VOLUME ONE HUNDRED AND TWENTY

VITAMINS AND HORMONES

Parathyroid Hormone

VOLUME ONE HUNDRED AND TWENTY

VITAMINS AND HORMONES

Parathyroid Hormone

Series Editor

GERALD LITWACK, PhD
Toluca Lake, North Hollywood, California

ELSEVIER

ACADEMIC PRESS
An imprint of Elsevier

Academic Press is an imprint of Elsevier
50 Hampshire Street, 5th Floor, Cambridge, MA 02139, United States
525 B Street, Suite 1650, San Diego, CA 92101, United States
The Boulevard, Langford Lane, Kidlington, Oxford OX5 1GB, United Kingdom
125 London Wall, London, EC2Y 5AS, United Kingdom

First edition 2022

ISBN: 978-0-323-99225-1
ISSN: 0083-6729

For information on all Academic Press publications
visit our website at https://www.elsevier.com/books-and-journals

Publisher: Zoe Kruze
Acquisitions Editor: Leticia M. Lima
Developmental Editor: Federico Paulo S. Mendoza
Production Project Manager: Vijayaraj Purushothaman
Cover Designer: Greg Harris

Typeset by STRAIVE, India

Contents

Contributors

Juan A. Ardura
Bone Physiopathology Laboratory, Department of Basic Medical Sciences, CEU San Pablo University, CEU Universities, Madrid, Spain

Rebecca Burnley
UCB Pharma, Slough, United Kingdom

Bernadette Byrne
Department of Life Sciences, Imperial College, London, United Kingdom

H. Herschel Conaway
Department of Physiology and Cell Biology, University of Arkansas for Medical Sciences, Little Rock, Arkansas, United States

Peter A. Friedman
Department of Pharmacology and Chemical Biology, Laboratory for GPCR Biology, University of Pittsburgh School of Medicine, Pittsburgh, PA, United States

Arancha R. Gortazar
Bone Physiopathology Laboratory, Department of Basic Medical Sciences, CEU San Pablo University, CEU Universities, Madrid, Spain

Ami Grunbaum
Calcium Research Laboratories and Department of Medicine, McGill University and McGill University Health Centre, Montreal, QC, Canada

Petra Henning
Department of Internal Medicine and Clinical Nutrition, Institute of Medicine, Sahlgrenska Osteoporosis Centre and Centre for Bone and Arthritis Research at the Sahlgrenska Academy, University of Gothenburg, Gothenburg, Sweden

Sara Heredero-Jiménez
Bone Physiopathology Laboratory, Department of Basic Medical Sciences, CEU San Pablo University, CEU Universities, Madrid, Spain

Takahisa Hiramitsu
Department of Endocrine Surgery, Nagoya Daini Red Cross Hospital, Nagoya, Japan

Toshihiro Ichimori
Department of Endocrine Surgery, Nagoya Daini Red Cross Hospital, Nagoya, Japan

Akemi Ito
Ito Bone Histomorphometry Institute, Niigata, Japan

Lisa Joedicke
UCB Pharma, Slough, United Kingdom

Rachelle W. Johnson
Department of Medicine, Division of Clinical Pharmacology, and Vanderbilt Center for Bone Biology, Vanderbilt University Medical Center, Nashville, TN, United States

Sang Wan Kim
Department of Internal Medicine, Seoul National University College of Medicine and Seoul Metropolitan Government Boramae Medical Center, Seoul, Republic of Korea

Richard Kremer
Calcium Research Laboratories and Department of Medicine, McGill University and McGill University Health Centre, Montreal, QC, Canada

Makoto Kuro-o
Division of Anti-aging Medicine, Center for Molecular Medicine, Jichi Medical University, Shimotsuke, Japan

Joon-Hyop Lee
Laboratory of Genomics and Translational Medicine; Department of Surgery, Gachon University College of Medicine, Incheon, Korea

Sihoon Lee
Laboratory of Genomics and Translational Medicine; Department of Internal Medicine, Gachon University College of Medicine, Incheon, Korea

Ulf H. Lerner
Department of Internal Medicine and Clinical Nutrition, Institute of Medicine, Sahlgrenska Osteoporosis Centre and Centre for Bone and Arthritis Research at the Sahlgrenska Academy, University of Gothenburg, Gothenburg, Sweden

Mariangela Librizzi
Dipartimento di Scienze e Tecnologie Biologiche, Chimiche e Farmaceutiche (STEBICEF), Università di Palermo, Palermo, Italia

Claudio Luparello
Dipartimento di Scienze e Tecnologie Biologiche, Chimiche e Farmaceutiche (STEBICEF), Università di Palermo, Palermo, Italia

Nikita Machado
Department of General Surgery, University Hospitals Conneaut, OH, Clinical Assistant Professor of Surgery, Case Western Reserve School of Medicine, Cleveland, OH, United States

Joanne Marks
Department of Neuroscience, Physiology & Pharmacology, University College London, London, United Kingdom

T. John Martin
St Vincent's Institute of Medical Research, University of Melbourne, St Vincent's Health, Melbourne, VIC, Australia

Eduardo Martín-Guerrero
Bone Physiopathology Laboratory, Department of Basic Medical Sciences, CEU San Pablo University, CEU Universities, Madrid, Spain

David McMillan
UCB Pharma, Slough, United Kingdom

Taro Murata
Department of Urology, Tokyo, Teishin Hospital, Tokyo, Japan

Masaki Nakamura
Department of Nephrology and Urology, NTT East Kanto Hospital, Tokyo, Japan

Kosaku Nitta
Department of Medicine, Kidney Center, Tokyo Women's Medical University, Tokyo, Japan

Manabu Okada
Department of Endocrine Surgery, Nagoya Daini Red Cross Hospital, Nagoya, Japan

Julie Rhoades
Department of Medicine, Division of Clinical Pharmacology, and Vanderbilt Center for Bone Biology, Vanderbilt University Medical Center; Department of Veterans Affairs, Tennessee Valley Healthcare System, Nashville, TN, United States

Kaushik Sarkar
Department of Life Sciences, Imperial College, London; UCB Pharma, Slough, United Kingdom

Masahiko Sasaki
Department of Urology, Tokyo, Teishin Hospital, Tokyo, Japan

Larry J. Suva
Department of Veterinary Physiology and Pharmacology, College of Veterinary Medicine and Biomedical Sciences, Texas Veterinary Medical Center 4466 Texas A&M University, College Station, TX, United States

Cheng-Chia Tang
Endocrine Unit, Massachusetts General Hospital, Harvard Medical School, Boston, MA, United States

Yoshihiro Tominaga
Department of Endocrine Surgery, Nagoya Daini Red Cross Hospital, Nagoya, Japan

Toshihide Tomosugi
Department of Endocrine Surgery, Nagoya Daini Red Cross Hospital, Nagoya, Japan

Ken Tsuchiya
Department Blood Purification, Kidney Center, Tokyo Women's Medical University, Tokyo, Japan

Robert J. Unwin
Department of Neuroscience, Physiology & Pharmacology; Department of Renal Medicine, University College London, London, United Kingdom

Pablo Urena
Division of Nephrology, Clinique du Landy, Saint Ouen, France

Marc N. Wein
Endocrine Unit, Massachusetts General Hospital, Harvard Medical School, Boston, MA, United States

Marta Westwood
UCB Pharma, Slough, United Kingdom

Scott M. Wilhelm
Department of General and Endocrine Surgery, University Hospitals Cleveland, OH, Professor of Surgery, Case Western Reserve School of Medicine, Cleveland, OH, United States

Michael Wright
UCB Pharma, Slough, United Kingdom

Aiji Yajima
Department of Anatomy, Cell Biology and Physiology, Indiana University, School of Medicine, Indianapolis, IN, United States; Department of Urology, Tokyo, Teishin Hospital; Department Blood Purification, Kidney Center, Tokyo Women's Medical University, Tokyo, Japan

Sung-Hee Yoon
Endocrine Unit, Massachusetts General Hospital, Harvard Medical School, Boston, MA, United States

Preface

Parathyroid hormone (PTH) is secreted by the parathyroid glands. There are four pea-sized parathyroid glands (two pairs on each side) located next to the two lobes of the thyroid gland. The parathyroid glands are the sources of PTH. PTH is comprised of 84 amino acid residues, and its activity is located in the first 20–30 amino acids. PTH derives from a pre-pro-hormone consisting of 115 amino acids.

Hyperparathyroidism (HPTM) is a disease that is on the rise; in the United States, this disease occurs in as many as 7 adults per 1000, mostly in the elderly, and, significantly, has a three- to fourfold greater incidence in women than in men. In the United States, about 100,000 persons develop HPTM each year; it is one of the most common hormonal disorders.

HPTM can result from different causes. There is a genetic form of familial isolated HPTM that results from mutations in one or more of three genes: *MWN1* (encoding the mRNA for the protein menin, a tumor suppressor), *CDC73* (encoding the mRNA for the parafibromin protein, a tumor suppressor), and *CASR* (encoding the mRNA for the calcium-sensing receptor protein expressed in the parathyroid glands and kidneys). In HPTM, the parathyroids become enlarged and, not uncommonly, a nonmalignant adenoma develops in one or more parathyroid glands. Given normally functioning kidneys, the adenoma can be removed surgically, thus reducing the unregulated secretion of PTH and reducing any hypercalcemia resulting from excessive secretion of PTH. Surgery can be risky in elderly patients with concomitant diseases. There are drugs (e.g., Cinacalcet) that can be taken orally to reduce the level of calcium in cases of HPTM. The increased use of these drugs has reduced the need for surgery in some cases, although the drugs generate serious gastrointestinal side effects. HPTM can be associated with chronic kidney disease (CKD), which occurs in 1 of 7 adults or 15% of the population, affecting about 37 million adults. In addition, the frequent administration of Lasix (furosemide), a loop diuretic used for hypertension (one factor that may be causative of CKD), can result in lowered vitamin D blood levels. This may occur because the final reaction in the kidney (1-hydroxylation of 25-hydroxyvitamin D_3) to produce active vitamin D (1,25-dihydroxyvitamin D_3 or calcitriol) undoubtedly is reduced in CKD. The diuretic decreases the water volume of blood and also depletes calcium and other metals, actions that increase the risk of developing HPTM in cases of CKD. HPTM, as indicated, is often associated with low levels of vitamin D in the blood. Vitamin D binds to the intestinal cell vitamin D receptor that induces the production of calbindin, a protein involved in the transport of calcium ions across the cells lining the intestine so that dietary calcium ions have difficulty, when circulating vitamin D is low, in reaching the blood supply.

Hypoparathyroidism can occur when the parathyroid glands are injured or during surgical removal of the thyroid gland wherein, sometimes, the parathyroids are accidentally removed. In addition, it is possible to be born without parathyroid glands. Hypoparathyroidism occurs in about 37 persons per 100,000 in the United States, and it is more prevalent in women than in men. The treatment of hypoparathyroidism usually involves oral calcium carbonate, vitamin D, magnesium, thiazide diuretic, and PTH that is injected subcutaneously at about 50 μg per day. Routinely, the dosage of PTH needs to be varied depending on information from blood analyses.

When the level of calcium ions is elevated in the blood, the calcium-sensing receptor of the parathyroid glands senses the elevated level (determined by the association constant of calcium ion for the receptor), resulting in the inhibition of the release of PTH and the stimulation of calcitonin secretion from the thyroid gland. The calcium-sensing receptor is a G-protein-coupled receptor expressed in the parathyroid gland and in the kidney. In the kidney, it regulates the tubular reabsorption of calcium. When the level of calcium ions in the blood is low, the calcium-sensing receptor activates the release of PTH from the parathyroid glands. Increased circulating PTH activates bone osteoclasts to solubilize and release calcium into the blood, resulting in the elevation of circulating calcium ions. Overactivity of this scenario can result in fragile bones and bone breaks.

Besides calcium, phosphate is an important factor. When the level of calcium ions increases in the blood, the levels of phosphate are reduced; conversely, when the levels of calcium fall, blood phosphate ions increase. Phosphate ions in serum form salts with calcium (calcium phosphate), and this results in decreased levels of free ionized calcium in the blood. PTH decreases phosphate reabsorption at the proximal convoluted tubule of the kidney. PTH also acts on the kidney to stimulate the 1-hydroxylation of 25-hydroxyvitamin D_3 to generate more of the active vitamin. Enhanced levels of active vitamin D bind to the intestinal vitamin D receptor and stimulate the absorption of calcium (and also phosphate) into the bloodstream. In the kidney, PTH stimulates calcium ion reabsorption and phosphate excretion, which result in increased calcium ions and decreased phosphate ions in the blood. The action of PTH on bone results in the release of calcium and phosphate, increasing their levels in the blood. It also results in the release of fibroblast growth factor 23 (FGF23) from bone, resulting in phosphate excretion in the kidney. FGF23 has negative effects on the action of vitamin D in the intestine and on the parathyroid gland to reduce the secretion of PTH. PTH has a half-life of about 5 min in the body. It is broken down in the lysosome.

The reviews in this book focus on all the current aspects of PTH, a collection that should appeal to a wide variety of researchers and practitioners.

The cover illustration represents part of the surface structure of a molecule of reductive dehalogenase enzyme. This is one structure from the "Molecule of the Month" series contributed to the Protein Data Bank by Dr. David S. Goodsell to whom I am indebted for facilitating the cover illustration. In the illustration, the cobamide cofactor is in turquoise and the iron-sulfur clusters are in red and yellow. The chlorine-containing substrate is not shown because the total surface is not included.

I am indebted to Paulo Mendoza and Leticia Lima of Elsevier who facilitated the organization of the chapters for this book. The production group in Chennai, India, expertly created the final version.

Gerald Litwack
North Hollywood, California
April 14, 2022

CHAPTER ONE

Structural pharmacology of PTH and PTHrP

Larry J. Suva[a] and Peter A. Friedman[b,*]
[a]Department of Veterinary Physiology and Pharmacology, College of Veterinary Medicine and Biomedical Sciences, Texas Veterinary Medical Center 4466 Texas A&M University, College Station, TX, United States
[b]Department of Pharmacology and Chemical Biology, Laboratory for GPCR Biology, University of Pittsburgh School of Medicine, Pittsburgh, PA, United States
*Corresponding author: e-mail address: pafl0@pitt.edu

Contents

Abstract

Parathyroid hormone (PTH) and PTH-related peptide (PTHrP) regulate extracellular phosphate and calcium homeostasis as well as bone remodeling. PTH is a classic endocrine peptide hormone whose synthesis and negative feedback by multiple factors control release from the parathyroid glands. PTHrP is ubiquitously expressed (pre- and postnatally) and acts in an autocrine/paracrine manner. This review considers the structural pharmacology and actions of PTH and PTHrP, biological consequences of inherited mutations, engineered analogs that illuminate similarities and differences in physiologic actions, and targeted therapeutic opportunities.

Parathyroid hormone (PTH) and PTH-related protein (PTHrP) regulate extracellular mineral-ion homeostasis and skeletal growth and bone turnover. Acting through a shared type 1 PTH receptor (PTHR), PTH assures adequate supplies of calcium and phosphate to support bone mineralization, while PTHrP regulates skeletal turnover, anabolism, and repair. PTH and PTHrP are peptide hormones. The PTH glands synthesize and secrete PTH that acts on a limited set of discrete targets, whereas PTHrP is widely expressed and exhibits broad autocrine or paracrine actions. In the following

Vitamins and Hormones, Volume 120
ISSN 0083-6729
https://doi.org/10.1016/bs.vh.2022.03.001

review, we focus our attention on structural aspects of PTH and PTHrP that account for their shared and distinct activities, inherited disorders arising from polymorphisms or mutations, the regulatory disturbances these cause and the challenges they create for clinical measurement, and how engineered alterations modify the pharmacokinetics of hormone action and functional specificity. Finally, we speculate on potential opportunities to expand the repertoire of experimentally or therapeutically valuable agents.

1. Evolutionary biology of PTH and PTHrP

PTH is encoded on a gene consisting of three exons (*PTH*) (Vasicek et al., 1983) mapped to the short arm of human chromosome 11 (11.p15.3), chr11:13,492,045-13,496,175. Comprising eight exons on chromosome 12 (12p11.22), the human PTHrP gene (*PTHLH*) enjoys a more complex genomic structure, transcribing three isoforms of 139, 141, and 173 residues (Fig. 1) (Philbrick et al., 1996). PTHrP(1–139) is transcribed from exons 3 and 4, PTHrP(1–141) from 3, 4, 7, and PTHrP(1–173) from 3, 4, and 6 (Yasuda, Banville, Hendy, & Goltzman, 1989). A single gene duplication event is believed to be responsible for the emergence of PTHrP (Danks & Trivett, 2002). The biosynthesis and peptide processing of PTH (Murray, Rao, Divieti, & Bringhurst, 2005) and PTHrP (Clemens et al., 2001; Philbrick et al., 1996) have been extensively reviewed and will not be covered here. In contrast, the human PTH gene consists of three exons,

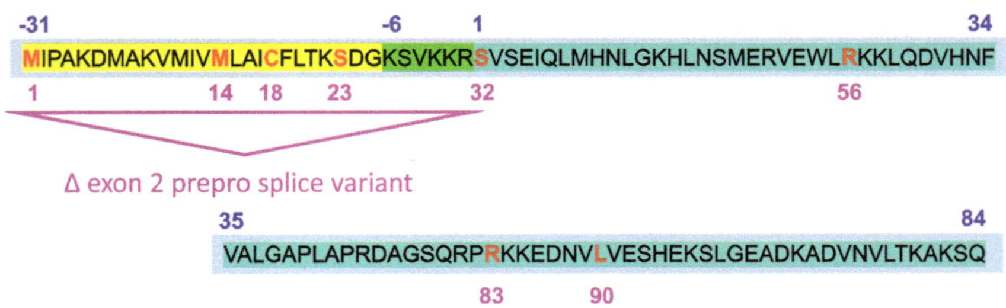

Fig. 1 Structural organization of human PTH showing described mutations. Primary amino acid sequence showing upstream pre (−31 to −7; yellow), pro (−6 to 0; green), and mature PTH (1 to 84, blue). By convention, the entire peptide is numbered from −31 to 84. Key sites are shown in blue above the sequence. Sites and locations of variants are counted from 1 starting at the amino terminus and are depicted in red. The magenta shows the large exon 2 deletion mutant that eliminates the prepro region. The individual mutations are described in the text and summarized with references in Table 1.

the second two of which encode a 25-residue "pre" signal sequence and a short, highly basic 6-residue "pro" sequence, followed by 84 amino acids that will form the mature hormone (Fig. 1).

New and emerging considerations of peptide variants provide insight into signaling mechanisms and clinical biochemistry. Eight PTH missense/nonsense mutations, along with one splicing mutation, have been described (Table 1). Despite the greater complexity of PTHrP, relatively few mutations have been reported, and those that have are clustered almost entirely with a severe skeletal phenotype. It may be that even modest clinical anomalies of mineral-ion metabolism permitted identifying the PTH mutations. In contrast, PTHrP mutations or variants may exist, but they either have a minor phenotype or have not been discovered because more common clinical screening for PTH levels does not extend to PTHrP. If the hormone mutation is active and alters interaction with PTHR, then there is likely no evolutionary advantage, and the mutation disappears. On the other hand, receptors are different and more likely to harbor activating/inactivating mutations that still interact with the ligand but are accompanied by altered signaling and potentially impact biological function. Those effects may be better tolerated evolutionarily.

Table 1 PTH mutations.

Location relative/absolute	Mutation	Dominance	PMID
−31/1	Met → Thr	Heterozygous	23873991
−18/14	Met1 → 4Lys	Heterozygous	28938448
−15/18	Cys → Arg	Heterozygous	2212001
−9/23	Ser → Pro, Ser → [a]	Homozygous Homozygous	10523031 22722080
1/32	Ser → Pro	Hemizygous	32421798
25/56	Arg → Cys	Homozygous	25891861
52/83	Arg → [a]	Hemizygous	18784115
59/90	Leu → Phe	Heterozygous	31433868

[a]Deletion.
Mutations collated by HGMD 2021.1 (Stenson et al., 2003). Additional details are included in the comprehensive review (Lee, Davaatseren, & Lee, 2020). Hemizygosity is synonymous with loss-of-heterozygosity.

2. PTH mutations

A variety of homozygous and heterozygous PTH coding region mutations have been identified. Interestingly, most of these are in the "pre" PTH sequence (Fig. 2). Here, we review the identified PTH variants starting at the amino-terminus of preproparathyroid hormone (preproPTH) and what is known about how they disorder mineral-ion metabolism.

Tomar et al. described a case of anomalous idiopathic hypoparathyroidism not attributable to calcium-sensing receptor autoantibodies (Tomar, Gupta, & Goswami, 2013). They identified a $Met^{-31}Thr$ replacement in PTH, causing a 6-base Met^{1}-Asp^{6} deletion due to introducing a BbsI restriction site within the prepro sequence.

A second methionine (Met^{14}) in the preproPTH signal sequence, i.e., within the pre element, is the locus of another heterozygous mutation, resulting in Familial Isolated Hypoparathyroidism. The described variant was predicted to interfere with the cleavage of preproPTH to proPTH (Cinque et al., 2017). In vitro studies demonstrated extensive cytoplasmic retention of the $Met^{14}Lys$ mutant PTH construct with reduced secretion consistent with deficient peptide processing compared to the wild-type construct that was extensively secreted to extracellular medium with little cellular retention.

Heterozygous Cys→Arg mutation at −15/18 within preproPTH was first described in 1990 by Arnold et al. (1990) in a subject previously identified with familial isolated hypoparathyroidism (Ahn, Antonarakis, Kronenberg, Igarashi, & Levine, 1986). The authors theorized that the underlying hypoparathyroidism in the affected individual was due to impaired glandular secretion. Subsequent work by others advanced the idea that the Cys18 mutant preproPTH served as a dominant-negative to

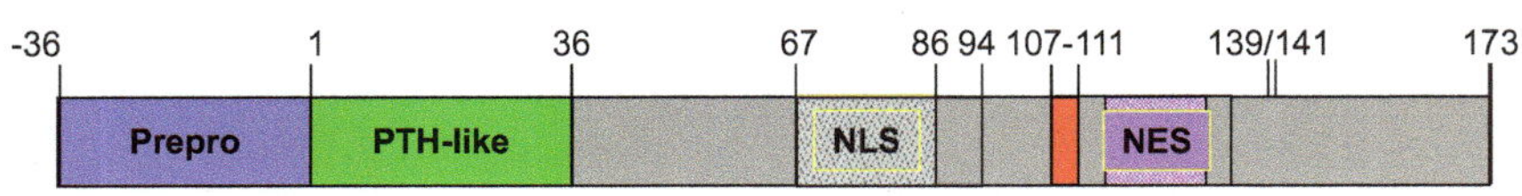

Fig. 2 Domains and amino acid sequence of human PTHrP (−36 to 173). Highlighted regions include the: prepro leader sequence (−36 to 0; blue), PTH-like region (1 to 36; green), putative nuclear localization signal (NLS; 67 to 93; yellow), osteostatin (107 to 111 red), and nuclear export signal (NES; 126 to 136; purple). Amino acid numbering of human PTHrP.

decrease assembly of the nascent protein with signal recognition particles in the endoplasmic reticulum and the translocation apparatus (Karaplis, Lim, Baba, Arnold, & Kronenberg, 1995). However, an abstract report of the same Cys→Arg mutation in a patient with sporadic idiopathic hypoparathyroidism (Lee et al., 2020) suggests that an alternate mechanism is at work. Datta et al. resolved the conundrum by finding that Cys→Arg mutant preproPTH is cytotoxic due to accumulation in the endoplasmic reticulum (Datta, Waheed, Shah, & Sly, 2007). Using a chemical chaperone rescued the phenotype, alleviated endoplasmic reticulum stress markers, and protected against cell death. Further, insofar as the cytotoxicity could be ascribed to a single allele, it explains the autosomal dominant inheritance pattern first noted by Arnold et al. (1990).

Two distinct homozygous Ser^{23} mutations in preproPTH (Table 1, Fig. 1) are the last known variants in the signal peptide. The first involves a T > C mutation changing the modified codon 23 from CTG to CCG, resulting in Pro replacing Ser at position 23 ($Ser^{23}Pro$) (Sunthornthepvarakul, Churesigaew, & Ngowngarmratana, 1999). Serum PTH levels were undetectable in affected consanguineous offspring of the heterozygous parents harboring the identical mutation supporting a pattern of autosomal recessive inheritance. Though not experimentally established, the authors proposed that the Ser > Pro mutation interferes with signal peptidase cleavage of preproPTH and failure to generate active hormone. The second described mutation involves a C > A substitution in exon 2, resulting in a premature stop codon (Dixon & Miller, 2018; Ertl, Stary, Streubel, Raimann, & Haeusler, 2012) and translation of non-functional proPTH.

Given stringent requirements for signal-particle binding (Akopian, Shen, Zhang, & Shan, 2013), it seems unlikely that mutations in proPTH would be tolerated. Indeed, no mutations in this region have been described as far as we know.

Parkinson and Thakker identified a subtle single-nucleotide change in exon 2 that displayed profound impact by converting a donor splice site to an endonuclease restriction motif, thereby excluding exon 2 consequent to the exon skipping event (Fig. 1) (Parkinson & Thakker, 1992). The presence of rare transcripts in nonspecific cells indicates a basal level of tissue-specific gene transcription at sites where they are generally active. Low-level pervasive transcription of tissue-specific genes has been termed ectopic or "illegitimate transcription" (Chelly, Concordet, Kaplan, & Kahn, 1989). Loss of exon 2 in the aberrant mRNA preproPTH transcript removes the initiation codon and the associated signal peptide sequence required for initiating PTH mRNA translation and translocation of the PTH peptide, respectively (Parkinson & Thakker, 1992).

Homozygous preproPTH mutations described thus far cause the catastrophic failure of peptide processing, resulting in negligible production of the biologically active hormone. In contrast, four downstream mutations identified in the mature hormone (Ser^{1}, Arg^{56}, Arg^{83}, Leu^{90}) disrupt function differently, and both are instructive about the structural determinants for PTH binding and signaling. Gild, Bullock, Luxford, Field, and Clifton-Bligh (2020) recently reported two sisters with a common inactivating PTH mutation at position 1/32 (Ser^{1}Pro). Serum PTH levels were low at birth and increased gradually as adults, as determined with various two-site assays. Such changes could result from parathyroid hyperplasia and a biological attempt to overcome diminished active hormone levels or arise from persistent hypocalcemia, as suggested by the authors. Previous comprehensive characterization of the amino-terminal determinants of PTH binding and receptor activation revealed complete intolerance by Ser^{1} for Pro substitution. However, Ala and Val were permissive (Gardella et al., 1991) thus generating abundant biologically inert hormone.

A different scenario attends homozygous Arg > Cys mutation at position 25/56 of PTH(1–84), as reported by Lee et al. (2015). Hypocalcemia and hyperphosphatemia characterize idiopathic hypoparathyroidism and pseudohypoparathyroidism. However, whereas idiopathic hypoparathyroidism is associated with diminished PTH levels, pseudohypoparathyroidism is attended by high PTH concentrations (Mantovani et al., 2018; Mantovani & Elli, 2021). The patients described by Lee were puzzling because they had symptoms of idiopathic hypoparathyroidism. Their circulating PTH levels were inconsistent and dependent on the assay employed. With an assay using a truncated amino-terminal detection antibody (PTH[13–34]), in conjunction with a common PTH(39–84) capture antibody, PTH levels were below normal. However, serum levels were modestly elevated with a PTH(1–34) detection antibody. Further complicating the picture, a PTH(1–84) assay that combines a PTH(1–3) detection antibody with the same PTH(39–84) capture antibody, serum PTH levels were markedly elevated. Results showing substantially elevated PTH levels would misleadingly suggest that hypocalcemia resulted from pseudohypoparathyroidism when, in fact, the biologically erroneous PTH levels are secondary to the altered affinity of the mutant PTH-Arg^{25}Cys peptide for the detection peptide. The contradictory and conflicting assay results underscore the clinical havoc that peptide mutations may cause. What accounts for the divergent assay results? Lee et al. suggest that $Arg^{25}Cys$ mutation lowers the affinity of the PTH(1–34) and PTH(13–34) detection antibodies. Further analysis is consistent with the view that the mutant PTH peptide has a lower affinity for the PTHR, compatible with the transient stimulation of cAMP compared to normal PTH. White et al., (2019) found that PTH-$Arg^{25}Cys$ exhibited only transient cAMP signaling and not the sustained activation typical of wild-type PTH. Further, the characteristic calcium allostery attending PTH action, where cAMP formation is positively influenced at elevated extracellular calcium concentrations, is absent in PTH-$Arg^{25}Cys$. The mutation disrupted interaction with acidic residues located in extracellular loop 1 of the PTHR, causing more rapid receptor recycling secondary to faster dissociation of the hormone/receptor complex. The mutant PTH-$Arg^{25}Cys$ also displays greater pH sensitivity, suggesting that this accounts for the absence of a sustained endosomal PTH cAMP synthesis (White et al., 2019).

A reported missense mutation of codon 83 from CGA to a stop codon (TGA) in a patient with parathyroid adenoma predicts premature truncation after the 52nd amino acid in PTH (Au et al., 2008). Surrounding parathyroid glands were normal and produced wild-type PTH. Accompanying in vitro studies demonstrated the formation of the amino-terminal PTH fragment but not full-length PTH, which along with sequencing results from peripheral-blood leukocytes, led the authors to conclude that the wild-type PTH allele had been deleted in the tumor. A prior analysis of carboxy-terminal determinants of hormone secretion revealed that translocation across the endoplasmic reticulum and signal sequence cleavage were impaired for PTH(1–52) (Lim, Gardella, Baba, Nussbaum, & Kronenberg, 1992). Insofar as the patient was only mildly hypocalcemic with normal serum phosphate levels supports the conclusion that the secreted fragments were broadly capable of maintaining mineral-ion homeostasis. Serum calcium levels normalized following tumor excision.

The final known PTH mutation was found in a screening analysis of childhood hypothyroidism and identified a novel variant at codon 59 (TTG>TTT). This heterozygous substitution replaced Phe for Leu ($Leu^{59}Phe$) (Wang et al., 2019). The biological activity of the mutant peptide was not characterized, so the mechanism underlying the pathology is unknown. Using PolyPhen to predict the effect of non-synonymous single nucleotide polymorphisms (nsSNPs), the authors propose that the missense variant was probably damaging. It is unclear why replacing an uncharged residue would not be tolerated. Leu-to-Phe exchange is generally considered permissive (Castro-Chavez, 2010; Jann, Werling, Chang, Haig, & Glass, 2008). Leu^{59} is perfectly conserved across primates (Malaivijitnond, Takenaka, Anukulthanakorn, & Cherdshewasart, 2002), bovine, rat, and porcine species (Suva et al., 1987), suggesting an important structural role. It may reside in an alpha-helical peptide region such that replacement with Phe and its large aromatic ring would disrupt tertiary protein structure.

3. PTHrP mutations

With nine exons and greater size than PTH, PTHrP offers further opportunities for coding errors to inflict damaging changes. However, relatively few PTHrP gene (*PTHLH*) mutations have been reported, and virtually all are associated with brachydactyly type E (Table 2). Unlike the point mutations found in PTH variants discussed above (Table 1), *PTHLH* mutations (HGMD 2021.1) include a total of 22 variants comprising 9 missense/nonsense mutations, 1 splicing variant (Reyes, Bravenboer, & Juppner, 2019), 2 small deletions (Jamsheer et al., 2016; Thomas-Teinturier et al., 2016), 3 gross deletions (Deshwar et al., 2019; Klopocki et al., 2010; Thomas-Teinturier et al., 2016), 4 gross insertions

Table 2 PTHrP mutations.

Location	Mutation	Dominance	Phenotype	PMID
−3	c.102-3A>G[a]	Heterozygous	BDE	30458061
15	Leu → Arg	Heterozygous	BDE	25801215
24	Gly → Ala	Heterozygous	Cleft lip/palate	27456059
42	Gln → Pro	Heterozygous	BDE	30835359
44	Leu → Pro	Heterozygous	BDE	20170896
56	R → [b]	Heterozygous	BDE	26763883
57	R → [b]	Heterozygous	BDE	29947179
60	Leu → Pro	Heterozygous	BDE	20170896
120	Lys → [b]	Heterozygous	BDE	20170896
178	[b] → Trp	Heterozygous	BDE	20170896

[a]HGVS designation.
[b]Deletion.
BDE = Brachydactyly type E. Mutations collated by HGMD 2021.1 (Stenson et al., 2003).

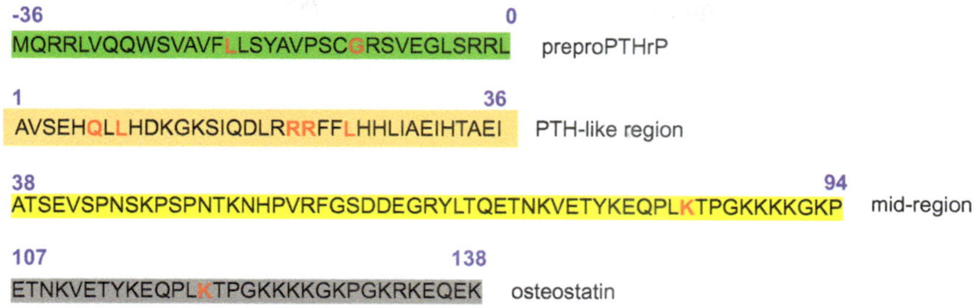

Fig. 3 hPTHrP post-translational peptides showing single residue mutations. Simple, single residue mutations in prepro, PTH-like, mid-region, and osteostatin PTHrP domains are indicated (red). A complete description, including complex insertion and deletion variants, is found in the text and summarized in Table 2.

(Collinson et al., 2010; Flottmann et al., 2016; Gray et al., 2014), 2 complex mutations (Maass et al., 2010), and 1 repeat (Gupta et al., 2008; Stenson et al., 2003). The location of single residue PTHrP variants is illustrated in Fig. 3.

A missense mutation changing Leu^{60} to Pro is associated with short stature (Klopocki et al., 2010). Using a chicken limb micromass culture assay, the authors confirmed diminished suppression of chondrocyte differentiation by the Leu^{60}Pro mutant compared to wild-type PTHrP as measured by alkaline phosphatase. Brachydactyly in humans with *PTHLH* disruptions resembles the short-limb chondrodysplasia of *Pthlh* knockout mice (Karaplis et al., 1994) and is attributable to premature birth growth plate closure.

A heterozygous *PTHLH* mutation in exon 4 converts the normal TGA stop codon to TGG, encoding Trp and allowing a large insertion uniquely affecting PTHrP(1–141), leading Reyes et al. (2019) to propose that PTHrP(1–141) and not PTHrP(1–139) is responsible for accelerated differentiation and premature growth plate closure.

A heterozygous two-nucleotide splice-site insertion trailing the prepro sequence of PTHrP shifts the open-reading frame, introducing 29 amino acids followed by a stop codon (Reyes et al., 2019). The mutation causes a unique origin for brachydactyly type E.

4. PTH and PTHrP pharmacology

PTH and PTHrP use the same G protein-coupled receptor, the PTH/PTHrP receptor (PTHR), to mediate their biological actions. Indeed, it has been only recently that the details of the mechanisms by which the two ligands bind to PTHR have begun to be unraveled (Cheloha, Gellman, Vilardaga, & Gardella, 2015; Ferrandon et al., 2009; Hattersley, Dean, Corbin, Bahar, & Gardella, 2015; Sutkeviciute, Clark, White, Gardella, & Vilardaga, 2019). The molecular details of PTHR ligand binding are discussed in more detail below. The current section will outline what is known regarding the conformational requirements for PTH and PTHrP that may inform the modes by which PTH and PTHrP act biologically to regulate calcium and bone metabolism when administered to humans.

Although acting through the same receptor, PTH and PTHrP have distinct biological roles (Cupp, Nayak, Adem, & Thomsen, 2013; Gensure, Gardella, & Juppner, 2005; Suva & Friedman, 2020). PTH, the primary product of the parathyroid gland, is a classical secreted 84-amino acid endocrine peptide hormone (PTH(1–84)) that regulates calcium and phosphate homeostasis via action on target cells in bone and kidney. PTH(1–34) (a synthetic non-physiologic peptide fragment of PTH) increases bone mineral density and bone strength in humans and is an effective treatment for postmenopausal osteoporosis (Black et al., 2003; Neer et al., 2001). In contrast, PTHrP acts in an autocrine/paracrine fashion and regulates cell proliferation and differentiation in various developing (Kronenberg, 2006) and post-natal tissues, including adult bone remodeling (Ansari et al., 2018; Martin, 2005).

4.1 PTH and PTHrP action is not identical

After the original cloning of PTHrP (Suva et al., 1987) and the demonstration of the high degree of sequence homology between PTH and PTHrP, there was much interest in the shared biologic actions of the two distinct gene products (Hock, Fonseca, Gunness-Hey, Kemp, & Martin, 1989; Juppner et al., 1988; Raisz, Simmons, Vargas, Kemp, & Martin, 1990). Perhaps the first inkling of the pharmacologic differences between the two evolutionarily-related proteins was uncovered in studies from the 1990s that involved the type 2 PTH receptor (PTH2R) (Usdin, Gruber, & Bonner, 1995). PTH2R is a poorly studied class B GPCR with 50% identity to PTHR at the amino acid level (Usdin et al., 1999). PTH2R mRNA expression demonstrates that it is most abundant in the brain but also expressed in the lung, pancreas, placenta, and testis with some limited expression in the cardiovascular system (Usdin et al., 1999).

From an evolutionary perspective, PTH2R may be part of a phylogenetically older regulatory system since zebrafish express a receptor more homologous to the PTH2R with no detectable PTHR-like gene (Bhattacharya, Yan, Postlethwait, & Rubin, 2011; Rubin et al., 1997). Protein expression and/or functional assays demonstrating specific activity such as PTH binding are limited, so controversy remains regarding precisely what tissues or cells are expressing PTH2R, and the physiologic function (if any) remains obscure.

Early studies stably expressing PTH2R in HEK293 cells demonstrated that despite the high degree of sequence homology between PTH and PTHrP in the N-terminal position 1–6 region (PTH position 1 is the only difference), PTHrP(1–34) does not interact with PTH2R, whereas PTH(1–34) does (Behar et al., 1996). The importance of the N-terminus of both PTH and PTHrP for interaction with the PTHR is well-documented (Nussbaum, Rosenblatt, & Potts Jr., 1980) because this region is an integral part of the so-called activation domain. These initial studies led to more interrogation of the differences in PTH and PTHrP action.

Unlike the PTHR, PTH2R interacts only weakly with PTHrP but has a high affinity for PTH, is strongly coupled to stimulation of cAMP accumulation, and weakly, in a cell-specific manner, to increase intracellular calcium concentration (Behar et al., 1996). Hoare, Clark, and Usdin (2000) identified the Tuberoinfundibular peptide of 39 residues (TIP39) as an endogenous PTH2R ligand that binds PTH2R with moderate affinity and does not appreciably activate PTHR. Despite the shared PTH2R action of PTH and TIP39, PTH and TIP39 elicit distinct responses via PTH2R (Bisello et al., 2004). These data further highlight the physiological regulation of PTH activity and how specific interactions mediate distinct physiologic responses.

4.2 PTH and PTHrP signaling

There is wide acceptance for the general model that PTHR amino-terminal sequences bind ligand carboxyl-terminal sequences with high affinity, and amino-terminal ligand sequences activate the receptor through interactions with the juxtamembrane portion of the receptor.

PTH and PTHrP are both directly implicated in bone remodeling (PTHR is abundant in osteoblasts and osteocytes in the bone microenvironment), and both mediate anabolic and catabolic effects on bone depending on the mode of administration (Suva & Friedman, 2020); pulsatile delivery is bone anabolic whereas continuous administration is catabolic. The synthetic N-terminal PTH(1–34) fragment stimulates a more prolonged increase in serum levels of calcium, 1,25-dihydroxy-vitamin-D, and bone resorption than PTHrP(1–36) fragments when delivered continuously (Horwitz et al., 2003, 2005), reflecting differences downstream of PTHR, assuming the interaction of the ligands with the receptor is consistent. This is indeed the case as PTH and PTHrP mediate their biological effects via identical heterotrimeric G protein (Gαβγ) pathways, primarily Gs/cAMP/protein kinase A (PKA) and Gq/PLC/Ca2+/PKC signaling, as well as G12/13/RhoA/phospholipase D and the mitogen-activated protein kinase (MAPK) signaling cascades (Chen, Wang, Hao, Hu, & Li, 2021; Cupp et al., 2013; Mantovani & Elli, 2019; Sutkeviciute et al., 2019). Current understanding attributes the specific biological actions of PTH and PTHrP to the ability of PTHR to adopt two distinct and active conformations that differ markedly in signaling duration and localization and thus distinct downstream activity (Feinstein et al., 2011; Okazaki et al., 2008; Sutkeviciute et al., 2019).

4.3 PTH and PTHrP peptides providing insight into physiology

The class B GPCRs including PTHR, are organized in two domains: one an extracellular domain involved in ligand-binding affinity and specificity and a second transmembrane domain required for receptor activation (Pal, Melcher, & Xu, 2012). The two-domain model supports information from peptide chemistry, suggesting the N-terminal domain of PTH interacts with the extracellular domain of the receptor and the PTH C-terminal domain with the transmembrane domain. Indeed, the mechanisms responsible for PTH and PTHrP binding and activation of PTHR fully support the two-site model involving initial rapid binding of the peptide C-terminal region (region 23–34) to the PTH1R N terminus, followed by a slower interaction of the peptide N terminus (region 1–14) with the so-called J-domain containing transmembrane α-helices that interconnect extracellular loops (Castro, Nikolaev, Palm, Lohse, & Vilardaga, 2005; Gardella & Juppner, 2001; Gensure et al., 2005). With the importance of PTH and PTHrP in the regulation of bone and calcium metabolism and the approval of PTH and PTHrP peptides for the treatment of osteoporosis, it seems appropriate to discuss how peptide chemistry has informed biology.

The anabolic effects of intermittent PTH(1–34) have been ascribed to PTHR-mediated increases in osteoblast number, resulting from the combination of increased differentiation of osteoblast precursors with a reduction in the apoptosis of mature, differentiated bone-forming osteoblasts (Jilka, Weinstein, Parfitt, & Manolagas, 2007). The catabolic effect of continuous PTH(1–34) has been associated with increased osteoblast/osteocyte-derived receptor activator of nuclear factor k-B ligand (RANKL) and decreased expression of the receptor activator of nuclear factor k-B ligand decoy protein osteoprotegerin (OPG), resulting in enhanced osteoclast formation and activity (Ma et al., 2001).

A wide variety of synthetic agonists/antagonists are described for the GPCR family of receptors. These molecules are characterized by a capacity to activate some or all of the entire signaling repertoire activated by endogenous ligands, likely associated with stabilization (or modifications) of distinct active receptor conformations capable of coupling to discrete signaling pathways (Kenakin, 2011; Urban et al., 2007). The idea that GPCR ligands can induce unique, ligand-specific receptor conformations resulting in differential activation of signal transduction pathways associated with that specific receptor is now widely accepted (Ghanouni et al., 2001; Kenakin, 1995).

Many PTH and PTHrP analogs promote differential PTHR signaling, including [Trp^1]PTHrP(1–36) and [Bpa1]PTHrP(1–36), that activate human PTHR-mediated Gs- but not Gq-signaling, ERK1/2 phosphorylation, β-arrestin recruitment, and receptor internalization (Bisello et al., 2002; Gesty-Palmer et al., 2009). Another ligand type, [Gly^1,Arg^{19}] PTH(1–28), stimulates Gs- but not Gq-signaling (Takasu, Gardella, Luck, Potts Jr., & Bringhurst, 1999). In vivo studies using PTH peptides administered in a pulsatile and, therefore, anabolic fashion have demonstrated that Gs-signaling is the predominant pathway leading to new bone formation and increased bone mass with little or no involvement of Gq (Hilliker, Wergedal, Gruber, Bettica, & Baylink, 1996; Rixon et al., 1994; Yang et al., 2007), although (Ogata et al., 2011) reported that PTHR-mediated Gq-signaling also contributes to anabolic effects of intermittently administered PTH(1–34). In addition, β-arrestin was shown to mediate PTH-stimulated bone formation without enhancing osteoclastogenesis in wild-type mice, but not in mice lacking β-arrestin (Ferrari et al., 2005). Considering these different conclusions, the exact PTHR signaling pathways participating in PTH and PTHrP responses have required further elucidation for some time. Therapeutically, PTHR-mediated signaling is well established in in vivo models of osteoporosis (ovariectomized rodents), with both PTH and PTHrP peptides increasing bone formation and bone mass without demonstrable elevations in osteoclastic bone resorption (Arlt et al., 2020; Fowler et al., 2012; Li, Liang, Shen, & Wronski, 1999). The precise mechanism underlying these actions was only recently resolved with the elucidation of the determinants of PTHR responsible for ligand selectivity and function (Sutkeviciute et al., 2019).

To determine the structural basis for molecular recognition and PTHR activation, Zhao and colleagues used cryo-electron microscopy (cryo-EM) to elucidate the near-atomic (3–4 Å) resolution structures of three active conformations of human PTHR bound to long-acting (LA)-PTH in complex with a stimulatory G protein (Zhao et al., 2019). This work was soon followed by elegant studies using the PTHR as a prototypic class B GPCR target, combined with molecular dynamics simulations and elastic network model-based methods, and uncovered a previously unknown site in PTHR that appears to modulate receptor action (Sutkeviciute et al., 2021). Other investigators have identified a small molecule (PCO371) currently in phase 1 as an orally available drug candidate for the treatment of hypoparathyroidism (Nishimura et al., 2020). PC0371 is an orally active small molecule PTHR agonist that dose-dependently increased serum calcium and decreased serum phosphate in thyroparathyroidectomized rats. The work of Sutkeviciute et al. (Sutkeviciute et al., 2021) may provide novel insight into discovering specific structures that modulate PTHR structure and function, providing an avenue to the first fully active small molecules that modulate PTHR signaling in the context of bone formation.

In summary, PTH and PTHrP regulate mineral-ion homeostasis and bone anabolism. Inherited gene mutations provide insights into vital structural determinants of hormone action. Multiple engineered PTH and PTHrP analogs are now available for clinical treatment of osteoporosis and related disorders. With high-resolution structures of the PTHR now in hand, novel peptides and small molecules will become available.

Acknowledgments

Original work cited herein was supported by grants R01DK105811 and R01GM140632 (P.A.F.) and CA166060, AA018282, and R01HD102909-01A1 (L.J.S.) from the National Institutes of Health.

References

Ahn, T. G., Antonarakis, S. E., Kronenberg, H. M., Igarashi, T., & Levine, M. A. (1986). Familial isolated hypoparathyroidism: A molecular genetic analysis of 8 families with 23 affected persons. *Medicine (Baltimore)*, *65*(2), 73–81.

Akopian, D., Shen, K., Zhang, X., & Shan, S. O. (2013). Signal recognition particle: An essential protein-targeting machine. *Annual Review of Biochemistry*, *82693-721*. https://doi.org/10.1146/annurev-biochem-072711-164732.

Ansari, N., Ho, P. W., Crimeen-Irwin, B., Poulton, I. J., Brunt, A. R., Forwood, M. R., et al. (2018). Autocrine and paracrine regulation of the murine skeleton by osteocyte-derived parathyroid hormone-related protein. *Journal of Bone and Mineral Research*, *33*(1), 137–153. https://doi.org/10.1002/jbmr.3291.

Arlt, H., Mullarkey, T., Hu, D., Baron, R., Ominsky, M. S., Mitlak, B., et al. (2020). Effects of abaloparatide and teriparatide on bone resorption and bone formation in female mice. *Bone Reports*, *13*, 100291. https://doi.org/10.1016/j.bonr.2020.100291.

Arnold, A., Horst, S. A., Gardella, T. J., Baba, H., Levine, M. A., & Kronenberg, H. M. (1990). Mutation of the signal peptide-encoding region of the preproparathyroid hormone gene in familial isolated hypoparathyroidism. *Journal of Clinical Investigation*, *86*(4), 1084–1087. https://doi.org/10.1172/JCI114811.

Au, A. Y., McDonald, K., Gill, A., Sywak, M., Diamond, T., Conigrave, A. D., et al. (2008). PTH mutation with primary hyperparathyroidism and undetectable intact PTH. *New England Journal of Medicine*, *359*(11), 1184–1186. https://doi.org/10.1056/NEJMc0802570.

Behar, V., Pines, M., Nakamoto, C., Greenberg, Z., Bisello, A., Stueckle, S. M., et al. (1996). The human PTH2 receptor: Binding and signal transduction properties of the stably expressed recombinant receptor. *Endocrinology*, *137*(7), 2748–2757. https://doi.org/10.1210/endo.137.7.8770894.

Bhattacharya, P., Yan, Y. L., Postlethwait, J., & Rubin, D. A. (2011). Evolution of the vertebrate pth2 (tip39) gene family and the regulation of PTH type 2 receptor (pth2r) and its endogenous ligand pth2 by hedgehog signaling in zebrafish development. *Journal of Endocrinology*, *211*(2), 187–200. https://doi.org/10.1530/JOE-10-0439.

Bisello, A., Chorev, M., Rosenblatt, M., Monticelli, L., Mierke, D. F., & Ferrari, S. L. (2002). Selective ligand-induced stabilization of active and desensitized parathyroid hormone type 1 receptor conformations. *Journal of Biological Chemistry*, *277*, 38524–38530.

Bisello, A., Manen, D., Pierroz, D. D., Usdin, T. B., Rizzoli, R., & Ferrari, S. L. (2004). Agonist-specific regulation of parathyroid hormone (PTH) receptor type 2 activity: Structural and functional analysis of PTH- and tuberoinfundibular peptide (TIP) 39-stimulated desensitization and internalization. *Molecular Endocrinology*, *18*(6), 1486–1498. https://doi.org/10.1210/me.2003-0487.

Black, D. M., Greenspan, S. L., Ensrud, K. E., Palermo, L., McGowan, J. A., Lang, T. F., et al. (2003). The effects of parathyroid hormone and alendronate alone or in combination in postmenopausal osteoporosis. *New England Journal of Medicine*, *349*(13), 1207–1215.

Castro, M., Nikolaev, V. O., Palm, D., Lohse, M. J., & Vilardaga, J. P. (2005). Turn-on switch in parathyroid hormone receptor by a two-step parathyroid hormone binding mechanism. *Proceedings of the National Academy of Sciences of the United States of America*, *102*(44), 16084–16089. https://doi.org/10.1073/pnas.0503942102.

Castro-Chavez, F. (2010). The rules of variation: Amino acid exchange according to the rotating circular genetic code. *Journal of Theoretical Biology*, *264*(3), 711–721. https://doi.org/10.1016/j.jtbi.2010.03.046.

Chelly, J., Concordet, J. P., Kaplan, J. C., & Kahn, A. (1989). Illegitimate transcription: Transcription of any gene in any cell type. *Proceedings of the National Academy of Sciences of the United States of America, 86*(8), 2617–2621. https://doi.org/10.1073/pnas.86.8.2617.

Cheloha, R. W., Gellman, S. H., Vilardaga, J. P., & Gardella, T. J. (2015). PTH receptor-1 signalling-mechanistic insights and therapeutic prospects. *Nature Reviews. Endocrinology, 11*(12), 712–724. https://doi.org/10.1038/nrendo.2015.139.

Chen, T., Wang, Y., Hao, Z., Hu, Y., & Li, J. (2021). Parathyroid hormone and its related peptides in bone metabolism. *Biochemical Pharmacology, 192*, 114669. https://doi.org/10.1016/j.bcp.2021.114669.

Cinque, L., Sparaneo, A., Penta, L., Mencarelli, A., Rogaia, D., Esposito, S., et al. (2017). Novel autosomal dominant PTH gene signal sequence mutation in a family with familial isolated hypoparathyroidism. *Journal of Clinical Endocrinology and Metabolism, 102*(11), 3961–3969. https://doi.org/10.1210/jc.2017-00250.

Clemens, T. L., Cormier, S., Eichinger, A., Endlich, K., Fiaschi-Taesch, N., Fischer, E., et al. (2001). Parathyroid hormone-related protein and its receptors: Nuclear functions and roles in the renal and cardiovascular systems, the placental trophoblasts and the pancreatic islets. *British Journal of Pharmacology, 134*(6), 1113–1136. https://doi.org/10.1038/sj.bjp.0704378.

Collinson, M., Leonard, S. J., Charlton, J., Crolla, J. A., Silve, C., Hall, C. M., et al. (2010). Symmetrical enchondromatosis is associated with duplication of 12p11.23 to 12p11.22 including PTHLH. *American Journal of Medical Genetics. Part A, 152A*(12), 3124–3128. https://doi.org/10.1002/ajmg.a.33567.

Cupp, M. E., Nayak, S. K., Adem, A. S., & Thomsen, W. J. (2013). Parathyroid hormone (PTH) and PTH-related peptide domains contributing to activation of different PTH receptor-mediated signaling pathways. *Journal of Pharmacology and Experimental Therapeutics, 345*(3), 404–418. https://doi.org/10.1124/jpet.112.199752.

Danks, J. A., & Trivett, M. K. (2002). Endocrinology and evolution: Lessons from comparative endocrinology. In J. A. H. Wass, & S. M. Shalet (Eds.), *Oxford textbook of endocrinology and diabetes* (pp. 9–13). Oxford: Oxford University Press.

Datta, R., Waheed, A., Shah, G. N., & Sly, W. S. (2007). Signal sequence mutation in autosomal dominant form of hypoparathyroidism induces apoptosis that is corrected by a chemical chaperone. *Proceedings of the National Academy of Sciences of the United States of America, 104*(50), 19989–19994. https://doi.org/10.1073/pnas.0708725104.

Deshwar, A. R., Spielmann, M., Vi, L., Mendoza-Londono, R., Dupuis, L., Stimec, J., et al. (2019). Disruption of the PTHLH regulatory landscape results in features consistent with hyperparathyroid disease. *American Journal of Medical Genetics. Part A, 179*(4), 663–667. https://doi.org/10.1002/ajmg.a.61071.

Dixon, J., & Miller, S. (2018). Successful pregnancies and reduced treatment requirement while breast feeding in a patient with congenital hypoparathyroidism due to homozygous c.68C > a null parathyroid hormone gene mutation. *BMJ Case Reports, 2018*. https://doi.org/10.1136/bcr-2017-223811.

Ertl, D. A., Stary, S., Streubel, B., Raimann, A., & Haeusler, G. (2012). A novel homozygous mutation in the parathyroid hormone gene (PTH) in a girl with isolated hypoparathyroidism. *Bone, 51*(3), 629–632. https://doi.org/10.1016/j.bone.2012.06.009.

Feinstein, T. N., Wehbi, V. L., Ardura, J. A., Wheeler, D. S., Ferrandon, S., Gardella, T. J., et al. (2011). Retromer terminates the generation of cAMP by internalized PTH receptors. *Nature Chemical Biology*, 7(5), 278–284. https://doi.org/10.1038/nchembio.545.

Ferrandon, S., Feinstein, T. N., Castro, M., Wang, B., Bouley, R., Potts, J. T., et al. (2009). Sustained cyclic AMP production by parathyroid hormone receptor endocytosis. *Nature Chemical Biology, 5*(10), 734–742. https://doi.org/10.1038/nchembio.206.

Ferrari, S., Pierroz, D., Glatt, V., Goddard, D., Bianchi, E., Lin, F., et al. (2005). Bone response to intermittent parathyroid hormone is altered in mice null for ß-arrestin2. *Endocrinology, 146*(4), 1854–1862.

Flottmann, R., Sowinska-Seidler, A., Lavie, J., Chateil, J. F., Lacombe, D., Mundlos, S., et al. (2016). Duplication of PTHLH causes osteochondroplasia with a combined brachydactyly type E/A1 phenotype with disturbed bone maturation and rhizomelia. *European Journal of Human Genetics, 24*(8), 1132–1136. https://doi.org/10.1038/ejhg.2015.266.

Fowler, T. W., McKelvey, K. D., Akel, N. S., Vander Schilden, J., Bacon, A. W., Bracey, J. W., et al. (2012). Low bone turnover and low BMD in down syndrome: Effect of intermittent PTH treatment. *PLoS One*, 7(8), e42967. https://doi.org/10.1371/journal.pone.0042967.

Gardella, T. J., Axelrod, D., Rubin, D., Keutmann, H. T., Potts, J. T., Jr., Kronenberg, H. M., et al. (1991). Mutational analysis of the receptor-activating region of human parathyroid hormone. *Journal of Biological Chemistry, 266*(20), 13141–13146.

Gardella, T. J., & Juppner, H. (2001). Molecular properties of the PTH/PTHrP receptor. *Trends in Endocrinology and Metabolism, 12*(5), 210–217. https://doi.org/10.1016/s1043-2760(01)00409-x.

Gensure, R. C., Gardella, T. J., & Juppner, H. (2005). Parathyroid hormone and parathyroid hormone-related peptide, and their receptors. *Biochemical and Biophysical Research Communications, 328*(3), 666–678. https://doi.org/10.1016/j.bbrc.2004.11.069.

Gesty-Palmer, D., Flannery, P., Yuan, L., Corsino, L., Spurney, R., Lefkowitz, R. J., et al. (2009). A β-arrestin-biased agonist of the parathyroid hormone receptor (PTH1R) promotes bone formation independent of G protein activation. *Science Translational Medicine, 1*(1). 1ra1 https://doi.org/10.1126/scitranslmed.3000071.

Ghanouni, P., Gryczynski, Z., Steenhuis, J. J., Lee, T. W., Farrens, D. L., Lakowicz, J. R., et al. (2001). Functionally different agonists induce distinct conformations in the G protein coupling domain of the beta 2 adrenergic receptor. *Journal of Biological Chemistry, 276*(27), 24433–24436. https://doi.org/10.1074/jbc.C100162200.

Gild, M. L., Bullock, M., Luxford, C., Field, M., & Clifton-Bligh, R. J. (2020). Congenital hypoparathyroidism associated with elevated circulating non-functional PTH due to novel PTH mutation. *Journal of Clinical Endocrinology and Metabolism, 105*(7). https://doi.org/10.1210/clinem/dgaa279.

Gray, M. J., van Kogelenberg, M., Beddow, R., Morgan, T., Wordsworth, P., Shears, D. J., et al. (2014). A new acro-osteolysis syndrome caused by duplications including PTHLH. *Journal of Human Genetics, 59*(9), 484–487. https://doi.org/10.1038/jhg.2014.58.

Gupta, A., Valimaki, V. V., Välimäki, M. J., Loyttyniemi, E., Richard, M., Bukka, P. L., et al. (2008). Variable number of tandem repeats polymorphism in parathyroid hormone-related protein as predictor of peak bone mass in young healthy Finnish males. *European Journal of Endocrinology of the European Federation of Endocrine Societies, 158*(5), 755–764. https://doi.org/10.1530/EJE-07-0886.

Hattersley, G., Dean, T., Corbin, B. A., Bahar, H., & Gardella, T. J. (2015). Binding selectivity of abaloparatide for PTH-type-1-receptor conformations and effects on downstream signaling. *Endocrinology, 157*(1), 141–149. https://doi.org/10.1210/en.2015-1726.

Hilliker, S., Wergedal, J. E., Gruber, H. E., Bettica, P., & Baylink, D. J. (1996). Truncation of the amino terminus of PTH alters its anabolic activity on bone in vivo. *Bone, 19*(5), 469–477. https://doi.org/10.1016/s8756-3282(96)00230-x.

Hoare, S. R., Clark, J. A., & Usdin, T. B. (2000). Molecular determinants of tuberoinfundibular peptide of 39 residues (TIP39) selectivity for the parathyroid hormone-2 (PTH2) receptor. N-terminal truncation of TIP39 reverses PTH2 receptor/PTH1 receptor binding selectivity. *Journal of Biological Chemistry*, *275*(35), 27274–27283. https://doi.org/10.1074/jbc.M003910200.

Hock, J. M., Fonseca, J., Gunness-Hey, M., Kemp, B. E., & Martin, T. J. (1989). Comparison of the anabolic effects of synthetic parathyroid hormone-related protein (PTHrP) 1-34 and PTH 1-34 on bone in rats. *Endocrinology*, *125*(4), 2022–2027. https://doi.org/10.1210/endo-125-4-2022.

Horwitz, M. J., Tedesco, M. B., Sereika, S. M., Hollis, B. W., Garcia-Ocana, A., & Stewart, A. F. (2003). Direct comparison of sustained infusion of human parathyroid hormone-related protein-(1-36) [hPTHrP-(1-36)] versus hPTH-(1-34) on serum calcium, plasma 1,25-dihydroxyvitamin D concentrations, and fractional calcium excretion in healthy human volunteers. *Journal of Clinical Endocrinology and Metabolism*, *88*(4), 1603–1609. https://doi.org/10.1210/jc.2002-020773.

Horwitz, M. J., Tedesco, M. B., Sereika, S. M., Syed, M. A., Garcia-Ocana, A., Bisello, A., et al. (2005). Continuous PTH and PTHrP infusion causes suppression of bone formation and discordant effects on 1,25$(OH)_2$ vitamin D. *Journal of Bone and Mineral Research*, *20*(10), 1792–1803. https://doi.org/10.1359/JBMR.050602.

Jamsheer, A., Sowinska-Seidler, A., Olech, E. M., Socha, M., Kozlowski, K., Pyrkosz, A., et al. (2016). Variable expressivity of the phenotype in two families with brachydactyly type E, craniofacial dysmorphism, short stature and delayed bone age caused by novel heterozygous mutations in the PTHLH gene. *Journal of Human Genetics*, *61*(5), 457–461. https://doi.org/10.1038/jhg.2015.172.

Jann, O. C., Werling, D., Chang, J. S., Haig, D., & Glass, E. J. (2008). Molecular evolution of bovine toll-like receptor 2 suggests substitutions of functional relevance. *BMC Evolutionary Biology*, *8288*. https://doi.org/10.1186/1471-2148-8-288.

Jilka, R. L., Weinstein, R. S., Parfitt, A. M., & Manolagas, S. C. (2007). Quantifying osteoblast and osteocyte apoptosis: Challenges and rewards. *Journal of Bone and Mineral Research*, *22*(10), 1492–1501. https://doi.org/10.1359/jbmr.070518.

Juppner, H., Abou-Samra, A. B., Unemo, S., Gu, W.-X., Potts, J. T., Jr., & Segre, G. V. (1988). The parathyroid hormone-like peptide associated with humoral hypercalcemia of malignancy and parathyroid hormone bind to the same receptor on the plasma membrane of ROS 17/2.8 cells. *Journal of Biological Chemistry*, *263*, 8557–8560.

Karaplis, A. C., Lim, S. K., Baba, H., Arnold, A., & Kronenberg, H. M. (1995). Inefficient membrane targeting, translocation, and proteolytic processing by signal peptidase of a mutant preproparathyroid hormone protein. *Journal of Biological Chemistry*, *270*(4), 1629–1635. https://doi.org/10.1074/jbc.270.4.1629.

Karaplis, A. C., Luz, A., Glowacki, J., Bronson, R. T., Tybulewicz, V. L., Kronenberg, H. M., et al. (1994). Lethal skeletal dysplasia from targeted disruption of the parathyroid hormone-related peptide gene. *Genes and Development*, *8*(3), 277–289. https://doi.org/10.1101/gad.8.3.277.

Kenakin, T. (1995). Pharmacological proteus? *Trends in Pharmacological Sciences*, *16*(8), 256–258. https://doi.org/10.1016/s0165-6147(00)89037-9.

Kenakin, T. (2011). Functional selectivity and biased receptor signaling. *Journal of Pharmacology and Experimental Therapeutics*, *336*(2), 296–302. https://doi.org/10.1124/jpet.110.173948.

Klopocki, E., Hennig, B. P., Dathe, K., Koll, R., de Ravel, T., Baten, E., et al. (2010). Deletion and point mutations of PTHLH cause brachydactyly type E. *American Journal of Human Genetics*, *86*(3), 434–439. https://doi.org/10.1016/j.ajhg.2010.01.023.

Kronenberg, H. M. (2006). PTHrP and skeletal development. *Annals of the New York Academy of Sciences*, *1068*, 1–13. https://doi.org/10.1196/annals.1346.002.

Lee, J. H., Davaatseren, M., & Lee, S. (2020). Rare PTH gene mutations causing parathyroid disorders: A review. *Endocrinology and Metabolism*, *35*(1), 64–70. https://doi.org/10.3803/EnM.2020.35.1.64.

Lee, S., Mannstadt, M., Guo, J., Kim, S. M., Yi, H. S., Khatri, A., et al. (2015). A homozygous [Cys25]PTH(1-84) mutation that impairs PTH/PTHrP receptor activation defines a novel form of hypoparathyroidism. *Journal of Bone and Mineral Research, 30*(10), 1803–1813. https://doi.org/10.1002/jbmr.2532.

Li, M., Liang, H., Shen, Y., & Wronski, T. J. (1999). Parathyroid hormone stimulates cancellous bone formation at skeletal sites regardless of marrow composition in ovariectomized rats. *Bone, 24*(2), 95–100. https://doi.org/10.1016/s8756-3282(98)00167-7.

Lim, S. K., Gardella, T. J., Baba, H., Nussbaum, S. R., & Kronenberg, H. M. (1992). The carboxy-terminus of parathyroid hormone is essential for hormone processing and secretion. *Endocrinology, 131*(5), 2325–2330. https://doi.org/10.1210/endo.131.5.1425431.

Ma, Y. L., Cain, R. L., Halladay, D. L., Yang, X., Zeng, Q., Miles, R. R., et al. (2001). Catabolic effects of continuous human PTH (1- -38) in vivo is associated with sustained stimulation of RANKL and inhibition of osteoprotegerin and gene-associated bone formation. *Endocrinology, 142*(9), 4047–4054. https://doi.org/10.1210/endo.142.9.8356.

Maass, P. G., Wirth, J., Aydin, A., Rump, A., Stricker, S., Tinschert, S., et al. (2010). A cis-regulatory site downregulates PTHLH in translocation t(8;12)(q13;p11.2) and leads to brachydactyly type E. *Human Molecular Genetics, 19*(5), 848–860. https://doi.org/10.1093/hmg/ddp553.

Malaivijitnond, S., Takenaka, O., Anukulthanakorn, K., & Cherdshewasart, W. (2002). The nucleotide sequences of the parathyroid gene in primates (suborder Anthropoidea). *General and Comparative Endocrinology, 125*(1), 67–78. https://doi.org/10.1006/gcen.2001.7735.

Mantovani, G., Bastepe, M., Monk, D., de Sanctis, L., Thiele, S., Usardi, A., et al. (2018). Diagnosis and management of pseudohypoparathyroidism and related disorders: First international consensus statement. *Nature Reviews. Endocrinology, 14*(8), 476–500. https://doi.org/10.1038/s41574-018-0042-0.

Mantovani, G., & Elli, F. M. (2019). Inactivating PTH/PTHrP signaling disorders. *Frontiers of Hormone Research, 51147-159*. https://doi.org/10.1159/000491045.

Mantovani, G., & Elli, F. M. (2021). PTH resistance. *Molecular and Cellular Endocrinology, 531*. 111311 https://doi.org/10.1016/j.mce.2021.111311.

Martin, T. J. (2005). Osteoblast-derived PTHrP is a physiological regulator of bone formation. *Journal of Clinical Investigation, 115*(9), 2322–2324. https://doi.org/10.1172/JCI26239.

Murray, T. M., Rao, L. G., Divieti, P., & Bringhurst, F. R. (2005). Parathyroid hormone secretion and action: Evidence for discrete receptors for the carboxyl-terminal region and related biological actions of carboxyl-terminal ligands. *Endocrine Reviews, 26*(1), 78–113. https://doi.org/10.1210/er.2003-0024.

Neer, R. M., Arnaud, C. D., Zanchetta, J. R., Prince, R., Gaich, G. A., Reginster, J. Y., et al. (2001). Effect of parathyroid hormone (1-34) on fractures and bone mineral density in postmenopausal women with osteoporosis. *New England Journal of Medicine, 344*(19), 1434–1441. https://doi.org/10.1056/NEJM200105103441904.

Nishimura, Y., Esaki, T., Isshiki, Y., Furuta, Y., Mizutani, A., Kotake, T., et al. (2020). Lead optimization and avoidance of reactive metabolite leading to PCO371, a potent, selective, and orally available human parathyroid hormone receptor 1 (hPTHR1) agonist. *Journal of Medicinal Chemistry, 63*(10), 5089–5099. https://doi.org/10.1021/acs.jmedchem.9b01743.

Nussbaum, S. R., Rosenblatt, M., & Potts, J. T., Jr. (1980). Parathyroid hormone•renal receptor interactions. Demonstration of two receptor-binding domains. *Journal of Biological Chemistry, 255*10183–25510187.

Ogata, N., Shinoda, Y., Wettschureck, N., Offermanns, S., Takeda, S., Nakamura, K., et al. (2011). $G\alpha_q$ signal in osteoblasts is inhibitory to the osteoanabolic action of parathyroid hormone. *Journal of Biological Chemistry, 286*(15), 13733–13740. https://doi.org/10.1074/jbc.M110.200196.

Okazaki, M., Ferrandon, S., Vilardaga, J. P., Bouxsein, M. L., Potts, J. T., Jr., & Gardella, T. J. (2008). Prolonged signaling at the parathyroid hormone receptor by peptide ligands targeted to a specific receptor conformation. *Proceedings of the National Academy of Sciences of the United States of America, 105*(43), 16525–16530. https://doi.org/10.1073/pnas.0808750105.

Pal, K., Melcher, K., & Xu, H. E. (2012). Structure and mechanism for recognition of peptide hormones by class B G-protein-coupled receptors. *Acta Pharmacologica Sinica, 33*(3), 300–311. https://doi.org/10.1038/aps.2011.170.

Parkinson, D. B., & Thakker, R. V. (1992). A donor splice site mutation in the parathyroid hormone gene is associated with autosomal recessive hypoparathyroidism. *Nature Genetics, 1*(2), 149–152. https://doi.org/10.1038/ng0592-149.

Philbrick, W. M., Wysolmerski, J. J., Galbraith, S., Holt, E., Orloff, J. J., Yang, K. H., et al. (1996). Defining the roles of parathyroid hormone-related protein in normal physiology. *Physiological Reviews, 76*(1), 127–173. https://doi.org/10.1152/physrev.1996.76.1.127.

Raisz, L. G., Simmons, H. A., Vargas, S. J., Kemp, B. E., & Martin, T. J. (1990). Comparison of the effects of amino-terminal synthetic parathyroid hormone-related peptide (PTHrP) of malignancy and parathyroid hormone on resorption of cultured fetal rat long bones. *Calcified Tissue International, 46*(4), 233–238. https://doi.org/10.1007/BF02555001.

Reyes, M., Bravenboer, B., & Juppner, H. (2019). A heterozygous splice-site mutation in PTHLH causes autosomal dominant shortening of metacarpals and metatarsals. *Journal of Bone and Mineral Research, 34*(3), 482–489. https://doi.org/10.1002/jbmr.3628.

Rixon, R. H., Whitfield, J. F., Gagnon, L., Isaacs, R. J., MacLean, S., Chakravarthy, B., et al. (1994). Parathyroid hormone fragments may stimulate bone growth in ovariectomized rats by activating adenylyl cyclase. *Journal of Bone and Mineral Research, 9*, 1179–1189. https://doi.org/10.1002/jbmr.5650090807.

Rubin, D., Chen, J., Zon, L., Fishman, M., Bergwitz, C., & Jüppner, H. (1997). Functional properties and tissue distribution of a parathyroid hormome type 2 receptor (Pth2r) in the zebrafish (Danio rerio). *Journal of Bone and Mineral Research, 12*, S369.

Stenson, P. D., Ball, E. V., Mort, M., Phillips, A. D., Shiel, J. A., Thomas, N. S., et al. (2003). Human gene mutation database (HGMD): 2003 update. *Human Mutation, 21*(6), 577–581. https://doi.org/10.1002/humu.10212.

Sunthornthepvarakul, T., Churesigaew, S., & Ngowngarmratana, S. (1999). A novel mutation of the signal peptide of the preproparathyroid hormone gene associated with autosomal recessive familial isolated hypoparathyroidism. *Journal of Clinical Endocrinology and Metabolism, 84*(10), 3792–3796. https://doi.org/10.1210/jcem.84.10.6070.

Sutkeviciute, I., Clark, L. J., White, A. D., Gardella, T. J., & Vilardaga, J. P. (2019). PTH/PTHrP receptor signaling, allostery, and structures. *Trends in Endocrinology and Metabolism, 30*(11), 860–874. https://doi.org/10.1016/j.tem.2019.07.011.

Sutkeviciute, I., Lee, J. Y., White, A. D., Maria, C. S., Pena, K. A., Savransky, S., et al. (2021). Precise druggability of the PTH type 1 receptor. *Nature Chemical Biology*. https://doi.org/10.1038/s41589-021-00929-w.

Suva, L. J., & Friedman, P. A. (2020). PTH and PTHrP actions on bone. *Handbook of Experimental Pharmacology, 262*, 27–46. https://doi.org/10.1007/164_2020_362.

Suva, L. J., Winslow, G. A., Wettenhall, R. E., Hammonds, R. G., Moseley, J. M., Diefenbach-Jagger, H., et al. (1987). A parathyroid hormone-related protein implicated in malignant hypercalcemia: Cloning and expression. *Science, 237*(4817), 893–896. https://doi.org/10.1126/science.3616618.

Takasu, H., Gardella, T. J., Luck, M. D., Potts, J. T., Jr., & Bringhurst, F. R. (1999). Amino-terminal modifications of human parathyroid hormone (PTH) selectively alter phospholipase C signaling via the type 1 PTH receptor: Implications for design of signal-specific PTH ligands. *Biochemistry, 38*(41), 13453–13460. https://doi.org/10.1021/bi990437n.

Thomas-Teinturier, C., Pereda, A., Garin, I., Diez-Lopez, I., Linglart, A., Silve, C., et al. (2016). Report of two novel mutations in PTHLH associated with brachydactyly type E and literature review. *American Journal of Medical Genetics. Part A*, *170*(3), 734–742. https://doi.org/10.1002/ajmg.a.37490.

Tomar, N., Gupta, N., & Goswami, R. (2013). Calcium-sensing receptor autoantibodies and idiopathic hypoparathyroidism. *Journal of Clinical Endocrinology and Metabolism*, *98*(9), 3884–3891. https://doi.org/10.1210/jc.2013-2158.

Urban, J. D., Clarke, W. P., von Zastrow, M., Nichols, D. E., Kobilka, B., Weinstein, H., et al. (2007). Functional selectivity and classical concepts of quantitative pharmacology. *Journal of Pharmacology and Experimental Therapeutics*, *320*(1), 1–13. https://doi.org/10.1124/jpet.106.104463.

Usdin, T. B., Gruber, C., & Bonner, T. I. (1995). Identification and functional expression of a receptor selectively recognizing parathyroid hormone, the PTH2 receptor. *Journal of Biological Chemistry*, *270*, 15455–15458.

Usdin, T. B., Hilton, J., Vertesi, T., Harta, G., Segre, G., & Mezey, E. (1999). Distribution of the parathyroid hormone 2 receptor in rat: Immunolocalization reveals expression by several endocrine cells. *Endocrinology*, *140*, 3363–3371. https://doi.org/10.1210/endo.140.7.6855.

Vasicek, T. J., McDevitt, B. E., Freeman, M. W., Fennick, B. J., Hendy, G. N., Potts, J. T., Jr., et al. (1983). Nulceotide sequence of the human parathyroid hormone gene. *Proceedings. National Academy of Sciences. United States of America*, *80*(8), 2127–2131. https://doi.org/10.1073/pnas.80.8.2127.

Wang, Y., Nie, M., Wang, O., Li, Y., Jiang, Y., Li, M., et al. (2019). Genetic screening in a large Chinese cohort of childhood onset hypoparathyroidism by next-generation sequencing combined with TBX1-MLPA. *Journal of Bone and Mineral Research*, *34*(12), 2254–2263. https://doi.org/10.1002/jbmr.3854.

White, A. D., Fang, F., Jean-Alphonse, F. G., Clark, L. J., An, H. J., Liu, H., et al. (2019). Ca^{2+} allostery in PTH-receptor signaling. *Proceedings of the National Academy of Sciences of the United States of America*, *116*(8), 3294–3299. https://doi.org/10.1073/pnas.1814670116.

Yang, D., Singh, R., Divieti, P., Guo, J., Bouxsein, M. L., & Bringhurst, F. R. (2007). Contributions of parathyroid hormone (PTH)/PTH-related peptide receptor signaling pathways to the anabolic effect of PTH on bone. *Bone*, *40*(6), 1453–1461. https://doi.org/10.1016/j.bone.2007.02.001.

Yasuda, T., Banville, D., Hendy, G. N., & Goltzman, D. (1989). Characterization of the human parathyroid hormone-like peptide gene. Functional and evolutionary aspects. *Journal of Biological Chemistry*, *264*(13), 7720–7725.

Zhao, L. H., Ma, S., Sutkeviciute, I., Shen, D. D., Zhou, X. E., de Waal, P. W., et al. (2019). Structure and dynamics of the active human parathyroid hormone receptor-1. *Science*, *364*(6436), 148–153. https://doi.org/10.1126/science.aav7942.

CHAPTER TWO

Salt inducible kinases and PTH1R action

Sung-Hee Yoon, Cheng-Chia Tang, and Marc N. Wein*
Endocrine Unit, Massachusetts General Hospital, Harvard Medical School, Boston, MA, United States
*Corresponding author: e-mail address: mnwein@mgh.harvard.edu

Contents

Abstract

Parathyroid hormone is a central regulator of calcium homeostasis. PTH protects the organism from hypocalcemia through its actions in bone and kidney. Recent physiologic studies have revealed key target genes for PTH receptor (PTH1R) signaling in these target organs. However, the complete signal transduction cascade used by PTH1R to accomplish these physiologic actions has remained poorly defined. Here we will review recent studies that have defined an important role for salt inducible kinases downstream of PTH1R in bone, cartilage, and kidney. PTH1R signaling inhibits the activity of salt inducible kinases. Therefore, direct SIK inhibitors represent a promising novel strategy to mimic PTH actions using small molecules. Moreover, a detailed understanding of the molecular circuitry used by PTH1R to exert its biologic effects will afford powerful new models to better understand the diverse actions of this important G protein coupled receptor in health and disease.

Vitamins and Hormones, Volume 120
ISSN 0083-6729
https://doi.org/10.1016/bs.vh.2022.04.008

1. Parathyroid hormone and PTH1R biology

Parathyroid hormone (PTH) is produced exclusively by the parathyroid gland (Lupp et al., 2010; Murray, Rao, Divieti, & Bringhurst, 2005). PTH is a 84-amino acid peptide whose major function is to protect the organism from hypocalcemia. When dropping calcium levels are sensed by the calcium-sensing receptor (CaSR) in the parathyroid gland, PTH will be released into the bloodstream. PTH rapidly mobilizes skeletal calcium stores via osteoclastic bone resorption, and stimulates renal tubular calcium reabsorption and renal synthesis of active vitamin D (1,25-dihydroxyvitamin D3 [1,25-$(OH)_2$D3] (Brown, 1983). Together, these physiologic actions of PTH work to maintain normocalcemia. Furthermore, PTH also regulates serum phosphate concentration by inhibiting tubular phosphate reabsorption in the kidneys. In bone, PTH exerts its action by primarily binding to and therefore activating PTH receptor1 (PTH1R) to initiate downstream signaling actions, leading to increased bone remodeling (Mannstadt, Jüppner, & Gardella, 1999). Chronic hyperparathyroidsim, a disease where parathyroid glands to over-produce PTH, leads to increased mobilization of calcium from the bone, loss of bone mass, and increased risks of fractures (Silva, Costa, Cusano, Kousteni, & Bilezikian, 2011; Wein & Kronenberg, 2018).

PTH1R is a class B G-protein coupled receptor (GPCR) family member. GPCRs contain seven membrane-spanning domains that connect intra-and extracellular loops. GPCRs are the biggest and most diverse family of membrane proteins in the human genome, and they are activated by a variety of ligands, transducing diverse extracellular stimuli to mediate cellular responses (Rosenbaum, Rasmussen, & Kobilka, 2009). In vertebrates, GPCRs are divided into five families based on their sequence and structural similarities, family A: rhodopsin, family B, secretin, family C, glutamate, adhesion, and Frizzled. GPCRs associate with three different intracellular signaling proteins: alpha subunit, beta subunit and gamma subunit. In unstimulated cells, GDP is bound to the alpha subunit. When a ligands bind to the GPCR, the G alpha subunit binds GTP and dissociates from the G gamma and beta subunits. The GTP-bound G alpha subunit and 'free' beta/gamma subunits can then activate downstream signaling proteins, and GTP hydrolysis terminates signal interaction.

Other family members of the class B GPCR family include calcitonin, calcitonin gene-related peptide (CGRP), receptors for glucagon, and glucagon-like peptides (GLPs) (Zhao et al., 2019). PTH1R is widely expressed in tissues, with the highest level of expression in kidney and in bone. PTH1R plays important roles in regulating mineral ion homeostasis, skeletal development, and bone metabolism. PTH binding to PTH1R receptor activates the G protein alpha subunits Gαs and Gαq, leading to production of 3′,5′-cyclic adenosine-5′-monophosphate (cAMP) (Jüppner et al., 1991). Elevated production of cAMP then leads to activation of protein kinase A (PKA). PTH1R also couples with phospholipase C via Gαq, leading to activation of PKC and increases intracellular Ca^{2+}. In addition, PTH1R can also couple to the Gα12/13-phospholipase D-transforming protein RhoA pathway and the β-arrestin-extracellular signal-regulated kinase (ERK1/2) pathway.

In humans, inactivating mutations in PTH1R results in Blomstrand chondrodysplasia, a genetic disease characterized by shortened limbs, increased bone density and advanced skeletal saturation (Jobert et al., 1998; Loshkajian et al., 1997). Activating mutation in PTH1R results in Jansen's metaphyseal chondrodysplasia (JMC), a rare form of short-limbed dwarfism characterized by abnormal endochondral bone formation and abnormalities in bone mineralization with autosomal dominant inheritance pattern (Schipani, Kruse, & Jüppner, 1995).

PTH1R is expressed on the surface of multiple mesenchymal lineage in bone including chondrocytes in growth plate, osteoblast precursors, osteoblasts, bone lining cells, and osteocytes (Wein & Kronenberg, 2018). Mice with conditional deletion of PTH1R in osteoblasts show reduced bone length and trabecular bone formation (Qiu et al., 2015). Conversely, mice with constitutively active PTH receptor (the same mutation found in humans with JMC) in osteoblasts or osteocytes show increased in trabecular bone mass (Calvi et al., 2001; Rhee et al., 2011). Parathyroid hormone-like related protein (PTHrP), which shares a structural similarity in the NH2-terminus as PTH, can also bind to PTH1R. In bone, paracrine acting PTHrP plays a crucial role in controlling hypertrophy of growth plate chondrocytes (Kronenberg, 2006). PTHrP also controls organogenesis during normal development and, in certain tumors, drives humoral hypercalcemia or malignancy (McCauley & Martin, 2012).

2. PTH as anabolic therapies for osteoporosis

Osteoporosis is a serious metabolic bone disease that is characterized by the deterioration of the microarchitecture and the mechanical properties, resulting in increased risks of fractures (Kanis, 2002). Worldwide, 1 in 3 women over the age of 50 years and 1 in 5 men will experience an osteoporotic fracture in their lifetime. Currently, osteoporosis affects 200 million people, and 10 million people in the US alone (Estell & Rosen, 2021; Sözen, Özışık, & Başaran, 2017; Wright et al., 2014). Fractures can significantly affect mobility, quality of life, and cause excessive mortality. Within the first year of a hip fracture, there is a 15–20% increase in mortality rate, and a 2.5-fold increase in risks of future fractures (Sözen et al., 2017). Every year, medical costs of osteoporotic fractures are approximately $14 billion dollars in the US and an additional $2 billion dollars for drug treatments (Blume & Curtis, 2011). Therefore, effective, novel and safe treatment options for osteoporosis represent an unmet medical need in our aging population.

The majority of currently-available osteoporosis medications work by blocking bone resorption by osteoclasts. In contrast, relatively few bone anabolic osteoporosis medications are used. While the initial physiologic actions of PTH are to stimulate osteoclastic bone formation, it has long been appreciated that PTH also can increase bone formation (Reeve et al., 1976). A potential teleologic explanation underlying this phenomenon is that PTH may have evolved actions to stimulate bone formation to protect the skeleton over time from calcium losses incurred by its initial actions. Intermittent subcutaneous injections of PTH amino acids 1–34 (Teriparatide, hPTH 1–34, TPTD) have been used as anabolic therapies to treat osteoporosis since its FDA approval in 2002. More recently, the PTHrP analog abaloparatide was also demonstrated to stimulate bone formation, increase bone mass, and reduce fractures in post-menopausal women with osteoporosis. Treatment with hPTH or abaloparatide leads to increased anabolic bone effects, bone mineral density (BMD), and prevents fractures in individuals who are at high risks (Estell & Rosen, 2021).

While effective, widespread use of PTH analogs for osteoporosis is limited to some degree by the need for once daily subcutaneous injections, potential risk of osteosarcoma, and occasional hypercalcemia. Moreover, intermittent PTH analog therapy also stimulates bone resorption over time, a phenomenon that may limit the therapeutic efficacy of these medications despite an initial 'anabolic window' where the bone anabolic effects appear to dominate (Silva et al., 2011). At present, development of 'next generation' osteoporosis therapeutics that exploit the ability of PTH to stimulate bone formation is limited by an incomplete knowledge of the cellular and molecular mechanisms used by PTH to stimulate bone formation. Osteocytes are the most abundant cell type in bone and likely central to the physiologic and pharmacologic actions of parathyroid hormone.

3. PTH action in osteocytes

Osteocytes, terminally-differentiated osteoblasts embedded in bone matrix, are the most abundant cell type in bone (Bonewald, 2011). Osteocytes are connected with other osteocytes or other bone cells on the surface through dendritic processes that extend from the cell bodies that travel through canaliculi. Through this "osteocyte network," osteocytes are able to communicate with other cells. Osteocytes produce paracrine-acting proteins, such as sclerostin and RANKL, that are important for bone remodeling. Sclerostin is a secreted inhibitor of WNT signaling, and osteocytes are the primary cell type responsible for sclerostin production. WNT signaling normally stimulates bone formation by osteoblasts. By locally blocking WNT actions in bone, sclerostin thus inhibits osteoblast function. Individuals lacking sclerostin show very high bone mass and resistance to fracture (Baron & Kneissel, 2013). Romosozumab is an anti-sclerostin antibody now available for treatment of post-menopausal women with osteoporosis (Baron & Gori, 2018). Osteocytes are also major source of RANKL (O'Brien, Nakashima, & Takayanagi, 2013), a crucial factor that stimulates osteoclast differentiation and function (Takayanagi, 2021). Thus, osteocytes express two key paracrine-acting factors that in turn regulate bone remodeling by osteoblasts (sclerostin) and osteoclasts (RANKL). As such, osteocytes can be viewed as 'conductors' of the "orchestra" of remodeling on bone surfaces.

Given the central role of osteocytes in controlling bone remodeling in response to parathyroid hormone, the question then emerges as to whether PTH regulates expression of these key paracrine-acting factors. Indeed, PTH potently controls sclerostin and RANKL expression in directions consistent with the net stimulatory effects of PTH on bone remodeling. PTH suppresses osteocytic sclerostin expression (Bellido et al., 2005; Keller & Kneissel, 2005), and PTH induces RANKL production by osteoblast lineage cells including osteocytes (Fu, Jilka, Manolagas, & O'Brien, 2002). While some debate exists as to the relative roles regarding PTH-induced changes in sclerostin and RANKL and the overall skeletal response to intermittent and chronic PTH (Wein, 2018), there is no doubt that PTH effects on osteocytes are central to the actions of this hormone on bone. For this reason, a detailed mechanistic understanding of the intracellular cascades used by PTH to enact these gene expression changes is likely to lead to important new advances in bone biology.

4. Salt inducible kinases

The AMP-activated protein kinase (AMPK) family of serine/threonine kinases have 14 members that play important roles in many biological processes including cellular energy metabolism and proliferation (Bright, Thornton, & Carling, 2009). There are three members in the SIK family in mammals, SIK1, SIK2 and SIK3. They earned their name because the mRNA expression of first family member (SIK1) was found to be *inducible* in the adrenal glands in rats by *salt* dietary intake (Wang, Takemori, Halder, Nonaka, & Okamoto, 1999). Rats were fed Na^{+} and K^{+}-enriched diet for 7 days and then were sacrificed and their adrenal glands were removed. A screen was performed to identify genes that were induced by high Na+/K+ in the adrenal gland, and SIK1 was identified as a novel cDNA clone encoding a 776-amino acid polypeptide sharing high similarity with AMPK/sucrose non-fermenting kinase (SNF1) (Wang et al., 1999). Other stimuli that induce SIK1 expression include adrenocorticotropin hormone signaling (Lin et al., 2001), glucagon signaling (Koo et al., 2005), excitable cell depolarization (Liu, Feldman, Machado, Vician, & Herschman, 2003), and circadian rhythms (Jagannath et al., 2013). In contrast, expression of closely-related family members SIK2 and SIK3 is *not* induced by salt intake; rather, these SIK isoforms are constitutively expressed and regulated at the post-translational level as detailed below. Both SIK2 and SIK3 are ubiquitously expressed in humans, where SIK2 shows the highest level of expression in adipose tissue and SIK3 shows the highest level of expression in brain. All three SIK isoforms are expressed in the bone, including osteoblast-lineage cells and osteoclasts (Wein, Foretz, Fisher, Xavier, & Kronenberg, 2018).

SIK cellular activity is predominantly regulated by two upstream acting kinases. Liver Kinase B1 LKB1 (encoded by the *Stk11* gene) phosphorylates the activation loop of all AMPK family kinases (Lizcano et al., 2004), thereby, promoting SIK activity. In contrast, cAMP-dependent protein kinase A (PKA) phosphorylates SIK isoforms at C-terminal residues outside of the kinase domain. PKA-dependent C-terminal SIK phosphorylation blocks cellular SIK activity via an allosteric mechanism involving recruitment of 14-3-3 chaperones and sequestration of SIKs from their cytoplasmic substrates (Sonntag, Vaughan, & Montminy, 2018). Thus, opposing actions of LKB1 and PKA regulate overall SIK cellular activity. At present, the best-characterized SIK substrates are class IIa histone deacetylases (HDACs) and cAMP-regulated transcriptional coactivators (CRTCs).

When phosphorylated by SIKs, these substrates are retained in the cytoplasm due to inducible 14-3-3 binding. SIK-mediated substrate phosphorylation is opposed by active dephosphorylation by yet-to-be-identified phosphatases. Thus, upon PKA-mediated SIK inhibition, SIK substrates quickly become dephosphorylated and can translocate into the nucleus. Nuclear class IIa HDACs block MEF2-driven gene expression (Haberland, Montgomery, & Olson, 2009), while nuclear CRTC proteins stimulate CREB-driven gene expression programs (Altarejos & Montminy, 2011).

5. Role of salt inducible kinases in responses to PTH in bone

SIK1 may negatively regulate osteoblast proliferation differentiation. Gene knockdown of SIK1 in primary osteoblasts led to increased in osteoblast differentiation and bone matrix mineralization in primary C2C12 cells in vitro (Kim et al., 2019). Lack of SIK1, CRTC1 becomes de-phosphorylated and gets translocated to the nucleus, where it stimulates the activity of CREB and induces genes that are associated with osteoblast differentiation, including *Id1* (Kim et al., 2019). Despite these cell culture findings, global SIK1-deficient mice have a modest bone phenotype. In addition, little evidence currently exists to suggest that SIK1 plays an important role downstream of PTH1R in bone cells.

Studies initially conceived to investigate how PTH signaling suppresses sclerostin expression in osteocyte have revealed a major role for SIK2 and SIK3 downstream of PTH1R in bone cells. Sclerostin expression in osteocytes is under the control of MEF2 family transcription factors (Collette et al., 2012; Leupin et al., 2007). Since class IIa HDAC proteins block MEF2-driven gene expression in many biologic systems (Haberland et al., 2009), the role of class IIa HDACs in MEF2-driven sclerostin production was investigated. PTH signaling promoting class IIa HDAC dephosphorylation and nuclear translocation, and HDAC4 and HDAC5 were shown to be needed for PTH-induced *Sost* suppression in vivo. Since SIKs are known class IIa HDAC kinases in other cell types, we asked whether these kinases were involved in PTH-mediated class IIa HDAC dephosphorylation. Indeed, in vitro SIK2 and SIK3 knockdown and SIK inhibitor treatment of cultured osteocytes reduced class IIa HDAC phosphorylation and reduced Sost expression. Moreover, PTH signaling rapidly leads to PKA-dependent SIK2/SIK3 phosphorylation which occurs at time points prior to class IIa HDAC dephosphorylation. Moreover, mice lacking SIK2 in osteocytes failed to show PTH-induced *Sost* suppression (Wein et al., 2016).

These results suggested a central role for SIK2 and SIK3 in PTH-induced *Sost* suppression in osteocytes. However, subsequent studies demonstrated that these kinases also control other key aspects of osteocytic PTH action. In addition to reducing Sost expression, small molecule SIK inhibitors stimulate RANKL expression via regulation of CRTC family transcriptional coactivators. Thus, SIKs control expression of both key paracrine-acting osteocyte derived regulates, sclerostin and RANKL. Unbiased transcriptomic profiling demonstrated remarkable molecular concordance between the effects of PTH and small molecule SIK inhibitors in cultured osteocytes. These results motivated further studies where mice were treated with the tool SIK inhibitor YKL-05-099. Similar to PTH, treatment with YKL-05-099 increased bone formation and boosted trabecular bone mass in mice (Wein et al., 2016), providing a key initial demonstrating for the bone anabolic potential of small molecule SIK inhibitors in vivo.

Subsequent genetic studies clearly demonstrated that SIK2 and SIK3 are the key isoforms that control PTH1R action in osteoblasts and osteocytes. Careful analysis of a complete allelic series of conditional knockout mice lacking individual or double combinations of SIK alleles in Dmp1-Cre lineage cells (mainly osteoblasts and osteocytes) revealed dramatic increases in trabecular bone mass when SIK2 and SIK3 are deleted, with minimal skeletal phenotypes of other single and double SIK mutant strains. Serum bone turnover markers, histology, and histomorphometry confirmed substantial increases in bone turnover in the absence of SIK2 and SIK3. Furthermore, bone transcriptomic profiling demonstrated that overall changes in gene expression patterns were remarkably similar in SIK2/SIK3 compound mice and animals expressing the active PTH1R found in JMC in osteoblast/osteocyte lineage cells (Nishimori, Lai, et al., 2019; Nishimori, O'Meara, et al., 2019). These findings clearly established that SIK2 and SIK3 (genes closely linked on orthologous regions of human chromosome 11 and mouse chromosome 9, likely arising from a gene duplication event) play a central role in skeletal remodeling in vivo. Importantly, these findings were confirmed in adult mice using a global inducible gene deletion model (Tang et al., 2021), results that further establish the therapeutic potential of targeting these kinases in postnatal mice. Importantly, global/inducible SIK2/SIK3 gene deletion led to dramatic changes in bone without overt extra-skeletal pathology. Recently, interest in developing 'next generation' SIK inhibitors for therapeutic indications such as cancer (Fan et al., 2021; Raab et al., 2021; Tesch et al., 2021) and inflammatory diseases (Darling & Cohen, 2021; van Gijsel-Bonnello et al., 2021) has piqued. Therefore, studying the effects of potent/specific 'next generation' SIK2/SIK3 inhibitors in bone may represent a pathway towards novel orally-available bone anabolic agents.

Interestingly, macrophage-derived exosomes may contain micro-RNAs (miR-5106) that target SIK2 and SIK3 mRNAs in adjacent mesenchymal lineage bone cells to promote osteoblast maturation (Xiong et al., 2021). Therefore, it is possible that hematopoietic/mesenchymal crosstalk in bone cells involves regulation of the PTH1R/SIK pathway. Additional studies have explored the role of SIKs in responses to PTH using in vitro cell culture models. For example, PTH treatment of adipose-derived stem cells promotes osteoblast maturation via induction of the CRTC target gene WNT4 (An et al., 2019). Therefore, it is likely that multiple downstream target genes of PTH1R/SIK action exist at different stages of osteoblast maturation that in concert regulate bone formation.

As discussed above, teriparatide and abaloparatide are PTH1R agonist peptides that are FDA-approved for treatment of post-menopausal osteoporosis. Despite the fact that both ligands engage the same receptor, it appears that abaloparatide stimulates less bone resorption and hypercalcemia than teriparatide, a property that may account for clinical differences between the two medications (Miller et al., 2016). Therefore, studies were performed to assess whether teriparatide and abaloparatide differ in their ability to regulate the PTH1R/SIK pathway and downstream RANKL expression in cultured bone cells. In carefully-designed washout studies, abaloparatide may be slightly less effective at stimulate RANKL than teriparatide (Makino et al., 2018). These results are consistent with clear data from heterologous expression systems demonstrating that teriparatide is better at promoting sustained PTH1R signaling and endocytosis than abaloparatide (Hattersley, Dean, Corbin, Bahar, & Gardella, 2016; Liu et al., 2019). That being said, the majority of studies comparing teriparatide and abaloparatide side-by-side reveal largely comparable biochemical and gene expression responses when tested in cells expressing endogenous levels of PTH1R (Le Henaff et al., 2019; Ricarte, Le Henaff, Kolupaeva, Gardella, & Partridge, 2018; Sato et al., 2021). Further kinetic washout-based studies are needed to assess potential differences in PTH1R-mediated regulation of SIK2/SIK3 phosphorylation and SIK substrate status in PTH1R-expressing target cells.

6. Role of salt inducible kinases in responses to PTH in cartilage

The PTHrP/PTH1R system plays a central role in controlling growth plate chondrocytes during endochondral bone formation (Kronenberg, 2003). The observation that MEF2 family transcription factors play a central role in controlling terminal chondrocyte hypertrophy (Arnold et al., 2007) prompted studies addressing the role of class IIa HDACs in chondrocyte maturation. Genetic interaction studies revealed a key role for HDAC4, a gene previously implicated in chondrocyte maturation (Vega et al., 2004), as a mediator of the effects of PTHrP (Nishimori, Lai, et al., 2019). Notably, transgenic mice overexpressing PTHrP in the growth plate show severe arrests in chondrocyte maturation, an effect that is completely abolished with concomitant HDAC4 gene deletion. Moreover, chronically elevated PTHrP signaling in the growth plate increases nuclear HDAC4 protein levels in chondrocytes (Nishimori, Lai, et al., 2019).

These results clearly establish class IIa HDACs as mediators of PTHrP actions in the growth plate, and lead to the inevitable question about whether SIKs control class IIa HDACs downstream of PTH1R action. SIK2 expression in growth plate chondrocytes is very low, with SIK3 being the major SIK isoform expressed in the growth plate. Global SIK3 gene deletion leads to a growth plate phenotype quite similar to what is seen with transgenic PTHrP overexpression (Sasagawa et al., 2012). PTHrP signaling in chondrocytes leads to PKA-dependent SIK3 phosphorylation, and chondrocyte specific SIK3 deletion recapitulates the growth plate defects seen in global SIK3 deficient mice. Remarkably, SIK3 gene deletion rescues perinatal lethality in mice lacking PTHrP. Finally, analysis compound mutant mice demonstrated that HDAC4 lies genetically downstream of SIK3 in controlling chondrocyte maturation (Nishimori, Lai, et al., 2019; Nishimori, O'Meara, et al., 2019). Human genetics has confirmed a key role for SIK3 in controlling chondrocyte maturation. Humans with rare homozygous SIK3 mutations demonstrate a JMC-like skeletal dyscrasia (Csukasi et al., 2018). Metabolomic analysis of chondrocytes SIK3-deficient mice demonstrated reductions in acetyl-coA that may also underlie defective maturation in these cells (Kosai et al., 2019). Taken together, these results demonstrate striking parallel models in chondrocytes and osteocytes. In chondrocytes, PTHrP signaling, via SIK3, blocks MEF2-driven expression of the hypertrophic gene expression program. In osteocytes, PTH signaling, via SIK2 and SIK3, blocks MEF2-driven sclerostin expression (Fig. 1).

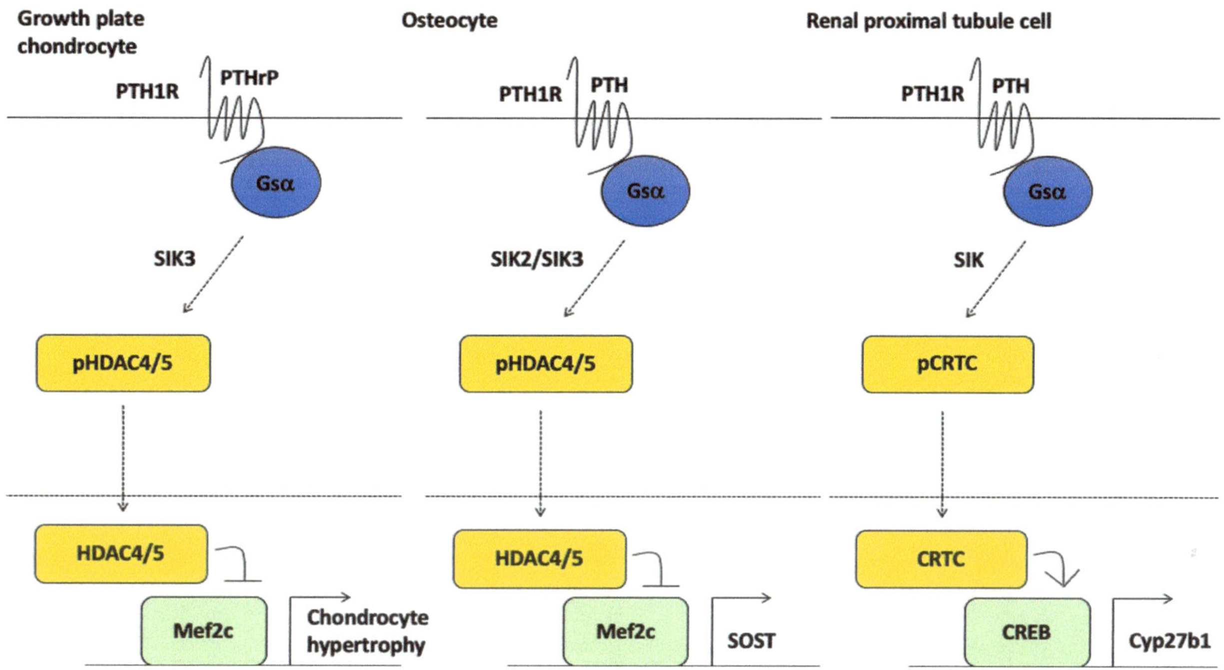

Fig. 1 Schematic model demonstrating a conserved role for SIKs downstream of PTH1R action in across cell types. Cyclic AMP generation leads to PKA-dependent SIK phosphorylation and inactivation. SIK inhibition reduces levels of SIK substrate phosphorylation. Upon dephosphorylation, SIK substrates translocate into the nucleus. Class IIa HDACs block MEF2-driven gene expression. In chondrocytes, MEF2 drives expression of the hypertrophic gene expression program which is blocked by PTHrP action. In osteocytes, MEF2 drives *SOST* expression which is blocked by PTH signaling. We hypothesize that PTH signaling in the proximal tubule stimulates *Cyp27b1* expression via a SIK/CRTC/CREB pathway, see text for details.

Recent studies have demonstrated a role for SIK3 in articular chondrocytes as well. PTHrP signaling in articular chondrocytes may prevent acquisition of a matrix catabolic state in the setting of inflammation and injury (Sampson et al., 2011). Accordingly, mice lacking SIK3 in chondrocytes are resistant to trauma-induced osteoarthritis. Treatment with pterosin B, an edible compound that degrades SIK3 protein, protects mice from injury-induced osteoarthritis (Yahara et al., 2016). Therefore, the robust body of evidence demonstrating an important role for SIK3 in growth plate chondrocyte maturation may have imminent therapeutic relevance in the setting of osteoarthritis.

7. Do salt inducible kinases participate in the renal actions of PTH?

Together with other hormones, such as FGF23 and 1,25-dihydroxy vitamin D, PTH contributes significantly to mineral metabolism by acting not only on bone, but also on kidneys. Three major actions of PTH in kidneys are (1) stimulation of Cyp27b1 expression; (2) inhibition of phosphate reabsorption; and (3) reduction in calcium excretion. As PTH secretion is stimulated by low serum calcium, all these actions work to protect the organism against hypocalcemia. At present, there are no published data regarding the role of SIKs in the renal actions of PTH. However, given the central role of these kinases downstream of PTH1R action in bone cells, it is highly likely that an analogous role in kidney may exist. Below, we will summarize the actions of PTH on distinct portions of the nephron, highlighted potential roles for SIKs in renal PTH biology along the way.

8. PTH stimulates Cyp27b1 expression

Vitamin D, obtained from ultraviolet rays of sunlight or diet, undergoes two-step hydroxylation to become biologically active and bind to vitamin D receptor (VDR). Vitamin D is first hydroxylated at C-25 by 25-hydroxylase (mainly Cyp2R1) in liver, then further metabolized to 1,25-dihydroxy-vitamin D (1,25-$(OH)_2$D) by 1-alpha-hydroxylase (Cyp27b1) in kidney. This cytochrome P450 enzyme Cyp27b1 is expressed in several tissues including kidney, bone, placenta, brain, and other endocrine organs (Bikle, Patzek, & Wang, 2018). However, extra-renal 1,25-dihydroxy-vitamin D production tends to be local (Adams & Hewison, 2012; Anderson & Atkins, 2008) and does not contribute significantly to the systemic 1,25-dihydroxy-vitamin D level and calcium homeostasis (Adams & Hewison, 2012). In addition, Cyp27b1 regulation by the canonical PTH receptor pathway via cAMP and PKA only occurs in the kidney (Adams et al., 2014; Van Der Meijden et al., 2016).

The role of PTH in the production of 1,25-dihydroxy-vitamin D (1,25-$(OH)_2D$) was first demonstrated in parathyroidectomized rats, whereas PTH administration rapidly induces the Cyp27b1 promoter (Garabedian, Holick, Deluca, & Boyle, 1972). This stimulatory effect of PTH on Cyp27b1 was shown in a discrete part of the nephron, the proximal tubule (Kawashima, Torikai, & Kurokawa, 1981). PTH exerts its action by binding to PTH receptor on proximal tubular epithelial cells, and subsequent increases in cAMP and PKA are predominantly responsible for inducing Cyp27b1 promoter activity (Armbrecht, Forte, Wongsurawat, Zenser, & Davis, 1984; Brenza & Deluca, 2000; Rost, Bikle, & Kaplan, 1981).

There are still unanswered questions in the exact molecular mechanism of how PTH stimulates Cyp27b1. However, an orphan nuclear receptor, Nr4a2, was shown to regulate PTH-induced Cyp27b1 expression in vitro (Zierold, Nehring, & DeLuca, 2007). Recently, Meyer et al. identified two kidney-specific, Cyp27b1 regulatory enhancers, Mettl1 (M1) and Mettl21 (M21) (Meyer et al., 2017). Among these, PTH induces Cyp27b1 expression via recruiting CREB (cAMP response element binding protein) to the M1 enhancer as demonstrated by Chromatin Immunoprecipitation sequencing (ChIP-Seq) and M1 knock-out and M1/M21 double knock-out mice. The important role for CREB in regulating renal Cyp27b1 expression raises the intriguing possibility that PTH1R/SIK/CRTC signaling may ultimately control these vital Cyp27b1 gene enhancers (Fig. 1). Recent evidence suggests that prolonged endosomal PTH1R signaling is required for renal Cyp27b1 induction (White et al., 2021). In light of this observation, if SIKs participate in renal Cyp27b1 induction, it will be of high interest to determine the intracellular spatial basis for this activity.

In addition, PTH also decreases renal 24-hydroxylase activity (Cyp24a1), which inactivates 1,25-$(OH)_2D$ by hydroxylation at C-24 (Shigematsu, Horiuchi, Ogura, Miyahara, & Suda, 1986), under cAMP-PKA pathway (Zierold, Reinholz, Mings, Prahl, & DeLuca, 2000). In a recent study, it was shown that PTH controls Cyp24a1 expression via a Cyp24a1 regulatory region, C24-DS1 (Meyer et al., 2019). All together, the actions of PTH in increasing Cyp27b1 expression and decreasing Cyp24a1 expression drive more 1,25-dihydroxy-vitamin D (1,25-$(OH)_2D$) production.

9. PTH decreases phosphate reabsorption in proximal tubule

Most of urinary phosphate is absorbed in proximal tubule, about 75% of which occurs in proximal convoluted tubule, and the rest is taken up by proximal straight tubule. In proximal tubule, Npt2a (sodium-phosphate co-transporter 2A; encoded by SLC34A1) in mice (and Npt2c in humans) are primarily responsible for divalent phosphate reabsorption (Blaine, Chonchol, & Levi, 2015). Its surface expression is stabilized by binding to a scaffold protein NHERF1 (Na^+/H^+ exchanger regulatory factor 1), which also assembles a protein complex associated with PTH receptor signaling, including the PTH receptor itself, phospholipase C (PLC), and actin, and stabilizes their expression as well at the apical membrane via cytoskeleton interaction (Hernando et al., 2002; Lederer, Khundmiri, & Weinman, 2003; Mahon & Segre, 2004; Murray, Lederer, & Khundmiri, 2015).

PTH regulates phosphate reabsorption in proximal tubule via Npt2a-NHERF1 complex. PTH activates PTH receptors on both apical and basolateral sides of proximal tubular epithelial cells, and phosphorylates NHERF1, instead of directly phosphorylating Npt2a (Deliot et al., 2005). This disrupts the binding of NHERF1 to Npt2a, which then becomes endocytosed via clathrin-coated pits and degraded in lysosomes (Pfister et al., 1998). Although there are some evidence that PTH decreases Npt2a mRNA transcription (Friedlaender et al., 2001; Kempson et al., 1995; Murray et al., 2013), rapid Npt2a internalization and downregulation occurs at the protein level without changing the mRNA abundance (Weinman, Steplock, Shenolikar, & Blanpied, 2011). As expected, in NHERF mutant mice and cells, Npt2a endocytosis is impaired. Interestingly, FGF23, another phosphaturic hormone, has also been shown to work through NHERF1 for its action (Sneddon et al., 2016). Phosphorylation sites of NHERF1 by PTH include Serine 77 and Threonine 95, located in the PDZ domain where Npt2a binding occurs (Weinman et al., 2010). Recently, it was also shown that Serine 290 is dephosphorylated by protein phosphatase 1 then subsequently re-phosphorylated by PTH, confirming reversible yet time-sensitive phosphorylation mechanism of PTH on phosphate transport (Zhang et al., 2019).

The inhibitory effect of PTH on phosphate reabsorption in proximal tubule is clearly demonstrated in many literatures in vitro and in vivo. Although there are several studies dissecting the signaling pathway, identifying the exact mechanism, and a potential role for SIKs, of how PTH signaling inhibits phosphate reabsorption in renal proximal tubule requires further research. Unlike 1,25-vitamin D regulation, the effects of PTH on phosphate reabsorption largely depend on post-transcriptional mechanisms. Since SIKs primarily control gene expression events in the nucleus, it is possible that SIKs control renal actions at the level of vitamin D activation but not phosphate handling.

10. PTH increases calcium reabsorption in distal tubule

Renal calcium reabsorption, in contrast to the two PTH actions described above, occurs at multiple sites throughout the nephron, reabsorbing ~98% of filtered calcium. Approximately 60% of filtered calcium is reabsorbed in proximal tubules, most of which is paracellular passive transport. Another site of calcium absorption is in the thick ascending limb, which absorbs 20–25% of filtered calcium. This process is mostly paracellular, regulated by a family of tight junction proteins called claudins (Moor & Bonny, 2016). There is also an evidence that PTH modulates calcium absorption in thick ascending limb through Claudin-14 (Sato et al., 2017). Lastly, distal tubule accounts for 5–10% of calcium reabsorption. Even though the relative contribution of calcium reabsorption in distal tubule may seem small, this process is highly regulated and the gene editing tools in vivo have demonstrated its significance (Hoenderop, De Pont, et al., 1999; Hoenderop et al., 2003; Ko, Choi, Oh, & Jeung, 2009; Loh et al., 2013).

Calcium reabsorption in distal tubule is an active, and transcellular process, involving 3 steps, (1) calcium entry into the cell at the apical surface, (2) intracellular buffering, (3) calcium exit at the basolateral membrane into the body. Calcium ions enter the cell via TRPV5 (transient receptor potential vanilloid 5), a calcium specific ion channel, and bind to calcium binding proteins, calbindin-D28k, and calmodulin if calbindin-D28k capacity is exceeded. These proteins serve as the buffers, and transport calcium to PMCA (plasma membrane calcium ATPase) or NCX1 (Na^{+}/Ca^{+2} Exchanger 1) at the basolateral membrane, for the calcium efflux.

PTH regulates calcium resorption in distal tubule via TRPV5. PTH increases the expression of TRPV5, at mRNA and protein level (Van Abel et al., 2005). Parathyroidectomy itself reduced TRPV5 transcript and protein abundance, which were rescued by PTH infusion in rats (Van Abel et al., 2005). PTH has also been shown to reduce caveolae-mediated

TRPV5 through PKC activation in vitro (Cha, Wu, & Huang, 2008). Groot et al. showed that T709 of TRPV5 becomes phosphorylated by PKA induced by PTH-cAMP pathway (De Groot et al., 2009; De Groot et al., 2011). This C-terminal part of TRPV5 containing T709 is where calmodulin binds to inactivate TRPV5 if excess Ca^{+2}. Thus, PTH induced phosphorylation at T709 diminishes calmodulin binding, which increases TRPV5 activity. The relative contribution of cAMP-PKA or PKC pathway is not as clear, but it is expected that both participate in the process of calcium absorption in distal tubule (Friedman, 1996; Hoenderop, De Pont, Bindels, & Willems, 1999). Moreover, as discussed previously, PTH increases 1,25-dihydroxy-vitamin D, which also increases TRPV5 and calbindin-D28k, enhancing calcium reabsorption in distal tubule (Hoenderop, Van Der Kemp, Urben, Strugnell, & Bindels, 2004). Since part of the distal tubule actions of PTH involve transcriptional regulation of TRPV5 expression, the possibility that PTH1R/SIK signaling controls TRPV5 expression via CRTC proteins remains open.

11. Summary and future directions

In conclusion, multiple established and ongoing lines of evidence demonstrate a key role for salt inducible kinases downstream of PTH1R in physiologically-important target tissues. PKA-mediated SIK regulation occurs in other GPCR signaling pathways (Darling & Cohen, 2021). Therefore, lessons learned from studying how SIKs participate in PTH actions may have broad implications for other physiologically-important cAMP-linked hormonal signaling modules. Currently, analysis of SIK action downstream of PTH has focused on two major classes of SIK2/SIK3 substrates, class IIa HDACs and CRTC proteins. However, it is highly likely that additional cell type-specific SIK substrates exist. For this reason, unbiased approaches to comprehensively map SIK-regulated phosphoproteins may illuminate new and important intracellular pathways used by these kinases in PTH1R-expressing target cells. Finally, the PTH receptor is a validated drug target for both osteoporosis and hypoparathyroidism. However, widespread use of PTH1R agonists is limited due to the need for daily subcutaneous injections. Orally-available small molecule SIK inhibitors bear tremendous potential for these indications among others. Information learned from detailed studies on the role of SIKs in bone remodeling will undoubtably inform future development of therapies that target this signaling pathway.

References

Adams, J. S., & Hewison, M. (2012). Extrarenal expression of the 25-hydroxyvitamin D-1-hydroxylase. *Archives of Biochemistry and Biophysics*, *523*(1), 95–102.

Adams, J. S., Rafison, B., Witzel, S., Reyes, R. E., Shieh, A., Chun, R., et al. (2014). Regulation of the extrarenal CYP27B1-hydroxylase. *The Journal of Steroid Biochemistry and Molecular Biology*, *144*, 22–27.

Altarejos, J. Y., & Montminy, M. (2011). CREB and the CRTC co-activators: Sensors for hormonal and metabolic signals. *Nature Reviews. Molecular Cell Biology*, *12*(3), 141–151.

An, Y., Zhao, J., Nie, F., Wu, Y., Xia, Y., & Li, D. (2019). Parathyroid hormone (PTH) promotes ADSC osteogenesis by regulating SIK2 and Wnt4. *Biochemical and Biophysical Research Communications*, *516*(2), 551–557.

Anderson, P. H., & Atkins, G. J. (2008). The skeleton as an intracrine organ for vitamin D metabolism. *Molecular Aspects of Medicine*, *29*(6), 397–406.

Armbrecht, H. J., Forte, L. R., Wongsurawat, N., Zenser, T. V., & Davis, B. B. (1984). Forskolin increases 1,25-Dihydroxyvitamin D3Production by rat renal slices in vitro*. *Endocrinology*, *114*(2), 644–649.

Arnold, M. A., Kim, Y., Czubryt, M. P., Phan, D., McAnally, J., Qi, X., et al. (2007). MEF2C transcription factor controls chondrocyte hypertrophy and bone development. *Developmental Cell*, *12*(3), 377–389.

Baron, R., & Gori, F. (2018). Targeting WNT signaling in the treatment of osteoporosis. *Current Opinion in Pharmacology*, *40*, 134–141.

Baron, R., & Kneissel, M. (2013). WNT signaling in bone homeostasis and disease: From human mutations to treatments. *Nature Medicine*, *19*(2), 179–192.

Bellido, T., Ali, A. A., Gubrij, I., Plotkin, L. I., Fu, Q., O'Brien, C. A., et al. (2005). Chronic elevation of parathyroid hormone in mice reduces expression of sclerostin by osteocytes: A novel mechanism for hormonal control of osteoblastogenesis. *Endocrinology*, *146*(11), 4577–4583.

Bikle, D. D., Patzek, S., & Wang, Y. (2018). Physiologic and pathophysiologic roles of extra renal CYP27b1: Case report and review. *Bone Reports*, *8*, 255–267.

Blaine, J., Chonchol, M., & Levi, M. (2015). Renal control of calcium, phosphate, and magnesium homeostasis. *Clinical Journal of the American Society of Nephrology*, *10*(7), 1257–1272.

Blume, S. W., & Curtis, J. R. (2011). Medical costs of osteoporosis in the elderly Medicare population. *Osteoporosis International*, *22*(6), 1835–1844.

Bonewald, L. F. (2011). The amazing osteocyte. *Journal of Bone and Mineral Research*, *26*(2), 229–238.

Brenza, H. L., & Deluca, H. F. (2000). Regulation of 25-Hydroxyvitamin D3 1α-hydroxylase gene expression by parathyroid hormone and 1,25-dihydroxyvitamin D3. *Archives of Biochemistry and Biophysics*, *381*(1), 143–152.

Bright, N. J., Thornton, C., & Carling, D. (2009). The regulation and function of mammalian AMPK-related kinases. *Acta Physiologica*, *196*(1), 15–26.

Brown, E. M. (1983). Four-parameter model of the sigmoidal relationship between parathyroid hormone release and extracellular calcium concentration in normal and abnormal parathyroid tissue*. *The Journal of Clinical Endocrinology and Metabolism*, *56*(3), 572–581.

Calvi, L. M., Sims, N. A., Hunzelman, J. L., Knight, M. C., Giovannetti, A., Saxton, J. M., et al. (2001). Activated parathyroid hormone/parathyroid hormone-related protein receptor in osteoblastic cells differentially affects cortical and trabecular bone. *Journal of Clinical Investigation*, *107*(3), 277–286.

Cha, S. K., Wu, T., & Huang, C. L. (2008). Protein kinase C inhibits caveolae-mediated endocytosis of TRPV5. *American Journal of Physiology. Renal Physiology*, *294*(5), F1212–F1221.

Collette, N. M., Genetos, D. C., Economides, A. N., Xie, L., Shahnazari, M., Yao, W., et al. (2012). Targeted deletion of Sost distal enhancer increases bone formation and bone mass. *Proceedings of the National Academy of Sciences of the United States of America*, *109*(35), 14092–14097.

Csukasi, F., Duran, I., Barad, M., Barta, T., Gudernova, I., Trantirek, L., et al. (2018). The PTH/PTHrP-SIK3 pathway affects skeletogenesis through altered mTOR signaling. *Science Translational Medicine*, *10*(459).

Darling, N. J., & Cohen, P. (2021). Nuts and bolts of the salt-inducible kinases (SIKs). *The Biochemical Journal*, *478*(7), 1377–1397.

De Groot, T., Kovalevskaya, N. V., Verkaart, S., Schilderink, N., Felici, M., Van Der Hagen, E. A. E., et al. (2011). Molecular mechanisms of calmodulin action on TRPV5 and modulation by parathyroid hormone. *Molecular and Cellular Biology*, *31*(14), 2845–2853.

De Groot, T., Lee, K., Langeslag, M., Xi, Q., Jalink, K., Bindels, R. J. M., et al. (2009). Parathyroid hormone activates TRPV5 via PKA-dependent phosphorylation. *Journal of the American Society of Nephrology*, *20*(8), 1693–1704.

Deliot, N., Hernando, N., Horst-Liu, Z., Gisler, S. M., Capuano, P., Wagner, C. A., et al. (2005). Parathyroid hormone treatment induces dissociation of type IIa Na+-P(i) cotransporter-Na+/H+ exchanger regulatory factor-1 complexes. *American Journal of Physiology. Cell Physiology*, *289*(1), C159–C167.

Estell, E. G., & Rosen, C. J. (2021). Emerging insights into the comparative effectiveness of anabolic therapies for osteoporosis. *Nature Reviews. Endocrinology*, *17*(1), 31–46.

Fan, D., Yang, H., Mao, W., Rask, P. J., Pang, L., Xu, C., et al. (2021). A novel salt inducible kinase 2 inhibitor, ARN-3261, sensitizes ovarian cancer cell lines and xenografts to carboplatin. *Cancers (Basel)*, *13*(3).

Friedlaender, M. M., Wald, H., Dranitzki-Elhalel, M., Zajicek, H. K., Levi, M., & Popovtzer, M. M. (2001). Vitamin D reduces renal NaPi-2 in PTH-infused rats: Complexity of vitamin D action on renal P(i) handling. *American Journal of Physiology. Renal Physiology*, *281*(3), F428–F433.

Friedman, P. A. (1996). Parathyroid hormone stimulation of calcium transport is mediated by dual signaling mechanisms involving protein kinase a and protein kinase. *Endocrinology*, *137*(1), 13–20.

Fu, Q., Jilka, R. L., Manolagas, S. C., & O'Brien, C. A. (2002). Parathyroid hormone stimulates receptor activator of NFkappa B ligand and inhibits osteoprotegerin expression via protein kinase a activation of cAMP-response element-binding protein. *The Journal of Biological Chemistry*, *277*(50), 48868–48875.

Garabedian, M., Holick, M. F., Deluca, H. F., & Boyle, I. T. (1972). Control of 25-hydroxycholecalciferol metabolism by parathyroid glands. *Proceedings of the National Academy of Sciences*, *69*(7), 1673–1676.

Haberland, M., Montgomery, R. L., & Olson, E. N. (2009). The many roles of histone deacetylases in development and physiology: Implications for disease and therapy. *Nature Reviews. Genetics*, *10*(1), 32–42.

Hattersley, G., Dean, T., Corbin, B. A., Bahar, H., & Gardella, T. J. (2016). Binding selectivity of abaloparatide for PTH-Type-1-receptor conformations and effects on downstream signaling. *Endocrinology*, *157*(1), 141–149.

Hernando, N., Deliot, N., Gisler, S. M., Lederer, E., Weinman, E. J., Biber, J., et al. (2002). PDZ-domain interactions and apical expression of type IIa Na/P(i) cotransporters. *Proceedings of the National Academy of Sciences of the United States of America*, *99*(18), 11957–11962.

Hoenderop, J. G. J., De Pont, J. J. H. H. M., Bindels, R. J. M., & Willems, P. H. G. M. (1999). Hormone-stimulated Ca2+ reabsorption in rabbit kidney cortical collecting system is cAMP-independent and involves a phorbol ester-insensitive PKC isotype. *Kidney International*, *55*(1), 225–233.

Hoenderop, J. G., van der Kemp, A. W., Hartog, A., van de Graaf, S. F., van Os, C. H., Willems, P. H., et al. (1999). Molecular identification of the apical Ca2+ channel in 1, 25-dihydroxyvitamin D3-responsive epithelia. *The Journal of Biological Chemistry*, *274*(13), 8375–8378.

Hoenderop, J. G. J., Van Der Kemp, A. W. C. M., Urben, C. M., Strugnell, S. A., & Bindels, R. J. M. (2004). Effects of vitamin D compounds on renal and intestinal Ca2+ transport proteins in 25-hydroxyvitamin D3-1α-hydroxylase knockout mice1. *Kidney International*, *66*(3), 1082–1089.

Hoenderop, J. G. J., Van Leeuwen, J. P. T. M., Van Der Eerden, B. C. J., Kersten, F. F. J., Van Derkemp, A. W. C. M., Mérillat, A.-M., et al. (2003). Renal Ca2+ wasting, hyperabsorption, and reduced bone thickness in mice lacking TRPV5. *The Journal of Clinical Investigation*, *112*(12), 1906–1914.

Jagannath, A., Butler, R., Godinho, S. I. H., Couch, Y., Brown, L. A., Vasudevan, S. R., et al. (2013). The CRTC1-SIK1 pathway regulates entrainment of the circadian clock. *Cell*, *154*(5), 1100–1111.

Jobert, A. S., Zhang, P., Couvineau, A., Bonaventure, J., Roume, J., Le Merrer, M., et al. (1998). Absence of functional receptors for parathyroid hormone and parathyroid hormone-related peptide in Blomstrand chondrodysplasia. *The Journal of Clinical Investigation*, *102*(1), 34–40.

Jüppner, H., Abou-Samra, A. B., Freeman, M., Kong, X. F., Schipani, E., Richards, J., et al. (1991). A G protein-linked receptor for parathyroid hormone and parathyroid hormone-related peptide. *Science*, *254*(5034), 1024–1026.

Kanis, J. A. (2002). Diagnosis of osteoporosis and assessment of fracture risk. *Lancet*, *359*(9321), 1929–1936.

Kawashima, H., Torikai, S., & Kurokawa, K. (1981). Localization of 25-hydroxyvitamin D3 1 alpha-hydroxylase and 24-hydroxylase along the rat nephron. *Proceedings of the National Academy of Sciences*, *78*(2), 1199–1203.

Keller, H., & Kneissel, M. (2005). SOST is a target gene for PTH in bone. *Bone*, *37*(2), 148–158.

Kempson, S. A., Lotscher, M., Kaissling, B., Biber, J., Murer, H., & Levi, M. (1995). Parathyroid hormone action on phosphate transporter mRNA and protein in rat renal proximal tubules. *The American Journal of Physiology*, *268*(4 Pt 2), F784–F791.

Kim, M. K., Kwon, J.-O., Song, M.-K., Kim, B., Kim, H., Lee, Z. H., et al. (2019). Salt-inducible kinase 1 regulates bone anabolism via the CRTC1-CREB-Id1 axis. *Cell Death & Disease*, *10*(11), 826.

Ko, S. H., Choi, K. C., Oh, G. T., & Jeung, E. B. (2009). Effect of dietary calcium and 1,25-(OH)2D3 on the expression of calcium transport genes in calbindin-D9k and -D28k double knockout mice. *Biochemical and Biophysical Research Communications*, *379*(2), 227–232.

Koo, S.-H., Flechner, L., Qi, L., Zhang, X., Screaton, R. A., Jeffries, S., et al. (2005). The CREB coactivator TORC2 is a key regulator of fasting glucose metabolism. *Nature*, *437*(7062), 1109–1111.

Kosai, A., Horike, N., Takei, Y., Yamashita, A., Fujita, K., Kamatani, T., et al. (2019). Changes in acetyl-CoA mediate Sik3-induced maturation of chondrocytes in endochondral bone formation. *Biochemical and Biophysical Research Communications*, *516*(4), 1097–1102.

Kronenberg, H. M. (2003). Developmental regulation of the growth plate. *Nature*, *423*(6937), 332–336.

Kronenberg, H. M. (2006). PTHrP and skeletal development. *Annals of the New York Academy of Sciences*, *1068*, 1–13.

Le Henaff, C., Ricarte, F., Finnie, B., He, Z., Johnson, J., Warshaw, J., et al. (2019). Abaloparatide at the same dose has the same effects on bone as PTH (1-34) in mice. *Journal of Bone and Mineral Research*.

Lederer, E. D., Khundmiri, S. J., & Weinman, E. J. (2003). Role of NHERF-1 in regulation of the activity of Na-K ATPase and sodium-phosphate co-transport in epithelial cells. *Journal of the American Society of Nephrology*, *14*(7), 1711–1719.

Leupin, O., Kramer, I., Collette, N. M., Loots, G. G., Natt, F., Kneissel, M., et al. (2007). Control of the SOST bone enhancer by PTH using MEF2 transcription factors. *Journal of Bone and Mineral Research*, *22*(12), 1957–1967.

Lin, X., Takemori, H., Katoh, Y., Doi, J., Horike, N., Makino, A., et al. (2001). Salt-inducible kinase is involved in the ACTH/cAMP-dependent protein kinase signaling in Y1 mouse adrenocortical tumor cells. *Molecular Endocrinology*, *15*(8), 1264–1276.

Liu, W., Feldman, J. D., Machado, H. B., Vician, L. J., & Herschman, H. R. (2003). Expression of depolarization-induced immediate early gene proteins in PC12 cells. *Journal of Neuroscience Research*, *72*(6), 670–678.

Liu, S., Jean-Alphonse, F. G., White, A. D., Wootten, D., Sexton, P. M., Gardella, T. J., et al. (2019). Use of backbone modification to enlarge the spatiotemporal diversity of parathyroid hormone Receptor-1 signaling via biased agonism. *Journal of the American Chemical Society*, *141*(37), 14486–14490.

Lizcano, J. M., Goransson, O., Toth, R., Deak, M., Morrice, N. A., Boudeau, J., et al. (2004). LKB1 is a master kinase that activates 13 kinases of the AMPK subfamily, including MARK/PAR-1. *The EMBO Journal*, *23*(4), 833–843.

Loh, N. Y., Bentley, L., Dimke, H., Verkaart, S., Tammaro, P., Gorvin, C. M., et al. (2013). Autosomal dominant hypercalciuria in a mouse model due to a mutation of the epithelial calcium channel, TRPV5. *PLoS One*, *8*(1), e55412.

Loshkajian, A., Roume, J., Stanescu, V., Delezoide, A. L., Stampf, F., & Maroteaux, P. (1997). Familial Blomstrand chondrodysplasia with advanced skeletal maturation: Further delineation. *American Journal of Medical Genetics*, *71*(3), 283–288.

Lupp, A., Klenk, C., Röcken, C., Evert, M., Mawrin, C., & Schulz, S. (2010). Immunohistochemical identification of the PTHR1 parathyroid hormone receptor in normal and neoplastic human tissues. *European Journal of Endocrinology*, *162*(5), 979–986.

Mahon, M. J., & Segre, G. V. (2004). Stimulation by parathyroid hormone of a NHERF-1-assembled complex consisting of the parathyroid hormone I receptor, phospholipase Cβ, and actin increases intracellular calcium in opossum kidney cells. *Journal of Biological Chemistry*, *279*(22), 23550–23558.

Makino, A., Takagi, H., Takahashi, Y., Hase, N., Sugiyama, H., Yamana, K., et al. (2018). Abaloparatide exerts bone anabolic effects with less stimulation of bone resorption-related factors: A comparison with Teriparatide. *Calcified Tissue International*, *103*(3), 289–297.

Mannstadt, M., Jüppner, H., & Gardella, T. J. (1999). Receptors for PTH and PTHrP: Their biological importance and functional properties. *The American Journal of Physiology*, *277*(5), F665–F675.

McCauley, L. K., & Martin, T. J. (2012). Twenty-five years of PTHrP progress: From cancer hormone to multifunctional cytokine. *Journal of Bone and Mineral Research*, *27*(6), 1231–1239.

Meyer, M. B., Benkusky, N. A., Kaufmann, M., Lee, S. M., Onal, M., Jones, G., et al. (2017). A kidney-specific genetic control module in mice governs endocrine regulation of the cytochrome P450 gene Cyp27b1 essential for vitamin D3 activation. *Journal of Biological Chemistry*, *292*(42), 17541–17558.

Meyer, M. B., Lee, S. M., Carlson, A. H., Benkusky, N. A., Kaufmann, M., Jones, G., et al. (2019). A chromatin-based mechanism controls differential regulation of the cytochrome

P450 gene Cyp24a1 in renal and non-renal tissues. *Journal of Biological Chemistry, 294*(39), 14467–14481.
Miller, P. D., Hattersley, G., Riis, B. J., Williams, G. C., Lau, E., Russo, L. A., et al. (2016). Effect of abaloparatide vs placebo on new vertebral fractures in postmenopausal women with osteoporosis: A randomized clinical trial. *JAMA, 316*(7), 722–733.
Moor, M. B., & Bonny, O. (2016). Ways of calcium reabsorption in the kidney. *American Journal of Physiology. Renal Physiology, 310*(11), F1337–F1350.
Murray, R. D., Holthouser, K., Clark, B. J., Salyer, S. A., Barati, M. T., Khundmiri, S. J., et al. (2013). Parathyroid hormone (PTH) decreases sodium-phosphate cotransporter type IIa (NpT2a) mRNA stability. *American Journal of Physiology. Renal Physiology, 304*(8), F1076–F1085.
Murray, R. D., Lederer, E. D., & Khundmiri, S. J. (2015). Role of PTH in the renal handling of phosphate. *AIMS Medical Science, 2*(3), 162–181.
Murray, T. M., Rao, L. G., Divieti, P., & Bringhurst, F. R. (2005). Parathyroid hormone secretion and action: Evidence for discrete receptors for the carboxyl-terminal region and related biological actions of carboxyl- terminal ligands. *Endocrine Reviews, 26*(1), 78–113.
Nishimori, S., Lai, F., Shiraishi, M., Kobayashi, T., Kozhemyakina, E., Yao, T. P., et al. (2019). PTHrP targets HDAC4 and HDAC5 to repress chondrocyte hypertrophy. *JCI Insight, 4*(5), e97903. https://doi.org/10.1172/jci.insight.97903.
Nishimori, S., O'Meara, M. J., Castro, C. D., Noda, H., Cetinbas, M., da Silva Martins, J., et al. (2019). Salt-inducible kinases dictate parathyroid hormone 1 receptor action in bone development and remodeling. *The Journal of Clinical Investigation, 129*(12), 5187–5203.
O'Brien, C. A., Nakashima, T., & Takayanagi, H. (2013). Osteocyte control of osteoclastogenesis. *Bone, 54*(2), 258–263.
Pfister, M. F., Ruf, I., Stange, G., Ziegler, U., Lederer, E., Biber, J., et al. (1998). Parathyroid hormone leads to the lysosomal degradation of the renal type II Na/pi cotransporter. *Proceedings of the National Academy of Sciences of the United States of America, 95*(4), 1909–1914.
Qiu, T., Xian, L., Crane, J., Wen, C., Hilton, M., Lu, W., et al. (2015). PTH receptor signaling in osteoblasts regulates endochondral vascularization in maintenance of postnatal growth plate. *Journal of Bone and Mineral Research, 30*(2), 309–317.
Raab, M., Rak, M., Tesch, R., Gasimli, K., Becker, S., Knapp, S., et al. (2021). The small-molecule inhibitor MRIA9 reveals novel insights into the cell cycle roles of SIK2 in ovarian cancer cells. *Cancers (Basel), 13*(15).
Reeve, J., Hesp, R., Williams, D., Hulme, P., Klenerman, L., Zanelli, J. M., et al. (1976). Anabolic effect of low doses of a fragment of human parathyroid hormone on the skeleton in postmenopausal osteoporosis. *Lancet, 1*(7968), 1035–1038.
Rhee, Y., Allen, M. R., Condon, K., Lezcano, V., Ronda, A. C., Galli, C., et al. (2011). PTH receptor signaling in osteocytes governs periosteal bone formation and intracortical remodeling. *Journal of Bone and Mineral Research, 26*(5), 1035–1046.
Ricarte, F. R., Le Henaff, C., Kolupaeva, V. G., Gardella, T. J., & Partridge, N. C. (2018). Parathyroid hormone (1-34) and its analogs differentially modulate osteoblastic RANKL expression via PKA/PP1/PP2A and SIK2/SIK3-CRTC3 signaling. *The Journal of Biological Chemistry*. https://doi.org/10.1074/Jbc.Ra118.004751.
Rosenbaum, D. M., Rasmussen, S. G. F., & Kobilka, B. K. (2009). The structure and function of G-protein-coupled receptors. *Nature, 459*(7245), 356–363.
Rost, C. R., Bikle, D. D., & Kaplan, R. A. (1981). In vitro stimulation of 25-hydroxycholecalciferol 1 alpha-hydroxylation by parathyroid hormone in chick kidney slices: Evidence for a role for adenosine 3′,5′-monophosphate. *Endocrinology, 108*(3), 1002–1006.

Sampson, E. R., Hilton, M. J., Tian, Y., Chen, D., Schwarz, E. M., Mooney, R. A., et al. (2011). Teriparatide as a chondroregenerative therapy for injury-induced osteoarthritis. *Science Translational Medicine*, *3*(101), 101ra193.

Sasagawa, S., Takemori, H., Uebi, T., Ikegami, D., Hiramatsu, K., Ikegawa, S., et al. (2012). SIK3 is essential for chondrocyte hypertrophy during skeletal development in mice. *Development*, *139*(6), 1153–1163.

Sato, T., Courbebaisse, M., Ide, N., Fan, Y., Hanai, J.-I., Kaludjerovic, J., et al. (2017). Parathyroid hormone controls paracellular Ca2+ transport in the thick ascending limb by regulating the tight-junction protein Claudin14. *Proceedings of the National Academy of Sciences*, *114*(16), E3344–E3353.

Sato, T., Verma, S., Khatri, A., Dean, T., Goransson, O., Gardella, T. J., et al. (2021). Comparable initial engagement of intracellular signaling pathways by parathyroid hormone receptor ligands Teriparatide, Abaloparatide, and long-acting PTH. *JBMR Plus*, *5*(5), e10441.

Schipani, E., Kruse, K., & Jüppner, H. (1995). A constitutively active mutant PTH-PTHrP receptor in Jansen-type metaphyseal chondrodysplasia. *Science*, *268*(5207), 98–100.

Shigematsu, T., Horiuchi, N., Ogura, Y., Miyahara, T., & Suda, T. (1986). Human parathyroid hormone inhibits renal 24-hydroxylase activity of 25-Hydroxyvitamin D3by a mechanism involving adenosine 3′,5′-monophosphate in rats*. *Endocrinology*, *118*(4), 1583–1589.

Silva, B. C., Costa, A. G., Cusano, N. E., Kousteni, S., & Bilezikian, J. P. (2011). Catabolic and anabolic actions of parathyroid hormone on the skeleton. *Journal of Endocrinological Investigation*, *34*(10), 801–810.

Sneddon, W. B., Ruiz, G. W., Gallo, L. I., Xiao, K., Zhang, Q., Rbaibi, Y., et al. (2016). Convergent signaling pathways regulate parathyroid hormone and fibroblast growth factor-23 action on NPT2A-mediated phosphate transport. *Journal of Biological Chemistry*, *291*(36), 18632–18642.

Sonntag, T., Vaughan, J. M., & Montminy, M. (2018). 14-3-3 proteins mediate inhibitory effects of cAMP on salt-inducible kinases (SIKs). *The FEBS Journal*, *285*(3), 467–480.

Sözen, T., Özışık, L., & Başaran, N.Ç. (2017). An overview and management of osteoporosis. *European Journal of Rheumatology and Inflammation*, *4*(1), 46–56.

Takayanagi, H. (2021). RANKL as the master regulator of osteoclast differentiation. *Journal of Bone and Mineral Metabolism*, *39*(1), 13–18.

Tang, C. C., Castro Andrade, C. D., O'Meara, M. J., Yoon, S. H., Sato, T., Brooks, D. J., et al. (2021). Dual targeting of salt inducible kinases and CSF1R uncouples bone formation and bone resorption. *eLife*, *10*.

Tesch, R., Rak, M., Raab, M., Berger, L. M., Kronenberger, T., Joerger, A. C., et al. (2021). Structure-based Design of Selective Salt-Inducible Kinase Inhibitors. *Journal of Medicinal Chemistry*, *64*(12), 8142–8160.

Van Abel, M., Hoenderop, J. G. J., Van Der Kemp, A. W. C. M., Friedlaender, M. M., Van Leeuwen, J. P. T. M., & Bindels, R. J. M. (2005). Coordinated control of renal Ca2+ transport proteins by parathyroid hormone. *Kidney International*, *68*(4), 1708–1721.

Van Der Meijden, K., Van Essen, H. W., Bloemers, F. W., Schulten, E. A. J. M., Lips, P., & Bravenboer, N. (2016). Regulation of CYP27B1 mRNA expression in primary human osteoblasts. *Calcified Tissue International*, *99*(2), 164–173.

van Gijsel-Bonnello, M., Darling, N. J., McDonald, L., Sime, M., Clark, J., Mezna, M., et al. (2021). Salt-inducible kinase (SIK) inhibition is protective in a mouse model of asthma. *bioRxiv*. 2021.2006.2007.447230.

Vega, R. B., Matsuda, K., Oh, J., Barbosa, A. C., Yang, X., Meadows, E., et al. (2004). Histone deacetylase 4 controls chondrocyte hypertrophy during skeletogenesis. *Cell*, *119*(4), 555–566.

Wang, Z.-N., Takemori, H., Halder, S. K., Nonaka, Y., & Okamoto, M. (1999). Cloning of a novel kinase (SIK) of the SNF1/AMPK family from high salt diet-treated rat adrenal1. *FEBS Letters*, *453*(1–2), 135–139.

Wein, M. N. (2018). Parathyroid hormone signaling in osteocytes. *JBMR Plus*, *2*(1), 22–30.

Wein, M. N., Foretz, M., Fisher, D. E., Xavier, R. J., & Kronenberg, H. M. (2018). Salt-inducible kinases: Physiology, regulation by cAMP, and therapeutic potential. *Trends in Endocrinology and Metabolism*, *29*(10), 723–735.

Wein, M. N., & Kronenberg, H. M. (2018). Regulation of bone remodeling by parathyroid hormone. *Cold Spring Harbor Perspectives in Medicine*, *8*(8).

Wein, M. N., Liang, Y., Goransson, O., Sundberg, T. B., Wang, J., Williams, E. A., et al. (2016). SIKs control osteocyte responses to parathyroid hormone. *Nature Communications*, 7, 13176.

Weinman, E. J., Steplock, D., Shenolikar, S., & Blanpied, T. A. (2011). Dynamics of PTH-induced disassembly of Npt2a/NHERF-1 complexes in living OK cells. *American Journal of Physiology. Renal Physiology*, *300*(1), F231–F235.

Weinman, E. J., Steplock, D., Zhang, Y., Biswas, R., Bloch, R. J., & Shenolikar, S. (2010). Cooperativity between the phosphorylation of Thr95 and Ser77 of NHERF-1 in the hormonal regulation of renal phosphate transport. *Journal of Biological Chemistry*, *285*(33), 25134–25138.

White, A. D., Pena, K. A., Clark, L. J., Maria, C. S., Liu, S., Jean-Alphonse, F. G., et al. (2021). Spatial bias in cAMP generation determines biological responses to PTH type 1 receptor activation. *Science Signaling*, *14*(703), eabc5944.

Wright, N. C., Looker, A. C., Saag, K. G., Curtis, J. R., Delzell, E. S., Randall, S., et al. (2014). The recent prevalence of osteoporosis and low bone mass in the United States based on bone mineral density at the femoral neck or lumbar spine. *Journal of Bone and Mineral Research*, *29*(11), 2520–2526.

Xiong, Y., Chen, L., Yan, C., Zhou, W., Yu, T., Sun, Y., et al. (2021). Correction to: M2 macrophagy-derived exosomal miRNA-5106 induces bone mesenchymal stem cells towards osteoblastic fate by targeting salt-inducible kinase 2 and 3. *Journal of Nanobiotechnology*, *19*(1), 88.

Yahara, Y., Takemori, H., Okada, M., Kosai, A., Yamashita, A., Kobayashi, T., et al. (2016). Pterosin B prevents chondrocyte hypertrophy and osteoarthritis in mice by inhibiting Sik3. *Nature Communications*, 7, 10959.

Zhang, Q., Xiao, K., Paredes, J. M., Mamonova, T., Sneddon, W. B., Liu, H., et al. (2019). Parathyroid hormone initiates dynamic NHERF1 phosphorylation cycling and conformational changes that regulate NPT2A-dependent phosphate transport. *Journal of Biological Chemistry*, *294*(12), 4546–4571.

Zhao, L.-H., Ma, S., Sutkeviciute, I., Shen, D.-D., Zhou, X. E., de Waal, P. W., et al. (2019). Structure and dynamics of the active human parathyroid hormone receptor-1. *Science*, *364*(6436), 148–153.

Zierold, C., Nehring, J. A., & DeLuca, H. F. (2007). Nuclear receptor 4A2 and C/EBPbeta regulate the parathyroid hormone-mediated transcriptional regulation of the 25-hydroxyvitamin D3-1alpha-hydroxylase. *Archives of Biochemistry and Biophysics*, *460*(2), 233–239.

Zierold, C., Reinholz, G. G., Mings, J. A., Prahl, J. M., & DeLuca, H. F. (2000). Regulation of the procine 1,25-dihydroxyvitamin D3-24-hydroxylase (CYP24) by 1,25-dihydroxyvitamin D3 and parathyroid hormone in AOK-B50 cells. *Archives of Biochemistry and Biophysics*, *381*(2), 323–327.

CHAPTER THREE

Physiological regulation of phosphate homeostasis

Joanne Marks[a] and Robert J. Unwin[a,b,*]
[a]Department of Neuroscience, Physiology & Pharmacology, University College London, London, United Kingdom
[b]Department of Renal Medicine, University College London, London, United Kingdom
*Corresponding author: e-mail address: robert.unwin@ucl.ac.uk

Contents

Abstract

Phosphate homeostasis is dependent on the interaction and coordination of four main organ systems: thyroid/parathyroids, gastrointestinal tract, bone and kidneys, and three key hormonal regulators, 1,25-hydroxyvitamin D3, parathyroid hormone and FGF23 with its co- factor klotho. Phosphorus is a critical nutritional element for normal cellular function, but in excess can be toxic to tissues, particularly the vasculature. As phosphate, it also has an important interaction and inter-dependence with calcium and calcium homeostasis sharing some of the same controlling hormones, although this is not covered in our article. We have chosen to provide a current overview of phosphate homeostasis only, focusing on the role of two major organ systems, the gastrointestinal tract and kidneys, and their contribution to the control of phosphate balance. We describe in some detail the mechanisms of intestinal and renal phosphate transport, and compare and contrast their regulation. We also consider a significant example

Vitamins and Hormones, Volume 120
ISSN 0083-6729
https://doi.org/10.1016/bs.vh.2022.04.007

of phosphate imbalance, with phosphate retention, which is chronic kidney disease; why consequent hyperphosphatemia is important, and some of the newer means of managing it.

Abbreviations

$1,25(OH)_2$vitD 1,25 dihydroxyvitamin D_3
CKD chronic kidney disease
FGF 23 fibroblast growth factor 23
Pi phosphate
PTH parathyroid hormone

1. Introduction

Phosphorus is one of the most abundant elements in the body and complexes with oxygen to form phosphate. Approximately 85% of phosphate (0.5–1 kg) is contained in bone as a constituent of crystalline hydroxyapatite; in other tissues it is found mainly in the intracellular compartment as a constituent of various organic compounds, including nucleic acids and membrane phospholipids. Phosphate is involved in both aerobic and an anaerobic metabolism, and in red blood cells, as 2,3-disphosphoglycerate (2,3-DPG), it has an essential role in oxygen delivery to tissues, and as ADP and ATP, it is a means of storing energy. While phosphate is a major intracellular anion, it is also present in plasma or serum. The normal serum phosphate concentration in adults is 0.81–1.45 mmol/L (2.5–4.5 mg/dL) (Burtis, Ashwood, Bruns, & Sawyer, 2008); in young children it is 30%–50% higher, depending on their age, probably because it is so essential for normal growth. Indeed, maintaining a stable extracellular phosphate concentration is essential for many biological processes, including skeletal development, conserving bone integrity, energy metabolism, cell signaling, and regulation of protein synthesis.

High serum phosphate levels are seen typically with chronic kidney disease (CKD), hyperparathyroidism, and metabolic or respiratory acidosis, whereas low serum phosphate levels can be due to excess alcohol, burns, starvation, and chronic treatment with some diuretics or bisphosphonates (Liamis, Milionis, & Elisaf, 2010). It is also important to recognize that phosphate can be present in biofluids in multiple ionized forms, depending on pH: in normal serum it exists primarily as divalent (HPO_4^{2-}) and monovalent ($H_2PO_4^{-1}$) phosphate; 25% is protein bound, 75% is ionized and freely filtered, some of which is complexed with Ca^{2+}, Mg^{2+} and Na^+.

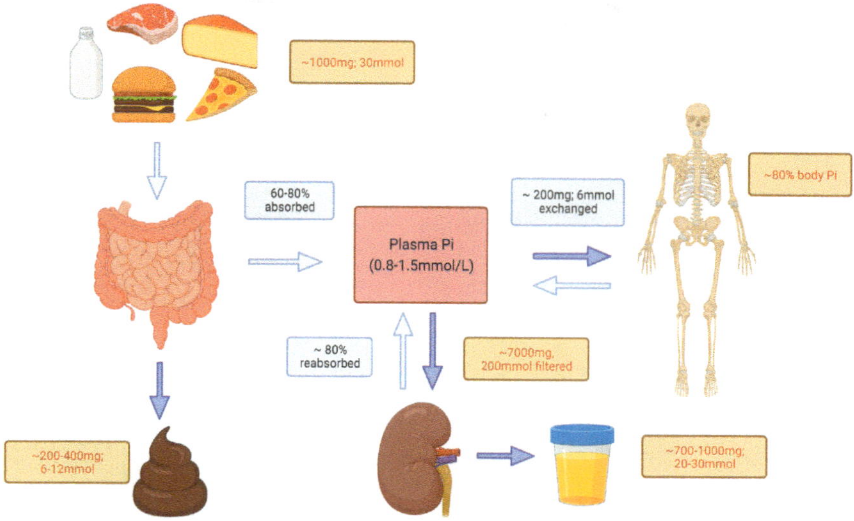

Fig. 1 Phosphate balance. The main organ systems responsible for body distribution of phosphate (Pi) and for maintaining phosphate balance. The main food sources high in phosphate include dairy and meat, and especially highly processed foodstuffs containing artificial flavorings and preservatives. Intestinal absorption tends not to be limiting. The figures given are for estimates of daily absorption, excretion and exchange. Phosphate in tissue exists mainly as PO_4^{3-} and in serum or plasma as HPO_4^{2-} or $H_2PO_4^{-}$, depending on pH.

Dietary intake of phosphate varies between 0.5 and 1.5 kg daily; its excretion in urine approximates net gastrointestinal absorption (*ca.* 60%–80%) to maintain phosphate balance. Phosphate in bone is a reservoir and can buffer changes in plasma and intracellular phosphate (Fig. 1). The physiological regulation of phosphate homeostasis is achieved by way of complex interactions between the kidney, intestine and bone, and several endocrine factors (Blau & Collins, 2015) (Fig. 2). Dietary phosphate depletion can occur, but it is normally compensated by renal phosphate conservation, unless the kidneys are themselves a primary source of phosphate loss.[a]

Any derangement in maintaining overall phosphate balance can have significant clinical consequences. The most commonly described are an associated increase in cardiovascular (CV) risk (Moe & Chen, 2008) and mortality (Stevens, Djurdjev, Cardew, Cameron, & Levin, 2004) in patients with phosphate overload caused by CKD, particularly in those patients receiving dialysis treatment (Copland, Komenda, Weinhandl,

[a] In chronic kidney disease (CKD), fractional phosphate excretion, which is normally <20%, increases significantly—see text for more detail (Bellasi et al., 2019).

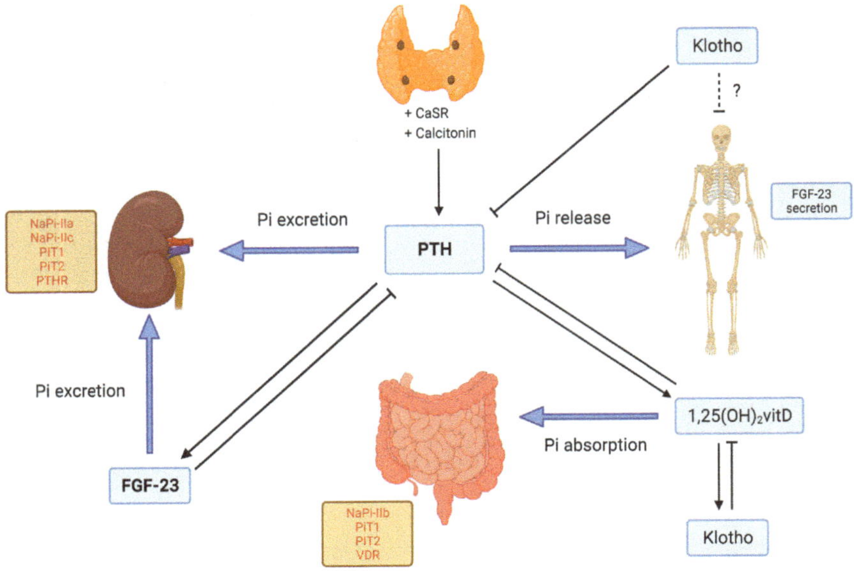

Fig. 2 Phosphate homeostasis. The main hormonal regulators, their sources, and the organ systems they act on to control phosphate (Pi) homeostasis, showing stimulation and inhibition. *Abbreviations:* CaSR, calcium sensing receptor; FGF23, fibroblast growth factor 23; PTHR, parathyroid hormone receptor; NaPi-IIa,b,c, and PiT-1,2 sodium-phosphate co-transporters (see text for details).

McCullough, & Morfin, 2016). A major complication of CKD is a rise in serum phosphate concentration leading to hyperphosphatemia, which drives increases in serum levels of parathyroid hormone (PTH) and fibroblast growth factor 23 (FGF23), and is associated with a decrease in active vitamin D levels, as well as the FGF23 co-receptor klotho. As GFR declines, detectable increases in FGF23 and PTH levels, and a decrease in klotho, occur *before* a measurable increase in serum phosphate concentration, making it difficult to know which factor may be the principal driver of these changes seen in CKD. This hormonal pattern is closely associated with the major complications of bone and cardiac disease, including vascular calcification, which occur in CKD (Neyra, Hu, & Moe, 2021). In early CKD, the physiological adaptation of elevated fibroblast FGF23 levels to maintain serum phosphate concentration is thought to be a direct cause of left ventricular hypertrophy (LVH), while hyperphosphatemia in late CKD stimulates vascular smooth muscle cells to undergo a phenotypic change that predisposes to calcification (McCullough, 2021).

The concept of phosphate toxicity in patients with normal renal function, particularly in those who consume a high proportion of processed food containing substantial amounts of highly bioavailable phosphate as additives and preservatives, is a subject of current concern and ongoing debate (Cooke, 2017; Erem & Razzaque, 2018; Rubio-Aliaga, 2020). In contrast, hypophosphatemia, often associated with acquired or genetic disorders that cause renal phosphate wasting, leads to important abnormalities of skeletal development and bone integrity, including rickets and osteomalacia in children, and osteomalacia in adults (Penido & Alon, 2012). Thus, by understanding and targeting the intricate processes involved in the maintenance of phosphate homeostasis, it is possible that newer therapeutic strategies may help to prevent the clinical manifestations and complications of phosphate imbalance.

2. An overview of phosphate homeostasis

Absorption of dietary phosphate from the gastrointestinal tract occurs predominately in the small intestine *via* both paracellular and transcellular routes, with the latter stimulated by the active form of vitamin D, 1,25 dihydroxyvitamin D_3 ($1,25(OH)_2$vitD). Extracellular phosphate concentrations are maintained by the hormones FGF23 and PTH, both of which inhibit renal phosphate reabsorption and increase the excretion of phosphate, and also control $1,25(OH)_2$vitD levels (Fig. 3)

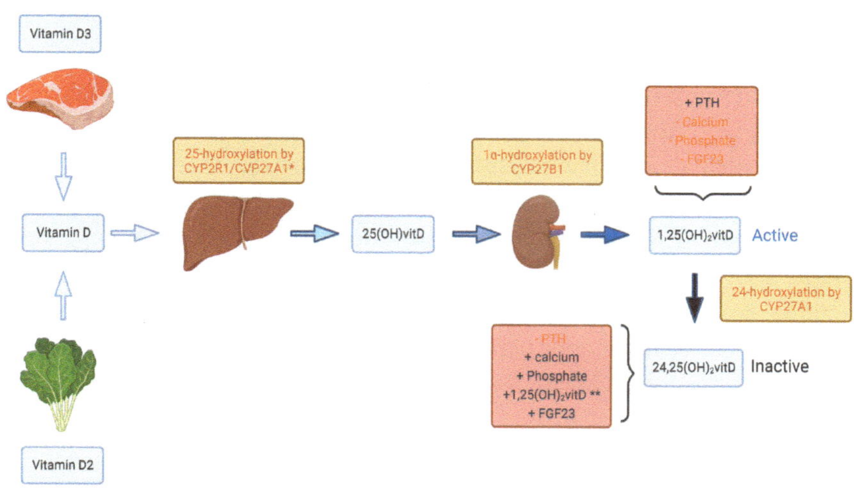

Fig. 3 Vitamin D metabolism. Vitamin D sources and the enzymes and factors, mainly PTH and FGF23 controlling its metabolism and illustrating the feedback control of active vitamin D synthesis (see text for details).

(Jacquillet & Unwin, 2019). The bone functions as a storage compartment for phosphate, containing approximately 85% of total body phosphate stores, in which it is complexed with calcium as hydroxyapatite crystals. PTH is involved in the regulation of bone turnover, resulting in the release of phosphate from the skeleton, while osteocytes and osteoblasts are responsible for the production of FGF-23 (Serna & Bergwitz, 2020).

3. Intestinal phosphate absorption

While it is recognized that the kidneys play a dominant role in the maintenance of extracellular phosphate balance through its excretion, it is also widely acknowledged that intestinal phosphate absorption is important and a promising therapeutic target for treating phosphate imbalance. However, we know less about the mechanisms involved in intestinal phosphate absorption than we do about the renal handling of phosphate, and it is still unclear how much phosphate absorption along the gut can be affected by changes in dietary intake of phosphate.

Studies performed in the 1970s on healthy volunteers documented that intestinal phosphate transport could be resolved into sodium-dependent and sodium-independent components, and that the human jejunum had a greater transport capacity than the ileum (Juan, Liptak, & Gray, 1976; Walton & Gray, 1979). These findings correlated with studies in rats that showed that the jejunum is also the dominant site for phosphate absorption in this species (McHardy & Parsons, 1956; Walling, 1977), and that transport at low phosphate concentrations is an active and saturable sodium-dependent process, but that at higher phosphate concentrations (>1 mmol/L) it is non-saturable and sodium-independent (Berner, Kinne, & Murer, 1976; Borowitz, Said, & Ghishan, 1986; Lee, Walling, & Corry, 1986). It is now accepted that these two pathways represent the transcellular and paracellular routes of phosphate absorption, respectively (Fig. 4) (Cozzolino, Ketteler, & Wagner, 2020; Marks, 2019; Saurette & Alexander, 2019).

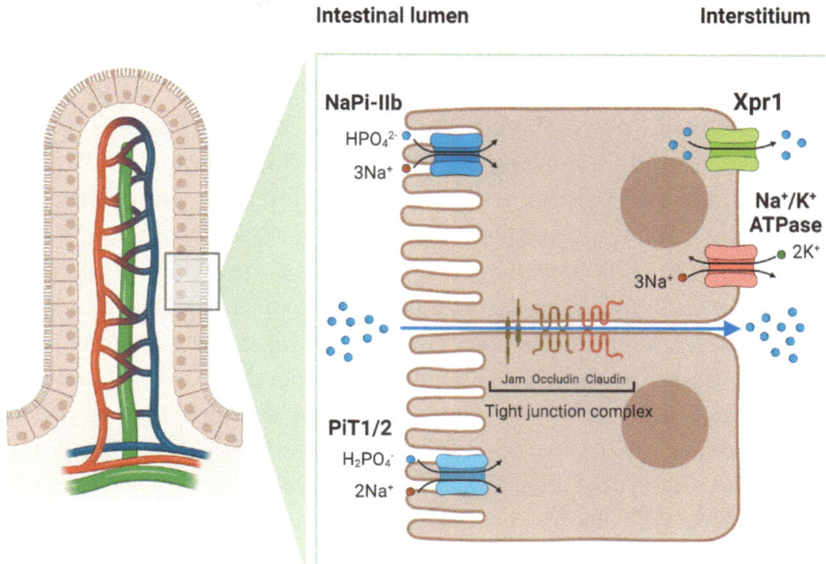

Fig. 4 An intestinal villous with two adjacent enterocytes showing the pathways and mechanisms for transcellular (NaPi-IIb and PiT1/2 for cell entry and putative Xrp1 for cell exit) and paracellular phosphate absorption regulated by tight junction proteins such as claudin 3 (see text for details).

4. Transcellular intestinal phosphate absorption and the role of NaPi-IIb

With the cloning of *SLC34A2* in 1998 (Hilfiker et al., 1998) and the finding that levels of the encoded protein NaPi-IIb can be regulated by dietary phosphate restriction and 1,25(OH)$_2$vitD (Hattenhauer, Traebert, Murer, & Biber, 1999; Katai et al., 1999), the focus of much of the subsequent research has been on clarifying the role of this protein in transcellular phosphate absorption. Kinetic studies of NaPi-IIb expressed in *Xenopus laevis* oocytes revealed the protein is a high-affinity sodium-dependent phosphate cotransporter with an apparent Km for phosphate of ~50 µmol/L and a Km for sodium of ~30 mmol/L (Hilfiker et al., 1998). The protein has a preference for HPO_4^{2-} and is electrogenic, transporting phosphate with a stoichiometry of 3:1 Na^+:HPO_4^{2-}, and is inhibited by phosphonoformic acid (PFA; foscarnet[b]) (Virkki, Biber, Murer, & Forster, 2007). Immunohistochemical studies consistently localize NaPi-IIb protein to the apical membrane of intestinal enterocytes (Giral et al., 2009; Hilfiker et al., 1998; Radanovic, Wagner, Murer, & Biber, 2005), although surprisingly there are significant differences in the regional profile for NaPi-IIb expression along the gastrointestinal tract of different rodent species, with expression occurring maximally in the ileum of mice (Marks et al., 2006; Radanovic et al., 2005) and in the jejunum of rats (Giral et al., 2009; Marks et al., 2006). (Humans are assumed to mirror more closely the regional pattern of phosphate absorption found in rats, rather than in mice.) Using conditions *in vitro* that favor sodium-dependent transcellular transport, uptake studies have correlated the profile for phosphate absorption in each rodent species to the regional expression of NaPi-IIb protein (Giral et al., 2009; Marks, Lee, Nadaraja, Debnam, & Unwin, 2015; Radanovic et al., 2005). In addition, by using tamoxifen-inducible *SLC34A2* knockout (KO) or intestinal specific *SLC34A2* KO models, it has been shown that ablation of this protein reduces transcellular phosphate absorption by 90% (Sabbagh et al., 2009) and 95% (Hernando et al., 2015), respectively. Importantly, when fed a standard phosphate diet, these mice have only moderately increased fecal phosphate excretion and can maintain normal serum phosphate levels due to an increase in renal phosphate reabsorption. This results in only a mild phenotype, despite the reduced transcellular intestinal phosphate absorption (Hernando et al., 2015). In contrast, when dietary phosphate is restricted, intestinal specific *SLC34A2* KO mice show elevated calcium excretion and lower bone mineral density, both indicative of increased bone resorption, leading to the suggestion that during long-term phosphate restriction, NaPi-IIb may be critical for bone health (Knöpfel et al., 2017).

4.1 Other sodium-dependent transporters involved in transcellular phosphate absorption

While it is apparent that under most circumstances NaPi-IIb is the rate-limiting step for transcellular phosphate absorption, other sodium-dependent phosphate cotransporters, such as members of the SLC20 (type III) family, PiT-1 (SLC20A1) and PiT-2 (SLC20A2), and sodium-independent mechanism(s), including a less well-defined H^+-dependent transporter, have

[b] Used as an early anti-viral to treat patients with HIV infection—hypophosphatemia was a commonly observed side effect.

also been identified (Candeal, Caldas, Guillén, Levi, & Sorribas, 2014, 2017). PiT1 and PiT2 are both sodium-dependent phosphate cotransporters with an apparent Km for phosphate of ~100 μmol/L and a Km for sodium of ~50 mmol/L. However, they preferentially transport H_2PO^{4-}, rather than HPO_4^{2-}, with a stoichiometry of 2:1 Na^+:$H_2PO_4^-$, and are not inhibited by PFA (Virkki et al., 2007). Unlike members of the SLC34 family, PiT proteins are ubiquitously expressed and present in epithelial and non-epithelial tissues (Forster, Hernando, Biber, & Murer, 2012). In the intestine they are localized to the apical membrane of enterocytes and there are regional and species differences in the profile of these proteins. In rats, both proteins are present in intestinal regions with a high absorptive capacity for phosphate (Candeal et al., 2017; Giral et al., 2009), while in mice, mRNA expression levels are similar throughout the small intestine, with the transcripts, particularly for PiT-1, found at a higher abundance in the distal colon (Pastor-Arroyo et al., 2020). Levels of PiT-1 and PiT-2 protein in rats have been shown to adapt to dietary phosphate intake (Aniteli et al., 2018; Candeal et al., 2017; Giral et al., 2009), although adaptation occurs with varying time-courses in the different intestinal segments (Marks, 2019). Using *SLC20A2* KO mice, Pastor-Arroyo et al. (2020) have recently demonstrated that the absence of this protein does not impact uptake in jejunal brush border membrane (BBM) vesicles from animals maintained on a normal phosphate diet, nor does it affect serum phosphate concentrations, or fecal or urinary phosphate excretion when the animals are maintained on a normal or low phosphate diet. However, following dietary phosphate restriction, *SLC20A2* KO mice have elevated levels of $1,25(OH)_2$vitD and are hypercalciuric (Pastor-Arroyo et al., 2020).

In a rat model of CKD, a reduction in NaPi-IIb-mediated transport has been reported that is partly compensated by increased activity of an alternative lower affinity transport pathway, possibly PiT-1-mediated (Ichida et al., 2020). Moreover, the pan-phosphate transporter inhibitor EOS789, which targets both NaPi-IIb *and* PiT-1/2, seems to be more effective than a selective NaPi-IIb inhibitor in controlling hyperphosphatemia in rats with CKD (Tsuboi et al., 2020), suggesting that SLC20 proteins may have a more significant role in transcellular intestinal phosphate absorption during CKD than had been thought up to now; however, how relevant these pre-clinical findings are to humans is still unclear. Intestinal phosphate absorption on a typical dietary intake did not differ in patients with moderate

CKD compared with controls, despite lower serum levels 1,25(OH)$_2$vitD (Stremke et al., 2021). Thus, unlike the kidneys, the intestine does not appear to compensate in humans by limiting phosphate absorption as kidney function declines, which had also been noted previously in a rat model of CKD (Marks et al., 2007).

4.2 Endocrine control of transcellular phosphate absorption

As in the kidney, the transcellular pathway for phosphate transport in the small intestine is under endocrine control. NaPi-IIb protein levels are upregulated by 1,25(OH)$_2$vitD administration, resulting in enhanced intestinal phosphate absorption (Hattenhauer et al., 1999; Marks et al., 2006). The effect of 1,25(OH)$_2$vitD in adults seems to occur in a non-transcriptional manner (Hattenhauer et al., 1999; Hernando, Gagnon, & Lederer, 2021; Marks et al., 2006; Xu, Inouye, Missey, Collins, & Ghishan, 2002), which contrasts with the effect in suckling rats, which is mediated partly by increased gene transcription (Xu et al., 2002). Recently, Hernando et al. (2021) have demonstrated that administration of 1,25(OH)$_2$vitD specifically increases transcellular, but not paracellular, phosphate absorption and that this effect is absent in *SLC34A2* KO mice. Taken together with the finding that the mRNA transcripts for PiT-1 and PiT-2 were unaltered, suggests that 1,25(OH)$_2$vitD enhances intestinal phosphate transport by a mechanism that exclusively involves NaPi-IIb.

Dietary phosphate is the other major regulator of transcellular phosphate absorption, with low dietary phosphate intake elevating circulating 1,25(OH)$_2$vitD levels and enhancing intestinal phosphate transport *via* NaPi-IIb (Capuano et al., 2005; Hattenhauer et al., 1999). However, administration of a low phosphate diet to vitamin-D receptor (VDR) KO mice and renal 1α-hydroxylase (*CYP27B1*) KO mice also increases NaPi-IIb protein expression, indicating the protein can be regulated by dietary phosphate alone and is independent of the 1,25(OH)$_2$vitD-VDR axis (Capuano et al., 2005; Segawa et al., 2004), although exactly how this occurs is still unclear.

Unlike control of NaPi-IIa in the kidney, the current view is that PTH and FGF23 may not affect intestinal Pi absorption directly, but instead act *via* feedback mechanisms resulting in altered 1,25(OH)$_2$vitD levels (Fig. 3). Elevated PTH has a stimulatory effect on 1,25(OH)$_2$vitD synthesis, while elevated FGF-23 inhibits the synthesis of 1,25(OH)2vitD; these alterations subsequently increase or decrease transcellular phosphate absorption along the small intestine, respectively, most likely as a result of 1,25 $(OH)_2$vitD-induced changes in NaPi-IIb protein levels (Saurette & Alexander, 2019).

4.3 Paracellular intestinal phosphate absorption

While under normal physiological conditions NaPi-IIb protein is responsible for most of the transcellular phosphate absorbed, there is growing evidence, mainly from *SLC34A2* KO mice, that this pathway does not play a dominant role in overall dietary phosphate absorption. For example, Sabbagh, Giral, Caldas, Levi, and Schiavi (2011) demonstrated that even when mice are maintained on a low phosphate diet, which should maximize NaPi-IIb-mediated transport, phosphate absorption is only ~50% less in *SLC34A2* KO mice compared with normal wildtype mice. Moreover, when maintained on a high phosphate diet, NaPi-IIb KO mice show either no significant increase (Ikuta et al., 2018), or even a decrease (Knöpfel et al., 2017), in fecal phosphate excretion compared with wild-type mice on the same diet. In both genotypes, a high dietary phosphate intake is accompanied by a significant increase in urinary phosphate excretion (Knöpfel et al., 2017), even though expression of NaPi-IIb would be decreased in wild-type mice on a high phosphate diet (Radanovic et al., 2005) or absent in the KO mice. Taken together these findings reflect the larger contribution of *passive* phosphate transport after consumption of a high phosphate load. This dominant role of the paracellular pathway is also evident from only partial correction of hyperphosphatemia in *SLC34A2* KO mice with CKD (Schiavi et al., 2012), and that normalization of serum phosphate in these animals requires treatment with a phosphate binder such as sevelamer.

Studies measuring phosphate concentrations in the bulk phase of intestinal contents in both rats and mice maintained on a normal phosphate diet have found that concentrations are in the millimolar range and exceed serum phosphate levels by several fold (Ikuta et al., 2018; Knöpfel et al., 2019; Marks et al., 2015). Therefore, luminal phosphate concentrations in this range would provide a significant gradient for paracellular absorption across the intestinal epithelium. Based on these findings, recent studies have started to characterize the paracellular pathway in more detail, including its regional profile, regulation, and potential mechanism(s). Studies using millimolar concentrations of phosphate, and in the presence or absence of sodium to evaluate paracellular phosphate transport in rats *in vivo*, have shown that the jejunum has the highest rate of total phosphate absorption and that ~70% can be attributed to the paracellular pathway (Marks et al., 2015). However, using Ussing chambers *ex vivo*, paracellular phosphate absorption was found to be similar in the rat jejunum and ileum, with only a marginally lower permeability found in the large intestine (Knöpfel et al., 2019). In the rat ileum paracellular absorption appears to slightly favor monovalent phosphate (Knöpfel et al., 2019). In mice, using the everted sac technique (Ikuta et al., 2018) or in Ussing chamber experiments (Knöpfel et al., 2019), paracellular absorption has also been shown to be similar in proximal and distal segments of the small intestine, and that absorption is not affected by NaPi-IIb deletion (Hernando, Gagnon, & Lederer, 2021; Ikuta et al., 2018). This apparent discrepancy between *in vivo* and *in vitro* estimates of paracellular permeability in different intestinal segments may be related to more recent findings on the role of the sodium-proton exchanger NHE3 and claudins in controlling paracellular permeability *in vivo* (see below).

4.4 Characterization and regulation of the paracellular pathway

Although the paracellular pathway is likely to be dominant for the absorption of dietary phosphate, the cellular mechanism involved is poorly characterized. It is unclear whether the tight junctions that form between adjacent intestinal enterocytes are truly passive and non-selectively permeable to phosphate, allowing its diffusion across the intestinal epithelium, or whether components of the tight junction complexes demonstrate a dynamic selectivity for phosphate. Potential candidates for selective phosphate permeability are the family of integral transmembrane proteins known as claudins, which have been shown to facilitate passive diffusion of several other ions across intestinal epithelia, including calcium (de Barboza, Guizzardi, & de Talamoni, 2015). These proteins form either epithelial barriers (sealing claudins) or paracellular channels (pore-forming claudins) and have distinct regional profiles of distribution along the intestine (Amasheh, Fromm, & Gunzel, 2010; Markov, Veshnyakova, Fromm, Amasheh, & Amasheh, 2010). Recent reports have described a correlation between claudin 3 and intestinal phosphate transport; in rat small intestine, this sealing claudin is highly expressed in the duodenum (Markov et al., 2010). Intestinal phosphate transport in the duodenum is significantly inhibited by diet-induced iron deficiency, an effect that is associated with the upregulation of claudin 3 mRNA and protein levels; in the jejunum, where diet-induced iron deficiency had no effect on phosphate absorption, claudin 3 remained unchanged (Asowata et al., 2021). In keeping with these findings, Hashimoto et al. (2020) have demonstrated that the VDR agonist lithocholic acid (LCA) suppresses claudin 3 levels in mouse ileum, resulting in increased phosphate permeability, and that in claudin 3 (*CLDN3*) KO mice, phosphate—but not calcium—absorption in the jejunum and ileum is significantly increased. These findings suggest that claudin 3 plays an important role in regulating paracellular phosphate absorption; however, given the complexity of the protein interactions involved in the formation and function of tight junctions, and the fact that LCA also decreased levels of occludin in the mouse ileum (Hashimoto et al., 2020), it is likely that other components of tight junction complexes also contribute to paracellular phosphate permeability. Although diet-induced iron deficiency and LCA have been shown to be novel regulators of paracellular phosphate permeability, $1,25(OH)_2$vitD does not affect paracellular absorption in mouse small intestine (Hernando, Pastor-Arroyo, et al., 2021). It is also worth noting that Ikuta et al. (2018) demonstrated alterations in mRNA and protein levels of both pore-forming and sealing claudins in the proximal and distal small intestine of *SLC34A2* KO mice, with the authors speculating that these changes might influence paracellular phosphate absorption. While this finding suggests that there may be crosstalk between the transcellular and paracellular pathways to optimize dietary phosphate absorption, functional studies by Knöpfel et al. (2019) failed to detect any alteration in paracellular transport in mice lacking NaPi-IIb. As already mentioned, another recent observation concerns the potential role of the sodium-proton exchanger NHE3 normally involved in intestinal sodium absorption and expressed at the apical membrane of enterocytes. It seems that changes in activity of this transporter can affect intracellular pH and that this can induce conformational changes in the tight junction proteins and alter paracellular permeability to phosphate (King et al., 2018).

While recent studies have increased our understanding of the processes involved in the absorption of dietary phosphate, observations in different species and from using different experimental techniques have sometimes provided contradictory and challenging information that requires further clarification. It is recognized that the intestinal paracellular pathway, with its high absorptive capacity, allows for large quantities of dietary phosphate to be absorbed and it is likely to be the dominant pathway under conditions of high dietary phosphate commonly found in typical western diets.

However, the complex combination of transcellular and paracellular mechanisms involved in intestinal phosphate absorption also provides the potential for manipulating phosphate homeostasis and developing novel strategies for treating disorders of phosphate balance (see later).

5. Renal phosphate reabsorption

Under normal conditions phosphate homeostasis is achieved predominately by balancing intestinal absorption of phosphate with its excretion *via* the kidneys. The latter occurs by a sodium-dependent transcellular pathway mediated by the renal SLC34 transporters, NaPi-IIa (SLC34A1) and NaPi-IIc (SLC34A3) (Fig. 5) (Weinman & Lederer, 2012); paracellular reabsorption is minor or insignificant (Kaufman & Hamburger, 1987). Kinetic studies of NaPi-IIa have shown the protein to be a sodium-dependent phosphate cotransporter with an apparent Km for phosphate of 50 μmol/L and a Km for sodium of 50 mmol/L. Like NaP-IIb in the intestine, the protein has a preference for HPO_4^{2-} and is electrogenic, transporting phosphate with a stoichiometry of 3:1 Na^+:HPO_4^{2-}, and is also inhibited by PFA (reviewed in Virkki et al., 2007). In contrast, NaPi-IIc is

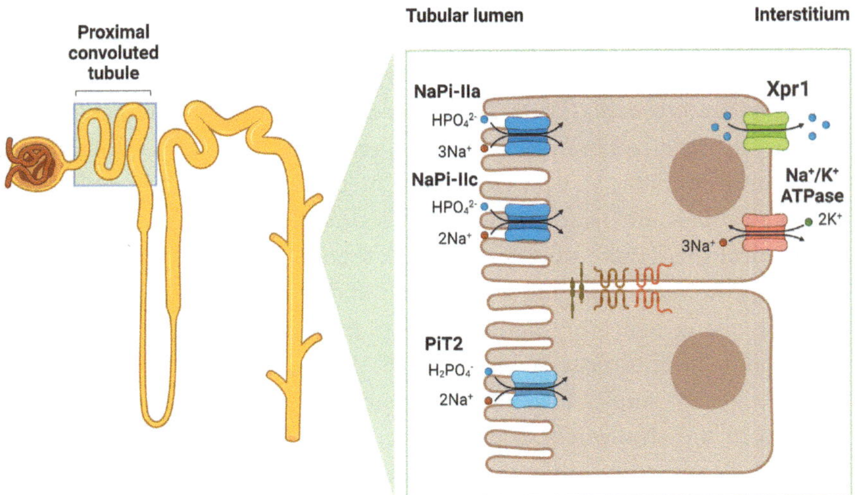

Fig. 5 A kidney nephron with two adjacent proximal tubular cells showing the transport mechanisms for transcellular reabsorption (NaPi-IIa and IIc and PiT2 for cell entry and putative Xrp1 for cell exit); paracellular transport of phosphate is negligible along the nephron, unlike the intestine (see text for details).

electroneutral and transports phosphate with a stoichiometry of 2:1 Na^+:HPO_4^{2-}, and has an apparent Km for phosphate of 70–80 μmol/L and a Km for sodium of 43–48 mmol/L; it also shows a preference for HPO_4^{2-} and is inhibited by PFA (Hernando, Gagnon, & Lederer, 2021).

Collectively, NaPi-IIa and NaPi-IIc are responsible for the reabsorption of ~80% of filtered phosphate. NaPi-IIa has been localized to the BBM throughout the proximal tubule (S1–S3 segments), although the highest protein levels are found in the S1 segment (Collins & Ghishan, 1994; Custer, Lotscher, Biber, Murer, & Kaissling, 1994). Generation of *SLC34A1* KO mice has shown that NaPi-IIa is responsible for ~70% of nephron phosphate absorption (Beck et al., 1998). In contrast, NaPi-IIc protein is only present in the S1 segment of the proximal tubule (Nowik et al., 2008), and while the protein is highly expressed in rodents during weaning, levels significantly diminish with age (Segawa et al., 2002). Studies using adult *SLC34A3* KO mice suggest that it plays a relatively minor role in phosphate homeostasis (Segawa et al., 2009) and accounts for ~30% of renal phosphate reabsorption when maximally expressed (Ohkido, Segawa, Yanagida, Nakamura, & Miyamoto, 2003; Segawa et al., 2002).

Genetic studies in humans have identified *SLC34A3* mutations as the cause of hereditary hypophosphatemic rickets with hypercalciuria (HHRH) (Bergwitz et al., 2006; Jaureguiberry, Carpenter, Forman, Jüppner, & Bergwitz, 2008). Numerous mutations have been described that ultimately result in intracellular accumulation of the protein. This lack of functional NaPi-IIc protein in the proximal tubule BBM leads to the hypophosphatemia, phosphaturia and osteomalacia characteristic of this disorder (Hernando, Gagnon, & Lederer, 2021). Mutations in *SLC34A1* have also been associated with altered phosphate homeostasis in humans, including hypophosphatemia and nephrolithiasis, infantile hypercalcemia and the renal Fanconi syndrome (Figueres, Beck-Cormier, Beck, & Marks, 2021). Of the 29 mutations that have been described, most are associated with trafficking defects (Wagner, Rubio-Aliaga, & Hernando, 2019). The clinical symptoms in patients with *SLC34A1* mutations may improve with maturity and on reaching adulthood, while problems persist into adulthood in patients with *SLC34A3* mutations. This led to speculation that NaPi-IIa may be the more important protein in early phases of life and that NaPi-IIc may be the dominant transporter in adult human kidney (Wagner et al., 2019), with the opposite being true for mice (Segawa et al., 2002). However, Gordon, Li, Doyle, Zaritsky, and Levine (2020) recently described a family with heterozygous inactivating mutations of both *SLC34A1* and *SLC34A3* and severe hypophosphatemia with rickets, implying a synergistic and non-redundant role for *SLC34A1* and *SLC34A3* in renal phosphate transport.

The roles of PiT1 and PiT2 in renal phosphate handling have also been investigated. Both proteins have been detected in the kidney, with localization of PiT2, but not PiT1, confirmed at the proximal tubule BBM (Villa-Bellosta & Sorribas, 2010). Under normal conditions, the protein is restricted to the S1 segment, whereas dietary phosphate restriction induces expression of PiT2 protein in all segments of the proximal tubule (Breusegem et al., 2009). Current experimental evidence suggests a minor role for this protein in renal phosphate reabsorption (Villa-Bellosta & Sorribas, 2010), which is supported by normophosphatemia observed in *SLC20A2* KO mice (Beck-Cormier et al., 2019), and in humans with *SLC20A2* inactivating mutations (Ramos et al., 2018).

5.1 Regulation of renal phosphate reabsorption

PTH, FGF23, and dietary phosphate are all considered to be the major regulators of NaPi-IIa protein levels. BBM protein levels of NaPi-IIa are reduced within minutes in response to PTH (Bacic et al., 2006; Kempson et al., 1995), within 2h in response to altered dietary phosphate load (Segawa et al., 2005; Villa-Bellosta et al., 2009) and within 4h in response to FGF23 (Küng et al., 2021). NaPi-IIc protein is also regulated by dietary phosphate (Segawa et al., 2005; Villa-Bellosta et al., 2008), PTH (Picard et al., 2010; Segawa et al., 2007) and FGF-23 (Segawa et al., 2003). Although, with the exception of FGF23, changes in NaPi-IIc protein expression occur over a longer time-course than NaPi-IIa, with BBM levels of NaPi-IIc reduced after 4h in response to PTH (Segawa et al., 2007), altered dietary phosphate load (Segawa et al., 2005; Villa-Bellosta et al., 2009).

PTH-induced internalization and degradation of NaPi-IIa and NaPi-IIc not only shows differences in their time-course, but the underlying cellular mechanisms also diverge. It has been documented that the rapid effect of PTH on NaPi-IIa occurs *via* clathrin-mediated endocytosis of the protein, with internalized transporters targeted from early to late endosomes and finally reaching lysosomes (Bacic et al., 2006; Pfister et al., 1998); while in contrast, degradation of NaPi-IIc does not take place in lysosomes (Picard et al., 2010; Segawa et al., 2005). A recent study has evaluated the time-course and routes of FGF23-induced internalization and degradation of NaPi-IIa and NaPi-IIc, and has shown that NaPi-IIa and NaPi-IIc are internalized concurrently into subapical vesicles that were positive for endosomal markers, indicating that both proteins are internalized *via* receptor-mediated endocytosis; however, in contrast to their route of degradation following PTH internalization, both proteins colocalized with lysosomal markers following FGF23 treatment (Küng et al., 2021).

The SLC20 transporter PiT-2 can also adapt to changes in dietary phosphate, although this occurs more slowly than with the SLC34 NaPi-IIa, but faster than NaPi-IIc (Villa-Bellosta et al., 2009). While it has been shown that PTH (Picard et al., 2010), FGF23 (Tomoe et al., 2010) and low dietary potassium (Breusegem et al., 2009) decrease, and metabolic acidosis increases (Villa-Bellosta & Sorribas, 2010) renal PiT2 protein levels, functional evidence for the impact of these changes on renal phosphate handling has not been shown. Rather, it has been suggested that SLC20 proteins may be involved in phosphate uptake for intracellular use, rather than playing any role in transcellular transport (Hernando, Gagnon, & Lederer, 2021).

One component of both intestinal and kidney transcellular phosphate transport that has remained unclear until now is the mechanism of epithelial cell basolateral membrane phosphate efflux and exit from the cell into blood. As shown in Figs. 4 and 5, the current model is that the phosphate exporter is Xpr1, originally identified as the receptor for murine leukemia retroviruses (Giovannini, Touhami, Charnet, Sitbon, & Battini, 2013), is the responsible transporter. However, its role in overall phosphate regulation is still unclear, although recent data suggest that its activity is controlled by one or more inositol hexakisphosphate kinases (IP6K) that generate signaling inositol pyrophosphates such as InsP7 and InsP8, which have both been shown to regulate intracellular phosphate levels *in vitro* (Li et al., 2020). Inhibiting generation of these signaling molecules *in vivo* has also been shown to reduce hyperphosphatemia and kidney damage in the rat model of adenine-induced CKD (Moritoh et al., 2021), opening up yet another potential therapeutic target to manage phosphate imbalance (see later).

To summarize, factors that regulate both renal and intestinal phosphate absorption include vitamin D, FGF23 with klotho, and also other phosphatonins such as MEPE (Aniteli et al., 2018; Dobbie, Unwin, Faria, & Shirley, 2007; Marks, Churchill, Debnam, & Unwin, 2008). Unlike the kidneys, in which phosphate reabsorption is downregulated by both PTH and FGF23 in CKD, the intestine does not seem to adapt in the same way. While CKD may decrease NaPi-IIb expression, dietary phosphate restriction has the opposite effect, and hyperphosphatemia will itself stimulate FGF23 and MEPE, and suppress active vitamin D, all of which result in no overall change in intestinal phosphate absorption (Aniteli et al., 2018; Marks et al., 2007; Stremke et al., 2021).

6. Phosphate imbalance

Table 1 lists some of the causes and associations of hyperphosphatemia and hypophosphatemia. The most common and clinically important cause of phosphate retention and hyperphosphatemia is CKD. The latter's

Table 1 Some listed examples of important clinical causes of hyperphosphatemia and hypophosphatemia, a major cause of hyperphosphatemia being chronic kidney disease (CKD; see text for details).

Hyperphosphatemia	Hypophosphatemia
Acute (AKI) or chronic kidney disease (CKD)	**Familial hypophosphatemic rickets**
Hypoparathyroidism	• Hereditary vitamin D-dependent (resistant) [*CYP27B1, CYP2R1, VDR*]
Acromegaly (GH/IGF1)	
Acute cellular shift or loss	• X-linked [PHEX – FGF23]
• Tumor lysis syndrome	• With hypercalciuria [SLC34A3]
• Rhabdomyolysis	Renal Fanconi syndrome(s) (drug-induced or familial]
• Lactic or ketoacidosis	Hyperparathyroidism
Exogenous load (e.g., laxatives)	Vitamin D deficiency (nutritional rickets)
FGF receptor inhibitors (used in cancer)	**Syndromic**
Vitamin D toxicity	• McCune-Albright [GNAS1 – FGF23]
Familial tumoral calcinosis [*GALNT3, FGF23*]	• Raine syndrome [FAM20C – FGF23]
Pseudo- (e.g., hypergobulinemia)	• With nephrolithiasis [SLC34A1, SLC9A3R1]
	• Opsismodysplasia [INPPL1]
	Acute
	Refeeding (insulin, glucagon), acute respiratory alkalosis, post-parathyroidectomy

well-established link to increased CV risk and disease makes preventing and controlling hyperphosphatemia an important clinical challenge in CKD. Although still elevated, the recommended clinical target for serum phosphate concentration is below 1.8 mmol/L (<5.5 mg/dL), which is difficult to achieve with the current clinical measures available that rely on oral phosphate binders in combination with dietary phosphate restriction to manage hyperphosphatemia in CKD (Sprague, Martin, & Coyne, 2021). These treatments are often associated with poor patient compliance due to the associated high pill burden and common gastrointestinal side effects, and the necessary dietary restrictions to reduce phosphate intake, and the ensuing risk of malnutrition (Cannata-Andía & Martin, 2016; Chan et al., 2017). Normalizing phosphate balance is rarely achievable in practice, as has demonstrating a clinical benefit of phosphate binders (Lioufas et al., 2021), and so has highlighted the need for new therapeutic approaches.

6.1 Targeting renal phosphate transport

The potential for directly targeting the process of renal phosphate reabsorption to prevent hyperphosphatemia in CKD has been made possible recently following discovery of a selective and orally bioavailable NaPi-IIa inhibitor, PF-06869206 (Filipski et al., 2018). Studies in the opossum kidney cell line have shown that the inhibitor induces a dose-dependent inhibition of sodium-dependent phosphate uptake (Thomas et al., 2020). Subsequent *in vivo* studies in both rats and mice corroborated that PF-06869206 causes a dose-dependent increase in fractional phosphate excretion and reduces plasma phosphate concentration, an effect that is absent in NaPi-IIa KO mice, confirming its target specificity (Clerin et al., 2020; Thomas et al., 2019, 2020). The effectiveness of the inhibitor in decreasing hyperphosphatemia in rodent models of CKD has also been demonstrated (Clerin et al., 2020; Thomas et al., 2019). However, this promising approach may be limited by differences in the relative contribution of NaPi-IIa and NaPi-IIc to renal phosphate reabsorption in humans and mice (Hernando, Gagnon, & Lederer, 2021). Exploratory studies in humans using PF-06869206 will be needed to determine whether NaPi-IIa is a viable clinical target.

6.2 Targeting intestinal phosphate transport

While there has been significant research on unravelling the mechanisms involved in intestinal phosphate absorption to identify new therapeutic targets, there is limited information on how this process adapts in response to a decline in GFR. It is assumed that the diminishing levels of 1,25 $(OH)_2$vitD that occur with progressive CKD would be accompanied by decreased intestinal phosphate absorption, an adaptation that would be expected to provide some compensation for the reduction in fractional phosphate excretion. However, given that 1,25$(OH)_2$vitD only impacts NaPi-IIb-mediated phosphate transport (Hernando, Pastor-Arroyo, et al., 2021) and that this pathway plays a relatively minor role in dietary phosphate absorption, it is unclear whether significant downregulation of NaPi-IIb occurs in CKD or would be beneficial. Studies in rats *in vivo* in models of CKD have confirmed that while serum 1,25$(OH)_2$vitD levels decrease, there is no net change in intestinal phosphate absorption (Marks et al., 2006; Turner et al., 2018; Vorland et al., 2020). As already mentioned, Stremke et al. (2021) have shown recently that fractional intestinal phosphate absorption did not differ in CKD patients compared with controls, despite lower serum 1,25$(OH)_2$vitD levels. Since these studies indicate that intestinal phosphate transport does not adapt or change significantly in CKD, it remains a suitable target to reduce hyperphosphatemia in CKD.

In contrast to the kidney, directly targeting intestinal phosphate transport has been studied in more detail and based on preclinical findings several compounds have been tested already in clinical trials. However, results so far have been mixed and this may reflect differences in the mechanisms responsible for phosphate absorption in rodents and humans. The first compound to be tested was the type II transport inhibitor, nicotinamide (Katai et al., 1999; Wu, Bacon, Al-Mahrouq, & Kempson, 1988). The compound was shown to inhibit intestinal phosphate transport *via* decreased jejunal NaPi-IIb protein levels and to prevent progressive hyperphosphatemia in rats with CKD (Eto, Miyata, Ohno, & Yamashita, 2005). Reports also suggested that relatively short-term treatment with nicotinamide of end-stage kidney disease (ESKD) patients on dialysis was effective at reducing serum phosphate levels (Lenglet et al., 2016). However, also using FGF23 levels as an index of phosphate retention, 12-months treatment with nicotinamide and lanthanum carbonate did not control hyperphosphatemia or reduce FGF23 levels in patients with moderate to severe CKD (Ix et al., 2019); these treatments have poor patient tolerance and significant gastrointestinal side-effects (He, Feng, Huo, Yang, & Liao, 2014; Ix et al., 2019; Lenglet et al., 2016).

ASP3325 is a small molecule inhibitor of NaPi-IIb reported initially to alleviate hyperphosphatemia in rat models of CKD, but unfortunately it failed to alter urinary or fecal phosphate excretion in healthy individuals or to reduce serum phosphate levels in ESKD (Larsson et al., 2018). A Phase 1b clinical study of another NaPi-IIb inhibitor, DS-2330b, also proved to be ineffective in controlling hyperphosphatemia in dialysis patients (Maruyama et al., 2021). This lack of efficacy may be due to a minor and negligible role of NaPi-IIb in humans; however, it is difficult to explain how niacin and nicotinamide have been shown to lower serum phosphate levels in ESKD, if their effect is not *via* NaPi-IIb. One possibility is that their phosphate lowering capacity may be via an action on other sodium phosphate cotransporters or absorptive pathways. In this context, a novel non-absorbed pan-inhibitor, EOS789, which targets NaPi-IIb, PIT-1 and PIT-2 in the intestine, has shown promising results in both normal and CKD rat models. It increased fecal ^{33}P excretion rates and reduced ^{33}P urinary excretion, suggesting reduced intestinal phosphate absorption (Tsuboi et al., 2020). In CKD rats this drug lowered serum phosphate, FGF23 and PTH levels, and also reduced ectopic calcification in the thoracic aorta and improved renal function (Tsuboi et al., 2020). A Phase 1b study has directly assessed the effectiveness of this compound for reducing intestinal phosphate absorption in dialysis patients using a radioisotope tracer method (Gallant et al., 2021). However, while short-term use of EOS789 appeared to be safe and well tolerated, the trend for reduced phosphate absorption was only evident following a *post hoc* analysis of the higher dose *versus* placebo, and follow-up confirmation of its effectiveness in humans is needed.

The efficiency of targeting the paracellular pathway for intestinal phosphate transport has also been tested in preclinical and clinical studies. In normal rats, tenapanor, a minimally absorbed inhibitor of NHE3, mentioned earlier, was shown to increase fecal phosphate excretion and reduce ^{33}P transfer from the gastrointestinal tract to plasma (Labonté et al., 2015). Using the 5/6th nephrectomy rat model of hyperphosphatemia, with active vitamin D (calcitriol) administration and a high phosphate diet to enhance vascular and soft tissue calcification, Labonté et al. also demonstrated that tenapanor reduced serum phosphate and FGF-23 levels, attenuated the decline in renal function and decreased vascular and soft tissue calcification associated with this model (Labonté et al., 2015). Later studies of tenapanor's mode of action suggested that intracellular proton retention and acidification caused by NHE3 inhibition affected tight junction proteins and reduced paracellular permeability to phosphate referred to earlier (King et al., 2018). In healthy volunteers, tenapanor increased fecal phosphate excretion and decreased urinary phosphate excretion (Johansson, Rosenbaum, Knutsson, & Leonsson-Zachrisson, 2017; King et al., 2018), and in a Phase 3 clinical trial it significantly reduced serum phosphate levels in dialysis patients (Block, Rosenbaum, Yan, & Chertow, 2019). More recent studies have started to explore the potential for using tenapanor as an add-on therapy with a phosphate binder. Studies in normal rats have shown that the effect of sevelamer, a widely used non-calcium-based phosphate binder, on urinary phosphate excretion is augmented by the addition of tenapanor (King et al., 2021). This approach has also proved successful in dialysis patients who have poor phosphate control when on a binder alone (Pergola, Rosenbaum, Yang, & Chertow, 2021; Shigematsu et al., 2021). Four weeks of treatment with a combination of a phosphate binder plus tenapanor resulted in a higher proportion of patients achieving a target phosphate level of <1.8 mmol/L (37.1% vs 21.8% on binder alone); an effect also associated with a reduction in FGF-23 levels (Pergola et al., 2021). This combined effect of using a phosphate binder together with tenapanor seems to be specific for the paracellular pathway, because intestinal phosphate absorption in dialysis patients treated with a combination of the pan-phosphate transporter inhibitor EOS789 and sevelamer was not significantly different from those treated with EOS789 alone (Gallant et al., 2021). Similarly, there was no additional phosphate-lowering effect when dialysis patients were treated with the NaPi-IIb inhibitor DS-2330b plus sevelamer (Maruyama et al., 2021), or when patients with moderate-to-severe CKD were treated with a combination of nicotinamide and a phosphate binder (Ix et al., 2019). These studies were based on the idea that binders can produce a low phosphate environment in the gastrointestinal tract and would enhance NaPi-IIb-mediated transport; however, while the patient numbers are small and treatment was for a relatively short time, the results demonstrate that targeting transcellular phosphate transport in humans is unlikely to be an effective therapeutic option.

Finally, although CKD has been discussed above in more detail as an important pathophysiological example of phosphate imbalance, this article is not intended to cover the pathophysiology of phosphate homeostasis and/or the treatment of other causes of hyperphosphatemia or hypophosphatemia, which topic has been covered in some recent reviews (Jacquillet & Unwin, 2019; Koumakis, Cormier, Roux, & Briot, 2021).

The maintenance of normal phosphate balance is a complex process involving a variety of hormones and bioactive factors, some well-established and understood, and perhaps others yet to be defined, all acting in concert on the kidneys, bone and gastrointestinal tract. While the main players are largely known, the conductor is still something of a mystery and how the known regulatory factors are orchestrated to maintain phosphate homeostasis. Indeed, it has been proposed that the gastrointestinal tract is a sensor of phosphate intake and the kidneys an effector, increasing or decreasing phosphate excretion accordingly, with much early attention focused on the validity of this concept and any potential mediators (Kumar, 2009; Thomas et al., 2016), rather than a sensing mechanism for phosphate. A mechanism of cellular phosphate sensing does seem both likely and essential. Evidence for this comes from the importance of phosphate as a nutritional source for plants and bacteria (Austin & Mayer, 2020). Perhaps phosphate transporters that are less directly involved in transcellular transport of phosphate such as PiT-1 (Michigami, Kawai, Yamazaki, & Ozono, 2018) or the more ubiquitously expressed, XrP1 with its phosphoinositol-dependent downstream signaling referred to earlier, are part of this process? We still need a better understanding of this phosphate sensing mechanism, which will no doubt have important future implications for both normal and disturbed phosphate homeostasis.

Conflict of interest

The authors declare that they have no competing financial interests. Robert Unwin is Emeritus Professor of Nephrology at UCL and currently working in Early Clinical Development, Early CVRM (Cardiovascular, Renal and Metabolism), R&D BioPharmaceuticals, AstraZeneca, Cambridge, UK.

References

Amasheh, S., Fromm, M., & Gunzel, D. (2010). Claudins of intestine and nephron—A correlation of molecular tight junction structure and barrier function. *Acta Physiologica, 201*(1), 133–140. https://doi.org/10.1111/j.1748-1716.2010.02148.x.

Aniteli, T. M., Siqueira, F. R., dos Reis, L. M., Dominguez, W. V., Oliveira, E. M. C., Castelucci, P., et al. (2018). Effect of variations in dietary Pi intake on intestinal Pi transporters (NaPi-IIb, PiT-1, and PiT-2) and phosphate-regulating factors (PTH, FGF-23, and MEPE). *Pflügers Archiv—European Journal of Physiology, 470*(4), 1–10. https://doi.org/10.1007/s00424-018-2111-6.

Asowata, E. O., Olusanya, O., Abaakil, K., Chichger, H., Srai, S. K. S., Unwin, R. J., et al. (2021). Diet-induced iron deficiency in rats impacts small intestinal calcium and phosphate absorption. *Acta Physiologica (Oxford, England)*, e13650. https://doi.org/10.1111/apha.13650.

Austin, S., & Mayer, A. (2020). Phosphate homeostasis—A vital metabolic equilibrium maintained through the INPHORS signaling pathway. *Frontiers in Microbiology, 11*, 1367. https://doi.org/10.3389/fmicb.2020.01367.

Bacic, D., Lehir, M., Biber, J., Kaissling, B., Murer, H., & Wagner, C. A. (2006). The renal Na^+/phosphate cotransporter NaPi-IIa is internalized via the receptor-mediated endocytic route in response to parathyroid hormone. *Kidney International, 69*(3), 495–503. https://doi.org/10.1038/sj.ki.5000148.

Beck, L., Karaplis, A. C., Amizuka, N., Hewson, A. S., Ozawa, H., & Tenenhouse, H. S. (1998). Targeted inactivation of Npt2 in mice leads to severe renal phosphate wasting, hypercalciuria, and skeletal abnormalities. *Proceedings of the National Academy of Sciences, 95*(9), 5372–5377. https://doi.org/10.1073/pnas.95.9.5372.

Beck-Cormier, S., Lelliott, C. J., Logan, J. G., Lafont, D. T., Merametdjian, L., Leitch, V. D., et al. (2019). Slc20a2, encoding the phosphate transporter PiT2, is an important genetic determinant of bone quality and strength. *Journal of Bone and Mineral Research, 34*(6), 1101–1114. https://doi.org/10.1002/jbmr.3691.

Bellasi, A., Micco, L. D., Russo, D., Simone, E. D., Iorio, M. D., Vigilante, R., et al. (2019). Fractional excretion of phosphate (FeP) is associated with end- stage renal disease patients with CKD 3b and 5. *Journal of Clinical Medicine, 8*(7), 1026. https://doi.org/10.3390/jcm8071026.

Bergwitz, C., Roslin, N. M., Tieder, M., Loredo-Osti, J. C., Bastepe, M., Abu-Zahra, H., et al. (2006). SLC34A3 mutations in patients with hereditary hypophosphatemic rickets with hypercalciuria predict a key role for the sodium-phosphate cotransporter NaPi-IIc in maintaining phosphate homeostasis. *The American Journal of Human Genetics, 78*(2), 179–192. https://doi.org/10.1086/499409.

Berner, W., Kinne, R., & Murer, H. (1976). Phosphate transport into brush-border membrane vesicles isolated from rat small intestine. *Biochemical Journal, 160*(3), 467–474. https://doi.org/10.1042/bj1600467.

Blau, J. E., & Collins, M. T. (2015). The PTH-vitamin D-FGF23 axis. *Reviews in Endocrine and Metabolic Disorders, 16*(2), 165–174. https://doi.org/10.1007/s11154-015-9318-z.

Block, G. A., Rosenbaum, D. P., Yan, A., & Chertow, G. M. (2019). Efficacy and safety of Tenapanor in patients with hyperphosphatemia receiving maintenance hemodialysis: A randomized phase 3 trial. *Journal of the American Society of Nephrology, 30*(4), 641–652. https://doi.org/10.1681/asn.2018080832.

Borowitz, S. M., Said, H. M., & Ghishan, F. K. (1986). The effects of metabolic acidosis on jejunal phosphate and glucose transport in weanling rats. *Pediatric Research, 20*(8), 763–767. https://doi.org/10.1203/00006450-198608000-00013.

Breusegem, S. Y., Takahashi, H., Giral-Arnal, H., Wang, X., Jiang, T., Verlander, J. W., et al. (2009). Differential regulation of the renal sodium-phosphate cotransporters NaPi-IIa, NaPi-IIc, and PiT-2 in dietary potassium deficiency. *AJP: Renal Physiology, 297*(2), F350–F361. https://doi.org/10.1152/ajprenal.90765.2008.

Burtis, C. A., Ashwood, E. R., Bruns, D. E., & Sawyer, B. G. (2008). *Tietz fundamentals of clinical chemistry*. Saunders Elsevier.

Candeal, E., Caldas, Y. A., Guillén, N., Levi, M., & Sorribas, V. (2014). Na^+-independent phosphate transport in Caco2BBE cells. *AJP: Cell Physiology, 307*(12), C1113–C1122. https://doi.org/10.1152/ajpcell.00251.2014.

Candeal, E., Caldas, Y. A., Guillén, N., Levi, M., & Sorribas, V. (2017). Intestinal phosphate absorption is mediated by multiple transport systems in rats. *American Journal of Physiology-Gastrointestinal and Liver Physiology, 312*(4), G355–G366. https://doi.org/10.1152/ajpgi.00244.2016.

Cannata-Andía, J. B., & Martin, K. J. (2016). The challenge of controlling phosphorus in chronic kidney disease. *Nephrology Dialysis Transplantation, 31*(4), 541–547. https://doi.org/10.1093/ndt/gfv055.

Capuano, P., Radanovic, T., Wagner, C. A., Bacic, D., Kato, S., Uchiyama, Y., et al. (2005). Intestinal and renal adaptation to a low-Pi diet of type II NaPi cotransporters in vitamin D receptor- and 1αOHase-deficient mice. *American Journal of Physiology-Cell Physiology, 288*(2), C429–C434. https://doi.org/10.1152/ajpcell.00331.2004.

Chan, S., Au, K., Francis, R. S., Mudge, D. W., Johnson, D. W., & Pillans, P. I. (2017). Phosphate binders in patients with chronic kidney disease. *Australian Prescriber, 40*(1), 9–14. https://doi.org/10.18773/austprescr.2017.002.

Clerin, V., Saito, H., Filipski, K. J., Nguyen, A. H., Garren, J., Kisucka, J., et al. (2020). Selective pharmacological inhibition of the sodium-dependent phosphate co- transporter NPT2a promotes phosphate excretion. *Journal of Clinical Investigation, 130*(12), 6510–6522. https://doi.org/10.1172/jci135665.

Collins, J. F., & Ghishan, F. K. (1994). Molecular cloning, functional expression, tissue distribution, and in situ hybridization of the renal sodium phosphate (Na^+/Pi) transporter in the control and hypophosphatemic mouse. *The FASEB Journal, 8*(11), 862–868. https://doi.org/10.1096/fasebj.8.11.8070635.

Cooke, A. (2017). Dietary food-additive phosphate and human health outcomes. *Comprehensive Reviews in Food Science and Food Safety, 16*(5), 906–1021. https://doi.org/10.1111/1541-4337.12275.

Copland, M., Komenda, P., Weinhandl, E. D., McCullough, P. A., & Morfin, J. A. (2016). Intensive hemodialysis, mineral and bone disorder, and phosphate binder use. *American Journal of Kidney Diseases, 68*(5), S24–S32. https://doi.org/10.1053/j.ajkd.2016.05.024.

Cozzolino, M., Ketteler, M., & Wagner, C. A. (2020). An expert update on novel therapeutic targets for hyperphosphatemia in chronic kidney disease: Preclinical and clinical innovations. *Expert Opinion on Therapeutic Targets, 24*(5), 1–12. https://doi.org/10.1080/14728222.2020.1743680.

Custer, M., Lotscher, M., Biber, J., Murer, H., & Kaissling, B. (1994). Expression of Na-P(i) cotransport in rat kidney: Localization by RT-PCR and immunohistochemistry. *American Journal of Physiology-Renal Physiology, 266*(5), F767–F774. https://doi.org/10.1152/ajprenal.1994.266.5.f767.

de Barboza, G. D., Guizzardi, S., & de Talamoni, N. T. (2015). Molecular aspects of intestinal calcium absorption. *World Journal of Gastroenterology, 21*(23), 7142–7154. https://doi.org/10.3748/wjg.v21.i23.7142.

Dobbie, H., Unwin, R. J., Faria, N. J. R., & Shirley, D. G. (2007). Matrix extracellular phosphoglycoprotein causes phosphaturia in rats by inhibiting tubular phosphate reabsorption. *Nephrology, Dialysis, Transplantation: Official Publication of the European Dialysis and Transplant Association—European Renal Association, 23*(2), 730–733. https://doi.org/10.1093/ndt/gfm535.

Erem, S., & Razzaque, M. S. (2018). Dietary phosphate toxicity: An emerging global health concern. *Histochemistry and Cell Biology, 150*(6), 711–719. https://doi.org/10.1007/s00418-018-1711-8.

Eto, N., Miyata, Y., Ohno, H., & Yamashita, T. (2005). Nicotinamide prevents the development of hyperphosphataemia by suppressing intestinal sodium-dependent phosphate transporter in rats with adenine-induced renal failure. *Nephrology Dialysis Transplantation, 20*(7), 1378–1384. https://doi.org/10.1093/ndt/gfh781.

Figueres, L., Beck-Cormier, S., Beck, L., & Marks, J. (2021). The complexities of organ crosstalk in phosphate homeostasis: Time to put phosphate sensing Back in the limelight. *International Journal of Molecular Sciences, 22*(11), 5701. https://doi.org/10.3390/ijms22115701.

Filipski, K. J., Sammons, M. F., Bhattacharya, S. K., Panteleev, J., Brown, J. A., Loria, P. M., et al. (2018). Discovery of orally bioavailable selective inhibitors of the sodium-phosphate cotransporter NaPi2a (SLC34A1). *ACS Medicinal Chemistry Letters, 9*(5), 440–445. https://doi.org/10.1021/acsmedchemlett.8b00013.

Forster, I. C., Hernando, N., Biber, J., & Murer, H. (2012). Phosphate transport kinetics and structure-function relationships of SLC34 and SLC20 proteins. *Current Topics in Membranes, 70*, 313–356. https://doi.org/10.1016/B978-0-12-394316-3.00010-7.

Gallant, K. M. H., Stremke, E. R., Trevino, L. L., Moorthi, R. N., Doshi, S., Wastney, M. E., et al. (2021). EOS789, a broad-spectrum inhibitor of phosphate transport, is safe with an indication of efficacy in a phase 1b randomized crossover trial in hemodialysis patients. *Kidney International, 99*(5), 1225–1233. https://doi.org/10.1016/j.kint.2020.09.035.

Giovannini, D., Touhami, J., Charnet, P., Sitbon, M., & Battini, J.-L. (2013). Inorganic phosphate export by the retrovirus receptor XPR1 in metazoans. *Cell Reports, 3*(6), 1866–1873. https://doi.org/10.1016/j.celrep.2013.05.035.

Giral, H., Caldas, Y., Sutherland, E., Wilson, P., Breusegem, S., Barry, N., et al. (2009). Regulation of rat intestinal Na-dependent phosphate transporters by dietary phosphate. *American Journal of Physiology. Renal Physiology, 297*(5), F1466–F1475. https://doi.org/10.1152/ajprenal.00279.2009.

Gordon, R. J., Li, D., Doyle, D., Zaritsky, J., & Levine, M. A. (2020). Digenic heterozygous mutations in SLC34A3 and SLC34A1 cause dominant hypophosphatemic rickets with hypercalciuria. *The Journal of Clinical Endocrinology & Metabolism, 105*(7), 2392–2400. https://doi.org/10.1210/clinem/dgaa217.

Hashimoto, N., Matsui, I., Ishizuka, S., Inoue, K., Matsumoto, A., Shimada, K., et al. (2020). Lithocholic acid increases intestinal phosphate and calcium absorption in a vitamin D receptor dependent but transcellular pathway independent manner. *Kidney International, 97*(6), 1164–1180. https://doi.org/10.1016/j.kint.2020.01.032.

Hattenhauer, O., Traebert, M., Murer, H., & Biber, J. (1999). Regulation of small intestinal Na-Pi type IIb cotransporter by dietary phosphate intake. *American Journal of Physiology—Gastrointestinal and Liver Physiology, 277*(4), G756–G762. https://doi.org/10.1152/ajpgi.1999.277.4.g756.

He, Y.-M., Feng, L., Huo, D.-M., Yang, Z.-H., & Liao, Y.-H. (2014). Benefits and harm of niacin and its analog for renal dialysis patients: A systematic review and meta-analysis. *International Urology and Nephrology, 46*(2), 433–442. https://doi.org/10.1007/s11255-013-0559-z.

Hernando, N., Gagnon, K., & Lederer, E. (2021). Phosphate transport in epithelial and nonepithelial tissue. *Physiological Reviews, 101*(1), 1–35. https://doi.org/10.1152/physrev.00008.2019.

Hernando, N., Myakala, K., Simona, F., Knöpfel, T., Thomas, L., Murer, H., et al. (2015). Intestinal depletion of NaPi-IIb/Slc34a2 in mice: Renal and hormonal adaptation. *Journal of Bone and Mineral Research, 30*(10), 1925–1937. https://doi.org/10.1002/jbmr.2523.

Hernando, N., Pastor-Arroyo, E. M., Marks, J., Schnitzbauer, U., Knöpfel, T., Bürki, M., et al. (2021). 1,25(OH)2 vitamin D3 stimulates active phosphate transport but not paracellular phosphate absorption in mouse intestine. *The Journal of Physiology, 599*(4), 1131–1150. https://doi.org/10.1113/jp280345.

Hilfiker, H., Hattenhauer, O., Traebert, M., Forster, I., Murer, H., & Biber, J. (1998). Characterization of a murine type II sodium-phosphate cotransporter expressed in mammalian small intestine. *Proceedings of the National Academy of Sciences, 95*(24), 14564–14569. https://doi.org/10.1073/pnas.95.24.14564.

Ichida, Y., Ohtomo, S., Yamamoto, T., Murao, N., Tsuboi, Y., Kawabe, Y., et al. (2020). Evidence of an intestinal phosphate transporter alternative to type IIb sodium-dependent phosphate transporter in rats with chronic kidney disease. *Nephrology Dialysis Transplantation, 36*(1), 68–75. https://doi.org/10.1093/ndt/gfaa156.

Ikuta, K., Segawa, H., Sasaki, S., Hanazaki, A., Fujii, T., Kushi, A., et al. (2018). Effect of Npt2b deletion on intestinal and renal inorganic phosphate (Pi) handling. *Clinical and Experimental Nephrology, 22*(3), 517–528. https://doi.org/10.1007/s10157-017-1497-3.

Ix, J. H., Isakova, T., Larive, B., Raphael, K. L., Raj, D. S., Cheung, A. K., et al. (2019). Effects of nicotinamide and lanthanum carbonate on serum phosphate and fibroblast growth factor-23 in CKD: The COMBINE trial. *Journal of the American Society of Nephrology, 30*(6), 1096–1108. https://doi.org/10.1681/asn.2018101058.

Jacquillet, G., & Unwin, R. J. (2019). Physiological regulation of phosphate by vitamin D, parathyroid hormone (PTH) and phosphate (Pi). *Pflügers Archiv*, *471*(1), 83–98. https://doi.org/10.1007/s00424-018-2231-z.

Jaureguiberry, G., Carpenter, T. O., Forman, S., Jüppner, H., & Bergwitz, C. (2008). A novel missense mutation in SLC34A3 that causes hereditary hypophosphatemic rickets with hypercalciuria in humans identifies threonine 137 as an important determinant of sodium-phosphate cotransport in NaPi-IIc. *American Journal of Physiology-Renal Physiology*, *295*(2), F371–F379. https://doi.org/10.1152/ajprenal.00090.2008.

Johansson, S., Rosenbaum, D. P., Knutsson, M., & Leonsson-Zachrisson, M. (2017). A phase 1 study of the safety, tolerability, pharmacodynamics, and pharmacokinetics of tenapanor in healthy Japanese volunteers. *Clinical and Experimental Nephrology*, *21*(3), 407–416. https://doi.org/10.1007/s10157-016-1302-8.

Juan, D., Liptak, P., & Gray, T. K. (1976). Absorption of inorganic phosphate in the human jejunum and its inhibition by Salmon calcitonin. *The Journal of Clinical Endocrinology & Metabolism*, *43*(3), 517–522. https://doi.org/10.1210/jcem-43-3-517.

Katai, K., Miyamoto, K., Kishida, S., Segawa, H., Nii, T., Tanaka, H., et al. (1999). Regulation of intestinal Na^+-dependent phosphate co-transporters by a low-phosphate diet and 1,25-dihydroxyvitamin D3. *The Biochemical Journal*, *343*(Pt. 3), 705–712.

Kaufman, J. S., & Hamburger, R. J. (1987). Lack of influence of volume flux on phosphate reabsorption in the proximal tubule. *Mineral and Electrolyte Metabolism*, *13*(3), 158–164.

Kempson, S. A., Lotscher, M., Kaissling, B., Biber, J., Murer, H., & Levi, M. (1995). Parathyroid hormone action on phosphate transporter mRNA and protein in rat renal proximal tubules. *American Journal of Physiology-Renal Physiology*, *268*(4), F784–F791. https://doi.org/10.1152/ajprenal.1995.268.4.f784.

King, A. J., Kohler, J., Fung, C., Jiang, Z., Quach, A., Kumaraswamy, P., et al. (2021). Combination treatment with tenapanor and sevelamer synergistically reduces urinary phosphorus excretion in rats. *American Journal of Physiology-Renal Physiology*, *320*(1), F133–F144. https://doi.org/10.1152/ajprenal.00137.2020.

King, A. J., Siegel, M., He, Y., Nie, B., Wang, J., Koo-McCoy, S., et al. (2018). Inhibition of sodium/hydrogen exchanger 3 in the gastrointestinal tract by tenapanor reduces paracellular phosphate permeability. *Science Translational Medicine*, *10*(456). https://doi.org/10.1126/scitranslmed.aam6474.

Knöpfel, T., Himmerkus, N., Günzel, D., Bleich, M., Hernando, N., & Wagner, C. A. (2019). Paracellular transport of phosphate along the intestine. *American Journal of Physiology-Gastrointestinal and Liver Physiology*, *317*(2), G233–G241. https://doi.org/10.1152/ajpgi.00032.2019.

Knöpfel, T., Pastor-Arroyo, E. M., Schnitzbauer, U., Kratschmar, D. V., Odermatt, A., Pellegrini, G., et al. (2017). The intestinal phosphate transporter NaPi-IIb (Slc34a2) is required to protect bone during dietary phosphate restriction. *Scientific Reports*, 7(1), 11018. https://doi.org/10.1038/s41598-017-10390-2.

Koumakis, E., Cormier, C., Roux, C., & Briot, K. (2021). The causes of hypo- and hyperphosphatemia in humans. *Calcified Tissue International*, *108*(1), 41–73. https://doi.org/10.1007/s00223-020-00664-9.

Kumar, R. (2009). Phosphate sensing. *Current Opinion in Nephrology and Hypertension*, *18*(4), 281–284. https://doi.org/10.1097/mnh.0b013e32832b5094.

Küng, C. J., Haykir, B., Schnitzbauer, U., Egli-Spichtig, D., Hernando, N., & Wagner, C. A. (2021). Fibroblast growth factor 23 leads to endolysosomal routing of the renal phosphate cotransporters NaPi-IIa and NaPi-IIc in vivo. *American Journal of Physiology-Renal Physiology*, *321*(6), F785–F798. https://doi.org/10.1152/ajprenal.00250.2021.

Labonté, E. D., Carreras, C. W., Leadbetter, M. R., Kozuka, K., Kohler, J., Koo-McCoy, S., et al. (2015). Gastrointestinal inhibition of sodium-hydrogen exchanger 3 reduces phosphorus absorption and protects against vascular calcification in CKD. *Journal of the American Society of Nephrology*, *26*(5), 1138–1149. https://doi.org/10.1681/asn.2014030317.

Larsson, T. E., Kameoka, C., Nakajo, I., Taniuchi, Y., Yoshida, S., Akizawa, T., et al. (2018). NPT-IIb inhibition does not improve hyperphosphatemia in CKD. *Kidney International Reports*, *3*(1), 73–80. https://doi.org/10.1016/j.ekir.2017.08.003.

Lee, D. B., Walling, M. W., & Corry, D. B. (1986). Phosphate transport across rat jejunum: Influence of sodium, pH, and 1,25-dihydroxyvitamin D3. *American Journal of Physiology-Gastrointestinal and Liver Physiology*, *251*(1), G90–G95. https://doi.org/10.1152/ajpgi.1986.251.1.g90.

Lenglet, A., Liabeuf, S., Esper, N. E., Brisset, S., Mansour, J., Lemaire-Hurtel, A.-S., et al. (2016). Efficacy and safety of nicotinamide in haemodialysis patients: The NICOREN study. In *Nephrology, dialysis, transplantation* Official Publication of the European Dialysis and Transplant Association-European Renal Association. https://doi.org/10.1093/ndt/gfw042. gfw042-10.

Li, X., Gu, C., Hostachy, S., Sahu, S., Wittwer, C., Jessen, H. J., et al. (2020). Control of XPR1-dependent cellular phosphate efflux by InsP8 is an exemplar for functionally-exclusive inositol pyrophosphate signaling. *Proceedings of the National Academy of Sciences*, *117*(7), 3568–3574. https://doi.org/10.1073/pnas.1908830117.

Liamis, G., Milionis, H. J., & Elisaf, M. (2010). Medication-induced hypophosphatemia: A review. *QJM: An International Journal of Medicine*, *103*(7), 449–459. https://doi.org/10.1093/qjmed/hcq039.

Lioufas, N. M., Pascoe, E. M., Hawley, C. M., Elder, G. J., Badve, S. V., Block, G. A., et al. (2021). Systematic review and meta-analyses of the effects of phosphate-lowering agents in nondialysis CKD. *Journal of the American Society of Nephrology*, *33*(1), 59–76. https://doi.org/10.1681/asn.2021040554.

Markov, A. G., Veshnyakova, A., Fromm, M., Amasheh, M., & Amasheh, S. (2010). Segmental expression of claudin proteins correlates with tight junction barrier properties in rat intestine. *Journal of Comparative Physiology B*, *180*(4), 591–598. https://doi.org/10.1007/s00360-009-0440-7.

Marks, J. (2019). The role of SLC34A2 in intestinal phosphate absorption and phosphate homeostasis. *Pflügers Archiv-European Journal of Physiology*, *471*(1), 165–173. https://doi.org/10.1007/s00424-018-2221-1.

Marks, J., Churchill, L. J., Debnam, E. S., & Unwin, R. J. (2008). Matrix extracellular phosphoglycoprotein inhibits phosphate transport. *Journal of the American Society of Nephrology*, *19*(12), 2313–2320. https://doi.org/10.1681/asn.2008030315.

Marks, J., Churchill, L. J., Srai, S. K., Biber, J., Murer, H., Jaeger, P., et al. (2007). Intestinal phosphate absorption in a model of chronic renal failure. *Kidney International*, *72*(2), 166–173. https://doi.org/10.1038/sj.ki.5002292.

Marks, J., Lee, G. J., Nadaraja, S. P., Debnam, E. S., & Unwin, R. J. (2015). Experimental and regional variations in Na^+-dependent and Na^+-independent phosphate transport along the rat small intestine and colon. *Physiological Reports*, *3*(1), e12281. https://doi.org/10.14814/phy2.12281.

Marks, J., Srai, S. K., Biber, J., Murer, H., Unwin, R. J., & Debnam, E. S. (2006). Intestinal phosphate absorption and the effect of vitamin D: A comparison of rats with mice. *Experimental Physiology*, *91*(3), 531–537. https://doi.org/10.1113/expphysiol.2005.032516.

Maruyama, S., Marbury, T. C., Connaire, J., Ries, D., Maxwell, W., & Rambaran, C. (2021). NaPi-IIb inhibition for hyperphosphatemia in CKD hemodialysis patients. *Kidney International Reports*, *6*(3), 675–684. https://doi.org/10.1016/j.ekir.2020.12.017.

McCullough, P. A. (2021). Phosphate control: The next frontier in dialysis cardiovascular mortality. *Cardiorenal Medicine*, *11*(3), 123–132. https://doi.org/10.1159/000516286.

McHardy, G. J. R., & Parsons, D. S. (1956). The absorption of inorganic phosphate from the small intestine of the rat. *Quarterly Journal of Experimental Physiology and Cognate Medical Sciences*, *41*(4), 398–409. https://doi.org/10.1113/expphysiol.1956.sp001211.

Michigami, T., Kawai, M., Yamazaki, M., & Ozono, K. (2018). Phosphate as a signaling molecule and its sensing mechanism. *Physiological Reviews*, *98*(4), 2317–2348. https://doi.org/10.1152/physrev.00022.2017.

Moe, S. M., & Chen, N. X. (2008). Mechanisms of vascular calcification in chronic kidney disease. *Journal of the American Society of Nephrology*, *19*(2), 213–216. https://doi.org/10.1681/asn.2007080854.

Moritoh, Y., Abe, S., Akiyama, H., Kobayashi, A., Koyama, R., Hara, R., et al. (2021). The enzymatic activity of inositol hexakisphosphate kinase controls circulating phosphate in mammals. *Nature Communications*, *12*(1), 4847. https://doi.org/10.1038/s41467-021-24934-8.

Neyra, J. A., Hu, M. C., & Moe, O. W. (2021). Klotho in clinical nephrology: Diagnostic and therapeutic implications. *Clinical Journal of the American Society of Nephrology*, *16*(1), 162–176. https://doi.org/10.2215/cjn.02840320.

Nowik, M., Picard, N., Stange, G., Capuano, P., Tenenhouse, H. S., Biber, J., et al. (2008). Renal phosphaturia during metabolic acidosis revisited: Molecular mechanisms for decreased renal phosphate reabsorption. *Pflügers Archiv-European Journal of Physiology*, *457*(2), 539–549. https://doi.org/10.1007/s00424-008-0530-5.

Ohkido, I., Segawa, H., Yanagida, R., Nakamura, M., & Miyamoto, K. (2003). Cloning, gene structure and dietary regulation of the type-IIc Na/Pi cotransporter in the mouse kidney. *Pflügers Archiv*, *446*(1), 106–115. https://doi.org/10.1007/s00424-003-1010-6.

Pastor-Arroyo, E. M., Knöpfel, T., Silva, P. H. I., Schnitzbauer, U., Poncet, N., Biber, J., et al. (2020). Intestinal epithelial ablation of Pit-2/Slc20a2 in mice leads to sustained elevation of vitamin D3 upon dietary restriction of phosphate. *Acta Physiologica*, *230*(2), e13526. https://doi.org/10.1111/apha.13526.

Penido, M. G. M. G., & Alon, U. S. (2012). Phosphate homeostasis and its role in bone health. *Pediatric Nephrology*, *27*(11), 2039–2048. https://doi.org/10.1007/s00467-012-2175-z.

Pergola, P. E., Rosenbaum, D. P., Yang, Y., & Chertow, G. M. (2021). A randomized trial of Tenapanor and phosphate binders as a dual-mechanism treatment for hyperphosphatemia in patients on maintenance dialysis (AMPLIFY). *Journal of the American Society of Nephrology*, *32*(6), 1465–1473. https://doi.org/10.1681/asn.2020101398.

Pfister, M. F., Ruf, I., Stange, G., Ziegler, U., Lederer, E., Biber, J., et al. (1998). Parathyroid hormone leads to the lysosomal degradation of the renal type II Na/Pi cotransporter. *Proceedings of the National Academy of Sciences*, *95*(4), 1909–1914. https://doi.org/10.1073/pnas.95.4.1909.

Picard, N., Capuano, P., Stange, G., Mihailova, M., Kaissling, B., Murer, H., et al. (2010). Acute parathyroid hormone differentially regulates renal brush border membrane phosphate cotransporters. *Pflügers Archiv-European Journal of Physiology*, *460*(3), 677–687. https://doi.org/10.1007/s00424-010-0841-1.

Radanovic, T., Wagner, C. A., Murer, H., & Biber, J. (2005). Regulation of intestinal phosphate transport I. Segmental expression and adaptation to low-Pi diet of the type IIb Na^+-Pi cotransporter in mouse small intestine. *American Journal of Physiology-Gastrointestinal and Liver Physiology*, *288*(3), G496–G500. https://doi.org/10.1152/ajpgi.00167.2004.

Ramos, E. M., Carecchio, M., Lemos, R., Ferreira, J., Legati, A., Sears, R. L., et al. (2018). Primary brain calcification: An international study reporting novel variants and associated phenotypes. *European Journal of Human Genetics*, *26*(10), 1462–1477. https://doi.org/10.1038/s41431-018-0185-4.

Rubio-Aliaga, I. (2020). Phosphate and kidney healthy aging. *Kidney and Blood Pressure Research, 45*(6), 802–811. https://doi.org/10.1159/000509831.

Sabbagh, Y., Giral, H., Caldas, Y. A., Levi, M., & Schiavi, S. C. (2011). Intestinal phosphate transport. *Advances in Chronic Kidney Disease, 18*(2), 85–90. https://doi.org/10.1053/j.ackd.2010.11.004.

Sabbagh, Y., O'Brien, S. P., Song, W., Boulanger, J. H., Stockmann, A., Arbeeny, C., et al. (2009). Intestinal Npt2b plays a major role in phosphate absorption and homeostasis. *Journal of the American Society of Nephrology, 20*(11), 2348–2358. https://doi.org/10.1681/asn.2009050559.

Saurette, M., & Alexander, R. T. (2019). Intestinal phosphate absorption: The paracellular pathway predominates? *Experimental Biology and Medicine, 244*(8), 646–654. https://doi.org/10.1177/1535370219831220.

Schiavi, S. C., Tang, W., Bracken, C., O'Brien, S. P., Song, W., Boulanger, J., et al. (2012). Npt2b deletion attenuates hyperphosphatemia associated with CKD. *Journal of the American Society of Nephrology, 23*(10), 1691–1700. https://doi.org/10.1681/asn.2011121213.

Segawa, H., Kaneko, I., Takahashi, A., Kuwahata, M., Ito, M., Ohkido, I., et al. (2002). Growth-related renal type II Na/Pi cotransporter. *Journal of Biological Chemistry, 277*(22), 19665–19672. https://doi.org/10.1074/jbc.m200943200.

Segawa, H., Kaneko, I., Yamanaka, S., Ito, M., Kuwahata, M., Inoue, Y., et al. (2004). Intestinal Na-Pi cotransporter adaptation to dietary Pi content in vitamin D receptor null mice. *American Journal of Physiology-Renal Physiology, 287*(1), F39–F47. https://doi.org/10.1152/ajprenal.00375.2003.

Segawa, H., Kawakami, E., Kaneko, I., Kuwahata, M., Ito, M., Kusano, K., et al. (2003). Effect of hydrolysis-resistant FGF23-R179Q on dietary phosphate regulation of the renal type-II Na/Pi transporter. *Pflügers Archiv-European Journal of Physiology, 446*(5), 585–592. https://doi.org/10.1007/s00424-003-1084-1.

Segawa, H., Onitsuka, A., Kuwahata, M., Hanabusa, E., Furutani, J., Kaneko, I., et al. (2009). Type IIc sodium-dependent phosphate transporter regulates calcium metabolism. *Journal of the American Society of Nephrology, 20*(1), 104–113. https://doi.org/10.1681/asn.2008020177.

Segawa, H., Yamanaka, S., Ito, M., Kuwahata, M., Shono, M., Yamamoto, T., et al. (2005). Internalization of renal type IIc Na-pi cotransporter in response to a high- phosphate diet. *American Journal of Physiology-Renal Physiology, 288*(3), F587–F596. https://doi.org/10.1152/ajprenal.00097.2004.

Segawa, H., Yamanaka, S., Onitsuka, A., Tomoe, Y., Kuwahata, M., Ito, M., et al. (2007). Parathyroid hormone-dependent endocytosis of renal type IIc Na-Pi cotransporter. *American Journal of Physiology-Renal Physiology, 292*(1), F395–F403. https://doi.org/10.1152/ajprenal.00100.2006.

Serna, J., & Bergwitz, C. (2020). Importance of dietary phosphorus for bone metabolism and healthy aging. *Nutrients, 12*(10), 3001. https://doi.org/10.3390/nu12103001.

Shigematsu, T., Une, Y., Ikejiri, K., Kanda, H., Fukagawa, M., & Akizawa, T. (2021). Therapeutic effects of add-on Tenapanor for hemodialysis patients with refractory hyperphosphatemia. *American Journal of Nephrology, 52*(6), 496–506. https://doi.org/10.1159/000516156.

Sprague, S. M., Martin, K. J., & Coyne, D. W. (2021). Phosphate balance and CKD-mineral bone disease. *Kidney International Reports, 6*(8), 2049–2058. https://doi.org/10.1016/j.ekir.2021.05.012.

Stevens, L. A., Djurdjev, O., Cardew, S., Cameron, E. C., & Levin, A. (2004). Calcium, phosphate, and parathyroid hormone levels in combination and as a function of dialysis duration predict mortality; evidence for the complexity of the association between mineral metabolism and outcomes. *Journal of the American Society of Nephrology, 15*(3), 770–779. https://doi.org/10.1097/01.asn.0000113243.24155.2f.

Stremke, E. R., Wiese, G. N., Moe, S. M., Wastney, M. E., Moorthi, R. N., & Gallant, K. M. H. (2021). Intestinal phosphorus absorption in moderate CKD and healthy adults determined using a radioisotopic tracer. *Journal of the American Society of Nephrology*, *32*(8), 2057–2069. https://doi.org/10.1681/asn.2020091340.

Thomas, L., Bettoni, C., Knöpfel, T., Hernando, N., Biber, J., & Wagner, C. A. (2016). Acute adaption to oral or intravenous phosphate requires parathyroid hormone. *Journal of the American Society of Nephrology*, 1–12. https://doi.org/10.1681/asn.2016010082.

Thomas, L., Xue, J., Murali, S. K., Fenton, R. A., Rieg, J. A. D., & Rieg, T. (2019). Pharmacological Npt2a inhibition causes phosphaturia and reduces plasma phosphate in mice with normal and reduced kidney function. *Journal of the American Society of Nephrology*, *30*(11), 2128–2139. https://doi.org/10.1681/asn.2018121250.

Thomas, L., Xue, J., Tomilin, V. N., Pochynyuk, O. M., Rieg, J. A. D., & Rieg, T. (2020). PF-06869206 is a selective inhibitor of renal Pi transport: Evidence from in vitro and in vivo studies. *American Journal of Physiology-Renal Physiology*, *319*(3), F541–F551. https://doi.org/10.1152/ajprenal.00146.2020.

Tomoe, Y., Segawa, H., Shiozawa, K., Kaneko, I., Tominaga, R., Hanabusa, E., et al. (2010). Phosphaturic action of fibroblast growth factor 23 in Npt2 null mice. *American Journal of Physiology-Renal Physiology*, *298*(6), F1341–F1350. https://doi.org/10.1152/ajprenal.00375.2009.

Tsuboi, Y., Ohtomo, S., Ichida, Y., Hagita, H., Ozawa, K., Iida, M., et al. (2020). EOS789, a novel pan-phosphate transporter inhibitor, is effective for the treatment of chronic kidney disease-mineral bone disorder. *Kidney International*, *98*(2), 343–354. https://doi.org/10.1016/j.kint.2020.02.040.

Turner, M. E., White, C. A., Hopman, W. M., Ward, E. C., Jeronimo, P. S., Adams, M. A., et al. (2018). Impaired phosphate tolerance revealed with an acute Oral challenge. *Journal of Bone and Mineral Research*, *33*(1), 113–122. https://doi.org/10.1002/jbmr.3294.

Villa-Bellosta, R., Barac-Nieto, M., Breusegem, S. Y., Barry, N. P., Levi, M., & Sorribas, V. (2008). Interactions of the growth-related, type IIc renal sodium/phosphate cotransporter with PDZ proteins. *Kidney International*, *73*(4), 456–464. https://doi.org/10.1038/sj.ki.5002703.

Villa-Bellosta, R., Ravera, S., Sorribas, V., Stange, G., Levi, M., Murer, H., et al. (2009). The Na^+-pi cotransporter PiT-2 (SLC20A2) is expressed in the apical membrane of rat renal proximal tubules and regulated by dietary Pi. *American Journal of Physiology-Renal Physiology*, *296*(4), F691–F699. https://doi.org/10.1152/ajprenal.90623.2008.

Villa-Bellosta, R., & Sorribas, V. (2010). Compensatory regulation of the sodium/phosphate cotransporters NaPi-IIc (SCL34A3) and Pit-2 (SLC20A2) during Pi deprivation and acidosis. *Pflügers Archiv-European Journal of Physiology*, *459*(3), 499–508. https://doi.org/10.1007/s00424-009-0746-z.

Virkki, L. V., Biber, J., Murer, H., & Forster, I. C. (2007). Phosphate transporters: A tale of two solute carrier families. *American Journal of Physiology-Renal Physiology*, *293*(3), F643–F654. https://doi.org/10.1152/ajprenal.00228.2007.

Vorland, C. J., Biruete, A., Lachcik, P. J., Srinivasan, S., Chen, N. X., Moe, S. M., et al. (2020). Kidney disease progression does not decrease intestinal phosphorus absorption in a rat model of chronic kidney disease–mineral bone disorder. *Journal of Bone and Mineral Research*, *35*(2), 333–342. https://doi.org/10.1002/jbmr.3894.

Wagner, C. A., Rubio-Aliaga, I., & Hernando, N. (2019). Renal phosphate handling and inherited disorders of phosphate reabsorption: An update. *Pediatric Nephrology*, *34*(4), 549–559. https://doi.org/10.1007/s00467-017-3873-3.

Walling, M. W. (1977). Intestinal Ca and phosphate transport: Differential responses to vitamin D3 metabolites. *American Journal of Physiology-Endocrinology and Metabolism*, *233*(6), E488. https://doi.org/10.1152/ajpendo.1977.233.6.e488.

Walton, J., & Gray, T. K. (1979). Absorption of inorganic phosphate in the human small intestine. *Clinical Science*, *56*(5), 407–412. https://doi.org/10.1042/cs0560407.

Weinman, E. J., & Lederer, E. D. (2012). PTH-mediated inhibition of the renal transport of phosphate. *Experimental Cell Research*, *318*(9), 1027–1032. https://doi.org/10.1016/j.yexcr.2012.02.037.

Wu, K. I., Bacon, R. A., Al-Mahrouq, H. A., & Kempson, S. A. (1988). Nicotinamide as a rapid- acting inhibitor of renal brush-border phosphate transport. *American Journal of Physiology-Renal Physiology*, *255*(1), F15–F21. https://doi.org/10.1152/ajprenal.1988.255.1.f15.

Xu, H., Inouye, M., Missey, T., Collins, J. F., & Ghishan, F. K. (2002). Functional characterization of the human intestinal NaPi-IIb cotransporter in hamster fibroblasts and Xenopus oocytes. *Biochimica et Biophysica Acta (BBA)—Biomembranes*, *1567*(1–2), 97–105. https://doi.org/10.1016/s0005-2736(02)00604-1.

CHAPTER FOUR

The parathyroid glands and parathyroid hormone: Insights from *PTH* gene mutations

Joon-Hyop Lee[a,b] and **Sihoon Lee**[a,c,*]

[a]Laboratory of Genomics and Translational Medicine, Gachon University College of Medicine, Incheon, Korea
[b]Department of Surgery, Gachon University College of Medicine, Incheon, Korea
[c]Department of Internal Medicine, Gachon University College of Medicine, Incheon, Korea
*Corresponding author: e-mail address: shleemd@gachon.ac.kr

Contents

Abstract

Nine mutations have been discovered in the parathyroid hormone (PTH) gene since it was initially sequenced in 1983. An autosomal dominant C18R mutation in the signal

Vitamins and Hormones, Volume 120
ISSN 0083-6729
https://doi.org/10.1016/bs.vh.2022.04.004

peptide was first reported in 1990, followed by an exon skipping mutation, leading to loss of exon 2 in 1992; the latter mutation prevents PTH biosynthesis, as exon 2 contains the initiation codon. The S23P and S23X mutations affecting the same residue were reported in 1999 and 2012, respectively, while in 2008, the somatic mutation, R83X, was detected in a parathyroid adenoma tissue sample from a patient with overt hyperparathyroidism. In 2013, the heterozygous p.Met1_Asp6del mutation was discovered incidentally in a case-control study, while another heterozygous mutation, M14K, was detected in the signal peptide 4 years later. In 2015, a homozygous R56C mutation was reported, and was the first hypoparathyroidism-causing mutation identified that affects the mature bioactive portion of PTH; this mutation has significantly contributed to the understanding of the molecular mechanisms involved in signal transduction through the PTH receptor. Recently, a novel homozygous S32P mutation was identified, which is also situated in the bioactive portion of PTH. The discovery of these nine mutations in the *PTH* gene and determination of the molecular mechanisms underlying their effects has provided deep insights into the synthesis, processing, and secretion of PTH. Future attempts to discover other such mutations will help elucidate as yet unknown functions of PTH, with potential clinical implications.

1. History of the parathyroid gland and parathyroid hormone

The parathyroid glands and parathyroid hormone (PTH) were unknown until just 140 years ago. Although there are several historical descriptions of the parathyroid glands, Sandström has been credited with their discovery and nomenclature. He first detected the parathyroid glands while examining a dog's neck, and later described their identification in other animals in 1880 (Sandstrom, 1880). Subsequently, he performed anatomical studies of 50 human corpses and named the new structure the "*glandulae parathyroidae*." As an anatomist, Sandström did not elucidate the functions of this new structure, beyond its macro and microscopic characteristics.

Although Sandström's discovery was published, it was not considered of significance for more than a decade until Eugene Gley, a French physiologist with an interest in the thyroid gland, came across the report. He had discovered the lower parathyroid glands in rabbits in 1891 and proceeded to study their function (Gley, 1891). Gley observed tetany and even death in thyroidectomized dogs, only when the glands described by Sandström were included in the surgical specimen (Gley, 1891). Similar events were observed in humans. Anton Wölfler, an assistant of the renowned surgeon, Theodor Billroth, described a patient experiencing muscle cramps and spastic twitching 3 days after receiving thyroid surgery (Wölfler, 1897). Ten of the 38 early thyroid operations conducted by Dr. Billroth had resulted in post-thyroidectomy tetany (Eiselsberg, 1890).

Jacob Erdheim, a Viennese Pathologist, was aware of the post-thyroidectomy complication that occurred among Billroth's patients. At the time, the cause of the complication was under debate; whether it was caused by removal of the thyroid, parathyroid, or even the combination of both, was not confirmed. Erdheim demonstrated that the culprit was loss of the parathyroid glands by complete microscopic examination of the whole neck area revealing their absence in three patients who died of post-thyroidectomy tetany (Erdheim, 1906).

The intense debate on how the parathyroid glands affected blood calcium levels and tetany persisted for 45 years after Sandström's discovery. James B. Collip, a Canadian biochemist, resolved this controversy in 1925 by successfully obtaining an acid extract from oxen parathyroid glands, which contained active hormones (Collip, 1925). He named the new hormone PTH and successfully treated a patient with tetany using his extract. Although the discovery of PTH allowed animal experiments to be conducted, it was insufficient to elucidate the mechanism underlying its activity. Despite Collip's discovery of PTH extract, not only was its long-term stability questionable, but inert peptide residues created during the HCL-extraction method also interfered with its activity. The development of novel methods to purify and characterize the hormone from the extract was necessary before the mechanism of action of PTH could be elucidated.

After a long period, including multiple attempts and failures to obtain pure PTH from the parathyroid glands, Gerald Aurbach overcame the shortcomings of Collip's extraction method buy successfully isolating PTH using phenol (Aurbach, 1959). Unlike acid solvents, which not only free the peptide from the glandular structure, but also cleave the active peptide itself, phenol (a strong organic solvent) releases the active peptide from the glands while maintaining its integrity. Aurbach's phenol extract produced stable and purifiable PTH, paving the way for production of sufficient stable hormone for use in future research. Subsequently in 1959, Aurbach (Aurbach, 1959) and Rasmussen and Craig (Rasmussen & Craig, 1959) separately isolated and purified PTH using countercurrent distribution methods. Amino acid sequencing of human PTH was first completed by Keutmann and colleagues in 1978 (Keutmann, Sauer, Hendy, O'Riordan, & Potts, 1978) and its nucleotide sequence was confirmed by Vasicek et al. (1983) (Fig. 1).

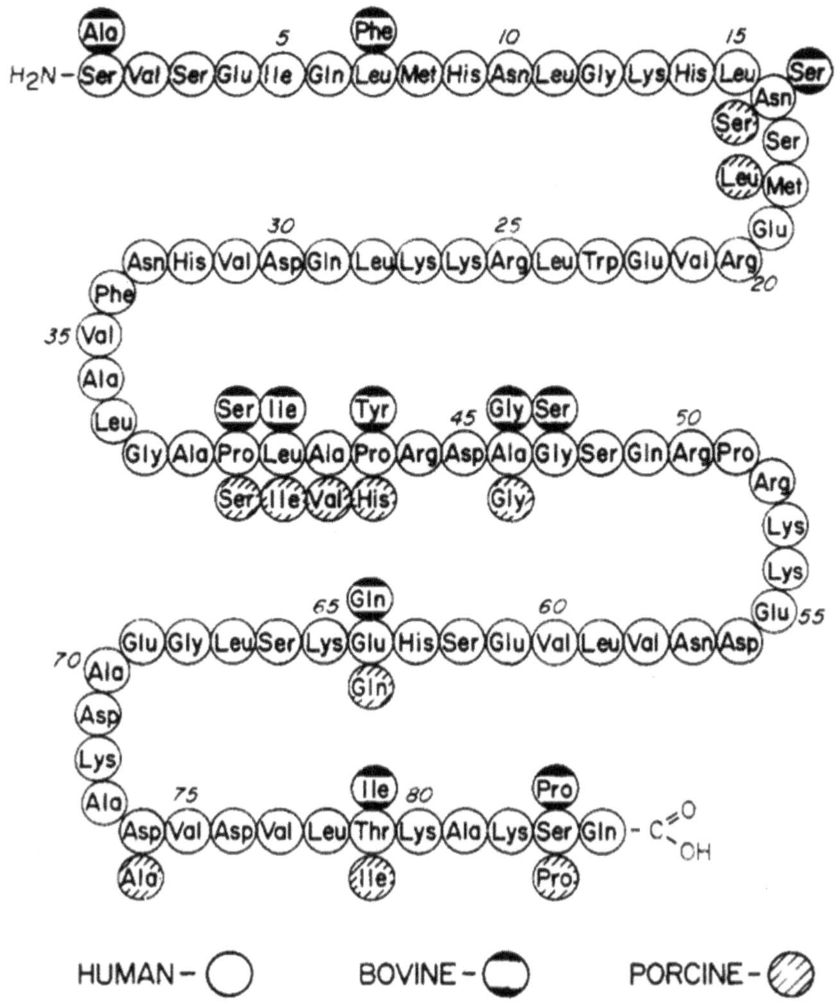

Fig. 1 The amino-acid sequences of human parathyroid hormone. The human hormone is shown by the backbone sequence; substitutions found in the bovine and porcine hormones arc indicated by circles alongside. *Adapted by Keutmann, H. T., Sauer, M. M., Hendy, G. N., O'Riordan, L. H., & Potts, J. T., Jr. (1978). Complete amino acid sequence of human parathyroid hormone.* Biochemistry, 17 *(26), 5723–5729. doi:10.1021/bi00619a019.*

2. Sequence and structure of the *PTH* gene

For many years, PTH biological activity was evaluated using bioassays that measured the rise in serum Ca^{2+} in parathyroidectomized animals. Such assays were time/labor consuming and costly. In 1969, Marcus and Aurbach developed a highly sensitive, specific, and reproducible system that could detect production of PTH-stimulated adenylate–cAMP (Marcus & Aurbach, 1969). This assay was invaluable for evaluation of the biological potency of numerous synthesized parathyroid peptides and subsequent resolution of the full PTH amino acid sequence. Identification of the PTH structure involved collaborative multi-disciplinary efforts by the Aurbach and Potts laboratories, along with many other scientists. Using the sequential degradation technique of Edman, both the Potts and Brewer groups reported the amino acid sequence of bovine PTH almost simultaneously in 1970 (Brewer & Ronan, 1970; Niall et al., 1970), followed by similar reports of the porcine (Sauer et al., 1974) and human (Brewer, Fairwell, Ronan, Sizemore, & Arnaud, 1972; Niall et al., 1974) hormones. There were several discrepancies between the structures proposed by the two research groups; however, through a series of elegant experiments, including nucleotide sequence analysis of the *PTH* gene coding region, the structures reported by the Potts group were verified as accurate (Keutmann, Niall, O'Riordan, & Potts, 1975; Vasicek et al., 1983). Thus, the synthetic peptides used today in parathyroid research are based on the hormone sequences originally reported and validated by the Potts group. Beyond resolving the amino acid sequence of these hormones, Potts and his colleagues synthesized and characterized a large number of peptides of varying length and amino acid sequence. The results of this work indicated that the biologically active portion of PTH resides in its NH_2-terminal domain, and suggested that a peptide comprising amino acids 1–29 could support the known activities of the full PTH sequence (amino acids 1–84) (Keutmann, Dawson, Aurbach, & Potts, 1972). This NH_2-terminal peptide increased serum Ca^{2+} in rats, stimulated prompt urinary excretion of cAMP and phosphate, and was a potent activator of cAMP production in the bone and kidney. This line of structure–function analysis led to the conclusion that PTH peptides of approximately 34–38 amino acids from the NH_2-terminus side are fully active *in vitro*, and have *in vivo* potencies approaching those of intact PTH (amino acids 1–84). In effect, truncating the peptides from the COOH-terminus side to lengths of 27 amino acids or less does not reduce the biological activity of the peptide. By contrast, the effect of deleting the first and second amino acids from the NH_2-terminus is profound. Despite binding to the PTH receptor with full avidity, a peptide missing the initial NH_2-terminal amino acid loses substantial activity, and elimination of the first two amino acids results in a complete loss of biological activity

(Tregear et al., 1973). Inevitably, these findings led to investigations of the biologic relevance of the COOH-terminal epitopes, which suggested that COOH-terminal sequences could stimulate distinct alkaline phosphatase activity (Murray, Rao, & Muzaffar, 1991). Further, receptors for COOH-terminal PTH fragments have been described on bone-derived cells (Divieti et al., 2001; Divieti, Geller, Suliman, Juppner, & Bringhurst, 2005).

The human *PTH* gene is a single-copy locus on the p-arm of chromosome 11 (11p15.2), spanning approximately 4kb of genomic DNA. It consists of three exons separated by two introns (Antonarakis et al., 1983; Vasicek et al., 1983). Exon 1 encodes a 5′ untranslated region (UTR), while exon 2 encodes six nucleotides of 5′ UTR, the entire signal peptide (pre-sequence) of 25 amino acids, and part of the pro-sequence. Exon 3 encodes the remainder of the pro-sequence, the entire full-length PTH (amino acids 1–84) and the 3′ UTR polyadenylation site. Codons for the first three amino acids and two nucleotides of the codon for the fourth amino acid, lysine, are in exon 2. The last nucleotide of the codon for the fourth amino acid and the last two amino acids, lysine and arginine, belong to exon 3. Both introns start with a GT dinucleotide and end with an AG dinucleotide, following the GT/AG rule (Fig. 2).

The human *PTH* gene has two transcription initiation sites, each one around 30bp downstream from different TATA boxes. Indeed, two *PTH* gene transcripts were synthesized when the rat pituitary cell line, GH4C1, was transfected with a plasmid containing the human *PTH* gene (Igarashi, Okazaki, Potter, Gaz, & Kronenberg, 1986). This genetic structure, including the exon/intron arrangement, is conserved and the nucleotide sequence encoding prepro-PTH shows a high degree of homology

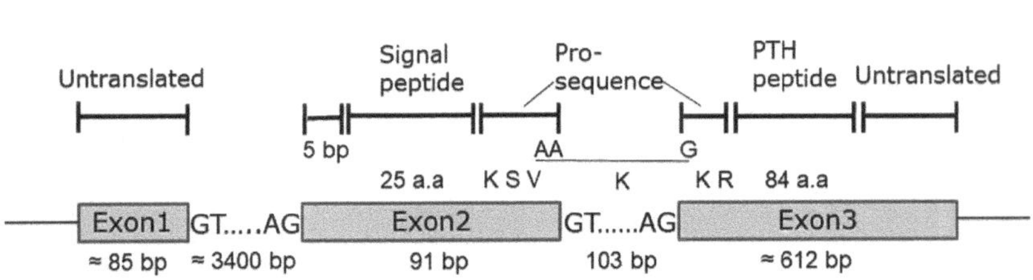

Fig. 2 Schematic structure of the human *PTH* gene. It consists of three exons and two introns. Exon 1 encodes most of the 5′ untranslated region (UTR). Exon 2 encodes five nucleotides of the 5′ UTR, the entire pre-sequence (signal peptide) of 25 amino-acids, and part of the pro-sequence (the first three of a total of six amino acids, K–S–V and two nucleotides (Ax2) of the fourth amino acid, K). Exon 3 encodes the remainder of the pro-sequence (the last nucleotide (G), of the fourth amino acid, K and the last two amino acids, K-R), the entire full-length *PTH* (1–84) and the 3′ UTR.

across numerous species (Gensure et al., 2004; He et al., 2002; Heinrich, Kronenberg, Potts, & Habener, 1984; Pinheiro et al., 2010; Weaver, Gordon, Kissil, Mead, & Kemper, 1984).

3. Synthesis, processing, and secretion of PTH

The *PTH* mRNA transcript is translated into prepro-PTH in ribosomes, which is then translocated through the rough endoplasmic reticulum (RER) and the trans-Golgi network, during which the pre and pro peptides are cleaved, leaving the active hormone molecules to be packaged and released in secretory granules, as required (Habener, Rosenblatt, & Potts, 1984).

Synthesis of PTH peptide is affected by serum 1α,25$(OH)_2D$, Ca^{2+}, and phosphate concentrations (Cantley, Russell, Lettieri, & Sherwood, 1985). The transcription rate of the *PTH* gene is reduced by binding of 1α,25 $(OH)_2D$ and the vitamin D receptor (VDR), which in turn binds to the vitamin D response element in the *PTH* gene promoter region. Serum Ca^{2+} affects the *PTH* gene indirectly by interacting with adenine- and uridine-rich elements (AREs) located in the 3' UTR, which regulate stability of the mRNA transcript (Naveh-Many, Sela-Brown, & Silver, 1999). Similarly, low serum phosphate level reduce the stability of the prepro-PTH encoding mRNA by interacting with the ARE (Levi et al., 2006).

After transcription, the nascent mRNA undergoes 5′-methyl capping, splicing, cleavage, and polyadenylation inside the nucleus (McKee & Silver, 2007). Depending on the serum Ca^{2+}-dependent activity of ARE-binding proteins, some mRNAs are stabilized for translation in ribosomes, while the remaining molecules await degradation, as they are destabilized through deadenylation and decapping in exosomes and processing bodies (Kumar & Thompson, 2011).

Many secreted proteins have signal peptides in their precursors that are critical in directing them through the cellular secretory pathway. The ribosome initiates translation of the signal sequence, which is a binding site for signal recognition particles (SRP). This complex binds to the SRP receptor (docking protein) located on the surface of the RER, thus docking the signal peptide to the outer border of the RER (Meyer, Krause, & Dobberstein, 1982). The signal sequence remains inserted into the membrane of the RER and is translocated, while the attached pro-mature peptide remains inside the RER. The signal sequence is cleaved by signal peptidase, an enzyme located on the surface membrane of the RER (Rapoport, 1990).

Signal peptides generally comprise three domains: a positively charged NH_2-terminal region, a hydrophobic central region, and a COOH-terminal polar region (von Heijne, 1985). The NH_2-terminal region, is crucial for translocation as it is related to the docking protein. The central hydrophobic area is the target for SRP (Walter & Blobel, 1981), and the COOH-terminal region influences signal peptidase cleavage activity (von Heijne, 1985).

The structural features of signal sequences are critical for the translocation of secretory peptides and for their cleavage by signal peptidase (von Heijne, 1983). The COOH-terminal region allows signal peptidases to adopt a hairpin structure near the cleavage site in the membrane. This hairpin structure has been proposed as required for insertion of the precursor into the RER membrane (Spiess & Lodish, 1986; Steiner, Quinn, Chan, Marsh, & Tager, 1980) and as an appropriate substrate for the signal peptidase (Inouye, Duffaud, & Inouye, 1986). Therefore, any changes near the cleavage site may result in the disruption of signal peptidase activity (Inouye et al., 1986).

The remaining pro-PTH peptide is then transported through a series of compartments, including the various cisternae of the Golgi complex, the trans-Golgi network, and secretory granules (Lippincott-Schwartz et al., 1990). The pro-peptide, which most likely aids effective translocation across the RER membrane, as well as functioning to maintain the inactivity of the mature PTH, is cleaved from the mature hormone, presumably in the trans-Golgi tubular network. Pro-PTH is not secreted from the parathyroid glands, and neither the pro-specific fragment, nor any of its possible derivative products, can be detected in its cells (Habener et al., 1984).

4. Chromosomal rearrangement of the *PTH* gene

DNA rearrangements, such as translocations, are associated with clonal oncogenesis, and frequently induce tumor-specific juxtaposition of proto-oncogenes with DNA sequences that are designed to activate transcription of other specific genes. This may lead to deregulated or overregulated proto-oncogene expression, resulting in oncogene activation.

4.1 *PTH* promotor/*cyclin D1* rearrangement

Molecular evidence for the presence of tumor-specific DNA rearrangement was found in parathyroid adenoma by detection of a *PRAD1/cyclin D1* fusion gene (Arnold et al., 1989; Arnold, Staunton, Kim, Gaz, & Kronenberg, 1988). This rearrangement led to separation of the 5' regulatory and non-coding exon 1 regions of *PTH* from the remainder of the gene, and placed non-*PTH* DNA adjacent to the *PTH* gene. Furthermore, each tumor cell had one intact *PTH* gene, responsible for the generation of PTH by the adenoma. Cloning of the breakpoint sequence in the *PTH* gene revealed that there was striking overexpression in the tumor and sequence analysis led to the discovery of what is now known as cyclin D1 (*CCND1*; also referred to as *PRAD 1*) (Fig. 3).

Parathyroid adenoma, with the well-defined *PTH/cyclin D1* rearrangement, has 11q13 breakpoints occurring between 1 and 15 kb upstream of cyclin D1 exon 1, with its exon-containing region retained and immediate promoter intact; in these rearrangements the cyclin D1 gene is attached to the 5′ regulatory region of the *PTH* gene, including the non-coding exon 1.

Over-expression of cyclin D1 in parathyroid tumors is hence directed by the DNA sequences upstream of the *PTH* gene. These findings provided

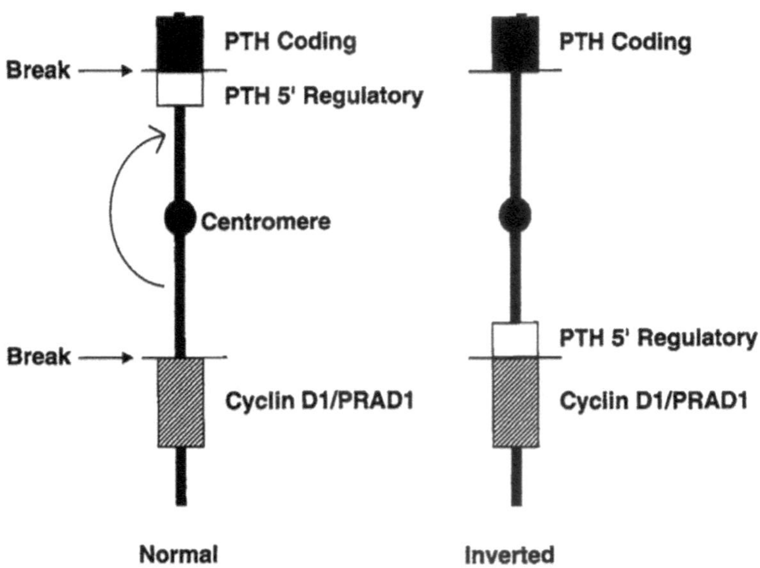

Fig. 3 Schematic diagram illustrating the DNA rearrangement involving the *PTH* gene and the *cyclin D1/PRAD1* gene in a subset of parathyroid adenomas. The chromosomal inversion event is deduced as the simplest cytogenetic mechanism consistent with the molecular details of this DNA rearrangement. *Adapted by Bilezikian, J. (Ed.). (2015).* The parathyroids—Basic and clinical concepts *(3rd ed.). London: Elsevier.*

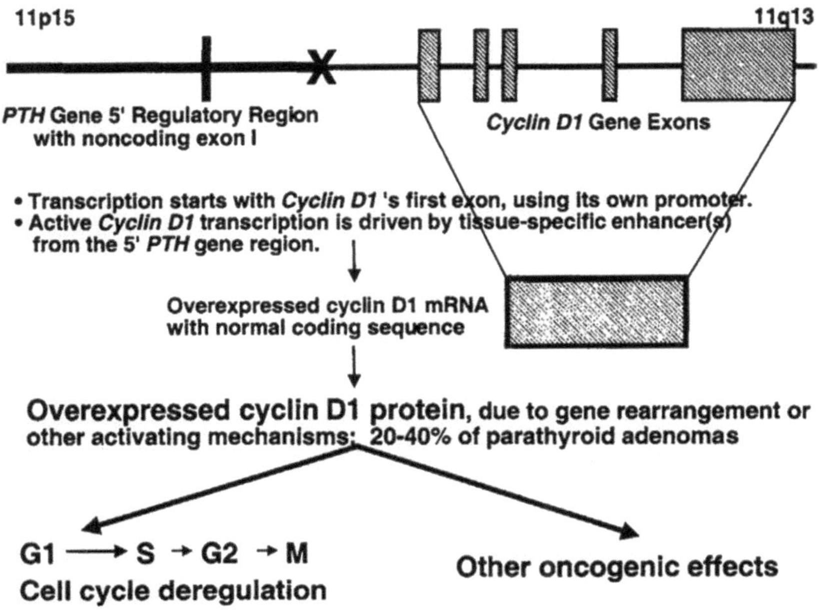

Fig. 4 Diagram of the directly observed molecular structure of the *PTH/cyclin D1* (PRAD1) DNA rearrangement and its functional consequences. *Adapted by Bilezikian, J. (Ed.) (2015).* The parathyroids—Basic and clinical concepts *(3rd ed.). London: Elsevier.*

clear proof of the presence of tissue-specific enhancer(s) of the *PTH* gene located in its 5' regulatory region (Mallya, Wu, Saria, Corrado, & Arnold, 2010), as confirmed by successful targeting of cyclin D1 transgene expression in mouse parathyroid tissue (Imanishi et al., 2001) (Fig. 4).

4.2 DNA rearrangement in the upstream regulatory region of the *PTH* gene

A DNA rearrangement in the upstream regulatory region of the *PTH* gene was also detected in an ovarian carcinoma with ectopic PTH production. The rearrangement replaced the control elements of the *PTH* gene with sequences that can interact with ovarian cell DNA-binding proteins, thereby leading to "ectopic" initiation of *PTH* expression (Nussbaum, Gaz, & Arnold, 1990). Moreover, amplification of the expression of the *PTH* gene with the upstream rearrangement was detected; i.e., it was present at several-fold increased copy numbers in ovarian cancer cells, almost definitely leading to increased PTH secretion (Nussbaum et al., 1990) (Fig. 5).

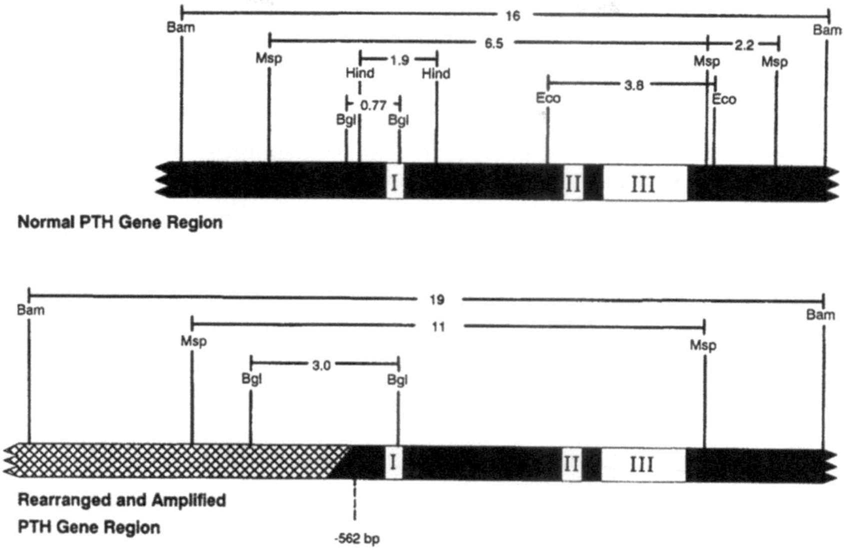

Fig. 5 Schematic diagram and restriction map of the rearranged and amplified *PTH* gene region (bottom) in the secreting ovarian tumor compared with the normal *PTH* gene region (top). *PTH* exons I, II, and III are represented by open bars; normally present introns and flanking regions are denoted by solid bars, and the DNA placed upstream of *PTH* by the tumor-specific rearrangement is denoted by the cross-hatched bar. Restriction sites are labeled, and fragment sizes are shown in kilobases. The precise location of the breakpoint of the rearrangement was narrowed to a segment between the HindIII site at −562 bp (562 base pairs upstream of the start of exon I) and the normally present BglII site 150 bp further upstream (no longer present in the rearranged gene). Bam, BamHI; Msp, MspI; Hind, HindIII; Bgl, BglII; Eco, EcoRI. *Adapted by Bilezikian, J. (Ed.) (2015).* The parathyroids—Basic and clinical concepts *(3rd ed.). London: Elsevier.*

In a pancreatic neuroendocrine neoplasm with ectopic production of PTH, no such DNA rearrangement was identified. Instead, there was evidence of transactivation of the intact *PTH* gene, with documented CpG hypomethylation of the *PTH* promoter in tumor tissue (VanHouten et al., 2006). More recently, a 2.5-kb noncoding RNA was found downstream of the *PTH* 3' region. This non-coding RNA was exclusively expressed in parathyroid adenoma, along with excess levels of *CYP24A1* and lack of *CaSR*, and was described in a retroperitoneal malignant fibrous histiocytoma with ectopic PTH production. Such co-occurrence may act as another driver of ectopic PTH-production during tumorigenesis (Uchida et al., 2017). Thus, the molecular mechanisms underlying ectopic PTH syndrome do not appear to be homogeneous.

5. Intragenic polymorphisms of the *PTH* gene

Several intragenic polymorphisms have been identified in the human *PTH* gene. These include a polymorphic tetranucleotide, $(AAAT)_n$, within intron 1 (Parkinson, Shaw, Himsworth, & Thakker, 1993), and several frequent single nucleotide polymorphisms, which are of clinical relevance and useful for genetic linkage analyses (Ahn, Antonarakis, Kronenberg, Igarashi, & Levine, 1986; Guo et al., 2010; Miric & Levine, 1992; Tenne et al., 2008; Tenne, McGuigan, Ahlborg, Gerdhem, & Akesson, 2010).

6. Mutations in the PTH signal peptide

6.1 Heterozygous PTH C18R mutation

In 1986, Dr. Ahn and colleagues conducted a study including 23 individual patients from eight families presenting with familial isolated hypoparathyroidism (FIH). The researchers conducted restriction endonuclease analysis of the genomic DNA, to determine whether the autosomal disease could be caused by any mutations within or near the *PTH* gene (Ahn et al., 1986). Their data demonstrated the absence of gross abnormalities, such as deletions, insertions, or rearrangements, in the affected individuals. Affected siblings in four families inherited different *PTH* alleles, a finding that hinted that the mutations causing FIH could be located outside the *PTH* gene region, in at least some subgroups of patients with the disease. Furthermore, concordance was observed between hypoparathyroidism inheritance and specific *PTH* alleles in affected members of two families, indicating that the condition could also be caused by alteration within or near the *PTH* gene. These results led the authors to conclude that FIH is a diverse set of diseases underpinned by genetic and molecular heterogeneity.

Four years later, the same group conducted further genetic analyses on one of the families included in their previous study (Ahn et al., 1986). The proband presented with hypocalcemic seizures and her serum PTH level was below normal. Her brother was also diagnosed with FIH, while another four siblings were biochemically and clinically unaffected. A T to C point mutation of the 52nd nucleotide in the prepro-PTH sequence was discovered in the proband (Arnold et al., 1990). This point mutation alters the 18th amino acid from cysteine to arginine (C18R), a change which affects the hydrophobic core of the signal sequence by introducing a positively-charged amino acid (arginine). Although the role of the hydrophobic core as a target of SRP had yet to be elucidated at the time of this discovery, the authors recognized the universal presence of a long stretch of hydrophobic amino acids in the signal sequence and it was known that introduction of charged residues into hydrophobic regions of signal sequences in prokaryotes often interfered with signal function (Emr, Hall, & Silhavy, 1980). Therefore, they hypothesized that the mutant protein would be processed ineffectively, as seen in the prokaryotic model, as the hydrophobic core is crucial for effective peptide translocation across the RER membrane. To test this hypothesis and determine whether the mutation was functionally silent, the authors constructed a prepro-PTH cDNA plasmid including the point mutation and compared the *in vitro* translated pro-PTH product with that from wild-type prepro-PTH cDNA. While both prepro-PTH RNAs were effectively translated, only the normal protein was efficiently processed into pro-PTH, suggesting that the point mutation in the signal sequence adversely affected interaction of the protein with the processing pathway. Although the authors demonstrated that the interaction of the peptide with the processing pathway was affected by the C18R mutation, their findings could not explain the how the autosomal dominant inheritance pattern was maintained. They proposed that the mutant peptide from the affected allele may obstruct the processing pathway by binding to the membrane, while it may be unable to complete the translocation process; however, this hypothesis was not supported by experimental evidence.

Four years later, the same group examined the biochemical effects of the C18R mutation and demonstrated that it impaired interaction among the nascent PTH protein, SRP, and the translocation apparatus (Karaplis, Lim, Baba, Arnold, & Kronenberg, 1995). Furthermore, signal peptidase could not effectively cleave the mutant signal sequence at the cleavage site. While this study provided experimental evidence that disruption of the hydrophobic core by the mutant signal sequence was responsible for a wide spectrum of translocation disturbances, the previously proposed hypothetical explanation that mutant products obstructed the pathway through membrane binding was neither proven nor refuted. Moreover, the same mutation was reported in a patient in Thailand with sporadic idiopathic hypoparathyroidism, while further functional analysis was not reported (Suprasdngsin, Wattanachal, & Preeysombat, 2001).

The molecular mechanism by which the C18R point mutation causes FIH was elucidated by Datta, Waheed, Shah, and Sly, (2007). Until this report, how a single mutant allele led to autosomal-dominant FIH was unexplained. Datta and colleagues transfected HEK293 cells to demonstrate that mutant peptides become trapped in the cells, most notably in the endoplasmic reticulum (ER). They hypothesized that the mutant peptides entrapped in the ER were products of the C18R allele, which exerted cytotoxic effects leading to apoptosis. The proapoptotic reaction was demonstrated by a marked increase in TUNEL and Annexin V staining and up-regulation of the proapoptotic transcription factor, CHOP, while increases in ER stress signals were evidenced by up-regulation of the ER stress-responsive proteins, BiP and PERK. These findings provided support for the hypothesis that the C18R single allele defect results in retention of the cytotoxic mutant protein in the ER, which can inflict cell injury and subsequent death via the apoptotic pathway. Further evidence for a causal link between RER stress and apoptosis was provided by a demonstration that introduction of a pharmacological chaperone (4-phenylbutyric acid), which reduces intracellular accumulation of the mutant peptide, resulted in reduced ER stress signals and apoptosis of the parathyroid cells. Hence, the Datta group provided a clear explanation of how the C18R mutation can lead to an autosomal dominant form of FIH (Hammer & McPhee, 2014).

6.2 Homozygous PTH S23P mutation

In 1999, Sunthornthepvarakul and colleagues in Thailand discovered a novel mutation in the signal sequence of the prepro-PTH peptide, which resulted in FIH (Sunthornthepvarakul, Churesigaew, & Ngowngarmratana, 1999). The proband was a child of a consanguineous marriage who presented with a seizure 7 days after birth, which was intractable to anticonvulsive agents. A diagnosis of FIH was made after two younger siblings presented with similar symptoms. The inheritance pattern was autosomal recessive and affected family members presented with severe hypocalcemia and hyperphosphatemia, with undetectable serum PTH levels. DNA sequencing revealed a mutation in the second exon of the prepro-*PTH* gene, comprising a T to C point mutation in the first nucleotide of the codon for the 23rd amino acid. This resulted in the 23rd serine being replaced by a proline in the 25 amino acid signal sequence of the prepro-PTH

(S23P). The mutation is situated in the crucial signal peptidase cleavage site; the peptidase exhibits strong preferences for specific amino acids in particular positions. The amino acid at position-1 from the cleavage site (25th amino acid of the signal sequence) must be small (alanine, serine, glycine, cysteine, threonine, or glutamine), while the residue at position -3 (23rd amino acid of the signal sequence) cannot be aromatic (phenylalanine, histidine, tyrosine, and tryptophan), charged (aspartic acid, glutamate, lysine, and arginine), or large polar (asparagine and glutamine), and there must be no proline in the region between these two positions (von Heijne, 1983). These two positions in the signal sequence must be intact for the signal peptidase to cleave prepro-PTH into pro-PTH peptides in the RER. This mutation was the first serine to proline change discovered at the 23rd amino acid in the signal sequence and, as proline is a strong helix-breaking residue, the authors hypothesized that this mutation could affect the signal function by hindering signal peptidase cleavage, and proposed that the uncleaved mutant signal peptide would be anchored to the membrane and degraded in the RER (Klausner & Sitia, 1990), which would explain the undetectable serum PTH levels and subsequent hypocalcemia and hyperphosphatemia in affected patients.

6.3 Homozygous PTH S23X mutation

An additional mutation in the *PTH* gene leading to FIH was discovered by Ertl, Stary, Streubel, Raimann, and Haeusler (2012). The initial presentation of the patient was similar to that of the case described by Sunthornthepvarakul, in that they presented with seizures at a very early stage, accompanied by severe hypocalcemia and hyperphosphatemia, and an autosomal recessive mode of inheritance (Sunthornthepvarakul et al., 1999). Mutation in the *PTH* gene was not initially considered, as measurement showed low but detectable serum PTH levels, unlike the previous case; however, further genetic studies revealed a C to A SNP that altered the same 23rd serine residue in the signal sequence reported by Sunthornthepvarakul into a stop codon, to induce a premature termination (S23X), resulting in non-translation of *PTH*. This mutation results in a truncated version of the peptide, without any cleavage sites for further processing; however, as the NH_2-terminus and hydrophobic core region of the signal peptide remain intact, the truncated product may still be translocated to the RER, despite being devoid of any biologically active residues of pro-PTH or mature PTH. Although the lack of production of bioactive PTH would be expected to lead to undetectable serum PTH levels in theory, the authors proposed transcriptional read-through of the mutant product as a potential explanation for the residual serum PTH detected in the patient (Ertl et al., 2012; Sunthornthepvarakul et al., 1999).

6.4 Heterozygous p.Met1_Asp6del PTH mutation

In 2004, Goswami et al. reported an extensive study of patients diagnosed with idiopathic hypoparathyroidism between 1998 and 2003 in India (Goswami et al., 2004). No mutations or other significant findings were initially observed in the *PTH* coding region among 49 patients compared with the 55 controls, contributing no new data regarding the mechanism underlying idiopathic hypoparathyroidism (Goswami et al., 2004). The same group conducted a case-control study in 2011 comprising 147 patients and 348 controls, primarily to search for mutations in the *CaSR* gene (Tomar, Gupta, & Goswami, 2013). One of the patients presented at 18 years of age with hypocalcemic symptoms and seizures for the past 3 years. She had no family history and her biochemical results indicated hypocalcemia, hyperphosphatemia, and low but detectable serum PTH. Restriction fragment length polymorphism analysis demonstrated a p.Met1_Asp6del mutation, due to a T to C substitution in the signal peptide region of *PTH*, specifically the translation initiation codon. No further data was presented in this report (Table 1).

6.5 Heterozygous M14K PTH mutation

In 2017, Cinque et al. conducted mutation analysis of the *CaSR*, *GCM2*, *GNA11*, and *PTH* genes to identify and characterize causal mutations in genomic DNA from FIH family members (Cinque et al., 2017). Biochemical analysis of one of the probands (which was undertaken because his mother and maternal grandmother presented with hypocalcemic symptoms) indicated hypocalcemia, hypophosphatemia, and extremely low serum PTH at 5 days after birth. Genomic DNA analysis identified a heterozygous substitution of T to A at the 41st nucleotide, resulting in a methionine to lysine substitution of the 14th amino acid of the signal peptide, encoded by the second exon of the *PTH* gene. This mutation lies within the hydrophobic core of the signal peptide and *in silico* analysis predicted, although

Table 1 List of patients with reported *PTH* mutations and their genetic features.

Index case (sex, age)	*PTH* mutation	Homozygous, heterozygous, or other	Familial	Reference
Male, infant	c.52T>C p.C18R	Heterozygous	Yes	Arnold et al. (1990), Datta et al. (2007)
Female	c.86+1G>C exon 2 skipped	Homozygous	Yes	Parkinson and Thakker (1992)
Female, 7 days old	c.67T>C p.S23P	Homozygous	Yes	Sunthornthepvarakul et al. (1999)
Female, 59 years old	c.247C>T p.R83X	Somatic; hemizygous	No	Au et al. (2008)
Female, 4 months old	c.68C>A p.S23X	Homozygous	Yes	Ertl et al. (2012)
Female, 18 years old	c.2T>C p.M1_D6del	Heterozygous	No	Tomar et al. (2013)
Male, early adolescence	c.166C>T p.R56C	Homozygous	Yes	Lee et al. (2015); White et al. (2019)
Male, 2 years old	c.41T>A p.M14K	Heterozygous	Yes	Cinque et al. (2017)
Female, at birth	c.94T>C p. S32P	Homozygous	Yes	Gild, Bullock, Luxford, Field, and Clifton-Bligh (2020)

PTH, parathyroid hormone.

inconsistently, that it would lead to impaired cleavage of prepro-PTH into pro-PTH. Furthermore, *in vitro* functional assays using HEK293 cells detected increased intracellular retention and impaired secretion into the medium of mutant M14K peptides compared with wild-type molecules. Addition of the pharmacological chaperone, 4-phenylbutyric acid, reduced retention of the intracellular mutant peptide 9-fold and increased its accumulation in the cell medium 15-fold in M14K mutant cells. These findings imply that the M14K mutation may act in similar a manner to the C18R mutation, by inducing ER stress and subsequent apoptosis (Arnold et al., 1990); both mutations are situated in the hydrophobic core region of the signal peptide (Fig. 6).

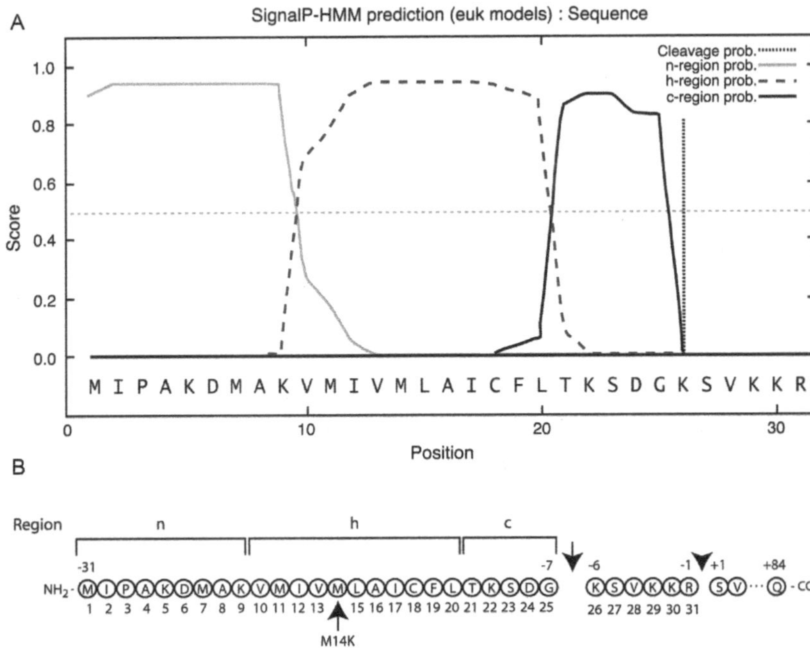

Fig. 6 Characteristics of the N-terminal region of preproPTH as a signal peptide and a summary of the SignalP predictions (A) A highly relevant probability (prob.) score was obtained, and the positions of the n-, h-, and c-regions were predicted, in addition to a signal peptide cleavage site between amino acids 25 and 26 using the SignalP hidden Markov model program (version 3.0). (B) Bracketed sequences indicate the predicted n-, h-, and c-regions. The signalase cleavage site is indicated by the arrow, and the pro-protein convertase cleavage site is indicated by the arrowhead. In addition, the M14K and the C18R mutations are indicated by the arrow. *Adapted by Cinque, L., Sparaneo, A., Penta, L., Mencarelli, A., Rogaia, D., Esposito, S., et al. (2017). Autosomal dominant PTH gene signal sequence mutation in a family with familial isolated hypoparathyroidism.* The Journal of Clinical Endocrinology and Metabolism, 102*(11), 3961–3969. doi:10.1210/jc.2017-00250; https://doi.org/10.1210/jc.2017-00250.*

7. *PTH* exon skipping mutation

7.1 Homozygous *PTH* exon 2 skipping mutation

In 1992, Parkinson and Thakker reported a case of FIH resulting from a consanguineous marriage (Parkinson & Thakker, 1992). Three of the children in the family experienced neonatal hypocalcemic seizures, and had undetectable serum PTH levels, while six unaffected relatives had normal biochemical results. The authors investigated the *PTH* gene in the patients to screen for abnormalities associated with *PTH* mRNA processing. The three exons and the four intron-exon boundaries that were identified by direct double-stranded DNA sequencing were amplified by PCR. A single-nucleotide G to C substitution in the first nucleotide of the second intron of the *PTH* gene was observed in patients, but not controls. This mutation changes the donor splice site sequence, GTAAGT, into a recognition site for the restriction endonuclease DdeI (CTAAG), leading the authors to suggest that the hypoparathyroidism may have resulted from abnormal *PTH* mRNA splicing. Further investigation revealed that the normal *PTH* cDNA consists of 258 bp, while the abnormal *PTH* cDNA was 168 bp. This difference in size corresponds precisely with the size of exon 2 (90 bp). Furthermore, nucleotide sequencing of the *PTH* cDNA from the patient revealed that exon 2 had been skipped and lost, resulting in exon 1 being merged onto exon 3. Since exon 2 contains the initiation codon (ATG), the resulting mutant allele cannot initiate the translation of *PTH* mRNA into the prepro-PTH protein and the translocation of the PTH peptide through the ER prior to secretion. Furthermore, this characteristic would only be expressed if both alleles were mutant, which explains the autosomal recessive mode of inheritance in the patients.

8. Mutations in the mature PTH peptide

8.1 Heterozygous PTH R83X mutation (R52X in mature PTH)

An interesting case report was published by Au and colleagues in 2008, describing a hypercalcemic patient with undetectable serum PTH (Au et al., 2008). The 59-year-old female patient presented with back pain. Further work-up showed ambivalent results, with hypercalcemia and undetectable serum PTH measured using two different types of assay, the Immulite assay (BioMediq DPC) and the Liaison N-tact assay (DiaSorin). Imaging studies indicated a soft tissue mass at the inferior pole region of the right thyroid, with increased uptake on sestamibi scan. Surgical pathology revealed a parathyroid adenoma that stained negative for PTH, whereas the adjacent parathyroid tissue stained positive on immunohistochemical analysis. Such conflicting findings led the authors to conduct exon sequencing, revealing a SNP causing a missense mutation of the 52nd residue (R52X) of PTH. The wild-type sequence was not detected in the adenoma; however, sequencing of DNA from peripheral blood leukocytes indicated that the mutation was heterozygous. The authors suggested that loss of heterozygosity resulted in absence of the wild-type *PTH* allele in the tumor, leading to production of biologically active, but truncated, NH_2-terminal PTH fragments, which could not be detected using conventional assays. Interestingly, the patient had a post-operative course including a decrease in serum Ca^{2+} and an increase in serum PTH to normal levels.

In 1992, Lim constructed a series of human prepro-PTH mutants with the COOH-terminal region truncated at various points along its length, to evaluate the biological function of the COOH -terminal region of PTH (Lim, Gardella, Baba, Nussbaum, & Kronenberg, 1992). When the peptide truncated at the 83rd amino acid, which would produce the same peptide as that reported in the study by Au et al. (2008), was transfected into COS cells, the truncated prepro-PTH was efficiently translocated; however, the truncated PTH was secreted in small amounts, which may either indicate that the pro-PTH molecule was inefficiently cleaved to truncated PTH, or that truncated pro-PTH peptides were destroyed, rather than being converted to truncated PTH.

8.2 Homozygous PTH R56C mutation (R25C in mature PTH)

A homozygous arginine to cysteine mutation at residue 25 (R25C) in the mature PTH (1–84) polypeptide was discovered in three siblings from a family affected by FIH (Lee et al., 2015). All three showed hypocalcemia, hyperphosphatemia, and each had increased or low-normal PTH levels, depending on the type of PTH assay employed (Fig. 7). Although the parents denied that their marriage was consanguineous, the family had lived on an isolated island off the western coast of South Korea for generations. Previous reports of the diverse mutations causing FIH described changes in the prepro leader region of the hormone, resulting in defective synthesis and secretion of the hormone; however, this mutation is positioned within the mature bioactive domain of PTH and does not affect its synthesis or secretion. This was further confirmed using two methods: first transfection of COS-7 cells with mutant PTH, and second by analysis of patient blood samples. The biological properties of ^{Cys25}PTH(1–34), including its capacity to bind to the PTH 1 receptor (PTH1R), and stimulate cAMP production, were three-fold lower in mutant, relative to wild type, PTH (1–34), based on testing in the human osteoblast-derived SaOS-2 cell line. Whereas one

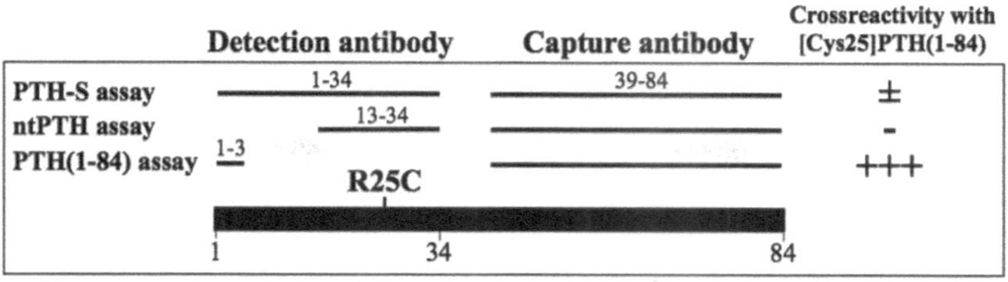

Fig. 7 Schematic representation of the PTH fragments used for affinity purification of the antibodies employed in different PTH assays. For ligand capture, all three PTH assays use similar antibodies that were affinity-purified with immobilized PTH(39–84). The detection antibodies were affinity-purified either with immobilized PTH(1–34) (PTH-S assay), PTH(13–34) (ntPTH assay), or PTH(1–3) (PTH(1–84) assay). Overall cross-reactivity with [Cys25]PTH(1–84) in the three assays is indicated by – and +. *Adapted by Lee, S., Mannstadt, M., Guo, J., Kim, S. M., Yi, H. S., Khatri, A., et al. (2015). A homozygous [Cys25] PTH (1-84) mutation that impairs PTH/PTHrP receptor activation defines a novel form of hypoparathyroidism.* Journal of Bone and Mineral Research, *30(10), 1803–1813. doi:10.1002/jbmr.2532; https://doi.org/10.1002/jbmr.2532.*

injection of a high dose of wild-type PTH (1–34) or ^{Cys25}PTH(1–34) caused a proportionate elevation in plasma and urine cAMP levels in mice, consecutive injections of ^{Cys25}PTH(1–34) resulted in small, but convincing, differences in serum Ca^{2+} and phosphate levels, relative to infusion of wild-type PTH (1–34).

Further, a unique and explicit interaction pattern between ^{Cys25}PTH(1-34) and PTH1R was identified, explaining the relatively high serum Ca^{2+} levels in patients, which helped prevent more severe hypocalcemia. It remains to be determined why the parathyroid glands in these patients did not secrete greater quantities of mutant PTH, to maintain completely normal serum Ca^{2+} levels. Some have proposed that the interaction between mutant PTH and PTHR1 could be intrinsically defective, and not amenable to amelioration by introduction of a larger volume of ligand (Lee et al., 2015).

The prolonged retention time of ligands on receptors caused by extracellular Ca^{2+}, which increases receptor activation duration and cAMP signaling, was discovered by White et al. 4 years later (White et al., 2019); however, this positive Ca^{2+} allostery was flawed in mutant ^{Cys25}PTH (1–34) (Fig. 8). Hence, Ca^{2+} allostery may have a meaningful role in the primary signaling processes that regulate calcium homeostasis. Specific functional analysis of ^{Cys25}PTH(1–34) in cellular endosomal cAMP production

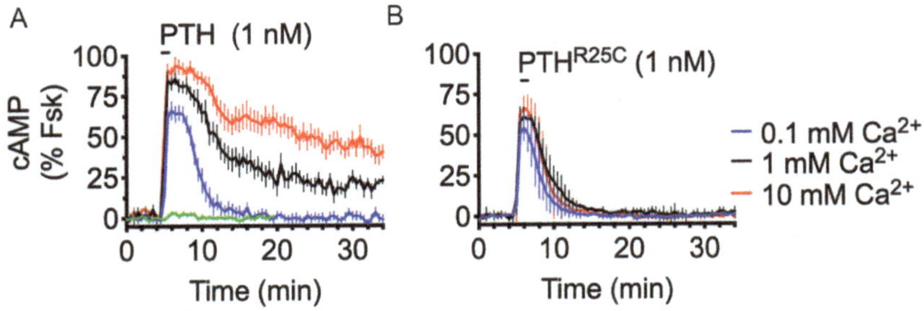

Fig. 8 Involvement of *PTH* R25 in interaction stabilization and receptor signaling shown by allosteric action of extracellular Ca2+ on the PTH signaling. (A) Averaged cAMP responses over 35 min in HEK293 cells stably expressing the *PTHR* stimulated with 1 nM PTH in the presence of a range of $[Ca^{2+}]$ measured by FRET. (B) Averaged cAMP production in response to the PTH^{R25C} under the same condition. cAMP time-course experiments in HEK cells revealed that the PTH^{R25C} completely lacks sensitivity to $[Ca^{2+}]$. *Adapted by White, A. D., Fang, F., Jean-Alphonse, F. G., Clark, L. J., An, H. J., Liu, H., et al. (2019). Ca(2+) allostery in PTH-receptor signaling.* Proceedings of the National Academy of Sciences of the United States of America, 116*(8), 3294–3299. doi:10.1073/pnas.1814670116; https://doi.org/10.1073/pnas.1814670116.*

and calcemic responses in patients will assist in determining the relevance of the association between the disease and endosomal PTHR signaling (Hammer & McPhee, 2014).

8.3 Homozygous PTH S32P mutation (S1P in mature PTH)

In 2020, Gild reported a proline to serine substitution of the 32nd residue of prepro-PTH (S32P), which is the first amino acid of mature PTH (S1P), in two sisters diagnosed with congenital hypoparathyroidism (Gild et al., 2020). Interestingly, although they showed low-normal serum PTH for many years, serum PTH concentrations increased markedly in both sisters when they reached their 30s or 40s, despite the patients remaining normocalcemic (Fig. 9). *In silico* analysis with Provean, SIFT, and Polyphen-2 predicted that the variant, ^{Pro1}PTH(1–84), would have damaging effects on PTH protein function. Moreover, Gardella et al. investigated ^{Pro1}PTH(1–84) protein function *in vitro* by expressing degenerate oligonucleotides in COS cells, followed by assays of PTH receptor-binding and cAMP production in rat osteosarcoma ROS17/2.8 cells

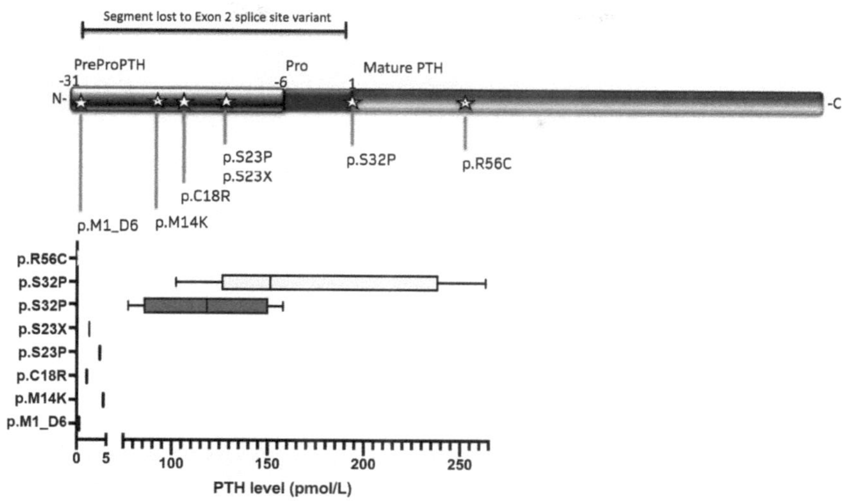

Fig. 9 Identified mutations in *PTH* with corresponding hormone levels. *Adapted by Gild, M. L., Bullock, M., Luxford, C., Field, M., & Clifton-Bligh, R. J. (2020). Congenital hypoparathyroidism associated with elevated circulating nonfunctional parathyroid hormone due to novel PTH mutation.* The Journal of Clinical Endocrinology and Metabolism, 105*(7). doi:10.1210/clinem/dgaa279; https://doi.org/10.1210/clinem/dgaa279.*

(Gardella et al., 1991). These studies revealed that ^{Pro1}PTH(1–84) did not bind to the PTH receptor or produce post-receptor cAMP. Taken together with the clinical phenotype in two patients, the ^{Pro1}PTH(1–84) variant was concluded to be biologically inactive and nonfunctional.

9. Conclusions and future directions

The parathyroid gland is the only human organ whose history has been recorded in detail, including its discovery, and the processes of determining its hormone function and its clinical applications. Only nine mutations have been discovered in PTH since its amino acid and nucleotide sequences were determined (Bilezikian, 2015; Lee, Davaatseren, & Lee, 2020) (Figs. 10 and 11). Diverse and intensive molecular genetic studies of rare diseases have deepened understanding of these conditions, raising the possibility of new insights into human physiology and molecular

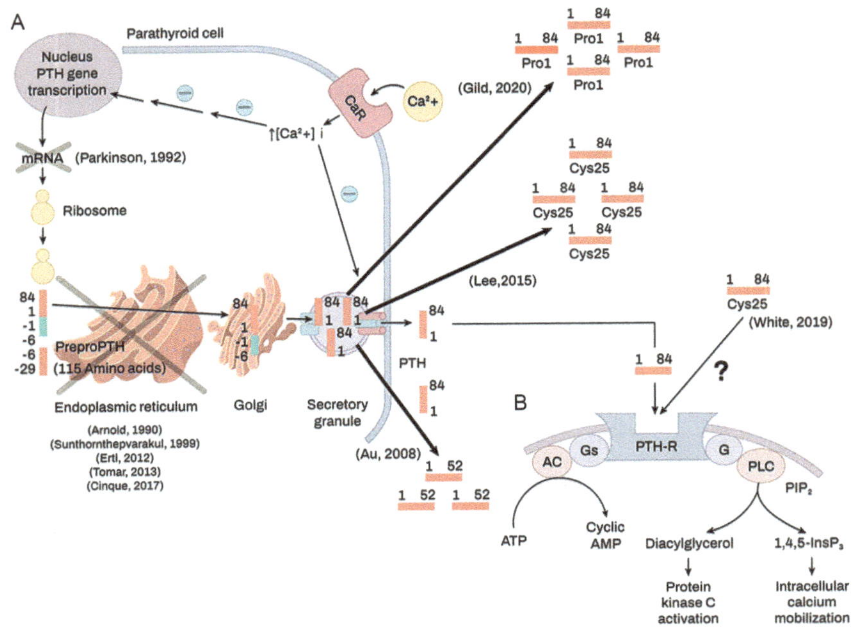

Fig. 10 Pathophysiological explanations of each mutation in *PTH* gene in each step. (A) Illustration of parathyroid hormone (PTH) synthesis and secretion from chief cells of parathyroid glands. In each step, pathophysiological explanations are noted with crossmarks. (B) Signal transduction pathways from the activated PTH receptor from its respective target tissues, i.e., bone and kidney. Note that differences in signal transduction pathways activated by either wild type *PTH* or Cys25*PTH*(1-84) are still to be investigated. CaR, calcium-sensing receptor; AC adenylate cyclase; PTH-R, PTH receptor; G, G protein; Gs, Gs alpha subunit; PLC, phospholipase C; PIP2, phosphatidylinositol 4,5-bisphosphate; ATP, adenosine triphosphate; AMP, adenosine monophosphate. *Adapted by Lee, J. H., Davaatseren, M., & Lee, S. (2020). Rare PTH gene mutations causing parathyroid disorders: A review.* Endocrinology and Metabolism (Seoul, Korea), 35*(1), 64–70. doi:10.3803/EnM.2020.35.1.64; https://doi.org/10.3803/EnM.2020.35.1.64.*

pathology. Similarly, while mutations in the *PTH* gene are mainly found in the rare diseases, idiopathic familial hypoparathyroidism and hyperparathyroidism, they have resulted in a broader understanding of the physiological function of normal parathyroid glands and deeper molecular genetic knowledge of the factors involved in their activities. This understanding raises the possibility for development of new treatments for more common diseases, such as osteoporosis and other metabolic bone and mineral diseases, and further enhances opportunities to apply previously hypothesized treatments.

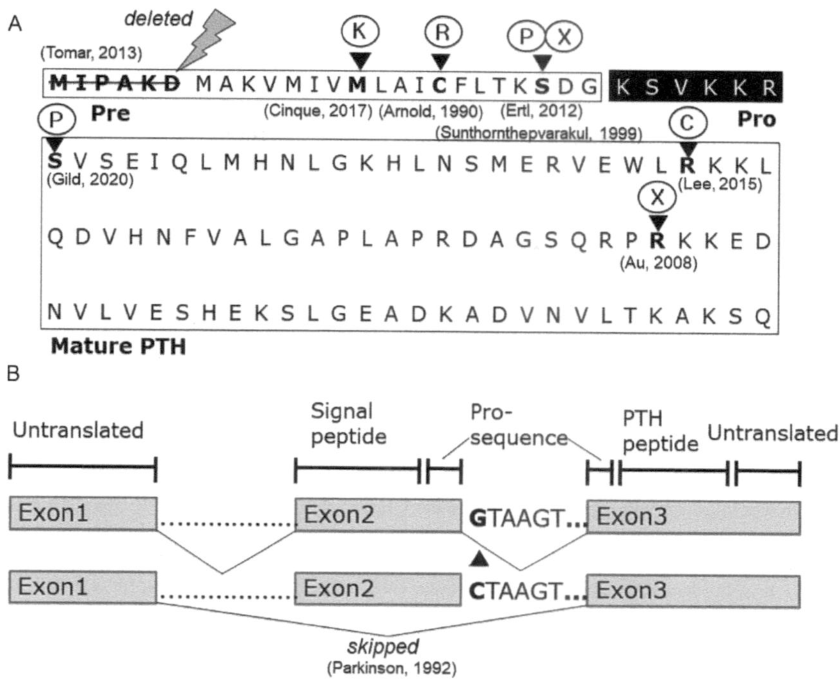

Fig. 11 Summary of *PTH* gene mutations. (A) Primary sequence of human parathyroid hormone (PTH) and organization of its domains. The pre- (signal sequence), pro- (black box), and mature regions (boxed) of PTH are labeled. The first six amino acids in the signal sequence are deleted in a patient with hypoparathyroidism. The arrowheads indicate the wild-type amino acids in pre-pro-PTH that are replaced by mutant amino acids or terminated. (B) Schematic presentation of the human parathyroid hormone (PTH) gene. The G to C nucleotide change at the beginning of intron 2 results in exon 2 being skipped in patients with autosomal dominant familial hypoparathyroidism; See (reference) for more details. *Adapted by Lee, J. H., Davaatseren, M., & Lee, S. (2020). Rare PTH gene mutations causing parathyroid disorders: A review.* Endocrinology and Metabolism (Seoul, Korea), 35*(1), 64–70. doi:10.3803/EnM.2020.35.1.64; https://doi.org/10.3803/EnM.2020.35.1.64.*

Acknowledgments

This chapter was supported by the Basic Science Research Program of the National Research Foundation of Korea Grant (to Dr. Joon-Hyop Lee [NRF-2019R1C1C1008384] and to Dr. Sihoon Lee [NRF-2019R1F1A1063459]).

Conflicts of interest

No potential conflict of interest relevant to this article was reported.

References

Ahn, T. G., Antonarakis, S. E., Kronenberg, H. M., Igarashi, T., & Levine, M. A. (1986). Familial isolated hypoparathyroidism: A molecular genetic analysis of 8 families with 23 affected persons. *Medicine (Baltimore)*, *65*(2), 73–81.

Antonarakis, S. E., Phillips, J. A., 3rd, Mallonee, R. L., Kazazian, H. H., Jr., Fearon, E. R., Waber, P. G., et al. (1983). Beta-globin locus is linked to the parathyroid hormone (PTH) locus and lies between the insulin and PTH loci in man. *Proceedings of the National Academy of Sciences of the United States of America*, *80*(21), 6615–6619. https://doi.org/10.1073/pnas.80.21.6615.

Arnold, A., Horst, S. A., Gardella, T. J., Baba, H., Levine, M. A., & Kronenberg, H. M. (1990). Mutation of the signal peptide-encoding region of the preproparathyroid hormone gene in familial isolated hypoparathyroidism. *Journal of Clinical Investigations*, *86*(4), 1084–1087. https://doi.org/10.1172/JCI114811.

Arnold, A., Kim, H. G., Gaz, R. D., Eddy, R. L., Fukushima, Y., Byers, M. G., et al. (1989). Molecular cloning and chromosomal mapping of DNA rearranged with the parathyroid hormone gene in a parathyroid adenoma. *Journal of Clinical Investigations*, *83*(6), 2034–2040. https://doi.org/10.1172/JCI114114.

Arnold, A., Staunton, C. E., Kim, H. G., Gaz, R. D., & Kronenberg, H. M. (1988). Monoclonality and abnormal parathyroid hormone genes in parathyroid adenomas. *New England Journal of Medicine*, *318*(11), 658–662. https://doi.org/10.1056/NEJM198803173181102.

Au, A. Y., McDonald, K., Gill, A., Sywak, M., Diamond, T., Conigrave, A. D., et al. (2008). PTH mutation with primary hyperparathyroidism and undetectable intact PTH. *New England Journal of Medicine*, *359*(11), 1184–1186. https://doi.org/10.1056/NEJMc0802570.

Aurbach, G. D. (1959). Isolation of parathyroid hormone after extraction with phenol. *Journal of Biological Chemistry*, *234*, 3179–3181.

Bilezikian, J. (Ed.). (2015). *The parathyroids—Basic and clinical concepts* (3rd ed.). London: Elsevier.

Brewer, H. B., Jr., Fairwell, T., Ronan, R., Sizemore, G. W., & Arnaud, C. D. (1972). Human parathyroid hormone: Amino-acid sequence of the amino-terminal residues 1-34. *Proceedings of the National Academy of Sciences of the United States of America*, *69*(12), 3585–3588. https://doi.org/10.1073/pnas.69.12.3585.

Brewer, H. B., Jr., & Ronan, R. (1970). Bovine parathyroid hormone: Amino acid sequence. *Proceedings of the National Academy of Sciences of the United States of America*, *67*(4), 1862–1869. https://doi.org/10.1073/pnas.67.4.1862.

Cantley, L. K., Russell, J., Lettieri, D., & Sherwood, L. M. (1985). 1,25-Dihydroxyvitamin D3 suppresses parathyroid hormone secretion from bovine parathyroid cells in tissue culture. *Endocrinology*, *117*(5), 2114–2119. https://doi.org/10.1210/endo-117-5-2114.

Cinque, L., Sparaneo, A., Penta, L., Mencarelli, A., Rogaia, D., Esposito, S., et al. (2017). Autosomal dominant PTH gene signal sequence mutation in a family with familial isolated hypoparathyroidism. *The Journal of Clinical Endocrinology and Metabolism*, *102*(11), 3961–3969. https://doi.org/10.1210/jc.2017-00250.

Collip, J. (1925). The extraction of a parathyroid hormone which will prevent or control parathyroid tetany and which regulates the level of blood calcium. *Journal of Biological Chemistry*, *63*, 395–438.

Datta, R., Waheed, A., Shah, G. N., & Sly, W. S. (2007). Signal sequence mutation in autosomal dominant form of hypoparathyroidism induces apoptosis that is corrected by a chemical chaperone. *Proceedings of the National Academy of Sciences of the United States of America*, *104*(50), 19989–19994. https://doi.org/10.1073/pnas.0708725104.

Divieti, P., Geller, A. I., Suliman, G., Juppner, H., & Bringhurst, F. R. (2005). Receptors specific for the carboxyl-terminal region of parathyroid hormone on bone-derived cells: Determinants of ligand binding and bioactivity. *Endocrinology*, *146*(4), 1863–1870. https://doi.org/10.1210/en.2004-1262.

Divieti, P., Inomata, N., Chapin, K., Singh, R., Juppner, H., & Bringhurst, F. R. (2001). Receptors for the carboxyl-terminal region of pth (1-84) are highly expressed in osteocytic cells. *Endocrinology*, *142*(2), 916–925. https://doi.org/10.1210/endo.142.2.7955.

Eiselsberg, A. (1890). Über Tetanie im anschlusse an Kropf-operationen. *Wien Klin Wochenschr*, *4*, 1–37.

Emr, S. D., Hall, M. N., & Silhavy, T. J. (1980). A mechanism of protein localization: The signal hypothesis and bacteria. *The Journal of Cell Biology*, *86*(3), 701–711. https://doi.org/10.1083/jcb.86.3.701.

Erdheim, J. (1906). *Über Epithelköperehenbelung bei Osteomalakie. Vol. 1*.

Ertl, D. A., Stary, S., Streubel, B., Raimann, A., & Haeusler, G. (2012). A novel homozygous mutation in the parathyroid hormone gene (PTH) in a girl with isolated hypoparathyroidism. *Bone*, *51*(3), 629–632. https://doi.org/10.1016/j.bone.2012.06.009.

Gardella, T. J., Axelrod, D., Rubin, D., Keutmann, H. T., Potts, J. T., Jr., Kronenberg, H. M., et al. (1991). Mutational analysis of the receptor-activating region of human parathyroid hormone. *Journal of Biological Chemistry*, *266*(20), 13141–13146.

Gensure, R. C., Ponugoti, B., Gunes, Y., Papasani, M. R., Lanske, B., Bastepe, M., et al. (2004). Identification and characterization of two parathyroid hormone-like molecules in zebrafish. *Endocrinology*, *145*(4), 1634–1639. https://doi.org/10.1210/en.2003-0964.

Gild, M. L., Bullock, M., Luxford, C., Field, M., & Clifton-Bligh, R. J. (2020). Congenital hypoparathyroidism associated with elevated circulating nonfunctional parathyroid hormone due to novel PTH mutation. *The Journal of Clinical Endocrinology and Metabolism*, *105*(7). https://doi.org/10.1210/clinem/dgaa279.

Gley, E. (1891). Sur les fonctions du corps thyroïde. *Comptes Rendus de Societe Biologique*, *43*, 841–847.

Goswami, R., Mohapatra, T., Gupta, N., Rani, R., Tomar, N., Dikshit, A., et al. (2004). Parathyroid hormone gene polymorphism and sporadic idiopathic hypoparathyroidism. *Journal of Clinical Endocrinology and Metabolism*, *89*(10), 4840–4845. https://doi.org/10.1210/jc.2004-0273.

Guo, Y., Zhang, L. S., Yang, T. L., Tian, Q., Xiong, D. H., Pei, Y. F., et al. (2010). IL21R and PTH may underlie variation of femoral neck bone mineral density as revealed by a genome-wide association study. *Journal of Bone Minerals & Research*, *25*(5), 1042–1048. https://doi.org/10.1359/jbmr.091040.

Habener, J. F., Rosenblatt, M., & Potts, J. T., Jr. (1984). Parathyroid hormone: Biochemical aspects of biosynthesis, secretion, action, and metabolism. *Physiological Reviews*, *64*(3), 985–1053. https://doi.org/10.1152/physrev.1984.64.3.985.

Hammer, G., & McPhee, S. (2014). *Pathophysiology of disease: An introduction to clinical medicine. Chapter 17, Disorders of the parathyroids & calcium & phosphorus metabolism* (7th ed.). New York: McGraw-Hills.

He, B., Tong, T. K., Hiou-Tim, F. F., Al-Akad, B., Kronenberg, H. M., & Karaplis, A. C. (2002). The murine gene encoding parathyroid hormone: Genomic organization, nucleotide sequence and transcriptional regulation. *Journal of Molecular Endocrinology*, *29*(2), 193–203. https://doi.org/10.1677/jme.0.0290193.

Heinrich, G., Kronenberg, H. M., Potts, J. T., Jr., & Habener, J. F. (1984). Gene encoding parathyroid hormone. Nucleotide sequence of the rat gene and deduced amino acid sequence of rat preproparathyroid hormone. *Journal of Biological Chemistry*, *259*(5), 3320–3329.

Igarashi, T., Okazaki, T., Potter, H., Gaz, R., & Kronenberg, H. M. (1986). Cell-specific expression of the human parathyroid hormone gene in rat pituitary cells. *Molecular and Cellular Biology*, *6*(5), 1830–1833. https://doi.org/10.1128/mcb.6.5.1830.

Imanishi, Y., Hosokawa, Y., Yoshimoto, K., Schipani, E., Mallya, S., Papanikolaou, A., et al. (2001). Primary hyperparathyroidism caused by parathyroid-targeted overexpression of cyclin D1 in transgenic mice. *Journal of Clinical Investigations*, *107*(9), 1093–1102. https://doi.org/10.1172/JCI10523.

Inouye, S., Duffaud, G., & Inouye, M. (1986). Structural requirement at the cleavage site for efficient processing of the lipoprotein secretory precursor of Escherichia coli. *Journal of Biological Chemistry*, *261*(24), 10970–10975.

Karaplis, A. C., Lim, S. K., Baba, H., Arnold, A., & Kronenberg, H. M. (1995). Inefficient membrane targeting, translocation, and proteolytic processing by signal peptidase of a mutant preproparathyroid hormone protein. *Journal of Biological Chemistry*, *270*(4), 1629–1635. https://doi.org/10.1074/jbc.270.4.1629.

Keutmann, H. T., Dawson, B. F., Aurbach, G. D., & Potts, J. T., Jr. (1972). A biologically active amino-terminal fragment of bovine parathyroid hormone prepared by dilute acid hydrolysis. *Biochemistry*, *11*(10), 1973–1979. https://doi.org/10.1021/bi00760a037.

Keutmann, H. T., Niall, H. D., O'Riordan, J. L., & Potts, J. T., Jr. (1975). A reinvestigation of the amino-terminal sequence of human parathyroid hormone. *Biochemistry*, *14*(9), 1842–1847. https://doi.org/10.1021/bi00680a006.

Keutmann, H. T., Sauer, M. M., Hendy, G. N., O'Riordan, L. H., & Potts, J. T., Jr. (1978). Complete amino acid sequence of human parathyroid hormone. *Biochemistry*, *17*(26), 5723–5729. https://doi.org/10.1021/bi00619a019.

Klausner, R. D., & Sitia, R. (1990). Protein degradation in the endoplasmic reticulum. *Cell*, *62*(4), 611–614. https://doi.org/10.1016/0092-8674(90)90104-m.

Kumar, R., & Thompson, J. R. (2011). The regulation of parathyroid hormone secretion and synthesis. *Journal of the American Society of Nephrology*, *22*(2), 216–224. https://doi.org/10.1681/ASN.2010020186.

Lee, J. H., Davaatseren, M., & Lee, S. (2020). Rare PTH gene mutations causing parathyroid disorders: A review. *Endocrinology and Metabolism (Seoul, Korea)*, *35*(1), 64–70. https://doi.org/10.3803/EnM.2020.35.1.64.

Lee, S., Mannstadt, M., Guo, J., Kim, S. M., Yi, H. S., Khatri, A., et al. (2015). A homozygous [Cys25] PTH (1-84) mutation that impairs PTH/PTHrP receptor activation defines a novel form of hypoparathyroidism. *Journal of Bone and Mineral Research*, *30*(10), 1803–1813. https://doi.org/10.1002/jbmr.2532.

Levi, R., Ben-Dov, I. Z., Lavi-Moshayoff, V., Dinur, M., Martin, D., Naveh-Many, T., et al. (2006). Increased parathyroid hormone gene expression in secondary hyperparathyroidism of experimental uremia is reversed by calcimimetics: Correlation with post-translational modification of the trans acting factor AUF1. *Journal of the American Society of Nephrology*, *17*(1), 107–112. https://doi.org/10.1681/ASN.2005070679.

Lim, S. K., Gardella, T. J., Baba, H., Nussbaum, S. R., & Kronenberg, H. M. (1992). The carboxy-terminus of parathyroid hormone is essential for hormone processing and secretion. *Endocrinology*, *131*(5), 2325–2330. https://doi.org/10.1210/endo.131.5.1425431.

Lippincott-Schwartz, J., Donaldson, J. G., Schweizer, A., Berger, E. G., Hauri, H. P., Yuan, L. C., et al. (1990). Microtubule-dependent retrograde transport of proteins into the ER in the presence of brefeldin A suggests an ER recycling pathway. *Cell*, *60*(5), 821–836. https://doi.org/10.1016/0092-8674(90)90096-w.

Mallya, S. M., Wu, H. I., Saria, E. A., Corrado, K. R., & Arnold, A. (2010). Tissue-specific regulatory regions of the PTH gene localized by novel chromosome 11 rearrangement breakpoints in a parathyroid adenoma. *Journal of Bone and Mineral Research*, *25*(12), 2606–2612. https://doi.org/10.1002/jbmr.187.

Marcus, R., & Aurbach, G. D. (1969). Bioassay of parathyroid hormone in vitro with a stable preparation of adenyl cyclase from rat kidney. *Endocrinology, 85*(5), 801–810. https://doi.org/10.1210/endo-85-5-801.

McKee, A. E., & Silver, P. A. (2007). Systems perspectives on mRNA processing. *Cell Research, 17*(7), 581–590. https://doi.org/10.1038/cr.2007.54.

Meyer, D. I., Krause, E., & Dobberstein, B. (1982). Secretory protein translocation across membranes-the role of the "docking protein". *Nature, 297*(5868), 647–650. https://doi.org/10.1038/297647a0.

Miric, A., & Levine, M. A. (1992). Analysis of the preproPTH gene by denaturing gradient gel electrophoresis in familial isolated hypoparathyroidism. *Journal of Clinical Endocrinology and Metabolism, 74*(3), 509–516. https://doi.org/10.1210/jcem.74.3.1740484.

Murray, T. M., Rao, L. G., & Muzaffar, S. A. (1991). Dexamethasone-treated ROS 17/2.8 rat osteosarcoma cells are responsive to human carboxylterminal parathyroid hormone peptide hPTH (53-84): Stimulation of alkaline phosphatase. *Calcified Tissue International, 49*(2), 120–123. https://doi.org/10.1007/BF02565133.

Naveh-Many, T., Sela-Brown, A., & Silver, J. (1999). Protein-RNA interactions in the regulation of PTH gene expression by calcium and phosphate. *Nephrology, Dialysis, Transplantation, 14*(4), 811–813. https://doi.org/10.1093/ndt/14.4.811.

Niall, H. D., Keutmann, H., Sauer, R., Hogan, M., Dawson, B., Aurbach, G., et al. (1970). The amino acid sequence of bovine parathyroid hormone I. *Hoppe-Seyler's Zeitschrift für Physiologische Chemie, 351*(12), 1586–1588.

Niall, H. D., Sauer, R. T., Jacobs, J. W., Keutmann, H. T., Segre, G. V., O'Riordan, J. L., et al. (1974). The amino-acid sequence of the amino-terminal 37 residues of human parathyroid hormone. *Proceedings of the National Academy of Sciences of the United States of America, 71*(2), 384–388. https://doi.org/10.1073/pnas.71.2.384.

Nussbaum, S. R., Gaz, R. D., & Arnold, A. (1990). Hypercalcemia and ectopic secretion of parathyroid hormone by an ovarian carcinoma with rearrangement of the gene for parathyroid hormone. *New England Journal of Medicine, 323*(19), 1324–1328. https://doi.org/10.1056/NEJM199011083231907.

Parkinson, D. B., Shaw, N. J., Himsworth, R. L., & Thakker, R. V. (1993). Parathyroid hormone gene analysis in autosomal hypoparathyroidism using an intragenic tetranucleotide (AAAT)n polymorphism. *Human Genetics, 91*(3), 281–284. https://doi.org/10.1007/BF00218273.

Parkinson, D. B., & Thakker, R. V. (1992). A donor splice site mutation in the parathyroid hormone gene is associated with autosomal recessive hypoparathyroidism. *Nature Genetics, 1*(2), 149–152. https://doi.org/10.1038/ng0592-149.

Pinheiro, P. L., Cardoso, J. C., Gomes, A. S., Fuentes, J., Power, D. M., & Canario, A. V. (2010). Gene structure, transcripts and calciotropic effects of the PTH family of peptides in Xenopus and chicken. *BMC Evolutionary Biology, 10*, 373. https://doi.org/10.1186/1471-2148-10-373.

Rapoport, T. A. (1990). Protein transport across the ER membrane. *Trends in Biochemical Sciences, 15*(9), 355–358. https://doi.org/10.1016/0968-0004(90)90076-n.

Rasmussen, H., & Craig, L. (1959). Purification of parathyroid hormone by use of counter-current distribution. *Journal of the American Chemical Society, 81*.

Sandstrom, I. (1880). Omen ny Kortel hos mennisken och atskilige baggdjur Lakarefore rings. *Upsala*.

Sauer, R. T., Niall, H. D., Hogan, M. L., Keutmann, H. T., O'Riordan, J. L., & Potts, J. T., Jr. (1974). The amino acid sequence of porcine parathyroid hormone. *Biochemistry, 13*(9), 1994–1999. https://doi.org/10.1021/bi00706a033.

Spiess, M., & Lodish, H. F. (1986). An internal signal sequence: The asialoglycoprotein receptor membrane anchor. *Cell, 44*(1), 177–185. https://doi.org/10.1016/0092-8674(86)90496-4.

Steiner, D. F., Quinn, P. S., Chan, S. J., Marsh, J., & Tager, H. S. (1980). Processing mechanisms in the biosynthesis of proteins. *Annals of the New York Academy of Sciences, 343*, 1–16. https://doi.org/10.1111/j.1749-6632.1980.tb47238.x.

Sunthornthepvarakul, T., Churesigaew, S., & Ngowngarmratana, S. (1999). A novel mutation of the signal peptide of the preproparathyroid hormone gene associated with autosomal recessive familial isolated hypoparathyroidism. *Journal of Clinical Endocrinology and Metabolism, 84*(10), 3792–3796. https://doi.org/10.1210/jcem.84.10.6070.

Suprasdngsin, C., Wattanachal, A., & Preeysombat, C. (2001). 128 Annual Research Abstract 164. In *Parathyroid hormone gene mutation in patient with hypoparathyroidism*. Bangkok: Mahidol University.

Tenne, M., McGuigan, F. E., Ahlborg, H., Gerdhem, P., & Akesson, K. (2010). Variation in the PTH gene, hip fracture, and femoral neck geometry in elderly women. *Calcified Tissue International, 86*(5), 359–366. https://doi.org/10.1007/s00223-010-9351-6.

Tenne, M., McGuigan, F., Jansson, L., Gerdhem, P., Obrant, K. J., Luthman, H., et al. (2008). Genetic variation in the PTH pathway and bone phenotypes in elderly women: Evaluation of PTH, PTHLH, PTHR1 and PTHR2 genes. *Bone, 42*(4), 719–727. https://doi.org/10.1016/j.bone.2007.12.005.

Tomar, N., Gupta, N., & Goswami, R. (2013). Calcium-sensing receptor autoantibodies and idiopathic hypoparathyroidism. *Journal of Clinical Endocrinology and Metabolism, 98*(9), 3884–3891. https://doi.org/10.1210/jc.2013-2158.

Tregear, G. W., Van Rietschoten, J., Greene, E., Keutmann, H. T., Niall, H. D., Reit, B., et al. (1973). Bovine parathyroid hormone: Minimum chain length of synthetic peptide required for biological activity. *Endocrinology, 93*(6), 1349–1353. https://doi.org/10.1210/endo-93-6-1349.

Uchida, K., Tanaka, Y., Ichikawa, H., Watanabe, M., Mitani, S., Morita, K., et al. (2017). An excess of CYP24A1, lack of CaSR, and a novel lncRNA near the PTH gene characterize an ectopic PTH-producing tumor. *Journal of the Endocrine Society, 1*(6), 691–711. https://doi.org/10.1210/js.2017-00063.

VanHouten, J. N., Yu, N., Rimm, D., Dotto, J., Arnold, A., Wysolmerski, J. J., et al. (2006). Hypercalcemia of malignancy due to ectopic transactivation of the parathyroid hormone gene. *The Journal of Clinical Endocrinology and Metabolism, 91*(2), 580–583. https://doi.org/10.1210/jc.2005-2095.

Vasicek, T. J., McDevitt, B. E., Freeman, M. W., Fennick, B. J., Hendy, G. N., Potts, J. T., Jr., et al. (1983). Nucleotide sequence of the human parathyroid hormone gene. *Proceedings of the National Academy of Sciences of the United States of America, 80*(8), 2127–2131. https://doi.org/10.1073/pnas.80.8.2127.

von Heijne, G. (1983). Patterns of amino acids near signal-sequence cleavage sites. *European Journal of Biochemistry, 133*(1), 17–21. https://doi.org/10.1111/j.1432-1033.1983.tb07424.x.

von Heijne, G. (1985). Signal sequences. The limits of variation. *Journal of Molecular Biology, 184*(1), 99–105. https://doi.org/10.1016/0022-2836(85)90046-4.

Walter, P., & Blobel, G. (1981). Translocation of proteins across the endoplasmic reticulum III. Signal recognition protein (SRP) causes signal sequence-dependent and site-specific arrest of chain elongation that is released by microsomal membranes. *Journal of Cellular Biology, 91*(2 Pt. 1), 557–561. https://doi.org/10.1083/jcb.91.2.557.

Weaver, C. A., Gordon, D. F., Kissil, M. S., Mead, D. A., & Kemper, B. (1984). Isolation and complete nucleotide sequence of the gene for bovine parathyroid hormone. *Gene, 28*(3), 319–329. https://doi.org/10.1016/0378-1119(84)90149-5.

White, A. D., Fang, F., Jean-Alphonse, F. G., Clark, L. J., An, H. J., Liu, H., et al. (2019). Ca(2+) allostery in PTH-receptor signaling. *Proceedings of the National Academy of Sciences of the United States of America, 116*(8), 3294–3299. https://doi.org/10.1073/pnas.1814670116.

Wölfler, A. (1897). Weitere Beiträge zur chirurgischen Behandlung des Kropfes. *Wiener Medicinische Wochenschrift, 28*, 758–760.

CHAPTER FIVE

Modulation of PTH1R signaling by an extracellular binding antibody

Kaushik Sarkar[a,b,†], Lisa Joedicke[b,¶], Marta Westwood[b,||], Rebecca Burnley[b,‡], Michael Wright[b,§], David McMillan[b], and Bernadette Byrne[a,*]

[a]Department of Life Sciences, Imperial College, London, United Kingdom
[b]UCB Pharma, Slough, United Kingdom
*Corresponding author: e-mail address: b.byrne@imperial.ac.uk

Contents

Abstract

Parathyroid hormone receptor 1 (PTH1R) is a class B G-protein coupled receptor with key roles in bone development. The receptor signals through both the G_s and G_q G-proteins as well as through β-arrestin in a G-protein independent manner.

† Present address: Adaptimmune, 60 Jubilee Avenue, Milton Park, Abingdon, Oxfordshire, OX14 4RX, United Kingdom.

‡ Present address: M3, Western Avenue, Milton Park, Abingdon, Oxfordshire, OX14 4SH, United Kingdom.

§ Present address: Vertex Pharmaceuticals, 86-88 Jubilee Avenue, Milton Park, Abingdon, Oxfordshire, OX14 4RW, United Kingdom.

¶ Present address: F. Hoffmann-La Roche, Pharma Research and Early Development (pRED), Grenzacherstrasse 124, 4070 Basel, Switzerland.

|| Present address: Pharmaron UK Ltd. Innovation Park, West Cl, Hertford Rd, Hoddesdon EN11 9FH, United Kingdom.

Vitamins and Hormones, Volume 120
ISSN 0083-6729
https://doi.org/10.1016/bs.vh.2022.04.002

Current treatments for bone disorders, such as osteoporosis, target the PTH1R but are suboptimal in their efficacy. Monoclonal antibodies represent a major growth area in therapeutics as a result of their superior specificity and long serum half-life. Here, we discovered antibodies against the extracellular domain (ECD) of PTH1R from a phage display library. One of these antibodies, ECD-ScFvhFc, binds PTH1R with high affinity and although it has little or no effect on G-protein dependent receptor signaling, it does reduce PTH1R mediated β-arrestin signaling. Hydrogen-deuterium exchange mass spectrometry (HDX-MS) demonstrated that the ECD-ScFvhFc binding site overlapped partially with that of the cognate ligand, PTH. The results of this study demonstrate the suitability of PTH1R as a target for therapeutic antibody development.

1. Introduction

1.1 Parathyroid hormone, parathyroid 1 receptor and bone development

Parathyroid hormone receptor 1 (PTH1R) is a member of the secretin family of class B G-protein coupled receptors (GPCRs), physiologically binding to the peptide ligands, parathyroid hormone (PTH) and parathyroid hormone related peptide (PTHrP). PTH1R is mainly expressed in the osteoblast cells of bone tissue and renal tubular cells of the kidney, that are major targets of both PTH and PTHrP (Datta & Abou-Samra, 2009). Despite binding to the same receptor, PTH and PTHrP exhibit low sequence identity (∼16%) (Schluter & Piper, 1998). PTH is mainly synthesized and expressed by the parathyroid gland, whereas PTHrP is produced by a range of cell types including skin, growth plate chondrocytes, smooth muscles, bone, kidney, neuronal and glial cells (Datta & Abou-Samra, 2009).

The role of PTH1R in bone development and bone cell differentiation has been extensively explored (Jobert et al., 1998; Lanske et al., 1999; Schipani, Kruse, & Jüppner, 1995) and mutations of the receptor are known to be associated with bone abnormalities (Nampoothiri et al., 2016; Saito et al., 2018). The N-terminal 34 residue fragment of human parathyroid hormone, PTH (1–34) (referred to clinically as teriparatide, Forsteo® or Forteo®) is an FDA-approved treatment for osteoporosis (Neer et al., 2001). Irregular administration of teriparatide lowers the risk of spinal fracture in osteoporosis patients although its effects on non-vertebral bones are much less pronounced (Neer et al., 2001). At a molecular level, treatment with teriparatide induces bone matrix production by the osteoblasts, which is likely to be responsible for the reduced spinal fracture risk. However, this treatment has significant side effects, including the induction of osteoclastogenesis by driving secretion of growth factors and cytokines such as RANKL. Chronic usage of Terparatide leads to bone resorption, even when it is administered on an irregular basis (Neer et al., 2001). Other side effects of teriparatide include hypercalcemia, nausea, headache, dizziness, and leg cramps (Neer et al., 2001), clearly demonstrating that this is a far from ideal therapeutic approach. More recently, a novel parathyroid hormone related peptide analogue, abaloparatide (TYMLOS®) received FDA-approval for treatment of osteoporosis (Leder et al., 2015). Data from studies on this molecule indicate that it increases bone density with a lower incidence of hypercalcemia than other treatments (Leder et al., 2019). Currently however, all available treatments for osteoporosis are far from optimal, and thus significant scope remains for the development of novel osteoporosis drugs targeting PTH1R.

1.2 The PTH1R

PTH1R is G-protein coupled receptor, characterized by the standard 7 transmembrane (TM) domains, an extracellular N-terminus and an intracellular C-terminus common to all GPCRs but as a class B receptor, it also contains a large soluble extracellular domain (ECD) (Assil, Qi, Arai, Shomali, & Abou-Samra, 2001). The mature PTH1R consists of 567 amino acids following cleavage of the signal peptide. The PTH1R ECD is encoded by residues 1–188 and contains 6 conserved cysteine residues, forming 3 disulfide bridges, and 4 N-linked glycosylation sites (Grauschopf et al., 2000). The N-terminal region of PTH binds into a pocket formed by the transmembrane domains of the receptor while the C-terminal domain binds into the ECD via a number of hydrophobic interactions (Fig. 1A and B). The insertion of the N-terminus of PTH into this binding pocket induces a conformational change in transmembrane helix 6, and this in turn allows binding and activation of the G-protein.

PTH1R, like other class B GPCRs, can bind to more than one G-protein subclass (Schwindinger et al., 1998) and thus can activate different intracellular signaling pathways. The amino terminal end of the third intra-cellular loop plays a key role in G_s or $G_{q/11}$ binding (Huang et al., 1996). Previously, it was shown that replacing the EKKY sequence in intracellular loop 2 of the receptor, with DSEL in Rat PTH1R, selectively inhibited G_{q11} binding, leaving G_s coupling and signaling unaffected (Iida-Klein et al., 1997) suggesting that this region is selective for G_α coupling.

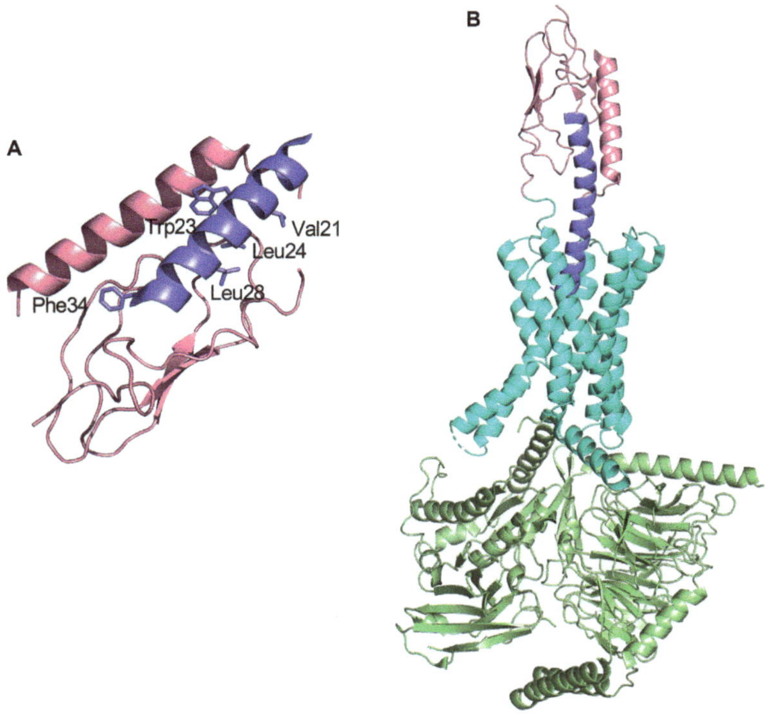

Fig. 1 Structures of the PTH1R (A) The ECD in complex with PTH_{15-34} (PDB: 3C4M) (Pioszak & Xu, 2008). The ECD is shown in pale pink and the PTH_{15-34} is in purple, both structures shown in ribbon representation. Key residues of the PTH_{15-34} involved in interaction with the ECD are indicated in stick representation. (B) The complete PTH1R in complex with G-protein and a PTH analogue (PDB: 6NBF) (Zhao et al., 2019). All elements of the complex are shown in ribbon representation with the ECD shown in pale pink, the PTH analogue shown in purple, the 7TM α-helices in cyan and the G-protein shown in light green.

Complete removal of the C-terminal tail of PTH1R, yields a receptor construct with elevated PTH-induced cAMP generation (Iida-Klein et al., 1997). This study along with others have shown that the C-terminal tail and intra-cellular loops of PTH1R not only play a major role in receptor activation and deactivation, but also fine tune receptor signaling through different G-protein subclasses.

1.3 PTH1R and biased agonism

Desensitization of ligand activated PTH1R is initiated by G protein-coupled receptor kinases (GRKs) which phosphorylate serines 489, 491, 492, 493, 495, 501 and 504, located at the N-terminal end of the intracellular

C-terminal domain. Upon phosphorylation the receptor preferentially binds to β-arrestin which initiates receptor internalization. Mutating all 7 serines to alanine completely abrogates phosphorylation and internalization (Tawfeek, Qian, & Abou-Samra, 2002). It is well established that the scaffolding properties of β-arrestins can bring elements of certain signaling pathways in close proximity to an internalized receptor (Luttrell et al., 2018). β-arrestin has been shown to play a role in receptor signaling through a range of molecules, including the mitogen-activated protein kinases (MAPK), extracellular signal–regulated kinases (ERK), c-Jun N-terminal kinase (JNK), and p38 as well as Akt, PI3 kinase, and RhoA (DeWire, Ahn, Lefkowitz, & Shenoy, 2007). A proline rich motif (P-XX-P) at the C-terminal tail of PTH1R plays a crucial role in β-arrestin mediated signaling via the MAPK pathway (Rey, Manen, Rizzoli, Caverzasio, & Ferrari, 2006).

The ability of PTH1R to signal through two distinct G-proteins as well as mediate G-protein independent β-arrestin signaling has paved the way for the development of novel therapeutics that act as "biased ligands," able to preferentially stimulate or inhibit only one signalling pathway (Rajagopal et al., 2011). A number of biased ligands for PTH1R have been described (Gesty-Palmer & Luttrell, 2011), including the Gs selective, PTH (1 − 31) (Whitfield & Morley, 1995), Gq selective PTH (3–34) (Jouishomme et al., 1992) and D-Trp^{12},Tyr^{34}-PTH (7–34) which acts as an inverse G_s agonist (Gardella et al., 1996), among others (Fig. 2). PTH1R-dependent signaling through β-arrestin results in bone formation without the activation of osteoclasts. A β-arrestin biased ligand D-Trp^{12}, Tyr^{34}-PTH (7–34) (PTH-βarr or PTH (7–34) has been shown to induce anabolic bone formation in vivo (Gesty-Palmer et al., 2006). However, further studies have shown that the β-arrestin biased agonist effects of PTH-βarr were not reproducible in human embryonic kidney (HEK293), Chinese hamster ovary (CHO) and U2OS cells (van der Lee et al., 2013). These disparate findings demonstrate that care must be taken when predicting the in vivo effects of PTH (7–34) from in vitro pharmacology.

1.4 Monoclonal antibodies

In recent years many monoclonal antibodies (mAbs) have entered the biotherapeutics market, offering new treatments for chronic and life-threatening diseases including rheumatoid arthritis (Malaviya & Östör, 2011) and cancer (Teo, Chew, & Phipps, 2016). The high specificity of mAbs means that they bind to target molecules very effectively and exhibit

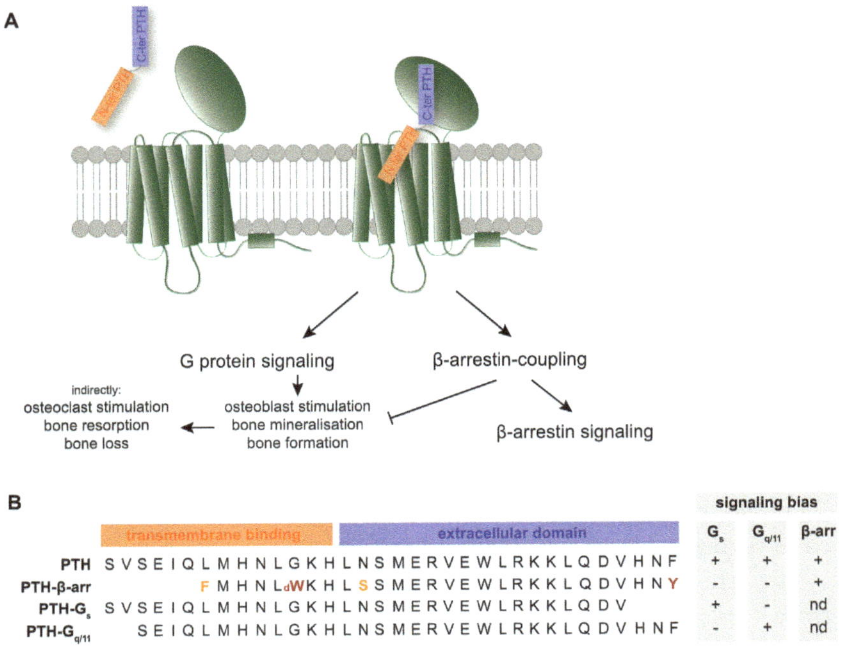

Fig. 2 (A) Summary of signalling through the PTH1R. PTH induced G-protein signaling results in a range of direct and indirect effects on bone formation and resorption. PTH induced β-arrestin coupling results in upregulation of β-arrestin signalling pathways and inhibition of G-protein signaling. (B) Differential effects of different PTH1R specific ligands. While PTH stimulates Gs, Gq and β-arrestin dependent signaling pathways, the other ligands are biased towards one signaling pathway. *Figure reproduced from Sarkar, K., Joedicke, L., Westwood, M., Burnley, R., Wright, M., McMillan, D., et al. (2019). Modulation of PTH1R signaling by an ECD binding antibody results in inhibition of β-arrestin 2 coupling.* Scientific Reports, *9(1), 14432. doi:10.1038/s41598-019-51016-z.*

low exposure to the CNS, reducing the risk of therapeutic side effects (Kaplon & Reichert, 2021). They also tend to have longer half-lives than small molecules meaning lower dosing frequencies. With the advent of mAb therapeutics, targeting cell-surface expressed receptors with exquisite specificity is now routine and GPCRs present a challenging yet important target class (Hutchings, 2020). There are a small number of FDA-approved GPCR-specific therapeutic mAbs. These include Mogalizumab which targets the C—C Motif Chemokine Receptor 4, used in the treatment of a range of adult T-cell lymphomas and acting by direct binding to immunoglobulin-γFc region receptor IIIa resulting in enhanced antibody dependent cytotoxicity of immune cells known to negatively affect the anti-tumor immune response. Another example is Erenumab, which targets the Calcitonin Gene-Related Peptide Receptor by acting as a direct antagonist of the receptor and is used in the treatment of migraine. There are also several additional GPCR-specific antibodies at various phases of early clinical development (Hutchings, 2020). Our recent research describes the discovery of antibodies against the PTH1R (Sarkar et al., 2019). One of these antibodies, termed ECD-scFvhFc, bound the receptor with high affinity, had little or no effect on PTH1R-cAMP signalling but did influence β-arrestin signalling, indicating that this antibody was acting like a biased ligand. Here we summarize the antibody discovery process and describe the characterization of the antibody PTH1R function.

2. Results

2.1 Antibody discovery overview

The PTH1R ECD (residues M1-G188), expressed in ExpiHEK293 cells as a fusion with a C-terminal Fc tag was purified to high homogeneity using Protein A affinity chromatography and the Fc tag removed (Fig. 3A). Isothermal calorimetry was used to confirm binding of PTH to the ECD with an affinity (~4 μM) in agreement with published values (Pioszak & Xu, 2008) (Fig. 3B). The process of screening and characterizing the antibodies is summarized in Fig. 4. The protein was then used to pull out antibodies from an in-house (UCB Pharma SA, unpublished) human naïve scFv phage library, with a diversity of approximately 1×10^{11} colony forming units (cfu), using phage display. This resulted in 411 phage particles, which upon sequencing yielded 154 unique sequences that differed in the V_H CDR3 region. Sequences with identical V_H CDR3 but differing in the V_L region were also included. These were cloned and expressed as (single-chain Fv with mouse Fc) scFvmFc TAP fragments in the ExpiHEK293 cells and the supernatant assayed for ability to bind cell surface expressed full length PTH1R binding using high-throughput flow cytometry. On the basis of the results from this analysis and the sequence diversity of the receptor binders, 18 antibodies were chosen for further characterization. scFv constructs tagged with a C-terminal human Fc tag were generated for each of the 18 antibodies. These were individually expressed in ExpiHEK293 cells, and purified using Protein A affinity chromatography. The individual proteins were checked for purity and used for a more detailed binding analysis using both flow cytometry (Fig. 5) and surface plasmon resonance (SPR).

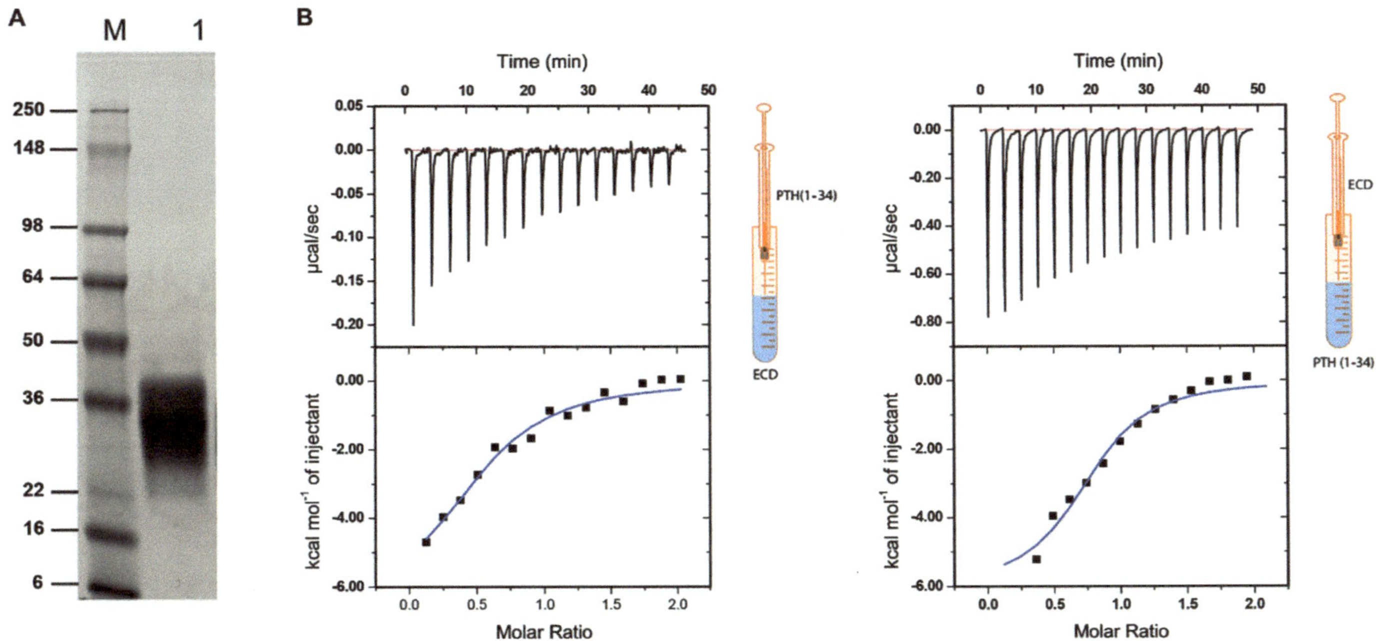

Fig. 3 Isolation and characterization of the PTH1R ECD (A) SDS-PAGE analysis of purified PTH1R ECD (Lane 1). The diffuse nature of the band is indicative of glycosylated protein. M = molecular weight markers. (B) ITC analysis of the PTH1R ECD. In the left panel, the ligand was titrated into a solution of the purified protein, while in the right panel the purified ECD was titrated into the ligand. The binding affinity of the ECD for PTH-(1–34) was approximated to 4 μM (titrating PTH (1–34) into PTH1R ECD) and 3.8 μM (titrating PTH1R ECD into PTH (1–34)). *Figure adapted from Sarkar, K., Joedicke, L., Westwood, M., Burnley, R., Wright, M., McMillan, D., et al. (2019). Modulation of PTH1R signaling by an ECD binding antibody results in inhibition of β-arrestin 2 coupling.* Scientific Reports, *9(1), 14432. doi:10.1038/s41598-019-51016-z.*

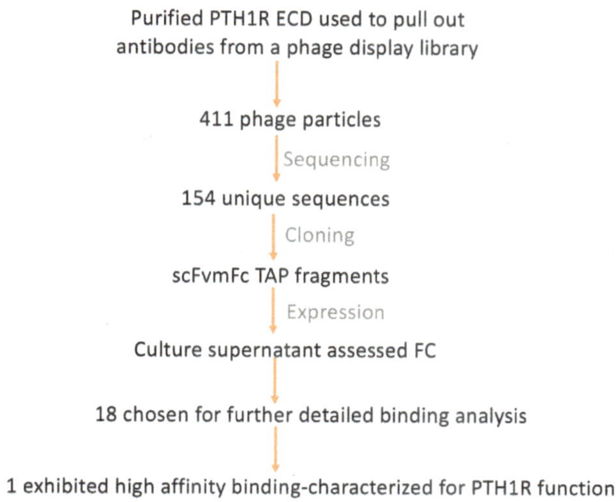

Fig. 4 Summary of the PTH1R antibody discovery and characterization process.

In order to assess antibody binding to surface expressed PTH1R, flow cytometry analysis was carried out in the absence (blue bars, Fig. 5B) and presence (red bars, Fig. 5B) of PTH. Only one of the antibodies, PG6, exhibited decreased fluorescence in the presence of the ligand indicating that it is outcompeted by the PTH for receptor binding. While flow cytometry is useful for providing qualitative analysis of receptor binding of the antibodies, SPR analysis can provide quantitative information. In this case the purified antibodies were immobilized on an SPR chip using their Fc tag and purified soluble PTH1R-ECD was flowed over. Most of the antibodies bound exhibited detectable binding to the ECD with affinities ranging from 4.2 nM to ~1 μM (data not shown). Taken together the flow cytometry and SPR results characterized the kinetics of binding of the antibody but also demonstrated that the antibody binds to both cell surface expressed full-length receptor and the purified ECD respectively. The antibody, ECD-ScFvhFc, with the highest binding affinity (4.2 nM) was further characterized in terms of its effects on PTH1R signaling.

2.2 ECD-ScFvhFc and PTH1R signaling

As mentioned above PTH1R can induce G-protein dependent signaling through both Gs and Gq G-proteins as well as G-protein independent signaling through β-arrestin. Thus, we assessed the effect of ECD-ScFvhFc

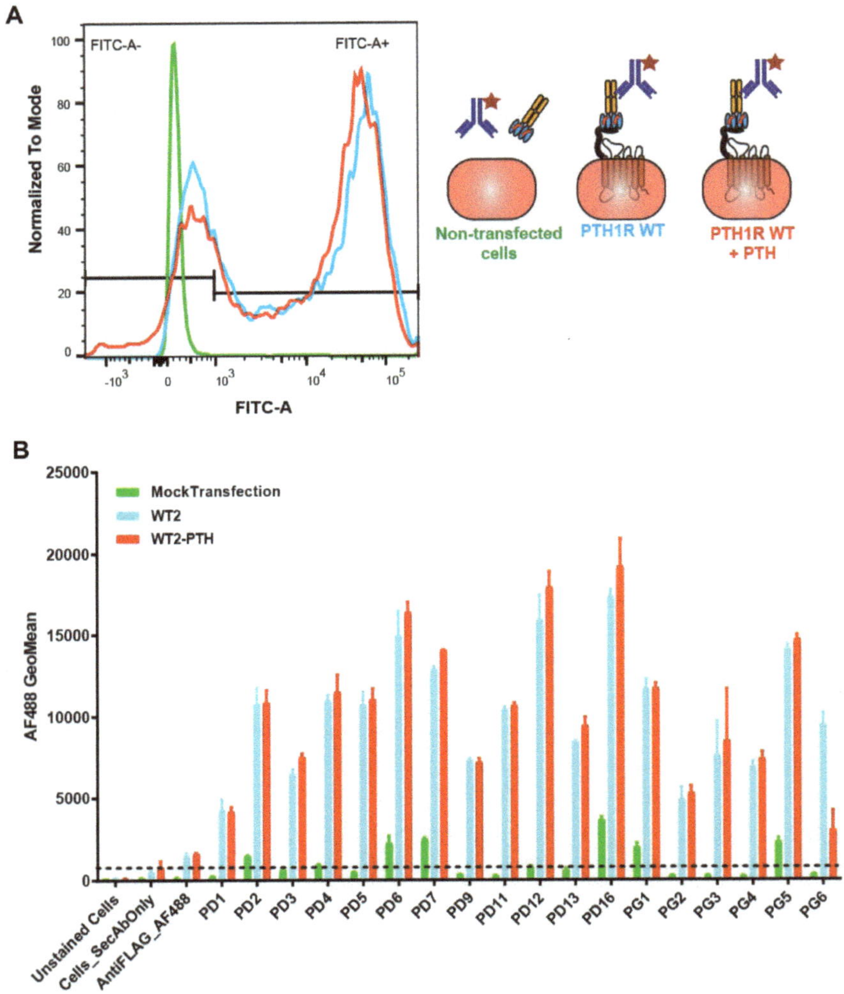

Fig. 5 (A) A representative overlay of flow cytometry data showing sample data for PD9 scFv-hFc staining of cells transfected with WT2 receptor (cyan) and non-transfected (green) cells. The overlay also shows the effect of PTH addition to the WT expressing cells (red trace). (B) A plot summarizing the geometric mean of fluorescence as shown by all selected ECD binding scFv-hFc molecules. Each data point represents the mean and standard deviation of triplicate measurements. The black dashed line represents the fluorescence levels from cells stained with secondary antibody only.

on both types of PTH1R mediated signaling. Signaling through Gs was assessed using a cAMP assay based on immunocompetition, with an increase in intracellular cAMP reported as a decrease in the homogeneous time-resolved fluorescence (HTRF) ratio (Fig. 6A). The analysis shows a

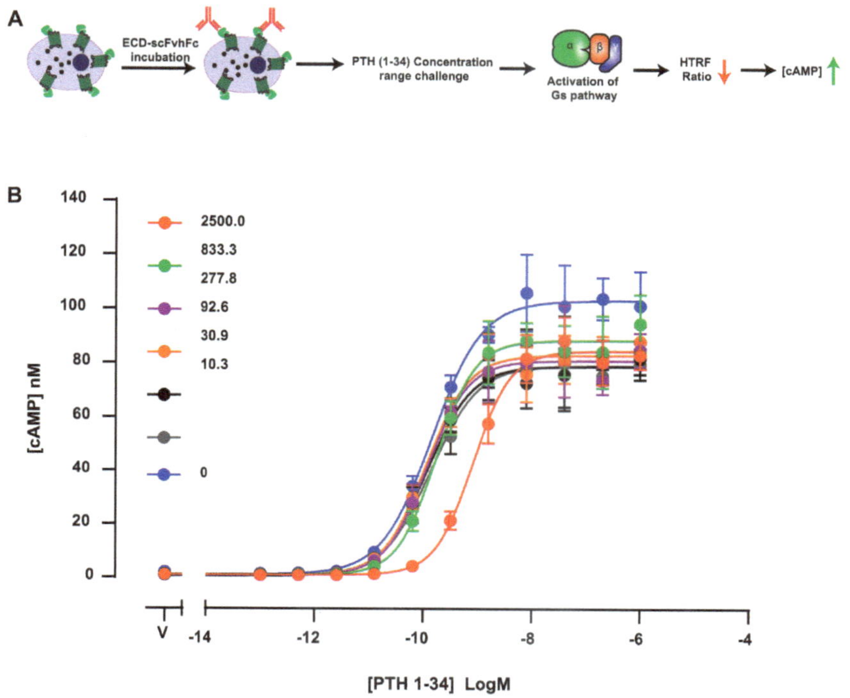

Fig. 6 Effect of ECD-scFvhFc on PTH induced cAMP production. (A) Schematic of the assay. Cells expressing the full length PTH1R were incubated with different concentrations of ECD-scFvhFc. The cells were then challenged with a range of PTH concentrations. This results in activation of the Gs pathway, with a decrease in the HTRF ratio indicative of an increase in cAMP concentration. (B) Assay results. The change in cAMP concentration in response to increasing concentrations of PTH. The different colored curves represent cAMP production by cells preincubated with different concentrations of ECD-scFvhFc molecule (nM, indicated by the key). *Figure reproduced from Sarkar, K., Joedicke, L., Westwood, M., Burnley, R., Wright, M., McMillan, D., et al. (2019). Modulation of PTH1R signaling by an ECD binding antibody results in inhibition of β-arrestin 2 coupling.* Scientific Reports, *9(1), 14432. doi:10.1038/s41598-019-51016-z.*

PTH dose-dependent increase in cAMP production in the presence of no ECD-ScFvhFc (blue 0 curve, Fig. 6B) with an EC_{50} of 1.43E-10M. Concentrations of antibody as high as 833nM have little effect on PTH-induced cAMP production. It is only at the highest concentration of antibody that a marked and significant shift in cAMP production is observable, with an increase in the EC_{50} to 8.27E-10M. Thus ECD-ScFvhFc reduces PTH induced PTH1R signaling through Gs, however, only at very high concentrations, much higher than those that would ever be experienced physiologically.

The effect of ECD-scFvhFc on PTH1R G-protein independent signaling through β-arrestin was assessed using an enzyme fragment complementation assay. The receptor-β-arrestin 2 complex forms a fully functional β-galactosidase which produces a luminescent product upon substrate addition (Fig. 7A). Addition of increasing concentrations of PTH to cells expressing full length PTH1R tagged with ProLink™ (PK) results in increasing amounts of β-arrestin coupling with a pEC50 of 8.30 ± 0.05 (Fig. 7B, blue curve). Addition of ECD-scFvhFc alone does not result in β-arrestin recruitment (Fig. 7B, red curve). Stimulating cells with 10nM PTH(1–34) in the presence of ECD-scFvhFc results in an antibody concentration-dependent reduction in β-arrestin recruitment (Fig. 7C, red curve), with an IC_{50} value of 102.3 ± 14.8 nM. No such reduction is seen in the presence of non-PTH1R specific mAb. These results demonstrate that the ECD-ScFvhFc antibody can act as a biased ligand, reducing β-arrestin coupling while having little or no effect on Gs signaling.

2.3 Characterization of antibody binding to full length receptor

Expression and isolation of integral membrane proteins for biophysical analysis is challenging (Gulezian et al., 2021; He, Wang, & Yan, 2014). There are several issues, with the main one being the requirement to remove the membrane protein from the native membrane environment (Sadaf, Cho, Byrne, & Chae, 2015). This is normally achieved by the use of detergents which both disrupt the interactions between the membrane protein and the membrane and associate with the hydrophobic regions of the membrane protein, providing an effective shield from the aqueous environment (Stetsenko & Guskov, 2017). However, the actual process of detergent based solubilization removes lipids that are often crucial for structure and function of membrane proteins. Recent advances have been made in the development of alternative membrane mimetic systems to standard detergents, with one key alternative styrene maleic acid (SMA), which punches holes in membranes cutting out membrane proteins together with the surrounding lipids (Knowles et al., 2009; Lee et al., 2016). The resultant SMA lipid particles (SMALPs) are shielded from the aqueous environment by the SMA polymer. The SMAs have been successfully applied to the study of other GPCRs (Routledge et al., 2020).

While it was possible to express the WT form of PTH1R as a His-tagged protein, purification resulted in very little protein, likely due to instability in

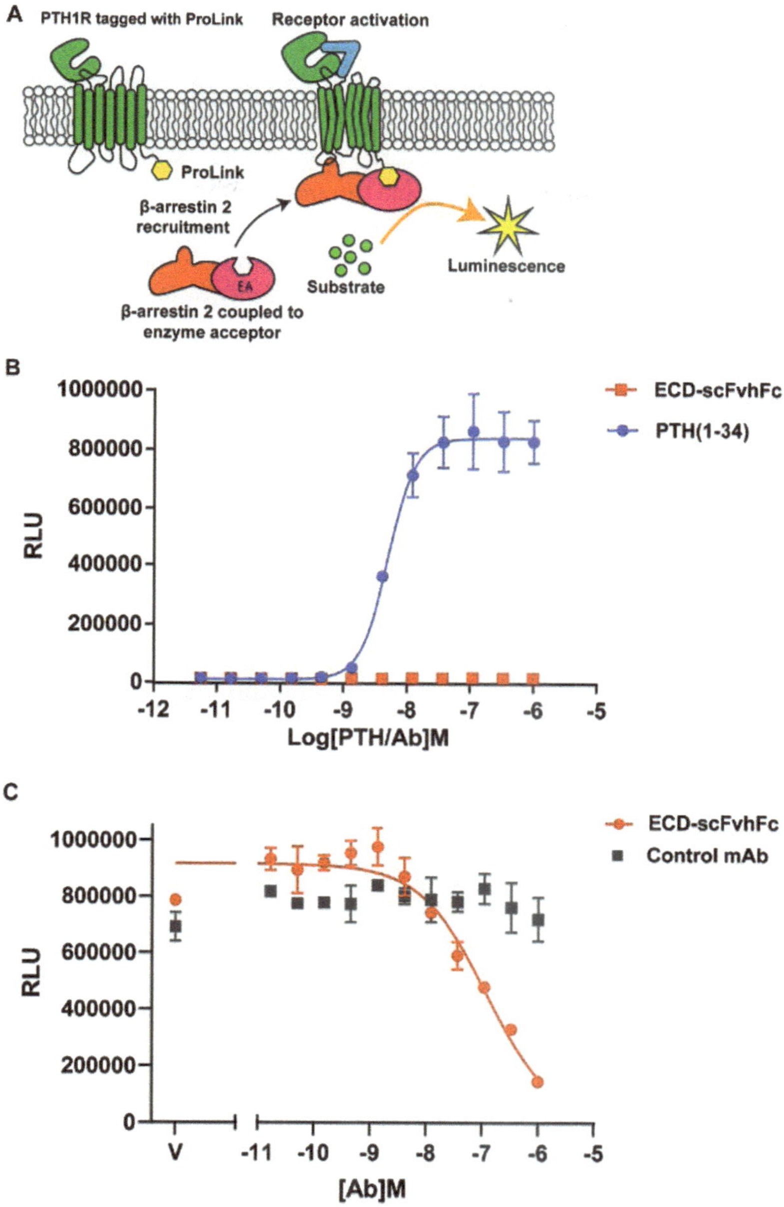

Fig. 7 Effect of ECD-scFvhFc on PTH induced β-arrestin coupling. (A) Assay set up. The PTH1R is expressed as a fusion with a C-terminal ProLink tag (enzyme donor), while the β-arrestin 2 is co-expressed with a β-galactosidase enzyme fragment (enzyme acceptor). Activation of PTH1R with PTH causes recruitment of the β-arrestin complex to the receptor and the enzyme acceptor and donor combine to form a fully functionally β-galactosidase enzyme. In the presence of substrate, the β-galactosidase produces a luminescent signal. (B) Luminescent signal produced by the addition of increasing concentrations of PTH alone (blue curve) or ECD-scFvhFC alone (red curve) to cells expressing the PTH1R and β-arrestin 2 constructs. (C) Luminescent signal produced by the addition of 10 nM PTH in combination with increasing concentrations of either ECD-ScFvhFc (red curve) or a non-PTH1R specific control mAb (grey).

solution and thus loss of protein through aggregation and precipitation. Conformational flexibility is a major contributor to protein instability thus we made a series of constructs truncated at either the N or C terminus or both termini. The construct that gave the best results in terms of yield of purified protein was termed ΔC29–491 which lacked the native signal sequence and the C-terminal 102 amino acid residues. The final expressed construct was tagged with a twin Strep tag at the N-terminus and a 10His tag at the C-terminus and also contained a HA signal sequence. The construct sequence included two tobacco etch virus protease cleavage sites to allow for tag removal after protein isolation. A control construct lacking the whole ECD as well as the C-terminal 102 amino acid residues, ΔCΔN171–491 was also generated (Fig. 8A). The two PTH1R proteins were expressed and then isolated to high homogeneity as SMALPs (Fig. 8B) and then used for SPR analysis of ECD-scFvhFc binding. The proteins were immobilized onto the SPR chip via an Anti-His tag polyclonal antibody. ECD-scFvhFc bound to the ΔC29–491 construct with a Kd$\sim$4nM (Fig. 5A). The ECD-scFvhFc did not bind to the ΔNΔC171–491 construct. Thus, the ECD-scFvhFc binds with high affinity to the ECD region of the PTH1R.

2.4 Mapping the ECD-scFvhFc binding site

From the findings shown above it was clear that the scFvhFc bound specifically to the ECD of PTH1R but it was not possible to identify the precise binding site. Hydrogen deuterium exchange (HDX) coupled to MS has emerged as a sensitive tool to monitor protein-protein interactions at peptide level resolution (Mistarz, Brown, Haselmann, & Rand, 2016). HDX-MS defines the solvent accessibility of different regions of a protein by monitoring the exchange of hydrogen to deuterium—the exchange reaction depends on intrinsic protein motions (Mistarz et al., 2016). H-bonding networks greatly reduce the rate and efficiency of HDX (Masson et al., 2019), whereas highly dynamic regions typically undergo a higher level of HDX (Konermann, Pan, & Liu, 2011). The most valuable information arises from differential HDX analysis (ΔHDX), which probes differences between two protein states (e.g., protein alone and protein in complex with another molecule). Such differences in deuterium uptake are mapped onto the protein structure enabling a fast readout of the protein interaction site. It proved impossible to carry out HDX-MS analysis of the PTH1R ΔC29–491 purified in SMA due to issues with aggregation

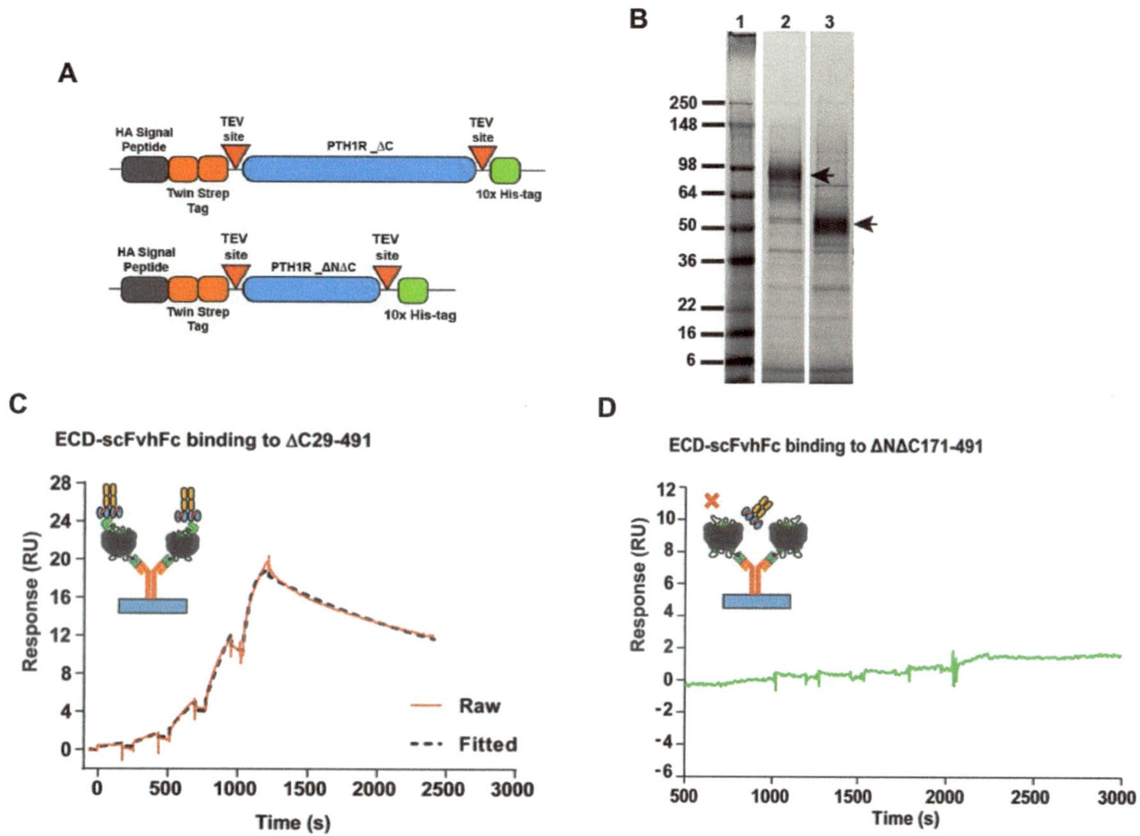

Fig. 8 ECD-scFvhFc binds to the PTH1R with high affinity. (A) Constructs used for expression and purification of the PTH1R. (B) SDS-PAGE analysis of purified PTH1R receptor protein. Lane 1, molecular markers; Lane 2, PTH1R ΔC29–491; Lane 3, PTH1R ΔNΔC171–491. The protein was isolated using Ni^{2+}-NTA affinity chromatography in the form of SMALPs. SPR analysis of ECD-scFvhFc binding to (C) PTH1R ΔC29–491 and (D) PTH1R ΔNΔC171–491.

observed during the low pH quenching step, so pure ECD, as described above, was used. The ECD was analyzed alone and in complex with 20 μM of the ECD-scFvhFc (expressed in Fab format). The 0.5 min time point deuteration measurement was used for statistical analysis and definition of the epitope, as it represented the largest changes in HDX dynamics between bound and unbound states. After labelling, the protein was digested using pepsin into peptides, and the difference in deuterium incorporation calculated per peptide. Peptides were identified covering 99% of the ECD sequence.

As described previously (Sarkar et al., 2019), overlapping peptides assigned to the N-terminus of the ECD, covering the sequence ^{30}DVMTKEEGIFLLHRAQA^{46}, were found to be more protected from exchange in the ECD-scFvhFc complex (Fig. 9A, ΔHDX% apo versus

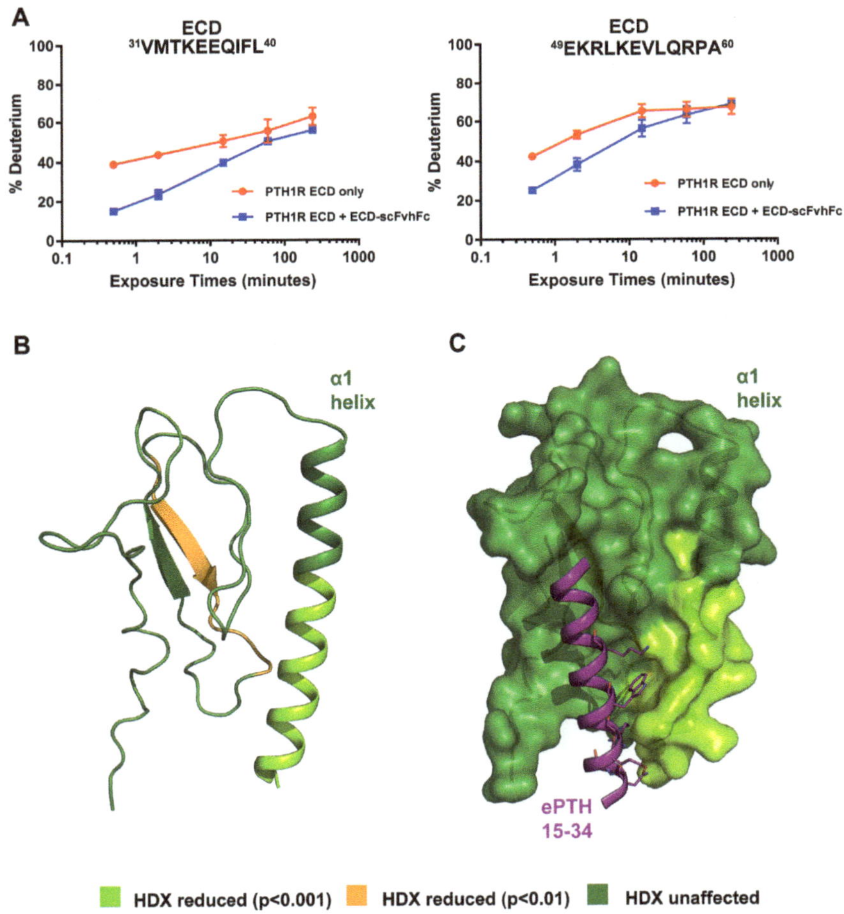

Fig. 9 See figure legend on opposite page.

bound $P<0.001$). Two further regions also showed reduced uptake of deuterium in the presence of the scFvhFc: ^{62}IM^{63} (these residues are not resolved in available crystal structures) and ^{125}EVVAVPCPDY^{134} (ΔHDX% apo versus bound $P<0.01$). Since the results reflect peptide-level resolution, it is not possible to discern precisely which of the residues within these regions are most important for the ECD-scFvhFc interaction. It is not possible to rule out that a specific protection may be due directly to binding or as an indirect effect e.g. the stabilization of adjacent structural elements. Mapping the HDX-MS results onto the crystal structure of PTH1R-ECD (PDB ID: 3C4M, Fig. 9B) reveals that the affected peptides cluster close to the PTH (1–34) binding site. Indeed, the region most affected by scFvhFc binding is the α1 helix, some residues of which are known to interact directly with PTH (1–34) (Fig. 9A,B). However, other residues known to be involved in the ECD-PTH (1–34) interaction are not identified as being protected here. This suggests a partial overlap of the ECD-scFvhFc epitope with the known PTH (1–34) binding site.

3. Discussion

GPCRs are major therapeutic targets (Santos et al., 2017) while mAbs represent a substantially growing area of biotherapeutics due to their enhanced efficacy and safety compared to small molecules (Datta & Abou-Samra, 2009; Kaplon & Reichert, 2021). In this chapter, we describe our efforts to discover functional antibodies against the functionally

Fig. 9 Mapping the ECD-scFvhFc binding site (A) Deuterium uptake plot showing the difference in apo (red) and ECD-scFvhFc bound PTH1R-ECD (blue). Peptides corresponding to the sequence 31VMTKEEQ——EVLQRPA60 exhibited slower exchange rates when bound to ECD-scFvhFc. (B) The putative epitope of ECD-scFvhFc is located on the α1-helix on the PTH1R-ECD and is highlighted in light green. The additional peptide showing slower HDX is depicted in orange, areas of the PTH1R-ECD that showed no change in HDX are colored dark green. (C), Surface representation of the PTH (15–34)-bound PTH1R-ECD reveals an overlap of the putative ECD-scFvhFc epitope and the PTH (15–34) binding site. Residues interacting with the PTH1R-ECD in that area are shown as sticks. The second peptide that showed reduced HDX (orange) is located opposite the PTH (15–34) binding site and does not overlap. The figure was prepared using PDB 6FJ3 (Ehrenmann et al., 2018). *The figure is reproduced from Sarkar, K., Joedicke, L., Westwood, M., Burnley, R., Wright, M., McMillan, D., et al. (2019). Modulation of PTH1R signaling by an ECD binding antibody results in inhibition of β-arrestin 2 coupling.* Scientific Reports, *9(1), 14432. doi:10.1038/s41598-019-51016-z.*

important ECD of PTH1R. A number of antibodies were successfully identified with one, ECD-scFvhFc, exhibiting similar, high affinity binding to both the ECD alone and a construct of the PTH1R containing both the ECD and the 7TM domains but lacking the long flexible C-terminal region, ΔC29–491, both solubilized and isolated in SMALPs. SMALPs have a number of advantages over detergent-based solubilization methods including reduced disruption of membrane protein-lipid interactions (Rothnie, 2016), and a greater suitability for SPR analysis as demonstrated here. The recent structure of the PTH1R was obtained in complex with a number of lipids, assigned cholesterol from the electron density (Zhao et al., 2019). This highlights the close association between the protein and the lipids and indicates a structural and functional role for these interactions. It is not possible to say which lipids were present in the SMALPs of PTH1R from the current study, although future lipidomics analysis could be used to identify which lipids co-purify with the protein and native mass spectrometry could reveal which of these lipids are forming direct interactions with the protein (Pyle et al., 2018, 2019). Mapping of the precise binding site of the ECD-ScFvhFc revealed overlap with the PTH binding site (Fig. 10A,B).

Importantly, the ECD-scFvhFc has clear functional effects on cell surface expressed WT PTH1R, significantly reducing PTH induced β-arrestin 2 coupling while having little effect on Gs signalling (Fig. 10C). Thus ECD-scFvhFc acts as a biased antagonist, specifically reducing the likelihood of the receptor adopting the conformation required for β-arrestin 2 interaction. As a result of this β-arrestin 2 antagonism, ECD-scFvhFc may extend PTH-induced G-protein dependent signaling. The precise molecular basis of the antagonism is unclear and would require further analysis but to the best of our knowledge the ECD-scFvhFc is the first reported antibody based negative-allosteric modulator of β-arrestin 2 coupling in PTH1R.

Such biased GPCR ligands are being becoming increasingly well characterized (Ferraino et al., 2021; Kalinkovich & Livshits, 2021; Neve et al., 2021). They have significant potential as therapeutics as a result of their very specific actions (Kenakin, 2011). There are other PTH1R ligands, derivatives of PTHrP, that exhibit similar biased function as ECD-scFvhFc (Fig. 10D) (Gesty-Palmer et al., 2006; Jouishomme et al., 1992). Notably, these biased PTH derivatives also inhibit PTH induced signalling through Gq, something not assessed as yet for ECD-scFvhFc but clearly of interest for future studies.

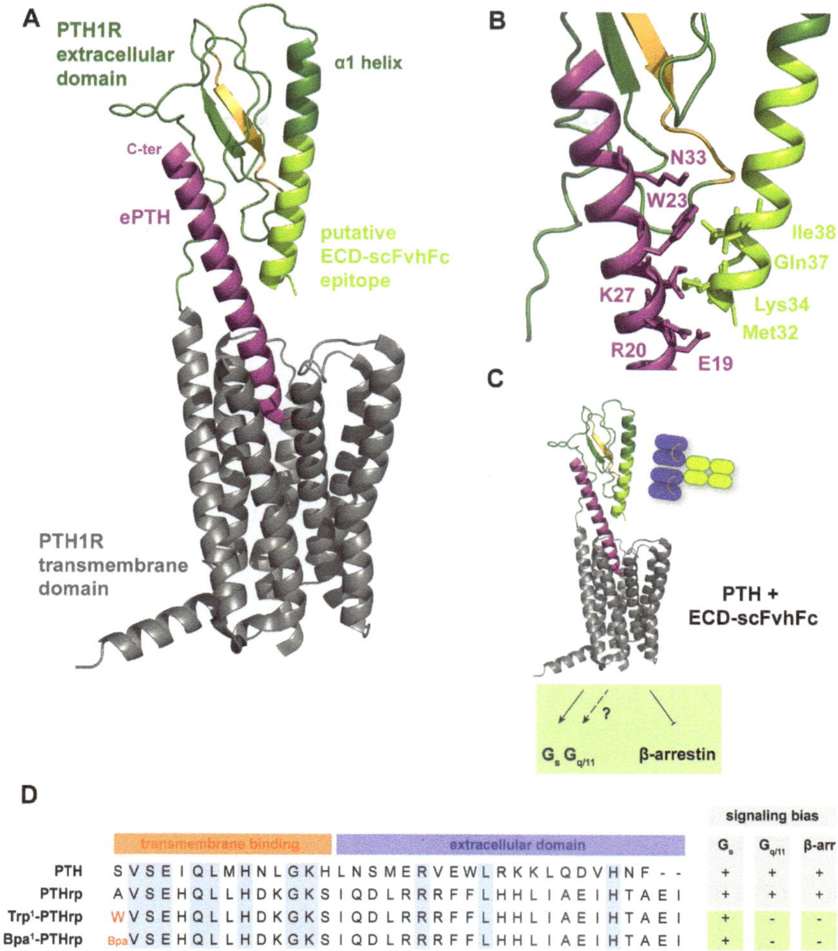

Fig. 10 Structure and function of ECD-scFvhFc interaction. (A) Schematic showing structure of PTH1R in complex with PTH. The TM domains are colored grey, PTH is colored purple and the ECD is colored according to the HDX-MS results; major ΔHDX, light green; minor ΔHDX, gold; no ΔHDX dark green. PDB (B) Zoomed in view of the interaction of PTH with the region of the ECD indicated to bind the ECD-scFvhFc with key side chains shown in stick representation. (C) Schematic highlighting that the ECD-scFvhFc has no effect on PTH induced PTH1R-Gs-protein signaling PTH1R function but does reduce b-arrestin recruitment. The effects of ECD-scFvhFc on Gq-protein signaling have yet to be characterized. (D) Example PTH ligands with details of their signaling bias.

PTH signaling eventually results in osteoblast stimulation, bone mineralization and bone formation but also triggers bone resorption via indirect activation of the RANKL and OPG pathways (Huang et al., 2004). β-arrestin recruitment to the receptor is thus essential for termination of G protein signaling and maintenance of a balance between bone formation and resorption (Gesty-Palmer & Luttrell, 2011). β-arrestin based PTH1R agonists are known to have a positive pharmacological effect (Gesty-Palmer & Luttrell, 2011) but further in vivo studies would be required to explore the physiological effect of β-arrestin recruitment antagonism in PTH1R. The ECD-scFvhFc however also represents a useful tool to explore the complexities of PTH1R signaling and also to potentially stabilize a key PTH1R conformation for structural analysis (Errasti-Murugarren, Bartoccioni, & Palacín, 2021).

In summary our study demonstrates the feasibility of obtaining functional antibodies against the PTH1R both as potential therapeutic agents and as tools for research.

Acknowledgments

Authors would like to thank Gill Holdsworth, Nicolas Basse, Sarfaraj Topia, Daniel Lightwood, Anthony ScottTucker, Jeff Kennedy, Oliver Mace, Catherine Simpson and Grace Smith for their contribution to this work. Authors would also like to thank Phil Stanley for statistical support.

References

Assil, I. Q., Qi, L. J., Arai, M., Shomali, M., & Abou-Samra, A. B. (2001). Juxtamembrane region of the amino terminus of the corticotropin releasing factor receptor type 1 is important for ligand interaction. *Biochemistry*, *40*(5), 1187–1195. https://doi.org/10.1021/bi001758y.

Datta, N. S., & Abou-Samra, A. B. (2009). PTH and PTHrP signaling in osteoblasts. *Cellular Signalling*, *21*(8), 1245–1254. https://doi.org/10.1016/j.cellsig.2009.02.012.

DeWire, S. M., Ahn, S., Lefkowitz, R. J., & Shenoy, S. K. (2007). β-Arrestins and cell signaling. *Annual Review of Physiology*, *69*(1), 483–510. https://doi.org/10.1146/annurev.physiol.69.022405.154749.

Ehrenmann, J., Schöppe, J., Klenk, C., Rappas, M., Kummer, L., Doré, A. S., et al. (2018). High-resolution crystal structure of parathyroid hormone 1 receptor in complex with a peptide agonist. *Nature Structural & Molecular Biology*, *25*(12), 1086–1092. https://doi.org/10.1038/s41594-018-0151-4.

Errasti-Murugarren, E., Bartoccioni, P., & Palacín, M. (2021). Membrane protein stabilization strategies for structural and functional studies. *Membranes*, *11*(2), 155. https://doi.org/10.3390/membranes11020155.

Ferraino, K. E., Cora, N., Pollard, C. M., Sizova, A., Maning, J., & Lymperopoulos, A. (2021). Adrenal angiotensin II type 1 receptor biased signaling: The case for "biased" inverse agonism for effective aldosterone suppression. *Cellular Signalling*, *82*, 109967. https://doi.org/10.1016/j.cellsig.2021.109967.

Gardella, T., Luck, M., Jensen, G., Schipani, E., Jr., Potts, J. T., & Juppner, H. (1996). Inverse agonism of amino-terminally truncated parathyroid hormone (PTH) and PTH-related peptide (PTHrP) analogs revealed with constitutively active mutant PTH/PTHrP receptors. *Endocrinology*, *137*, 3936–3941. https://doi.org/10.1210/endo.137.9.8756569.

Gesty-Palmer, D., Chen, M., Reiter, E., Ahn, S., Nelson, C. D., Wang, S., et al. (2006). Distinct β-Arrestin- and G protein-dependent pathways for parathyroid hormone receptor-stimulated ERK1/2 activation. *The Journal of Biological Chemistry*, *281*(16), 10856–10864. https://doi.org/10.1074/jbc.m513380200.

Gesty-Palmer, D., & Luttrell, L. M. (2011). 'Biasing' the parathyroid hormone receptor: A novel anabolic approach to increasing bone mass? *British Journal of Pharmacology*, *164*(1), 59–67. https://doi.org/10.1111/j.1476-5381.2011.01450.x.

Grauschopf, U., Lilie, H., Honold, K., Wozny, M., Reusch, D., Esswein, A., et al. (2000). The N-terminal fragment of human parathyroid hormone receptor 1 constitutes a hormone binding domain and reveals a distinct disulfide pattern. *Biochemistry*, *39*(30), 8878–8887. https://doi.org/10.1021/bi0001426.

Gulezian, E., Crivello, C., Bednenko, J., Zafra, C., Zhang, Y., Colussi, P., et al. (2021). Membrane protein production and formulation for drug discovery. *Trends in Pharmacological Sciences*, *42*(8), 657–674. https://doi.org/10.1016/j.tips.2021.05.006.

He, Y., Wang, K., & Yan, N. (2014). The recombinant expression systems for structure determination of eukaryotic membrane proteins. *Protein & Cell*, *5*(9), 658–672. https://doi.org/10.1007/s13238-014-0086-4.

Huang, Z., Chen, Y., Pratt, S., Chen, T.-H., Bambino, T., Nissenson, R. A., et al. (1996). The N-terminal region of the third intracellular loop of the parathyroid hormone (PTH)/PTH-related peptide receptor is critical for coupling to cAMP and inositol phosphate/Ca^{2+} signal transduction pathways. *Journal of Biological Chemistry*, *271*(52), 33382–33389. https://doi.org/10.1074/jbc.271.52.33382.

Huang, J. C., Sakata, T., Pfleger, L. L., Bencsik, M., Hallloran, B. P., Bikle, D. D., et al. (2004). PTH differentially regulates expression of RANKL and OPG. *Journal of Bone and Mineral Research*, *19*, 235–244. https://doi.org/10.1359/jbmr.0301226.

Hutchings, C. J. (2020). A review of antibody-based therapeutics targeting G protein-coupled receptors: An update. *Expert Opinion on Biological Therapy*, *20*(8), 1–11. https://doi.org/10.1080/14712598.2020.1745770.

Iida-Klein, A., Guo, J., Takemura, M., Drake, M. T., Potts, J. T., Abou-Samra, A., et al. (1997). Mutations in the second cytoplasmic loop of the rat parathyroid hormone (PTH)/PTH-related protein receptor result in selective loss of PTH-stimulated phospholipase C activity. *Journal of Biological Chemistry*, *272*(11), 6882–6889. https://doi.org/10.1074/jbc.272.11.6882.

Jobert, A. S., Zhang, P., Couvineau, A., Bonaventure, J., Roume, J., Merrer, M. L., et al. (1998). Absence of functional receptors for parathyroid hormone and parathyroid hormone-related peptide in Blomstrand chondrodysplasia. *Journal of Clinical Investigation*, *102*(1), 34–40. https://doi.org/10.1172/jci2918.

Jouishomme, H., Whitfield, J., Chakravarthy, B., Durkin, J., Gagnon, L., Issacs, R., et al. (1992). The protein kinase-C activation domain of the parathyroid hormone. *Endocrinology*, *130*, 53–60.

Kalinkovich, A., & Livshits, G. (2021). Biased and allosteric modulation of bone cell-expressing G protein-coupled receptors as a novel approach to osteoporosis therapy. *Pharmacological Research*, *171*, 105794. https://doi.org/10.1016/j.phrs.2021.105794.

Kaplon, H., & Reichert, J. M. (2021). Antibodies to watch in 2021. *MAbs*, *13*(1), 1860476. https://doi.org/10.1080/19420862.2020.1860476.

Kenakin, T. (2011). Functional selectivity and biased receptor signaling. *Journal of Pharmacology and Experimental Therapeutics*, *336*(2), 296–302. https://doi.org/10.1124/jpet.110.173948.

Knowles, T. J., Finka, R., Finka, R., Smith, C., Smith, C., Lin, Y.-P., et al. (2009). Membrane proteins solubilized intact in lipid containing nanoparticles bounded by styrene maleic acid copolymer. *Journal of the American Chemical Society*, *131*(22), 7484–7485. https://doi.org/10.1021/ja810046q.

Konermann, L., Pan, J., & Liu, Y.-H. (2011). Hydrogen exchange mass spectrometry for studying protein structure and dynamics. *Chemical Society Reviews*, *40*(3), 1224–1234. https://doi.org/10.1039/c0cs00113a.

Lanske, B., Amling, M., Neff, L., Guiducci, J., Baron, R., & Kronenberg, H. M. (1999). Ablation of the PTHrP gene or the PTH/PTHrP receptor gene leads to distinct abnormalities in bone development. *Journal of Clinical Investigation*, *104*(4), 399–407. https://doi.org/10.1172/jci6629.

Leder, B. Z., O'Dea, L. S. L., Zanchetta, J. R., Kumar, P., Banks, K., McKay, K., et al. (2015). Effects of abaloparatide, a human parathyroid hormone-related peptide analog, on bone Mineral density in postmenopausal women with osteoporosis. *The Journal of Clinical Endocrinology & Metabolism*, *100*(2), 697–706. https://doi.org/10.1210/jc.2014-3718.

Leder, B. Z., Zapalowski, C., Hu, M., Hattersley, G., Lane, N. E., Singer, A. J., et al. (2019). Fracture and bone mineral density response by baseline risk in patients treated with abaloparatide followed by alendronate: Results from the phase 3 ACTIVExtend trial. *Journal of Bone and Mineral Research*, *34*(12), 2213–2219. https://doi.org/10.1002/jbmr.3848.

Lee, S. C., Knowles, T. J., Postis, V. L. G., Jamshad, M., Parslow, R. A., Lin, Y.-P., et al. (2016). A method for detergent-free isolation of membrane proteins in their local lipid environment. *Nature Protocols*, *11*(7), 1149–1162. https://doi.org/10.1038/nprot.2016.070.

Luttrell, L. M., Wang, J., Plouffe, B., Smith, J. S., Yamani, L., Kaur, S., et al. (2018). Manifold roles of β-arrestins in GPCR signaling elucidated with siRNA and CRISPR/Cas9. *Science Signaling*, *11*(549), eaat7650. https://doi.org/10.1126/scisignal.aat7650.

Malaviya, A. P., & Östör, A. J. K. (2011). Rheumatoid arthritis and the era of biologic therapy. *Inflammopharmacology*, *20*(2), 59–69. https://doi.org/10.1007/s10787-012-0123-y.

Masson, G. R., Burke, J. E., Ahn, N. G., Anand, G. S., Borchers, C., Brier, S., et al. (2019). Recommendations for performing, interpreting and reporting hydrogen deuterium exchange mass spectrometry (HDX-MS) experiments. *Nature Methods*, *16*(7), 595–602. https://doi.org/10.1038/s41592-019-0459-y.

Mistarz, U. H., Brown, J. M., Haselmann, K. F., & Rand, K. D. (2016). Probing the binding interfaces of protein complexes using gas-phase H/D exchange mass spectrometry. *Structure/Folding and Design*, *24*(2), 310–318. https://doi.org/10.1016/j.str.2015.11.013.

Nampoothiri, S., Fernández-Rebollo, E., Yesodharan, D., Gardella, T. J., Rush, E. T., Langman, C. B., et al. (2016). Jansen metaphyseal chondrodysplasia due to heterozygous H223R-PTH1R mutations with or without overt hypercalcemia. *The Journal of Clinical Endocrinology & Metabolism*, *101*(11), 4283–4289. https://doi.org/10.1210/jc.2016-2054.

Neer, R. M., Arnaud, C. D., Zanchetta, J. R., Price, R., Gaich, G. A., Reginster, J.-Y., et al. (2001). Effect of parathyroid hormone (1-34) on fractures and bone mineral density in postmenopausal women with osteoporosis. *The New England Journal of Medicine*, *344*, 1434–1441.

Neve, J. D., Barlow, T. M. A., Tourwé, D., Bihel, F., Simonin, F., & Ballet, S. (2021). Comprehensive overview of biased pharmacology at the opioid receptors: Biased ligands and bias factors. *RSC Medicinal Chemistry*, *12*(6), 828–870. https://doi.org/10.1039/d1md00041a.

Pioszak, A. M., & Xu, H. E. (2008). Molecular recognition of parathyroid hormone by its G protein-coupled receptor. *Proceedings of the National Academy of Sciences of the United States of America*, *105*, 5034–5039. https://doi.org/10.1073/pnas.0801027105.

Pyle, E., Guo, C., Hofmann, T., Schmidt, C., Ribiero, O., Politis, A., et al. (2019). Protein-lipid interactions stabilize the oligomeric state of BOR1p from *Saccharomyces cerevisiae*. *Analytical Chemistry*, *91*(20), 13071–13079. https://doi.org/10.1021/acs.analchem.9b03271.

Pyle, E., Kalli, A. C., Amillis, S., Hall, Z., Lau, A. M., Hanyaloglu, A. C., et al. (2018). Structural lipids enable the formation of functional oligomers of the eukaryotic purine symporter UapA. *Cell Chemical Biology*, *25*(7), 840–848.e4. https://doi.org/10.1016/j.chembiol.2018.03.011.

Rajagopal, S., Ahn, S., Rominger, D. H., Gowen-MacDonald, W., Lam, C. M., Dewire, S. M., et al. (2011). Quantifying ligand bias at seven transmembrane receptors. *Molecular Pharmacology*, *80*, 367–377.

Rey, A., Manen, D., Rizzoli, R., Caverzasio, J., & Ferrari, S. L. (2006). Proline-rich motifs in the parathyroid hormone (PTH)/PTH-related protein receptor C terminus mediate scaffolding of c-Src with β-Arrestin2 for ERK1/2 activation. *Journal of Biological Chemistry*, *281*(50), 38181–38188. https://doi.org/10.1074/jbc.m606762200.

Rothnie, A. J. (2016). Detergent-free membrane protein purification. *Methods in Molecular Biology (Clifton, N.J.)*, *1432*, 261–267. https://doi.org/10.1007/978-1-4939-3637-3_16.

Routledge, S. J., Jamshad, M., Little, H. A., Lin, Y.-P., Simms, J., Thakker, A., et al. (2020). Ligand-induced conformational changes in a SMALP-encapsulated GPCR. *Biochimica et Biophysica Acta. Biomembranes*, *1862*(6), 183235. https://doi.org/10.1016/j.bbamem.2020.183235.

Sadaf, A., Cho, K. H., Byrne, B., & Chae, P. S. (2015). Chapter four amphipathic agents for membrane protein study. *Methods in Enzymology*, *557*, 57–94. https://doi.org/10.1016/bs.mie.2014.12.021.

Saito, H., Noda, H., Gatault, P., Bockenhauer, D., Loke, K. Y., Hiort, O., et al. (2018). Progression of Mineral ion abnormalities in patients with Jansen metaphyseal chondrodysplasia. *The Journal of Clinical Endocrinology & Metabolism*, *103*(7), 2660–2669. https://doi.org/10.1210/jc.2018-00332.

Santos, R., Ursu, O., Gaulton, A., Bento, A. P., Donadi, R. S., Bologa, C. G., et al. (2017). A comprehensive map of molecular drug targets. *Nature*, *16*(1), 19–34. https://doi.org/10.1038/nrd.2016.230.

Sarkar, K., Joedicke, L., Westwood, M., Burnley, R., Wright, M., McMillan, D., et al. (2019). Modulation of PTH1R signaling by an ECD binding antibody results in inhibition of β-arrestin 2 coupling. *Scientific Reports*, *9*(1), 14432. https://doi.org/10.1038/s41598-019-51016-z.

Schipani, E., Kruse, K., & Jüppner, H. (1995). A constitutively active mutant PTH-PTHrP receptor in Jansen-type metaphyseal chondrodysplasia. *Science (New York, N.Y.)*, *268*(5207), 98–100.

Schluter, K.-D., & Piper, H. M. (1998). Cardiovascular actions of parathyroid hormone and parathyroid hormone-related peptide. *Cardiovascular Research*, *37*, 34–41.

Schwindinger, W. F., Fredericks, J., Watkins, L., Robinson, H., Bathon, J. M., Pines, M., et al. (1998). Coupling of the PTH/PTHrP receptor to multiple G-proteins. *Endocrine*, *8*(2), 201–209. https://doi.org/10.1385/endo:8:2:201.

Stetsenko, A., & Guskov, A. (2017). An overview of the top ten detergents used for membrane protein crystallization. *Crystals*, 7(7), 197. https://doi.org/10.1002/elsc.201400187.

Tawfeek, H. A., Qian, F., & Abou-Samra, A. B. (2002). Phosphorylation of the receptor for PTH and PTHrP is required for internalization and regulates receptor signaling. *Molecular Endocrinology*, *16*, 1–13. https://doi.org/10.1210/mend.16.1.0760.

Teo, E. C., Chew, Y., & Phipps, C. (2016). A review of monoclonal antibody therapies in lymphoma. *Critical Reviews in Oncology/Hematology*, *97*, 72–84. https://doi.org/10.1016/j.critrevonc.2015.08.014.

van der Lee, M. M. C., Verkaar, F., Wat, J. W. Y., van Offenbeek, J., Timmerman, M., Voorneveld, L., et al. (2013). β-Arrestin-biased signaling of PTH analogs of the type 1 parathyroid hormone receptor. *Cellular Signalling*, *25*(2), 527–538. https://doi.org/10.1016/j.cellsig.2012.11.012.

Whitfield, J. F., & Morley, P. (1995). Small bone-building fragments of parathyroid hormone: New therapeutic agents for osteoporosis. *Trends in Pharmacological Sciences*, *16*(11), 382–386. https://doi.org/10.1016/s0165-6147(00)89079-3.

Zhao, L.-H., Ma, S., Sutkeviciute, I., Shen, D.-D., Zhou, X. E., de Waal, P. W., et al. (2019). Structure and dynamics of the active human parathyroid hormone receptor-1. *Science*, *364*(6436), 148–153. https://doi.org/10.1126/science.aav7942.

CHAPTER SIX

Parathyroid hormone-related protein (PTHrP) and malignancy

Ami Grunbaum and Richard Kremer*
Calcium Research Laboratories and Department of Medicine, McGill University and McGill University Health Centre, Montreal, QC, Canada
*Corresponding author: e-mail address: richard.kremer@mcgill.ca

Contents

Abstract

PTHrP (parathyroid hormone related protein) is an important mediator of malignancy-related tumor progression and hypercalcemia that shares considerable homology and functionality with parathyroid hormone. In this chapter, we review what has been elucidated to date regarding PTHrP's role in malignancies. Starting with a review of calcium metabolism and regulation, we then summarize the discovery and structure of PTHrP and development of sensitive immunoassays for specific measurement. Subsequently, we explore its role in tumor progression, with emphasis on the primary tumor as well as

Vitamins and Hormones, Volume 120
ISSN 0083-6729
https://doi.org/10.1016/bs.vh.2022.03.002

skeletal and non-osseus metastases. We then consider the clinical implications of PTHrP in cancer before concluding with a discussion of both established and potential treatments for malignancy associated hypercalcemia and bone metastases.

1. Background

PTHrP was discovered during the search for a mediator of hypercalcemia in the setting of various malignancies in which PTH-like effects were noted despite there not being any measurable circulating PTH (Burtis et al., 1987; Strewler et al., 1987; Suva et al., 1987). Hypercalcemia is a relatively common clinical phenomenon, of which over 90% of cases are due to either primary hyperparathyroidism (PHPT) or malignancies (Gastanaga et al., 2016; Stewart, 2005). In fact, it is estimated that hypercalcemia will affect between 20% and 30% of cancer patients at some point during their disease course (Stewart, 2005). While the most common cause of hypercalcemia is PHPT, the most common in the inpatient setting and for severe cases (above 3 mmol/L) are malignancies of which as many as 80% have evidence of elevated PTHrP (Horwitz, 2018).

1.1 Calcium metabolism and regulation

Calcium is an essential element that is involved in many physiological processes as a second messenger, playing an important role in cellular stimulus-response reactions. Examples of calcium's critical functions include muscle contraction, nerve conduction and blood coagulation. Arguably calcium's most well-known role is in bone formation and remodeling where, in addition to its intra-cellular signaling role, it is also responsible for maintaining bone strength and integrity.

Calcium is the most abundant mineral in the human body, of which over 99% (approximately 1 kg) is found as calcium-phosphate salts in bone and teeth (Favus & Goltzman, 2013). Blood plasma levels are tightly controlled at 2.2–2.6 mmol/L, while intracellular concentrations are on the order of 0.2 μmol/L. Plasma calcium is primarily found in two forms—protein bound and free. Free (or ionized) calcium is the physiologically active form and makes up approximately 45% of blood calcium. Under normal physiologic conditions, free circulating calcium levels are tightly regulated by the interactions of parathyroid hormone (PTH) and 1,25 dihydroxyvitamin D_3 (1,25D).

The major effectors for promoting physiologic increases in blood calcium levels are PTH and 1,25D (Favus & Goltzman, 2013). PTH, secreted by the parathyroid glands, increases blood calcium by (1) stimulating calcium release from bone—bone resorption, (2) stimulating renal calcium reabsorption and phosphate excretion in the loop of Henle and distal convoluted tubules, and (3) stimulating 1-alpha-hydroxylase, the renal enzyme responsible for converting 25 hydroxyvitamin D_3 (25D) to the active vitamin D compound 1,25D. In turn, 1,25D (1) increases calcium and phosphate absorption in the small intestine, (2) decreases renal excretion of calcium and phosphate, and (3) with the participation of PTH leads to increased calcium release from the bone. Increases in calcium as well as 1,25D exert negative feedback on PTH secretion.

1.1.1 Bone metabolism

In bone, the Parathyroid Hormone Receptor (PTH1R) is localized on the bone-producing osteoblast cells which can bind with and be activated by both PTH and PTHrP to indirectly enhance osteoclastic bone resorption by enhancing the production of the cytokine RANKL (receptor activator of nuclear factor (NF) kappaB ligand), a member of the tumor necrosis factor (TNF) family (Rouleau, Mitchell, & Goltzman, 1990). Under the influence of RANKL stimulation, osteoclasts then proliferate and differentiate from their mononuclear precursors, eventually fusing to form multinucleated osteoclasts. These can then be activated to form bone-resorbing osteoclasts. Under normal physiologic conditions, osteoprotegerin (OPG) (a soluble decoy receptor of RANKL) can prevent excessive activation of osteoclastic bone resorption.

1.1.2 Renal regulation

As with PTH, an elevation in the circulating level of PTHrP interacting with PTH1R results in phosphate wasting and calcium retention by the kidney, along with increased urinary excretion of cAMP. In the renal proximal tubules, stimulation of adenylate cyclase causes internalization of the type II Na+/Pi- (inorganic phosphate) co-transporter leading to decreased phosphate reabsorption and phosphaturia (Keusch et al., 1998). About 20% of filtered calcium is reabsorbed in the cortical thick ascending limb of the loop of Henle (CTAL) and 15% in the distal convoluted tubule (DCT). Both PTH and PTHrP bind to the PTH1R via a homologous cAMP-mediated mechanism, enhancing calcium reabsorption. In the proximal tubule, PTH and PTHrP can both bind to the PTH1R and stimulate 25-hydroxyvitamin

D 1α hydroxylase (1a(OH)ase). However, PTHrP appears to have less of a stimulatory effect on 1α(OH) ase (Horwitz et al., 2005). This has a significant impact on 1,25D levels in PTHrP producing tumors (discussed further below).

1.2 Discovery of PTHrP

The association between elevated blood calcium and neoplastic disease was first noted in the 1920s (Zondek, Petrow, & Siebert, 1923), when it was postulated that release of calcium from bone by the direct osteolytic action of malignant cells was responsible for the hypercalcemia associated with malignancy. However, hypercalcemia was often accompanied by hypophosphatemia, the hallmark of hyperparathyroidism (HPT), and also occurred in the absence of skeletal metastases. In Albright (1941), Albright hypothesized that a malignant tumor releasing a systemically active factor may be the cause of this hypercalcemic derangement (Goltzman, Stewart, & Broadus, 1981; Nissenson, Abbott, Teitelbaum, Clark, & Arnaud, 1981; Rodan et al., 1983). This hypothesis was raised when a patient with a renal carcinoma, metastatic to bone, presented with hypercalcemia and hypophosphatemia in the absence of any obvious parathyroid dysfunction. The patient's biochemical abnormalities were temporarily normalized following tumor reduction but became apparent again as the tumor resumed its growth. It was reasoned that since hyperphosphatemia did not accompany hypercalcemia despite evidence of osteolysis, perhaps the tumor was secreting a phosphaturic substance like parathyroid hormone (PTH). Although no bioassayable hormone was detected in the tumor, the important observation had been made that malignancy-associated hypercalcemia (MAH) could have a humoral basis and that the factor might be related to PTH. The search for a novel protein with PTH-like bioactivity intensified using several bioassays that had been developed for PTH (Goltzman et al., 1981; Nissenson et al., 1981; Rodan et al., 1983). Using both in vivo and in vitro approaches, PTH-like bioactivity was identified in the blood and in tumor extracts from patients with MAH. These same bioassays were used in various combinations during the purification of a novel protein, initially named parathyroid-hormone-like peptide (PLP), from tumor tissue (Burtis et al., 1987; Moseley et al., 1987; Strewler et al., 1987). When the gene was cloned and demonstrated to share structural homology with PTH, the protein was renamed parathyroid-hormone-related protein, or PTHrP (Mangin, Ikeda, Dreyer, & Broadus, 1989; Suva et al., 1987).

The term humoral hypercalcemia of malignancy (HHM) was later coined to distinguish the syndrome caused by a circulating factor, now known to be PTHrP, from hypercalcemia caused by the direct lytic action of tumor cells in bone. The demonstration that neutralizing antibodies to PTHrP could reduce or eliminate the hypercalcemia in animal models of HHM (Kukreja et al., 1988; Henderson, Shustik, et al., 1990) provided strong evidence that this was indeed the causal factor of this syndrome. HHM was associated primarily with renal cell carcinomas and squamous cell cancers originating in a variety of primary sites, whereas hypercalcemia due to local osteolysis was commonly associated with metastatic breast disease and with hematologic malignancies. Nevertheless, it is now clear that the "direct" lytic action of tumor cells is also mediated in part by PTHrP, and that PTHrP released by these tumors may also act on a systemic basis.

1.3 Cloning of PTHrP

Despite many unsuccessful attempts to identify the PTH-like factor(s) in the blood of hypercalcemic patients using immunoassays directed at PTH (Powell, Singer, Murray, Minkin, & Potts Jr, 1973) and later the lack of expression of PTH mRNA in tumors of hypercalcemic cancer patients (Simpson et al., 1983), the search for this elusive factor continued. Using several bioassays, PTH-like bioactivity was identified in the blood (Goltzman et al., 1981) and in tumor extracts from patients with MAH (Nissenson et al., 1981; Rodan et al., 1983). Using similar bioassays, a fragment of the PTH-like protein was purified (Burtis et al., 1987; Moseley et al., 1987; Strewler et al., 1987), and soon after complementary DNA (cDNA) clones encoding PTHrP were isolated (Ikeda et al., 1995; Strewler et al., 1987; Suva et al., 1987).

1.4 Structure of PTHrP

1.4.1 Gene structure

The PTH and PTHLH genes respectively encoding PTH and PTHrP have been localized to the short arms of human chromosomes 11 and 12 and suggest a common ancestral origin (Goltzman, Hendy, & Banville, 1989). Similarities in their structural organization and in the functional properties of their amino termini provide further support for the hypothesis that PTH and PTHLH are members of a single gene family. PTHLH is transcribed from at least three promoters and undergoes differential splicing, giving rise to heterogenous mRNA species. Its cDNA encodes a prototypical secretory protein with predicted mature isoforms of 139, 141, and 173 amino acids that exhibit a high degree of homology within the first 13 amino-acids to the N-terminus of PTH. This permits its binding with a G-protein coupled PTH receptor, the type 1 PTH/PTHrP receptor (PTH1R) (Jüppner et al., 1991). Interestingly, the three isoforms were shown to be expressed and secreted by both normal and tumor tissues (Campos, Zhang, & Drucker, 1994) but PTHrP1-139 and 1-141 were considerably more abundant than PTHrP1-173 in non-neoplastic tissues, whereas an increase in PTHrP 1-173 was observed in a large proportion of neoplastic tissues. Conservation of PTHrP is evident with high homology across a variety of mammalian species (Burtis, 1992). This, along with similarities to the PTH gene structure organization, suggest that the two genes probably arose from a common ancestor through gene duplication (Fig. 1).

1.4.2 Functional domains of PTHrP

1.4.2.1 Interactions with cell surface receptors (via amino-terminal domain)

The first 13 amino-terminal amino acids are highly homologous with human PTH, and the initial 34 amino-terminal residues of both PTH

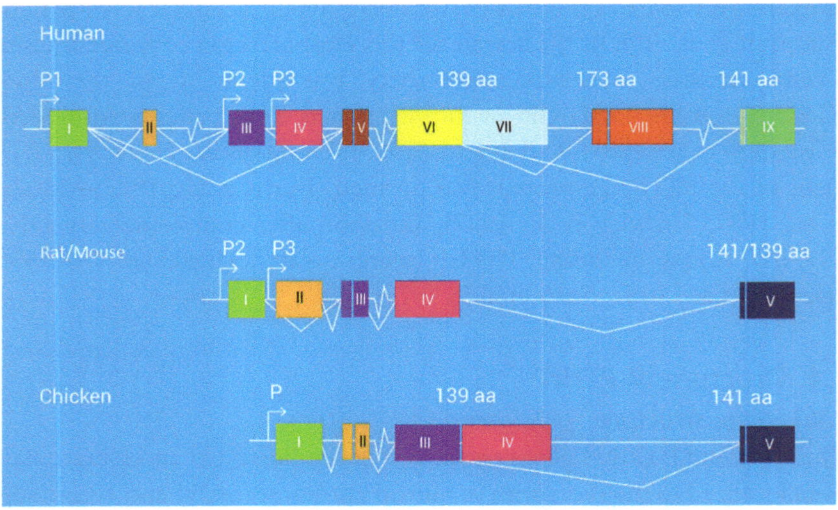

Fig. 1 Gene structure of PTHLH. Alternative splicing gives rise to three separate human PTHrP isoforms which differ at their carboxyl-terminal ends and contain either 139, 141, or 173 amino acids. The mRNAs for each of these isoforms are commonly expressed in various normal and cancer tissues in humans, but the full characterization of their tissue distribution, processing and function remains to be established. In contrast to the human *PTHLH* gene organization, the gene organization in other species is relatively simple with only one isoform per species.

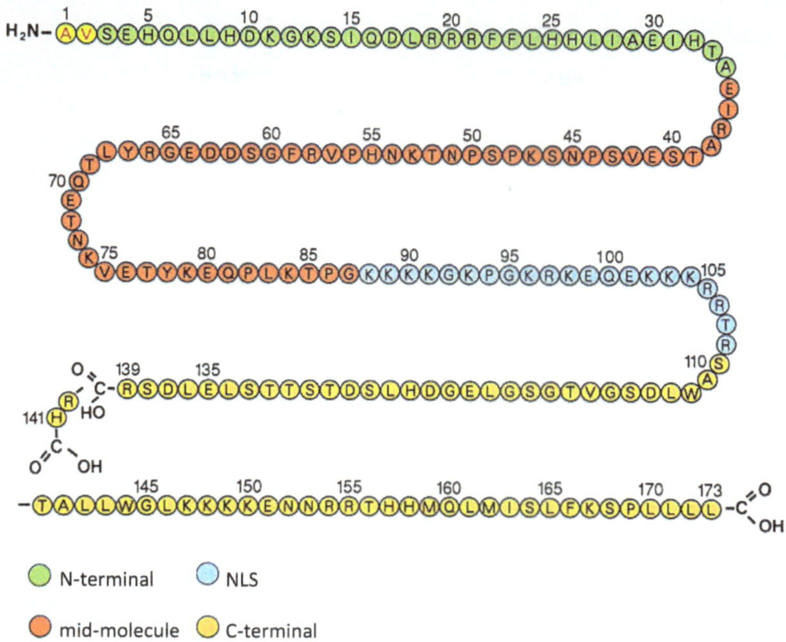

Fig. 2 Amino acid sequence and functional domains of human PTHrP. Human PTHrP arises from alternative splicing as three possible mature isoforms of 139, 141 or 173 amino acids. As detailed in the text, the PTHrP molecule has several functional domains (amino terminal, mid molecule, nuclear localization sequence and carboxy terminal). The initial 34 amino acids share 3-D structural similarity with PTH and are essential for binding and activation of the PTH1R (NLS—Nuclear Localization Sequence; PTH1R—PTH receptor)

and PTHrP contain the functional elements necessary for binding and interacting with the membrane spanning PTH receptor (PTH1R) (Fig. 2).

More specifically, residues 15–34 interact with the receptor's extracellular domain to permit binding of the hormones to the receptor (Lee et al., 1994). Residues 1–14 interact with the transmembrane helices and extracellular connecting loops of PTH1R to induce the conformational changes required for initiation of down-stream signaling (Castro, Nikolaev, Palm, Lohse, & Vilardaga, 2005; Gardella et al., 1994).

PTH1R is a transmembrane G-protein coupled receptor (GPCR). It contains a seven-membrane-spanning "serpentine" domain, a large extracellular ligand-binding domain and an intracellular COOH-terminal domain.

The receptor couples to Gs and Gq leading to activation of the protein kinase A (PKA) and protein kinase C (PKC) pathways (Mannstadt, Jüppner, & Gardella, 1999). As for other GPCRs, PTH1R undergoes cyclical activation, desensitization, and internalization (Weinman, Hall, Friedman, Liu-Chen, & Shenolikar, 2006). After ligand binding and endocytosis, the receptor is either recycled to the cell membrane or targeted for degradation.

PTH1R signaling via the cAMP pathway leads to PKA activation and phosphorylation of the cyclic AMP response element binding protein (CREB) (Yang et al., 2007). There are differences between PTH and PTHrP activations of cAMP responses, including a differential capacity for stimulating 1,25D production. It has been demonstrated that the truncated moieties PTH1-34 and PTHrP1-36 bind to PTH1R with similar association kinetics. However, the receptor-ligand dissociation is of shorter duration with PTHrP1-36 (Ferrandon et al., 2009). This indicates that PTH1R interaction with PTHrP is less stable. Consequently, cAMP signaling response is not as sustained with PTHrP (Dean, Vilardaga, Potts Jr, & Gardella, 2008). It has been shown that in comparison to PTH1-34, PTHrP1-36 has a reduced ability to increase 1,25D blood levels (Horwitz et al., 2005). This may in part explain why 1,25D levels are not as elevated in PTHrP associated malignancies as compared to PHPT.

In the brain, it has been reported that N-terminal and C-terminal sequences of PTHrP act through distinct signaling mechanisms via cAMP/PKA and changes in intra-cellular calcium respectively (Fukayama, Tashjian Jr, Davis, & Chisholm, 1995), suggesting that the C-terminal domain of PTHrP acts through a distinct receptor. Others have reported the presence of cell surface binding proteins for carboxyl terminal fragments of PTHrP on skeletal cells (Orloff & Stewart, 1995). To date however, the molecular nature of these receptors and binding proteins remains unclear.

1.4.2.2 Other functional domains of PTHrP

Whereas the functional association of PTHrP's N-terminal fragment (residues 1–36) with PTH1R and its signaling mechanism has been extensively studied, the activities and interactions of the mid and C-terminal regions with putative receptors are not as well understood. It is the functionality of these non-amino terminal regions of PTHrP that provide many of the distinctive functions of PTHrP as compared to PTH.

1.4.2.3 Mid-molecule portion (residues 35–84)

Promotes calcium transport across placental membrane, enabling skeletal calcium deposition in the developing fetus. This activity is independent of PTH1R, although to date, no receptor has been linked to this activity (Care et al., 1990; Kovacs et al., 1996; Wu et al., 1996). This activity appears to be mediated through cytosolic calcium influx and phospholipase C signaling (Orloff et al., 1996).

1.4.2.4 Nuclear localization sequence (residues 87–107)

In murine studies, the intracrine action of PTHrP appears to be mediated through residues 87–107 in the mid-region of the protein (Henderson et al., 1995). This region encodes a nuclear localization sequence (NLS) that is both necessary and sufficient to direct the passage of PTHrP to the nuclear compartment. The mature PTHrP peptide can originate from a conventional translational site permitting extracellular secretion or a non-conventional start site that disrupts its secretion and directs the product to the nucleus (Nguyen, He, & Karaplis, 2001). Nuclear translocation of PTHrP is required to prevent severe growth retardation and early lethality in mice (Miao et al., 2008).

The biological impact of nucleolar PTHrP on cell function was demonstrated in vitro where chondrocytes expressing wild-type PTHrP were protected from apoptosis induced by serum deprivation, whereas cells expressing the protein without the NLS were not. Subsequently, the nuclear localization of PTHrP was also found to be protective in studies of prostate (Dougherty et al., 1999) and breast cancer cells (Sepulveda, Shen, & Falzon, 2002).

Functionally, intranuclear PTHrP seems to delay apoptosis (Henderson et al., 1995) and to increase cell proliferation. Indeed, a series of studies (Zhang et al., 2010) using knockin mice which express PTHrP(1-86) but not the NLS (nor the carboxy terminal region) have shown that intranuclear PTHrP causes increased phosphorylation and inactivation of the retinoblastoma gene, resulting in increased cell cycle progression and self-renewal. This may provide a molecular basis for the phenotypic changes observed (including growth failure, premature osteoporosis, reduced hematopoiesis, altered energy metabolism, and ultimately premature death) (Miao et al., 2008). Whether this intracrine action promotes tumor progression in PTHrP expressing tumors remains to be determined.

1.4.2.5 COOH-terminal fragments (residues 107 and beyond)

The pentapeptide PTHrP107-11 (known as osteostatin) has been shown to inhibit osteoclastic bone resorption in culture (Fenton et al., 1991) and in mice (Cornish, Callon, Nicholson, & Reid, 1997). The effect of C-terminal fragments on osteoblast activity is controversial, with investigations demonstrating both proliferative (Cornish et al., 1999) and inhibitory (Seitz, Zhang, Simmons, & Cooper, 1995) actions on cultured osteoblasts.

1.4.2.6 Detection of PTHrP

Shortly after the discovery and characterization of PTHrP, assays were developed to detect this factor in the circulation (Budayr et al., 1989; Grill et al., 1991; Henderson, Bernier, D'Amour, & Goltzman, 1990; Kao, Klee, Taylor, & Heath 3rd, 1990; Ratcliffe, Norbury, Stott, Heath, & Ratcliffe, 1991). The majority of these assays were based on the recognition of discrete regions of the PTHrP molecule by specific antibodies. The early PTHrP assays lacked the superiority of two-site immunoassays, also called sandwich assays, to measure circulating PTHrP with both specificity and sensitivity. One of the first two-site PTHrP immunoassays developed (Burtis et al., 1990) was able to characterize PTHrP in a variety of cancer conditions with and without hypercalcemia. This immunoradiometric (IRMA) assay used two polyclonal antibodies, a capture antibody raised against PTHrP37-74, and a signal antibody raised against PTHrP1-36, making the assay quite specific for moieties containing at least the first 74 amino acids of the molecule. The assay sensitivity of 1.0 pmol/L was sufficient to detect PTHrP in approximately 50% of healthy volunteers with a mean of 1.9 ± 1.6 pmol/L and did not cross-react with PTH1-84 or C-terminal fragments of PTHrP.

Burtis' group developed a number of immunoassays targeting specific regions of PTHrP, revealing that patients with MAH contain both N and C-terminal reactive fragments in circulation (Burtis et al., 1990; Burtis, Dann, Gaich, & Soifer, 1994). Using an alternative sandwich assay targeting the N-Terminal (1-36) and C-Terminal (109-136) regions, little or no immunoreactivity was detected, suggesting that the full length PTHrP1-141 molecule is rapidly metabolized.

Discrepancies in results between assays may be related to the variable molecular nature of circulating forms of PTHrP, either due to differences in tumor production or clearance, especially in the presence of altered hepatic and/or renal function in cancer patients. For example, PTHrP is cleared by the kidney (Imamura et al., 1991) and patients with renal insufficiency have elevated levels of PTHrP C-terminal fragments without elevation of N-terminal fragments (Burtis et al., 1990).

PTHrP may have value in prognostic evaluation. Supporting this argument, PTHrP assays have generally shown positive correlations between circulating PTHrP levels and more advanced stages of cancer (Kremer, Sebag, et al., 1996; Kremer, Shustik, Tabak, Papavasiliou, & Goltzman, 1996), and several studies have reported that elevation of serum PTHrP levels was associated with a worse prognosis (Hiraki et al., 2002; Pecherstorfer et al., 1994; Truong, de Edwardes, Papavasiliou, Goltzman, & Kremer, 2003).

To date, measurement of serum PTHrP has not proven to be a useful method for early tumor detection, possibly because of the molecular heterogeneity of PTHrP released into the circulation and the limited sensitivity of commercial assays unable to adequately measure normal levels of PTHrP. Almost invariably, current assays only detect PTHrP in the advanced stages of cancer when hypercalcemia is already present (Kremer, Sebag, et al., 1996; Kremer, Shustik, et al., 1996; Rankin, Grill, & Martin, 1997). Furthermore, the finding of hypercalcemia in association with suppressed PTH using highly sensitive and specific assays to measure intact PTH has often been taken as presumptive evidence of elevated circulating PTHrP as the cause of the hypercalcemia, in lieu of direct measurement of PTHrP.

In order to address the difficulty of isolating the most clinically informative PTHrP fragments from the complex and heterogenous mixtures found in circulation, several groups have attempted to develop assays that permit simultaneous measurements of multiple moieties. Theoretically, mass spectrometric (MS) based methods could do so, although the analytical sensitivities of the MS methods previously described (Lu et al., 2002; Washam et al., 2013) have been insufficient to quantify circulating PTHrP concentrations typically found in the early stage of cancers. A more recently developed method for LC-MS/MS detection of immune-enriched PTHrP in plasma with a lower detection limit of 0.6 pmol/L is promising (Kushnir et al., 2016). Recently, an ultrasensitive multiplexed two-site immunoassay capable of simultaneously measuring several circulating forms of PTHrP with a sensitivity of 150 aM (approximately 1000-fold greater than current immunoradiometric assays) has been reported (Otieno, Krause, Jones, Kremer, & Rusling, 2016). This assay accurately measures normal levels of PTHrP in healthy volunteers and elevated levels in hypercalcemic cancer patients. Prospective measurements are underway in early stage cancer patients and normocalcemic cancer patients at more advanced stages to determine its usefulness as a prognostic biomarker and to monitor tumor recurrence.

2. Characterization of MAH

2.1 Differences between PTHrP and PTH manifestations

With respect to ion homeostasis in the developing fetus, PTHrP is necessary for normal calcium homeostasis (Kovacs et al., 1996). In adults, PTH assumes the role of primary hormone responsible for calcium and phosphorus homeostasis whereas serum PTHrP concentrations are very low to undetectable. When neoplasms hypersecrete PTHrP, it mimics the effects of PTH on bone (i.e., bone resorption and calcium release) and kidney (i.e., calcium reabsorption and phosphate excretion) with the resulting hypercalcemia suppressing endogenous PTH secretion and hypophosphatemia (Fig. 3).

Despite the striking similarities between the biochemical manifestations of HPT and MAH that precipitated Albright's hypothesis, it subsequently became apparent that there were also a number of important differences between these syndromes.

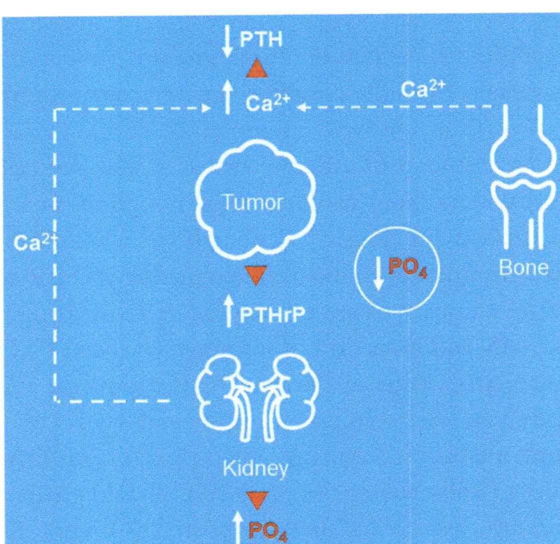

Fig. 3 PTHrP's actions and resulting biochemical changes. Tumor production and release of PTHrP leads to (1) bone resorption with calcium release into circulation and (2) increased renal calcium reabsorption and phosphate excretion. The resultant hypercalcemia suppresses parathyroid production and release of PTH.

Early attempts at the differential diagnosis of these hypercalcemic disorders identified not only subnormal concentrations of immunoreactive PTH (Powell et al., 1973) and no detectable PTH mRNA in the tumors (Simpson et al., 1983), but also a mild hypokalemic alkalosis in patients with MAH. In contrast, elevated PTH and mild metabolic acidosis were associated with HPT (Lafferty, 1966). In addition, circulating levels of 1,25D were reported low in PTHrP-associated hypercalcemia whereas in patients with HPT, 1,25D is often elevated (Stewart et al., 1980). Potential explanations for this apparent discrepancy are the very high calcium levels often seen in MAH that are known to inhibit the 1a(OH)ase enzyme (Bikle, 2014) or possibly unknown inhibitors of 1a(OH)ase released by tumor cells. However, the occurrence of low 1,25D levels in MAH is not widely accepted and low, normal or even high levels have been reported in association with this syndrome (Schweitzer, Hamdy, Frölich, Zwinderman, & Papapoulos, 1994). Indeed, in a large series of hypercalcemic patients with various malignancies and elevated PTHrP levels, the mean circulating levels of 1,25D were normal (Donovan et al., 2015). Another important difference between MAH and HPT are the histomorphometric changes observed in bone biopsy specimens and the reported uncoupling of bone resorption and bone formation processes in hypercalcemic cancer patients (Stewart et al., 1982). Similar results were obtained in tumor bearing rats producing PTHrP prior to the development of hypercalcemia (Yamato et al., 1995). Nevertheless, a study of the administration of an N-terminal fragment of PTHrP to human volunteers suggested that PTHrP has similar bioactivity to that of PTH (Fraher et al., 1992).

The differential effect of PTH compared to PTHrP on suppressing bone formation may be explained by the lack of variation in physiological concentrations of tumor-produced PTHrP. Infusion studies in humans demonstrate that both PTH and PTHrP produce inhibition of formation and stimulation of resorption when administered continuously (Horwitz et al., 2005). At least part of the osteoblast stimulating activity of PTH and therefore, likely of PTHrP, occurs via inhibition of sclerostin, an antagonist of the osteoblast growth factor Wnt (Bellido, Saini, & Pajevic, 2013). Whether this inhibition occurs differently when PTHrP is administered intermittently versus continuously remains to be determined.

2.2 Pathophysiologies of MAH (PTHrP associated and unassociated)

There are 4 widely accepted mechanisms by which malignancies cause hypercalcemia (Stewart, 2005):

2.2.1 Humoral hypercalcemia of malignancy (HHM)

By far the most common cause of MAH, PTHrP is responsible for approximately 80% of MAH (Horwitz, 2018). PTHrP is released systemically by malignant tumors, although it is also a normal gene product expressed in a variety of tissues. Under normal physiological conditions, PTHrP participates in the development of various tissues, facilitates transplacental calcium transport, and promotes calcium transfer to breast milk (Wysolmerski, 2012). PTHrP is produced and released by a variety of tumors, including squamous cell carcinomas such as lung, head and neck, renal, bladder, ovarian and breast cancers.

In addition to PTHrP, a number of other circulating proinflammatory cytokines (including EGF, GCSF, TNF-alpha, IL-1, IL-5, IL-6, IL-8 and IL-11) have been implicated in HHM to a more limited degree (Horwitz & Stewart, 2003; Nagai, Kyakumoto, & Sato, 2000).

While HHM is typically a late finding in advanced metastatic disease associated with a poor and short-term prognosis (Bhandari et al., 2019), breast cancer patients experience a somewhat longer median survival of 3–4.5 months (de Wit & Cleton, 1994). It is also classically described as an endocrine feature of paraneoplastic syndromes (Pelosof & Gerber, 2010).

2.2.2 Osteolytic metastases with local release of cytokines

The second most frequent cause of MAH (approximately 20%), osteolytic metastases are most commonly associated with breast cancer and myeloma (Chen, Sosnoski, & Mastro, 2010; Roodman, 2004).

These lesions are caused by factors produced by tumor cells and the bone micro-environment that stimulate osteoclasts and inhibit osteoblasts (Galson, Silbermann, & Roodman, 2012; Terpos, Christoulas, & Gavriatopoulou, 2018). Adhesion of tumor cells to bone marrow stromal cells stimulates RANKL expression and OPG inhibition by osteoblasts. In combination with a number of abundant inflammatory cytokines (including IL-6, TNFα, IL-1, IL-3, macrophage inflammatory protein-1α (MIP-1α) MIP-1β, decoy receptor 3(DcR3) and activin), this leads to pathologic osteolysis (Pearse et al., 2001).

Myeloma and breast tumor cells invading the skeleton produce PTHrP within the local bone microenvironment (Akhtari, Mansuri, Newman, Guise, & Seth, 2008) via the release of localized factors. In turn, PTHrP further stimulates RANKL release, resulting in a vicious cycle (Fig. 4). The local release of RANKL and various cytokines facilitates bone remodeling and resultant hypercalcemia (Roodman, 2010).

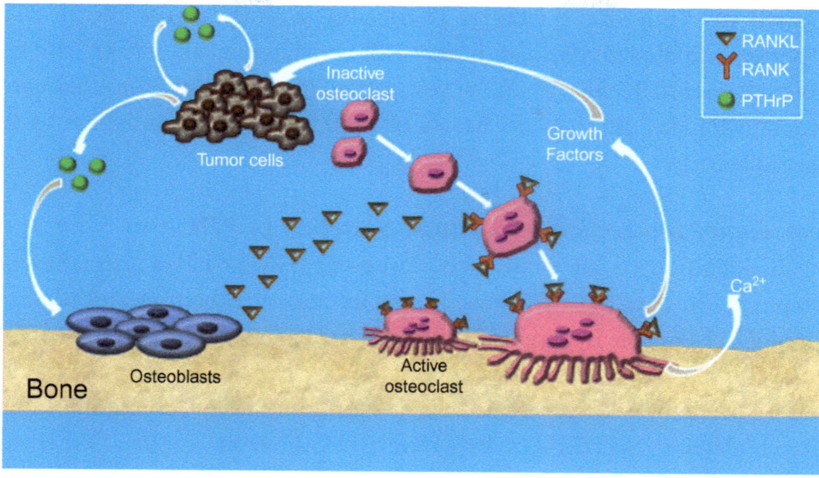

Fig. 4 Vicious cycle hypothesis of bone destruction in metastatic cancer. PTHrP released by tumor cells induces expression and release of RANKL from osteoblasts, which binds to RANK on dormant osteoclasts precursors, leading to their transformation into multinucleated, bone resorbing osteoclasts. Bone resorption releases calcium and bone-derived growth factors from the mineralized bone matrix that further stimulate tumor cell expansion and PTHrP release in a positive feedback loop.

2.2.3 Tumor production of 1,25D

The site of conversion of 25D to its active metabolite, 1,25D, by the 1a(OH) ase enzyme is known not to be restricted to the kidney. Although the kidney is still recognized as the primary site of circulating 1,25D production in vivo, several extrarenal sites of 1a(OH)ase activity have been identified. These include cells of the hematopoietic and immune systems, as well as cells in many other tissues that may give rise to solid tumors associated with MAH. In contrast to the systemic role played by kidney-derived 1,25D in calcium homeostasis, locally produced hormone is thought to regulate cell proliferation and differentiation in normal cells (Holick, 1999) and could modulate neoplastic transformation.

Whereas low circulating levels are typically found in the majority of hypercalcemic cancer patients, increased 1,25D is a common cause of hypercalcemia in patients with lymphoma (Breslau, McGuire, Zerwekh, Frenkel, & Pak, 1984; Edwards & Silberman, 2018; Jacobson, Bringhurst, Harris, Weitzman, & Aisenberg, 1989; Roodman, 1997). This increase in circulating 1,25D is often seen in the presence of renal impairment, suggesting an extrarenal source of the hormone. Furthermore, solid tumors such as the human small cell lung cancer cell line NCI H82 have been reported to synthesize a vitamin D metabolite with similar biochemical properties and bioactivity to authentic 1,25D (Mawer et al., 1994). The systemic increase in 1,25D may be accompanied by release of lymphoid cytokines and PTHrP (known mediators of MAH) and may contribute to the hypercalcemia.

2.2.4 Tumor production and ectopic secretion of PTH

A well-documented, albeit very rare cause of HHM, PTH is normally only secreted by the parathyroid glands. In rare instances, authentic ectopic secretion of PTH (ectopic hyperparathyroidism) has been found in a variety of histological types of tumors including lung, ovary, thyroid and thymus, and has been associated with hypercalcemia (Nussbaum, Gaz, & Arnold, 1990; Rizzoli, Aubert, Sappino, & Bonjour, 1994; Rizzoli, Pache, Didierjean, Bürger, & Bonjour, 1994; Strewler, Budayr, Clark, & Nissenson, 1993; Yoshimoto et al., 1989).

3. Role of PTHrP in tumor progression

As detailed above, tumor-derived PTHrP has an important impact on calcium homeostasis. In addition, PTHrP has other roles in the context of tumor progression.

3.1 Expression and regulation of PTHrP in tumors

It is apparent that the majority of patients with MAH and skeletal metastases have elevated blood concentrations of PTHrP, particularly in the context of breast cancer (Grill et al., 1991). Elevated circulating PTHrP concentrations in both the presence and absence of hypercalcemia have also been noted in hematological malignancies including non-Hodgkin's lymphoma (Burtis et al., 1990; Dodwell, Abbas, Morton, & Howell, 1991; Henderson, Bernier, et al., 1990; Kao et al., 1990; Kremer, Shustik, et al., 1996) and multiple myeloma (Budayr et al., 1989; Kao et al., 1990; Truong et al., 2003).

Elevated expression of PTHrP in primary tumor tissues has been associated with less favorable prognoses. PTHrP expression in prostate cancer tissues from patients without hypercalcemia was found to correlate with tumor stage (Deftos, 2000). Similarly, PTHrP expression in tumor samples appears to be correlated in lung cancers (Hidaka, Nishimura, & Nagao, 1998), renal carcinomas (Iwamura et al., 1999), colorectal tumors (Nishihara et al., 1999), breast cancers (Assaker et al., 2019) and pancreatic cancers (Pitarresi et al., 2021).

PTHrP tissue levels in breast cancer have been associated with progression of disease and bone metastases (Yoshida et al., 2000). More recently, in a large series of newly diagnosed triple negative breast cancer patients, Assaker et al. demonstrated that tissue expression of PTHrP correlated with decreased overall survival and progression of CNS (central nervous system) metastases (Assaker et al., 2019). In contrast, a retrospective analysis of surgically-resected breast tumor tissues found PTHrP to be associated with a more favorable outcome (Henderson et al., 2001; Henderson et al., 2006). The discrepancies in these findings may be due to differences in the patient populations studied, as Henderson's group excluded patients with more severe disease on neo-adjuvant therapy for whom tissue was unavailable.

Interestingly, PTHrP expression in skeletal metastases of breast malignancies was found to be significantly greater than in the primary tumors (Powell et al., 1991; Southby et al., 1990), supporting the hypothesis that PTHrP has an important role in metastatic spread and expansion into bone. This theory is further detailed in Section 3.2.

Regulation of PTHrP expression and production in normal cells was first investigated in studies of normal human keratinocytes (Kremer et al., 1991) that revealed PTHrP to be positively regulated by calcium and mitogenic stimuli, and negatively regulated by 1,25D. Supporting the role for calcium-modulation of PTHrP, Kim et al. documented a positive correlation between the calcium-sensing receptor (CaSR) and PTHrP mRNA expression in the MMTV-PyMT mouse model and in human samples (Kim et al., 2016). In addition, they demonstrated that ablation of the CaSR inhibited PTHrP expression and slowed tumor growth.

Regulatory sequences responding positively to growth factors and negatively to 1,25D in the promoter region of the rat PTHrP gene were identified within 1.2KB of the transcriptional start site. Further studies localized the presence of vitamin D response elements (VDRE) in the promoter region of the gene (Kremer, Sebag, et al., 1996). Similarly, in normal human mammary epithelial cells, both EGF and insulin-like growth factor I (IGF-I) stimulate PTHrP expression and production (Sebag, Henderson, Goltzman, & Kremer, 1994), whereas TGFβ (transforming growth factor β) has antagonistic effects to EGF (Ferrari, Rizzoli, & Bonjour, 1994). The induction of PTHrP expression by mitogenic stimuli has been demonstrated in a number of cell types including primary rat aortic smooth cells (Hongo et al., 1991), non-malignant established cell lines from normal human keratinocytes (Henderson, Sebag, Rhim, Goltzman, & Kremer, 1991), normal human cervical epithelial cells (Kremer, Woodworth, & Goltzman, 1996), and normal human melanocytes (El Abdaimi, Papavasiliou, Goltzman, & Kremer, 2000). In transformed cells, PTHrP is under negative regulation by dexamethasone and testosterone in rat Leydig cells (Liu, Goltzman, & Rabbani, 1993), and positive regulation by phorbol esters (Deftos, Gazdar, Ikeda, & Broadus, 1989; Rodan, Wesolowski, Ianacone, Thiede, & Rodan, 1989), calcitonin (Deftos, Hogue-Angeletti, Chalberg, & Tu, 1989), cyclic AMP (Chan, Strewler, King, & Nissenson, 1990), TGFβ (Kiriyama et al., 1993), and calcium (Brandt, Pandol, & Deftos, 1991; Henderson et al., 1991; Rizzoli & Bonjour, 1989). Several studies also indicate that PTHrP expression increases in parallel to tumor progression. In normal human keratinocytes (Henderson et al., 1991) and normal human prostate cells (Kremer, Goltzman, Amizuka, Webber, & Rhim, 1997), PTHrP is overexpressed following activation by the ras oncogene. In the pYMT model of breast tumor progression (Guy, Cardiff, & Muller, 1992), PTHrP expression progressively increases with tumor stage from hyperplasia to carcinoma (Li et al., 2011). One of the hallmarks of cancer is the growth of cells independently of mitogenic stimuli (Hanahan & Weinberg, 2011). We examined this potential mechanism in ras-transformed keratinocytes (Henderson et al., 1991) and demonstrated that PTHrP is expressed constitutively in these cells, in contrast to normal or established keratinocytes. Furthermore, these cells exhibited resistance to the negative regulation of PTHrP by 1,25D.

Overall, these studies have disclosed important new information about the expression and regulation of PTHrP.

3.2 Relationship of PTHrP with the bone microenvironment

Several animal cancer models associated with PTHrP production have been reported in the context of skeletal metastases and breast cancer (Guise et al., 1996; Mundy, 2002). In these models, PTHrP has been described as a predominant mediator of the bone metastatic process.

In the context of PTHrP regulation, the developmental transcription factor Gli2 (part of the Hedgehog signaling network that induces PTHrP expression) has been identified as a positive regulator of PTHrP and osteolysis in metastatic human breast cancer cells (Sterling et al., 2006) and was shown to be necessary for TGFβ activation of PTHrP (Johnson et al., 2011). In turn, PTHrP exerts downstream effects via CXCR4 (C-X-C chemokine receptor type 4) (Li et al., 2011), a GPCR (G-protein coupled receptor). CXCR4 is expressed in a variety of cancer cells and interacts with its ligand CXCL12

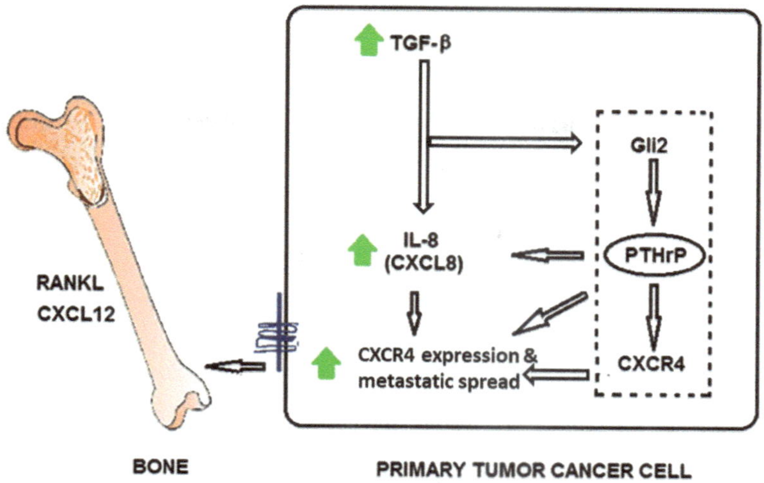

Fig. 5 Targeting of tumor cells to bone via the PTHrP–CXCR4–CXCL12 axis. Tumor production of PTHrP is stimulated by the Gli2 signaling molecule under the control of TGF-β. PTHrP upregulates tumor-surface expression of CXCR4. CXCL12, the natural ligand of CXCR4, is highly expressed in bone as well as other potential metastatic sites (including lung and liver) and acts as a chemoattractant of circulating tumor cells, facilitating metastatic seeding.

(C-X-C motif chemokine 12). In addition, PTHrP and TGFβ can co-stimulate tumor cell production of IL-8, further enhancing CXCR4 expression. CXCL12 is produced in large quantities in the bone microenvironment where it favors the homing of tumor cells to bone (Hirbe, Morgan, & Weilbaecher, 2010). It has been demonstrated that PTHrP upregulates the expression of CXCR4 (Li et al., 2011). Thus, we propose that the PTHrP-CXCR4-CXCL12 axis (Fig. 5) contributes to the bone metastatic process.

MAH is strongly linked to bone metastases, with the majority of MAH patients having evidence of bone involvement and elevated circulating levels of PTHrP** (Henderson, Bernier, et al., 1990; Henderson, Shustik, et al., 1990). Certain malignancies, notably breast, lung, and renal carcinomas which are frequently associated with hypercalcemia also exhibit a strong propensity for spreading to bone. This had been proposed by Paget in the landmark 1889 Lancet article as the "seed to soil hypothesis" (Paget, 1989), namely that metastatic tumor cells will metastasize to sites where the local microenvironment is favorable. Paget's theory is directly relatable to the present context of PTHrP and the bone microenvironment, whereby in addition to the expression of PTHrP systemically by primary tumors, the hormone is also expressed locally with paracrine activities by metastases that have colonized the skeleton (Powell et al., 1991; Southby et al., 1990; Vargas et al., 1992).

Following homing to bone, osteotropic tumor cells adapt to the bone marrow space by producing factors that allow development and progression in their new microenvironment, in part by inducing osteolysis (Waning & Guise, 2014). As poetically phrased by Paget, bone provides a fertile environment for growth of tumors that are metastatic to the skeleton with a well-developed blood supply (from the hematopoietic marrow) as a source of nutrients and numerous growth factors which can provide a growth advantage. The bone matrix is a unique storage site of immobilized growth factors such as TGF-β, insulin-like growth factor (IGF) I and II, fibroblast growth factor (FGF)-1 and -2, and platelet-derived growth factors (PDGF). Tumor-derived PTHrP induces osteolysis and the release of such active growth factors from the matrix, which further facilitate tumor growth. In turn, growth factors such as TGF-β (Kakonen et al., 2002; Yin et al., 1999), as well as calcium per se, following their local release during osteoclastic resorption, may also further stimulate PTHrP production in a self-propagating vicious cycle (Fig. 4). This enhancement of PTHrP production by tumors in the bone microenvironment may explain the observation that breast cancer tissue from skeletal metastases may be more frequently positive for PTHrP expression than from other sites (Powell et al., 1991) and that patients with PTHrP negative primary breast cancers may have PTHrP positive metastatic lesions (Henderson et al., 2006). PTHrP from skeletal metastases may also contribute to tumor growth, survival and adhesion to extracellular bone matrix (Shen & Falzon, 2003).

3.3 Extra-skeletal actions of PTHrP in malignancy

In a series of studies using the pYMT model of breast tumor progression characterized by rapid progression to metastases in less than 11 weeks, Li et al. (2011) demonstrated that PTHrP is critical for tumor initiation, growth, and lung metastases, and is involved with both the PTHrP-CXCR4-CXCL12 axis (Li et al., 2011) and AKT cell survival pathway (Zhang et al., 2009). Furthermore, a monoclonal antibody raised against PTHrP1–33 administered to nude mice transplanted with the human breast cancer cell line MDAMB435 into the mammary fat pad significantly reduced tumor growth and the occurrence of lung metastases (Li et al., 2011). In support of this novel role of PTHrP in tumor initiation and progression, a large genome-wide association study (GWAS) identified PTHrP as one of three critical breast cancer susceptibility genes (Ghoussaini et al., 2012).

Pitarresi et al. (2021) have recently demonstrated the intimate role PTHrP has in pancreatic cancer progression. In silico genomic analyses supported by PDAC (pancreatic ductal adenocarcinoma) cell line studies revealed that in PDAC patients, PTHLH (PTHrP encoding gene) mRNA expression is often amplified, and that PTHLH gene copy number is enriched in metastases. The group went on to demonstrate that PTHLH gene deletion reduced tumor burden and abolished metastases while extending overall survival in a mouse model of PDCA. Further supporting the role of PTHrP in PDAC, anti-PTHrP monoclonal neutralizing antibodies reduced both primary tumor load and metastatic outgrowth. Ultimately, in the absence of PTHrP, tumors were unable to undergo EMT (epithelial-mesenchymal transition) and could not initiate the metastatic cascade. These studies highlight PTHrP's role as a driver of cancer progression and its potential as a therapeutic target.

Similarly, PTHrP overproduction has been reported in the A375 metastatic melanoma model (Huang et al., 2014). Using a PTHrP knockout approach and intra-cardiac transplantation of cancer cells, Huang et al. demonstrated that PTHrP plays a predominant role in the metastatic process within and outside of the skeleton. As such, it appears that PTHrP regulates cell invasion in vivo through alteration of the actin cytoskeleton.

Cancer cachexia is a complex multifactorial syndrome involving tumor and host-derived signaling factors and abnormal metabolism that alters the homeostatic mechanisms of energy balance in the body resulting in muscle mass deterioration (Henderson, Makhijani, & Mace, 2018). Cancer cachexia includes the participation of various different molecules such as cytokines, hormones, neuropeptides, neurotransmitters, and tumor-derived factors (Cariuk et al., 1997; Tisdale, 2002). Studies in animal models of lung cancer and skeletal metastases (Hashimoto et al., 2007; Iguchi, Onuma, Sato, Sato, & Ogata, 2001) indicate that tumor-produced PTHrP is involved in cachexia and accumulation of orixogenic peptides such as neuropeptide Y in the hypothalamus. Administration of monoclonal antibodies against PTHrP ameliorated the condition with a concomitant reduction in expression of the hypothalamic orixogenic peptides. The aforementioned A375 melanoma model also demonstrated reversal of weight loss and muscle wasting both by PTHrP knockout as well as administration of antibodies against PTHrP (Huang et al., 2014). Finally, to put the role of PTHrP in cachexia into clinical context, in a year-long study on a longitudinal observational cohort of over 600 patients with a variety of cancers (including respiratory, liver, pancreatobiliary, gastroesophageal and genitourinary), serum PTHrP levels predicted cancer-associated cachexia as assessed by BMI (body mass index) adjusted weight loss independent of hypercalcemia, inflammation, tumor burden, and other comorbidities (Hong et al., 2016).

PTHrP's role in progression and survival in other cancers such as prostate cancer, renal cell carcinoma, colon adenocarcinoma and hematological malignancies is addressed in the preceding section on expression of PTHrP in tumors.

4. Clinical implications of MAH

4.1 Clinical manifestations of hypercalcemia

The adult human body contains approximately 1000 g of calcium of which 99% is stored as hydroxyapatite crystal in bone (Favus & Goltzman, 2013). Circulating concentrations of calcium are tightly regulated, with blood levels in the range of 2.12–2.62 mmol/L. Of the total calcium in circulation, approximately 50% is in the biologically active ionized form (1.16–1.31 mmol/L) (Table 1).

Malignancy-associated PTHrP overproduction leading to measurable circulating levels of the hormone is a late finding in the advanced stages of cancer. The associated clinical findings are due to PTHrP's role in precipitating a hypercalcemic crisis. The pathological symptoms of hypercalcemia are dependent on both the rate and magnitude of elevation in circulating levels. In comparison to the more insidious rise in calcium levels linked to

Table 1 Clinical findings associated with hypercalcemia of malignancy.

Organ system	Clinical abnormalities
Neurologic	Lethargy, altered mental status (confusion to coma)
Cardiovascular	Hypertension, ECG changes (shortened QT interval, abnormal ST segment), arrhythmias
Musculoskeletal	Muscle weakness, bone pain, pathological fractures, diminished bone density
Gastrointestinal	Dysmotility, nausea, vomiting, constipation, anorexia
Renal	Nephrogenic diabetes insipidus, polyurea/polydipsia, dehydration, renal vasoconstriction, distal renal tubular acidosis, renal insufficiency, nephrocalcinosis/nephrolithiasis

the relatively slow rise and progression of PTH levels in primary hyperparathyroidism, MAH related PTHrP can cause a rapid and severe blood calcium elevation, with total levels that are often above 3 mmol/L.

Tight control of the 1000-fold gradient of calcium concentrations between the extracellular and intracellular spaces is integral for the maintenance of the plasma membrane's electrochemical potential. Hypercalcemia raises the potential across cell membranes, thus increasing the depolarization threshold of muscle cells (skeletal, smooth and cardiac) and neurons (Horwitz, 2018). This leads to the clinical manifestations of hypercalcemia involving multiple organ systems, including neurologic, cardiovascular, musculoskeletal, gastro-intestinal and renal.

Neurologic symptoms can manifest as lethargy and altered mental status, ranging from confusion to coma (Inzucchi, 2004). Cardiac conduction abnormalities include Q-T interval and ST-segment changes that can result in arrhythmias (Diercks, Shumaik, Harrigan, Brady, & Chan, 2004; Nishi, Barbagelata, Atar, Birnbaum, & Tuero, 2006), while hyperpolarization of vascular cells can cause systemic hypertension. In a similar vein, decreased striated skeletal and smooth intestinal muscle cell contractility can lead to skeletal muscle weakness and gastrointestinal dysmotility with resulting nausea, vomiting, constipation and anorexia. Hypercalcemia also has direct effects at the level of the nephron, preventing water reabsorption (Caruana & Buckalew Jr, 1988; Lins, 1978). This can lead to nephrogenic diabetes insipidus with resulting polyuria, renal vasoconstriction, distal renal tubular acidosis, dehydration and renal insufficiency. Prolonged or chronic hypercalcemia can lead to renal calcium deposits and nephrocalcinosis/nephrolithiasis. Finally, PTHrP's pathologic activation of osteoclast bone resorption (via osteoblast-mediated RANKL secretion as outlined above) can result in bone pain and pathological fractures.

4.2 Diagnostic approach and investigation of hypercalcemia

As mentioned previously, hypercalcemia is a relatively common clinical phenomenon most commonly associated with primary hyperparathyroidism or active malignancies. By the time a malignancy has advanced sufficiently to cause symptomatic hypercalcemia, the diagnosis is usually evident. Nevertheless, a step-wise algorithmic approach for the confirmation and investigation of hypercalcemia is recommended to establish true hypercalcemia while ruling-out other causes, avoid missing concomitant pathologies and ensuring appropriate management (Fig. 6).

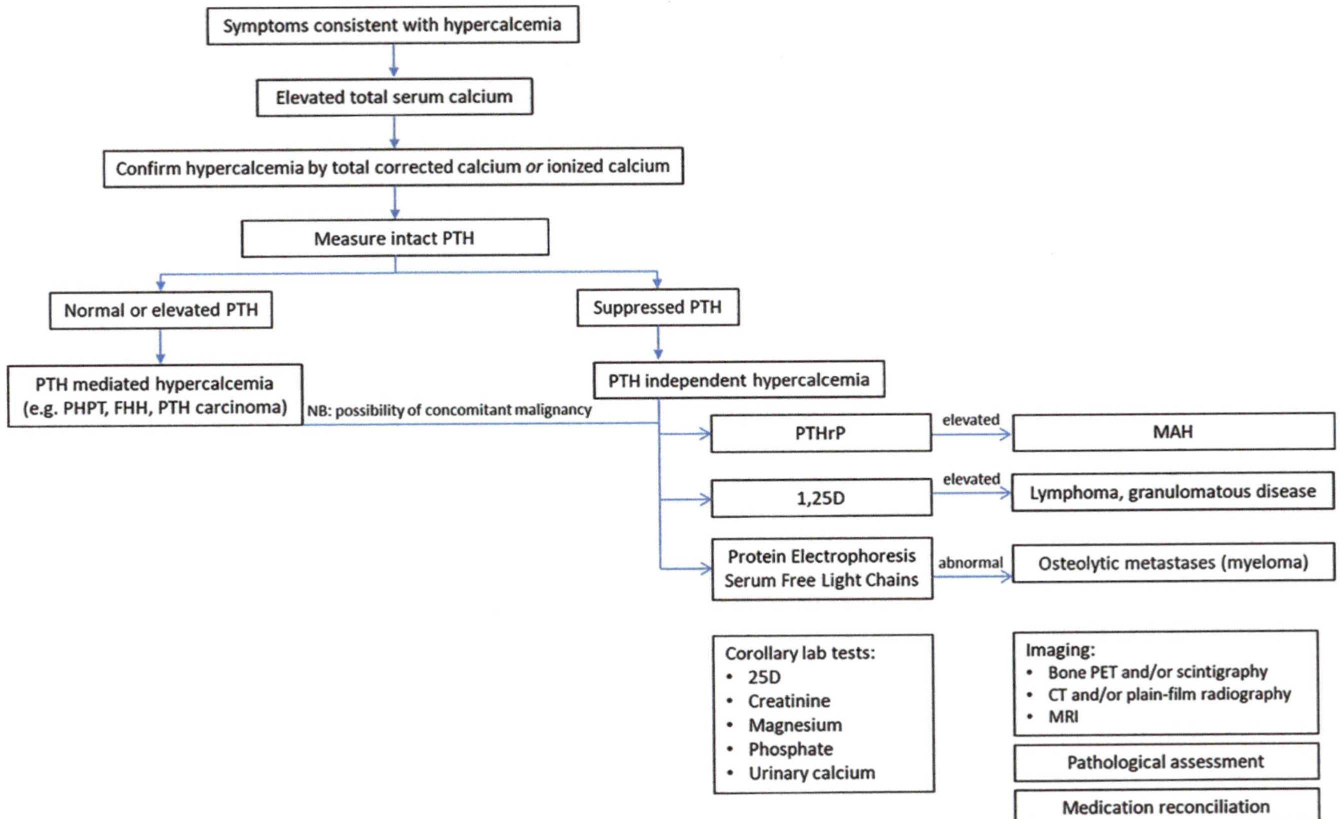

Fig. 6 Diagnostic and investigational algorithm of hypercalcemia (FHH—familial hypocalciuric hypercalcemia; MAH—malignancy associated hypercalcemia; PTH—parathyroid hormone; PTHrP—PTH related protein; PHPT—primary hyperparathyroidism; 25D—25 hydroxyvitamin D_3; 1,25D—1,25 dihydroxyvitamin D_3).

As discussed above, total serum calcium consists of protein-bound (predominantly to albumin) and ionized forms. It is the ionized (or free) form that is physiologically active. Generally, clinical laboratories measure and report total calcium levels. In pathologic states accompanied by altered albumin levels (e.g., low albumin levels in protein-losing enteropathies or nephropathies, malnutrition or acute inflammatory conditions; elevated protein levels in severe dehydration), total calcium levels may be misleading (factitious hypercalcemia associated with elevated protein levels, or overlooked true-hypercalcemia in patients with hypoalbuminemia). To overcome this, most clinical laboratories employ correction formulas to adjust for abnormal serum albumin levels. The most widely used of these is the Payne formula (Smith, Wilson, & Schneider, 2018): Adjusted calcium (mmol/L) = Total calcium (mmol/L) + 0.02 [40 − serum albumin (g/L)]. This and other correction formulae were derived by determining the linear regression relationship of serum calcium to albumin concentration in healthy patients. However, corrected calcium measurements are often unreliable in critically ill patients with extreme variations in pH and protein levels. In such circumstances, direct measurements of ionized calcium are recommended if available to both diagnose and guide management of hypercalcemia.

After confirming a true physiologic hypercalcemia, it is advisable to measure PTH levels to determine whether the hypercalcemia is PTH mediated or independent. The most common PTH mediated cause of hypercalcemia is primary hyperparathyroidism, but FHH (familial hypocalciuric hypercalcemia)—a benign condition caused by mutations in the calcium sensing receptor—or rare instances of malignant PTH production are possible. PTH suppression is indicative of a PTH-independent pathology. However, due to the frequency of PHPT in the general population, it is possible that it may coexist with MAH and make the differential diagnosis more challenging. In such scenarios, consider investigation for PTH-independent pathologies such as MAH.

At the discretion of the clinical investigators, subsequent investigations of PTH-independent hypercalcemia may include:

- PTHrP—Commonly associated with MAH including HHM, but often not available within the turn-around-time required for clinical management. From a management perspective, the degree of PTHrP elevation may predict response to bisphosphonates and mortality (Gurney, Grill, & Martin, 1993; Ling, A'Hern, & Hardy, 1995; Pecherstorfer et al., 1994; Wimalawansa, 1994). As described in the therapeutic modalities section below, in the setting of elevated circulating PTHrP, response to antiresorptives may be blunted and of limited efficacy. The few PTHrP assays that are currently available for clinical use are based on IRMA technology. Typical values associated with past and current versions of these assays are summarized in Table 2.

- 1,25D—Of limited utility, but possibly useful in the differential diagnosis of some lymphomas and granulomatous diseases such as tuberculosis and sarcoidosis (associated with elevations in 1,25D).
- Protein electrophoreses (serum and/or urine) and serum free light chains—Abnormal studies are often associated with osteolytic metastases arising from multiple myeloma.
- 25D—It is important to be cognizant of patients' vitamin D status (as measured by 25D blood levels) because of its importance in maintaining bone health and prevention of fractures. Due to the frequency of primary hyperparathyroidism in the general population, it is to be expected that the two disorders (i.e., hypercalcemia related to PHPT and MAH) will often co-exist.
- Urinary calcium levels can be useful in the management of hypercalcemia in order to avoid longer-term renal complications such as nephrocalcinosis.
- Low serum phosphate levels are suggestive of PTH-like activity.
- Low magnesium levels will blunt PTH receptor response to PTH and PTHrP and may mask the magnitude of hypercalcemia.
- It is also important to consider the patient's renal function (clinically estimated by serum creatinine) as PTH fragments may accumulate in kidney dysfunction. A significant proportion of inactive fragments lacking the first few N-terminus amino-acids are detected by the second generation of PTH assays currently used by most laboratories, resulting in elevated measurements that may lead to an erroneous diagnosis of hyperparathyroidism (Kritmetapak & Pongchaiyakul, 2019).
- Review the patient's medications (e.g., lithium, thiazide diuretics, vitamin D and calcium supplements) and consider limiting their use in order to mitigate their hypercalcemic effects.
- Imaging: In MAH, whole body assessment by a multimodal approach (plain-film radiography, whole-body scintigraphy, CT, MRI, and PET) is recommended (Cecchin et al., 2011). Imaging studies should evaluate each organ potentially involved in the pathogenesis of hypercalcemia. If available, 18F-fluoro-2-deoxyglucose (FDG)-PET/CT is more sensitive and accurate than bone scintigraphy for detection of

Table 2 Commercially available PTHrP assays.

Commercial assay	References	Detection limit	Control values (healthy subjects)	Cancer patients hypercalcemic	Cancer patients normocalcemic
Incstar Corporation	Truong et al. (2003)	0.2 pmol/L	<1 pmol/L	0.2–13.8 pmol/L	N/A
Stillwater, MN					
Recognizes PTHrP1-86					
Nichols Institutes Diagnostic	Truong et al. (2003)	0.3 pmol/L	<1 pmol/L	N/A	N/A
San Juan, Capistrano, CAL	Rizzoli et al. (1999)	0.3 pmol/L	<2.5 pmol/L	3.9–5.9 pmol/L	N/A
Recognizes PTHrP1-72	Donovan et al. (2015)	0.3 pmol/L	<2 pmol/L	3.3–7.2 pmol/L	N/A
Beckman Coulter	Otieno, Krause, Jones, Kremer, and Rusling (2016)	0.3 pmol/L	<1 pmol/L	0.3–7.0 pmol/L	<1 pmol/L
Brea, CAL					
Recognizes PTHrP1-86 with DSL8100 assay					
Mitsubishi LSI (Medicine Corporation) Japan Recognizes PTHrP1-87	Ikeda et al. (1994)	0.5 pmol/L	<1.1 pmol/L	3.7–18.9 pmol/L	0.6–1.3 pmol/L

active bone metastases from breast cancer (Hansen et al., 2021; Rong et al., 2013). A number of radionuclide tracers are available for localization of specific tumors (e.g., Ga-68 dotatate for neuroendocrine tumors) (Shell et al., 2018).

– Pathological assessment: As described in the preceding section (*Expression of PTHrP in tumors*), immune-histological staining of resected tumor lesions has been studied in the context of PTHrP producing malignancies and may be of added value when deliberating complex clinical cases of hypercalcemia in which circulating PTHrP levels are not elevated despite the abnormal presence of PTHrP in malignant tissues.

4.3 Therapeutic modalities for MAH and skeletal metastases

4.3.1 Clinically approved therapies and their limitations

The biological actions of PTHrP on bone and kidney have a central role in the therapeutic approach to patients with MAH. The mainstay of treatment for patients presenting with severe hypercalcemia is the combination of rehydration and anti-resorptive agents including bisphosphonates and monoclonal antibodies against RANKL (such as denosumab). Anti-resorptives work by inhibiting the skeletal component of hypercalcemia but are often less effective in hypercalcemic cancer patients with elevated PTHrP levels (Walls, Ratcliffe, Howell, & Bundred, 1994). Indeed, the proportion of recurrent hypercalcemia following calcium normalization is higher than in patients with MAH and normal levels of PTHrP (Grill et al., 1992; Gurney et al., 1993; Walls et al., 1994; Wimalawansa, 1994). A possible explanation of this resistance is the loss of responsiveness of PTHrP production by bone metastases to bisphosphonates, as local production of PTHrP may diminish the efficacy of these drugs on bone. PTHrP has a direct effect on RANKL production by osteoblasts (Roodman, 2004), as well as direct effects on tumor cell proliferation (Li et al., 2011; Maioli & Fortino, 2004). In addition, while anti-resorptives have direct effects on bone metabolism, they have no effect in inhibiting PTHrP's role in inducing renal reabsorption of calcium (Onuma et al., 2005; Rizzoli et al., 1999). Cancer patients suffering from skeletal metastases and/or MAH may respond to higher dosages administered at shorter intervals. Decreasing PTHrP production through alternate approaches would be of significant benefit for these patients using novel targeted therapies directed at PTHrP.

4.3.2 Investigational approaches to controlling PTHrP production and action

Anti-PTHrP neutralizing antibodies have been shown to be effective in reducing skeletal metastases and hypercalcemia (Guise et al., 2002; Iguchi et al., 1996), presumably by inhibiting PTHrP's actions at both levels. Targeting PTHrP in skeletal metastases using specific monoclonal antibodies to PTHrP is supported by pre-clinical studies in animal models of breast, prostate and lung metastases to bone (Guise et al., 1996; Li et al., 2022; Miki et al., 2004; Rabbani, Gladu, Harakidas, Jamison, & Goltzman, 1999), the three most common human malignancies. Additional support for this approach comes from clinical studies indicating that cancer patients overexpress PTHrP at the bone metastatic site (Powell et al., 1991), a feature that is likely to exacerbate the vicious cycle occurring in bone once cancer cells invade the skeleton. Emerging evidence in pre-clinical models of breast and lung cancer support the principle that combination therapies targeting both PTHrP with monoclonal antibodies and the bone microenvironment with the potent bisphosphonate zoledronic acid are superior to either agent administered separately (Camirand, Fadhil, Luco, Ochietti, & Kremer, 2013; Luco et al., 2013; Yamada et al., 2009).

Previously, low-calcemic analogs of 1,25D were investigated as potential therapeutics given their ability to inhibit PTHrP production without stimulating hypercalcemia via gastrointestinal or renal absorption of calcium or bone resorption. In cell and animal studies, these analogs demonstrated an ability to reverse MAH arising from PTHrP's effects (Rabbani, Gladu, Liu, & Goltzman, 1995).

An alternative approach for diminishing PTHrP overproduction in malignancy employed antisense technology. In this approach, engineered oligonucleotides complementary to PTHrP mRNA sequences are inserted pharmacologically and delivered to cells in order to arrest the translation to proteins, thus inhibiting endogenous PTHrP expression and diminishing hypercalcemia. The use of antisense PTHrP (Rabbani et al., 1995) or antisense furin (Liu, Amizuka, Goltzman, & Rabbani, 1995; Liu, Goltzman, & Rabbani, 1995) to reduce bioactive PTHrP in MAH models in vivo led to diminished hypercalcemia, decreased tumor cell proliferation and prolonged survival of the host animals.

Theoretically, these approaches can be used in the management of cancer patients exhibiting MAH secondary to excess PTHrP production. To date however, none of these investigational pharmaceuticals have shown success in human clinical trials, nor have they been approved by regulatory agencies.

4.3.3 *Potential non-skeletal roles for anti-PTHrP therapies*

Although skeletal metastases are associated with skeletal related events (SREs) such as pain, nerve compression and fractures (Sullivan, 1996), rapid disease progression usually involves extra-skeletal sites resulting in metastatic spread to vital organs that is resistant to most therapeutic regimens. This is particularly true for breast cancer expressing the triple negative phenotype. In a study of a model of breast tumor progression, PTHrP was critical for the initiation, growth and metastases of tumor to lungs (Li et al., 2011). Targeting PTHrP was proposed as a potential new therapy. Indeed, administration of monoclonal antibodies against PTHrP reduced breast tumor growth and lung metastases in breast cancer (Li et al., 2011) and reduced disease spread at multiple sites including liver, lungs and bones in a human melanoma model (Huang et al., 2014). Similarly, in a model of renal cancer, anti-PTHrP antibodies slowed tumor progression and its metastatic spread (Massfelder et al., 2004). Numerous in vitro studies support a role of PTHrP in growth and invasion in breast, colon, prostate and giant tumor cell of bone (GCT) cancers (Downs, Burton, Araiza, Hastings, & Deftos, 2011; Li et al., 2011; Mak, Cowan, Turcotte, Singh, & Ghert, 2011; Mula, Bhatia, & Falzon, 2010). Taken together, these studies support the rationale of targeting PTHrP as a novel approach to cancer treatment, possibly in the most refractory types such as triple negative breast cancer.

5. Conclusion

The existence of a PTH-like substance was first considered by Albright over 80 years ago. Since then, considerable progress has been made—from the discovery and cloning of the PTHrP molecule, to the elucidation of its role in fetal development and its pathological mechanisms and manifestations in malignancy. Advances in applying this knowledge to cancer therapeutics has lagged behind but is now emerging as an important consideration in the treatment of resistant malignancy associated hypercalcemia and possibly as a target for skeletal as well as non-skeletal metastases for both solid and hematological malignancies.

Acknowledgments

The authors acknowledge the support of the Canadian Institute for Health Research (CIHR) MOP142287 and the United States Department of Defense W81XWH-15-1-0723

Disclaimer

Richard Kremer (RK) is the inventor of the therapeutic monoclonal anti-PTHrP antibodies (PTHrP, its isoforms and antagonist thereto in the diagnosis and treatment of disease. United States Patent 850 1929. 2013/08/06) and authorized for use by Biochrom Pharma. RK is Founder and President of Biochrom Pharma. Ami Grunbaum declares no potential conflicts of interest.

References

Akhtari, M., Mansuri, J., Newman, K. A., Guise, T. M., & Seth, P. (2008). Biology of breast cancer bone metastasis. *Cancer Biology & Therapy*, *7*(1), 3–9. https://doi.org/10.4161/cbt.7.1.5163.

Albright, F. (1941). Case records of the Massachusetts General Hospital (Case 27461). *The New England Journal of Medicine*, *225*, 789–791.

Assaker, G., Camirand, A., Abdulkarim, B., Omeroglu, A., Deschenes, J., Joseph, K., et al. (2019). PTHrP, a biomarker for CNS metastasis in triple-negative breast cancer and selection for adjuvant chemotherapy in node-negative disease. *JNCI Cancer Spectrum*, *4*(1), pkz063. https://doi.org/10.1093/jncics/pkz063.

Bellido, T., Saini, V., & Pajevic, P. D. (2013). Effects of PTH on osteocyte function. *Bone*, *54*(2), 250–257. https://doi.org/10.1016/j.bone.2012.09.016.

Bhandari, S., Kumar, R., Tripathi, P., Chan, A., Mudra, S., & Redman, R. (2019). Outcomes of hypercalcemia of malignancy in patients with solid cancer: A national inpatient analysis. *Medical Oncology (Northwood, London, England)*, *36*(10), 90. https://doi.org/10.1007/s12032-019-1315-8.

Bikle, D. D. (2014). Vitamin D metabolism, mechanism of action, and clinical applications. *Chemistry & Biology*, *21*(3), 319–329. https://doi.org/10.1016/j.chembiol.2013.12.016.

Brandt, D. W., Pandol, S. J., & Deftos, L. J. (1991). Calcium-stimulated parathyroid hormone-like protein secretion: potentiation through a protein kinase-C pathway. *Endocrinology*, *128*(6), 2999–3004. https://doi.org/10.1210/endo-128-6-2999.

Breslau, N. A., McGuire, J. L., Zerwekh, J. E., Frenkel, E. P., & Pak, C. Y. (1984). Hypercalcemia associated with increased serum calcitriol levels in three patients with lymphoma. *Annals of Internal Medicine*, *100*(1), 1–6. https://doi.org/10.7326/0003-4819-100-1-1.

Budayr, A. A., Nissenson, R. A., Klein, R. F., Pun, K. K., Clark, O. H., Diep, D., et al. (1989). Increased serum levels of a parathyroid hormone-like protein in malignancy-associated hypercalcemia. *Annals of Internal Medicine*, *111*(10), 807–812. https://doi.org/10.7326/0003-4819-111-10-807.

Burtis, W. J. (1992). Parathyroid hormone-related protein: structure, function, and measurement. *Clinical Chemistry*, *38*(11), 2171–2183.

Burtis, W. J., Brady, T. G., Orloff, J. J., Ersbak, J. B., Warrell, R. P., Jr., Olson, B. R., et al. (1990). Immunochemical characterization of circulating parathyroid hormone-related protein in patients with humoral hypercalcemia of cancer. *The New England Journal of Medicine*, *322*(16), 1106–1112. https://doi.org/10.1056/NEJM199004193221603.

Burtis, W. J., Dann, P., Gaich, G. A., & Soifer, N. E. (1994). A high abundance midregion species of parathyroid hormone-related protein: Immunological and chromatographic characterization in plasma. *The Journal of Clinical Endocrinology and Metabolism*, *78*(2), 317–322. https://doi.org/10.1210/jcem.78.2.8106619.

Burtis, W. J., Wu, T., Bunch, C., Wysolmerski, J. J., Insogna, K. L., Weir, E. C., et al. (1987). Identification of a novel 17,000-dalton parathyroid hormone-like adenylate cyclase-stimulating protein from a tumor associated with humoral hypercalcemia of malignancy. *The Journal of Biological Chemistry*, *262*(15), 7151–7156.

Camirand, A., Fadhil, I., Luco, A. L., Ochietti, B., & Kremer, R. B. (2013). Enhancement of taxol, doxorubicin and zoledronate anti-proliferation action on triple-negative breast cancer cells by a PTHrP blocking monoclonal antibody. *American Journal of Cancer Research*, *3*(5), 500–508.

Campos, R. V., Zhang, L., & Drucker, D. J. (1994). Differential expression of RNA transcripts encoding unique carboxy-terminal sequences of human parathyroid hormone-related peptide. *Molecular Endocrinology (Baltimore, Md.)*, *8*(12), 1656–1666. https://doi.org/10.1210/mend.8.12.7708054.

Care, A. D., Abbas, S. K., Pickard, D. W., Barri, M., Drinkhill, M., Findlay, J. B., et al. (1990). Stimulation of ovine placental transport of calcium and magnesium by mid-molecule fragments of human parathyroid hormone-related protein. *Experimental Physiology*, *75*(4), 605–608. https://doi.org/10.1113/expphysiol.1990.sp003437.

Cariuk, P., Lorite, M. J., Todorov, P. T., Field, W. N., Wigmore, S. J., & Tisdale, M. J. (1997). Induction of cachexia in mice by a product isolated from the urine of cachectic cancer patients. *British Journal of Cancer*, *76*(5), 606–613. https://doi.org/10.1038/bjc.1997.433.

Caruana, R. J., & Buckalew, V. M., Jr. (1988). The syndrome of distal (type 1) renal tubular acidosis. Clinical and laboratory findings in 58 cases. *Medicine*, *67*(2), 84–99. https://doi.org/10.1097/00005792-198803000-00002.

Castro, M., Nikolaev, V. O., Palm, D., Lohse, M. J., & Vilardaga, J. P. (2005). Turn-on switch in parathyroid hormone receptor by a two-step parathyroid hormone binding mechanism. *Proceedings of the National Academy of Sciences of the United States of America*, *102*(44), 16084–16089. https://doi.org/10.1073/pnas.0503942102.

Cecchin, D., Motta, R., Zucchetta, P., Bui, F., Basso, S. M., & Lumachi, F. (2011). Imaging studies in hypercalcemia. *Current Medicinal Chemistry*, *18*(23), 3485–3493. https://doi.org/10.2174/092986711796642607.

Chan, S. D., Strewler, G. J., King, K. L., & Nissenson, R. A. (1990). Expression of a parathyroid hormone-like protein and its receptor during differentiation of embryonal carcinoma cells. *Molecular Endocrinology (Baltimore, Md.)*, *4*(4), 638–646. https://doi.org/10.1210/mend-4-4-638.

Chen, Y. C., Sosnoski, D. M., & Mastro, A. M. (2010). Breast cancer metastasis to the bone: Mechanisms of bone loss. *Breast Cancer Research: BCR*, *12*(6), 215. https://doi.org/10.1186/bcr2781.

Cornish, J., Callon, K. E., Lin, C., Xiao, C., Moseley, J. M., & Reid, I. R. (1999). Stimulation of osteoblast proliferation by C-terminal fragments of parathyroid hormone-related protein. *Journal of Bone and Mineral Research: the Official Journal of the American Society for Bone and Mineral Research*, *14*(6), 915–922. https://doi.org/10.1359/jbmr.1999.14.6.915.

Cornish, J., Callon, K. E., Nicholson, G. C., & Reid, I. R. (1997). Parathyroid hormone-related protein-(107-139) inhibits bone resorption in vivo. *Endocrinology*, *138*(3), 1299–1304. https://doi.org/10.1210/endo.138.3.4990.

de Wit, S., & Cleton, F. J. (1994). Hypercalcemia in patients with breast cancer: A survival study. *Journal of Cancer Research and Clinical Oncology*, *120*(10), 610–614. https://doi.org/10.1007/BF01212816.

Dean, T., Vilardaga, J. P., Potts, J. T., Jr., & Gardella, T. J. (2008). Altered selectivity of parathyroid hormone (PTH) and PTH-related protein (PTHrP) for distinct conformations of the PTH/PTHrP receptor. *Molecular Endocrinology (Baltimore, Md.)*, *22*(1), 156–166. https://doi.org/10.1210/me.2007-0274.

Deftos, L. J. (2000). Prostate carcinoma: Production of bioactive factors. *Cancer*, *88*(12 Suppl), 3002–3008. https://doi.org/10.1002/1097-0142(20000615)88:12+<3002::aid-cncr16>3.3.co;2-t.

Deftos, L. J., Gazdar, A. F., Ikeda, K., & Broadus, A. E. (1989). The parathyroid hormone-related protein associated with malignancy is secreted by neuroendocrine tumors. *Molecular Endocrinology (Baltimore, Md.)*, *3*(3), 503–508. https://doi.org/10.1210/mend-3-3-503.

Deftos, L. J., Hogue-Angeletti, R., Chalberg, C., & Tu, S. (1989). PTHrP secretion is stimulated by CT and inhibited by CgA peptides. *Endocrinology*, *125*(1), 563–565. https://doi.org/10.1210/endo-125-1-563.

Diercks, D. B., Shumaik, G. M., Harrigan, R. A., Brady, W. J., & Chan, T. C. (2004). Electrocardiographic manifestations: Electrolyte abnormalities. *The Journal of Emergency Medicine*, *27*(2), 153–160. https://doi.org/10.1016/j.jemermed.2004.04.006.

Dodwell, D. J., Abbas, S. K., Morton, A. R., & Howell, A. (1991). Parathyroid hormone-related protein(50-69) and response to pamidronate therapy for tumour-induced hypercalcaemia. *European Journal of Cancer (Oxford, England: 1990)*, *27*(12), 1629–1633. https://doi.org/10.1016/0277-5379(91)90431-c.

Donovan, P. J., Achong, N., Griffin, K., Galligan, J., Pretorius, C. J., & McLeod, D. S. (2015). PTHrP-mediated hypercalcemia: causes and survival in 138 patients. *The Journal of Clinical Endocrinology and Metabolism*, *100*(5), 2024–2029. https://doi.org/10.1210/jc.2014-4250.

Dougherty, K. M., Blomme, E. A., Koh, A. J., Henderson, J. E., Pienta, K. J., Rosol, T. J., et al. (1999). Parathyroid hormone-related protein as a growth regulator of prostate carcinoma. *Cancer Research*, *59*(23), 6015–6022.

Downs, T. M., Burton, D. W., Araiza, F. L., Hastings, R. H., & Deftos, L. J. (2011). PTHrP stimulates prostate cancer cell growth and upregulates aldo-keto reductase 1C3. *Cancer Letters*, *306*(1), 52–59. https://doi.org/10.1016/j.canlet.2011.02.027.

Edwards, C. M., & Silberman, R. (2018). Myeloma bone disease and other hematological malignancies. In J. P. Bilezikian (Ed.), *Primer on the metabolic bone diseases and disorders of mineral metabolism* (9th ed., p. 760). Hoboken, NJ: Wiley-Blackwell.

El Abdaimi, K., Papavasiliou, V., Goltzman, D., & Kremer, R. (2000). Expression and regulation of parathyroid hormone-related peptide in normal and malignant melanocytes. *American Journal of Physiology. Cell Physiology*, *279*(4), C1230–C1238. https://doi.org/10.1152/ajpcell.2000.279.4.C1230.

Favus, M. J., & Goltzman, D. (2013). Regulation of calcium and magnesium. In C. J. Rosen (Ed.), *Primer on the metabolic bone diseases and disorders of mineral metabolism* (8th ed., pp. 173–179). John Wiley & Sons, Inc.

Fenton, A. J., Kemp, B. E., Hammonds, R. G., Jr., Mitchelhill, K., Moseley, J. M., Martin, T. J., et al. (1991). A potent inhibitor of osteoclastic bone resorption within a highly conserved pentapeptide region of parathyroid hormone-related protein; PTHrP[107-111]. *Endocrinology*, *129*(6), 3424–3426. https://doi.org/10.1210/endo-129-6-3424.

Ferrandon, S., Feinstein, T. N., Castro, M., Wang, B., Bouley, R., Potts, J. T., et al. (2009). Sustained cyclic AMP production by parathyroid hormone receptor endocytosis. *Nature Chemical Biology*, *5*(10), 734–742. https://doi.org/10.1038/nchembio.206.

Ferrari, S. L., Rizzoli, R., & Bonjour, J. P. (1994). Effects of epidermal growth factor on parathyroid hormone-related protein production by mammary epithelial cells. *Journal of Bone and Mineral Research: the Official Journal of the American Society for Bone and Mineral Research*, *9*(5), 639–644. https://doi.org/10.1002/jbmr.5650090508.

Fraher, L. J., Hodsman, A. B., Jonas, K., Saunders, D., Rose, C. I., Henderson, J. E., et al. (1992). A comparison of the in vivo biochemical responses to exogenous parathyroid hormone-(1-34) [PTH-(1-34)] and PTH-related peptide-(1-34) in man. *The Journal of Clinical Endocrinology and Metabolism*, *75*(2), 417–423. https://doi.org/10.1210/jcem.75.2.1322424.

Fukayama, S., Tashjian, A. H., Jr., Davis, J. N., & Chisholm, J. C. (1995). Signaling by N- and C-terminal sequences of parathyroid hormone-related protein in hippocampal neurons. *Proceedings of the National Academy of Sciences of the United States of America, 92*(22), 10182–10186. https://doi.org/10.1073/pnas.92.22.10182.

Galson, D. L., Silbermann, R., & Roodman, G. D. (2012). Mechanisms of multiple myeloma bone disease. *BoneKEy Reports, 1*, 135. https://doi.org/10.1038/bonekey.2012.135.

Gardella, T. J., Jüppner, H., Wilson, A. K., Keutmann, H. T., Abou-Samra, A. B., Segre, G. V., et al. (1994). Determinants of [Arg2]PTH-(1-34) binding and signaling in the transmembrane region of the parathyroid hormone receptor. *Endocrinology, 135*(3), 1186–1194. https://doi.org/10.1210/endo.135.3.8070362.

Gastanaga, V. M., Schwartzberg, L. S., Jain, R. K., Pirolli, M., Quach, D., Quigley, J. M., et al. (2016). Prevalence of hypercalcemia among cancer patients in the United States. *Cancer Medicine, 5*(8), 2091–2100. https://doi.org/10.1002/cam4.749.

Ghoussaini, M., Fletcher, O., Michailidou, K., Turnbull, C., Schmidt, M. K., Dicks, E., et al. (2012). Genome-wide association analysis identifies three new breast cancer susceptibility loci. *Nature Genetics, 44*(3), 312–318. https://doi.org/10.1038/ng.1049.

Goltzman, D., Hendy, G. N., & Banville, D. (1989). Parathyroid hormone-like peptide: Molecular characterization and biological properties. *Trends in Endocrinology and Metabolism: TEM, 1*(1), 39–44. https://doi.org/10.1016/1043-2760(89)90029-5.

Goltzman, D., Stewart, A. F., & Broadus, A. E. (1981). Malignancy-associated hypercalcemia: Evaluation with a cytochemical bioassay for parathyroid hormone. *The Journal of Clinical Endocrinology and Metabolism, 53*(5), 899–904. https://doi.org/10.1210/jcem-53-5-899.

Grill, V., Ho, P., Body, J. J., Johanson, N., Lee, S. C., Kukreja, S. C., et al. (1991). Parathyroid hormone-related protein: Elevated levels in both humoral hypercalcemia of malignancy and hypercalcemia complicating metastatic breast cancer. *The Journal of Clinical Endocrinology and Metabolism, 73*(6), 1309–1315. https://doi.org/10.1210/jcem-73-6-1309.

Grill, V., Murray, R. M., Ho, P. W., Santamaria, J. D., Pitt, P., Potts, C., et al. (1992). Circulating PTH and PTHrP levels before and after treatment of tumor induced hypercalcemia with Pamidronate Disodium (APD). *The Journal of Clinical Endocrinology and Metabolism, 74*(6), 1468–1470. https://doi.org/10.1210/jcem.74.6.1592895.

Guise, T. A., Yin, J. J., Taylor, S. D., Kumagai, Y., Dallas, M., Boyce, B. F., et al. (1996). Evidence for a causal role of parathyroid hormone-related protein in the pathogenesis of human breast cancer-mediated osteolysis. *The Journal of Clinical Investigation, 98*(7), 1544–1549. https://doi.org/10.1172/JCI118947.

Guise, T. A., Yin, J. J., Thomas, R. J., Dallas, M., Cui, Y., & Gillespie, M. T. (2002). Parathyroid hormone-related protein (PTHrP)-(1-139) isoform is efficiently secreted in vitro and enhances breast cancer metastasis to bone in vivo. *Bone, 30*(5), 670–676. https://doi.org/10.1016/s8756-3282(02)00685-3.

Gurney, H., Grill, V., & Martin, T. J. (1993). Parathyroid hormone-related protein and response to pamidronate in tumour-induced hypercalcaemia. *Lancet (London, England), 341*(8861), 1611–1613. https://doi.org/10.1016/0140-6736(93)90756-7.

Guy, C. T., Cardiff, R. D., & Muller, W. J. (1992). Induction of mammary tumors by expression of polyomavirus middle T oncogene: A transgenic mouse model for metastatic disease. *Molecular and Cellular Biology, 12*(3), 954–961. https://doi.org/10.1128/mcb.12.3.954-961.1992.

Hanahan, D., & Weinberg, R. A. (2011). Hallmarks of cancer: The next generation. *Cell, 144*(5), 646–674. https://doi.org/10.1016/j.cell.2011.02.013.

Hansen, J. A., Naghavi-Behzad, M., Gerke, O., Baun, C., Falch, K., Duvnjak, S., et al. (2021). Diagnosis of bone metastases in breast cancer: Lesion-based sensitivity of dual-time-point FDG-PET/CT compared to low-dose CT and bone scintigraphy. *PLoS One, 16*(11), e0260066. https://doi.org/10.1371/journal.pone.0260066.

Hashimoto, H., Azuma, Y., Kawasaki, M., Fujihara, H., Onuma, E., Yamada-Okabe, H., et al. (2007). Parathyroid hormone-related protein induces cachectic syndromes without directly modulating the expression of hypothalamic feeding-regulating peptides. *Clinical Cancer Research: An Official Journal of the American Association for Cancer Research, 13*(1), 292–298. https://doi.org/10.1158/1078-0432.CCR-06-1487.

Henderson, J. E., Amizuka, N., Warshawsky, H., Biasotto, D., Lanske, B. M., Goltzman, D., et al. (1995). Nucleolar localization of parathyroid hormone-related peptide enhances survival of chondrocytes under conditions that promote apoptotic cell death. *Molecular and Cellular Biology, 15*(8), 4064–4075. https://doi.org/10.1128/MCB.15.8.4064.

Henderson, J., Bernier, S., D'Amour, P., & Goltzman, D. (1990). Effects of passive immunization against parathyroid hormone (PTH)-like peptide and PTH in hypercalcemic tumor-bearing rats and normocalcemic controls. *Endocrinology, 127*(3), 1310–1318. https://doi.org/10.1210/endo-127-3-1310.

Henderson, M., Danks, J., Moseley, J., Slavin, J., Harris, T., McKinlay, M., et al. (2001). Parathyroid hormone-related protein production by breast cancers, improved survival, and reduced bone metastases. *Journal of the National Cancer Institute, 93*(3), 234–237. https://doi.org/10.1093/jnci/93.3.234.

Henderson, M. A., Danks, J. A., Slavin, J. L., Byrnes, G. B., Choong, P. F., Spillane, J. B., et al. (2006). Parathyroid hormone-related protein localization in breast cancers predict improved prognosis. *Cancer Research, 66*(4), 2250–2256. https://doi.org/10.1158/0008-5472.CAN-05-2814.

Henderson, S. E., Makhijani, N., & Mace, T. A. (2018). Pancreatic cancer-induced cachexia and relevant mouse models. *Pancreas, 47*(8), 937–945. https://doi.org/10.1097/MPA.0000000000001124.

Henderson, J., Sebag, M., Rhim, J., Goltzman, D., & Kremer, R. (1991). Dysregulation of parathyroid hormone-like peptide expression and secretion in a keratinocyte model of tumor progression. *Cancer Research, 51*(24), 6521–6528.

Henderson, J. E., Shustik, C., Kremer, R., Rabbani, S. A., Hendy, G. N., & Goltzman, D. (1990). Circulating concentrations of parathyroid hormone-like peptide in malignancy and in hyperparathyroidism. *Journal of Bone and Mineral Research: the Official Journal of the American Society for Bone and Mineral Research, 5*(2), 105–113. https://doi.org/10.1002/jbmr.5650050203.

Hidaka, N., Nishimura, M., & Nagao, K. (1998). Establishment of two human small cell lung cancer cell lines: the evidence of accelerated production of parathyroid hormone-related protein with tumor progression. *Cancer Letters, 125*(1-2), 149–155. https://doi.org/10.1016/s0304-3835(97)00503-x.

Hiraki, A., Ueoka, H., Bessho, A., Segawa, Y., Takigawa, N., Kiura, K., et al. (2002). Parathyroid hormone-related protein measured at the time of first visit is an indicator of bone metastases and survival in lung carcinoma patients with hypercalcemia. *Cancer, 95*(8), 1706–1713. https://doi.org/10.1002/cncr.10828.

Hirbe, A. C., Morgan, E. A., & Weilbaecher, K. N. (2010). The CXCR4/SDF-1 chemokine axis: A potential therapeutic target for bone metastases? *Current Pharmaceutical Design, 16*(11), 1284–1290. https://doi.org/10.2174/138161210791034012.

Holick, M. F. (1999). Vitamin D: Photobiology, metabolism, machanism of action and clinical applications. In M. J. Favus (Ed.), *Primer on the metabolic bone diseases and disorders of mineral metabolism* (pp. 92–98). Baltimore: Lippincott Williams and Wilkins.

Hong, N., Yoon, H. J., Lee, Y. H., Kim, H. R., Lee, B. W., Rhee, Y., et al. (2016). Serum PTHrP predicts weight loss in cancer patients independent of hypercalcemia, inflammation, and tumor burden. *The Journal of Clinical Endocrinology and Metabolism, 101*(3), 1207–1214. https://doi.org/10.1210/jc.2015-3785.

Hongo, T., Kupfer, J., Enomoto, H., Sharifi, B., Giannella-Neto, D., Forrester, J. S., et al. (1991). Abundant expression of parathyroid hormone-related protein in primary rat aortic smooth muscle cells accompanies serum-induced proliferation. *The Journal of Clinical Investigation*, *88*(6), 1841–1847. https://doi.org/10.1172/JCI115505.

Horwitz, M. J. (2018). Non-parathyroid hypercalcemia. In J. P. Bilezikian (Ed.), *Primer on the metabolic bone diseases and disorders of mineral metabolism* (9th ed., p. 639). Hoboken, NJ: Wiley-Blackwell.

Horwitz, M. J., & Stewart, A. F. (2003). Humoral hypercalcemia of malignancy. In M. J. Favus (Ed.), *Primer on the metabolic bone diseases and disorders of mineral metabolism* (5th ed., pp. 246–250). Philadelphia, PA: Lippincott Williams & Wilkins.

Horwitz, M. J., Tedesco, M. B., Sereika, S. M., Syed, M. A., Garcia-Ocaña, A., Bisello, A., et al. (2005). Continuous PTH and PTHrP infusion causes suppression of bone formation and discordant effects on 1,25(OH)2 vitamin D. *Journal of Bone and Mineral Research: the Official Journal of the American Society for Bone and Mineral Research*, *20*(10), 1792–1803. https://doi.org/10.1359/JBMR.050602.

Huang, D. C., Yang, X. F., Ochietti, B., Fadhil, I., Camirand, A., & Kremer, R. (2014). Parathyroid hormone-related protein: Potential therapeutic target for melanoma invasion and metastasis. *Endocrinology*, *155*(10), 3739–3749. https://doi.org/10.1210/en.2013-1803.

Iguchi, H., Onuma, E., Sato, K., Sato, K., & Ogata, E. (2001). Involvement of parathyroid hormone-related protein in experimental cachexia induced by a human lung cancer-derived cell line established from a bone metastasis specimen. *International Journal of Cancer*, *94*(1), 24–27. https://doi.org/10.1002/ijc.1425.

Iguchi, H., Tanaka, S., Ozawa, Y., Kashiwakuma, T., Kimura, T., Hiraga, T., et al. (1996). An experimental model of bone metastasis by human lung cancer cells: The role of parathyroid hormone-related protein in bone metastasis. *Cancer Research*, *56*(17), 4040–4043.

Ikeda, K., Inoue, D., Okazaki, R., Kikuchi, T., Ogata, E., & Matsumoto, T. (1995). Parathyroid hormone-related peptide in hypercalcemia associated with adult T cell leukemia/lymphoma: Molecular and cellular mechanism of parathyroid hormone-related peptide overexpression in HTLV-I-infected T cells. *Mineral and Electrolyte Metabolism*, *21*(1-3), 166–170.

Ikeda, K., Ohno, H., Hane, M., Yokoi, H., Okada, M., Honma, T., et al. (1994). Development of a sensitive two-site immunoradiometric assay for parathyroid et al., hormone-related peptide: evidence for elevated levels in plasma from patients with adult T-cell leukemia/lymphoma and B-cell lymphoma. *The Journal of Clinical Endocrinology and Metabolism*, *79*(5), 1322–1327. https://doi.org/10.1210/jcem.79.5.7962324.

Imamura, H., Sato, K., Shizume, K., Satoh, T., Kasono, K., Ozawa, M., et al. (1991). Urinary excretion of parathyroid hormone-related protein fragments in patients with humoral hypercalcemia of malignancy and hypercalcemic tumor-bearing nude mice. *Journal of Bone and Mineral Research: the Official Journal of the American Society for Bone and Mineral Research*, *6*(1), 77–84. https://doi.org/10.1002/jbmr.5650060113.

Inzucchi, S. E. (2004). Understanding hypercalcemia. Its metabolic basis, signs, and symptoms. *Postgraduate Medicine*, *115*(4), 69–76. https://doi.org/10.3810/pgm.2004.04.1486.

Iwamura, M., Wu, W., Muramoto, M., Ohori, M., Egawa, S., Uchida, T., et al. (1999). Parathyroid hormone-related protein is an independent prognostic factor for renal cell carcinoma. *Cancer*, *86*(6), 1028–1034. https://doi.org/10.1002/(sici)1097-0142(19990915)86:6<1028::aid-cncr19>3.0.co;2-g.

Jacobson, J. O., Bringhurst, F. R., Harris, N. L., Weitzman, S. A., & Aisenberg, A. C. (1989). Humoral hypercalcemia in Hodgkin's disease. *Clinical and Laboratory Evaluation. Cancer*, *63*(5), 917–923. https://doi.org/10.1002/1097-0142(19890301)63:5<917::aid-cncr2820630522>3.0.co;2-e.

Johnson, R. W., Nguyen, M. P., Padalecki, S. S., Grubbs, B. G., Merkel, A. R., Oyajobi, B. O., et al. (2011). TGF-beta promotion of Gli2-induced expression of parathyroid hormone-related protein, an important osteolytic factor in bone metastasis, is independent of canonical Hedgehog signaling. *Cancer Research*, *71*(3), 822–831. https://doi.org/10.1158/0008-5472.CAN-10-2993.

Jüppner, H., Abou-Samra, A. B., Freeman, M., Kong, X. F., Schipani, E., Richards, J., et al. (1991). A G protein-linked receptor for parathyroid hormone and parathyroid hormone-related peptide. *Science (New York, N.Y.)*, *254*(5034), 1024–1026. https://doi.org/10.1126/science.1658941.

Kakonen, S. M., Selander, K. S., Chirgwin, J. M., Yin, J. J., Burns, S., Rankin, W. A., et al. (2002). Transforming growth factor-beta stimulates parathyroid hormone-related protein and osteolytic metastases via Smad and mitogen-activated protein kinase signaling pathways. *The Journal of Biological Chemistry*, *277*(27), 24571–24578. https://doi.org/10.1074/jbc.M202561200.

Kao, P. C., Klee, G. G., Taylor, R. L., & Heath, H., 3rd. (1990). Parathyroid hormone-related peptide in plasma of patients with hypercalcemia and malignant lesions. *Mayo Clinic Proceedings*, *65*(11), 1399–1407. https://doi.org/10.1016/s0025-6196(12)62163-6.

Keusch, I., Traebert, M., Lötscher, M., Kaissling, B., Murer, H., & Biber, J. (1998). Parathyroid hormone and dietary phosphate provoke a lysosomal routing of the proximal tubular Na/Pi-cotransporter type II. *Kidney International*, *54*(4), 1224–1232. https://doi.org/10.1046/j.1523-1755.1998.00115.x.

Kim, W., Takyar, F. M., Swan, K., Jeong, J., VanHouten, J., Sullivan, C., et al. (2016). Calcium-sensing receptor promotes breast cancer by stimulating intracrine actions of parathyroid hormone-related protein. *Cancer Research*, *76*(18), 5348–5360. https://doi.org/10.1158/0008-5472.CAN-15-2614.

Kiriyama, T., Gillespie, M. T., Glatz, J. A., Fukumoto, S., Moseley, J. M., & Martin, T. J. (1993). Transforming growth factor beta stimulation of parathyroid hormone-related protein (PTHrP): A paracrine regulator? *Molecular and Cellular Endocrinology*, *92*(1), 55–62. https://doi.org/10.1016/0303-7207(93)90074-t.

Kovacs, C. S., Lanske, B., Hunzelman, J. L., Guo, J., Karaplis, A. C., & Kronenberg, H. M. (1996). Parathyroid hormone-related peptide (PTHrP) regulates fetal-placental calcium transport through a receptor distinct from the PTH/PTHrP receptor. *Proceedings of the National Academy of Sciences of the United States of America*, *93*(26), 15233–15238. https://doi.org/10.1073/pnas.93.26.15233.

Kremer, R., Goltzman, D., Amizuka, N., Webber, M. M., & Rhim, J. S. (1997). ras Activation of human prostate epithelial cells induces overexpression of parathyroid hormone-related peptide. *Clinical Cancer Research: An Official Journal of the American Association for Cancer Research*, *3*(6), 855–859.

Kremer, R., Karaplis, A. C., Henderson, J., Gulliver, W., Banville, D., Hendy, G. N., et al. (1991). Regulation of parathyroid hormone-like peptide in cultured normal human keratinocytes. Effect of growth factors and 1,25 dihydroxyvitamin D3 on gene expression and secretion. *The Journal of Clinical Investigation*, *87*(3), 884–893. https://doi.org/10.1172/JCI115094.

Kremer, R., Sebag, M., Champigny, C., Meerovitch, K., Hendy, G. N., White, J., et al. (1996). Identification and characterization of 1,25-dihydroxyvitamin D3-responsive repressor sequences in the rat parathyroid hormone-related peptide gene. *The Journal of Biological Chemistry*, *271*(27), 16310–16316. https://doi.org/10.1074/jbc.271.27.16310.

Kremer, R., Shustik, C., Tabak, T., Papavasiliou, V., & Goltzman, D. (1996). Parathyroid-hormone-related peptide in hematologic malignancies. *The American Journal of Medicine*, *100*(4), 406–411. https://doi.org/10.1016/S0002-9343(97)89515-0.

Kremer, R., Woodworth, C. D., & Goltzman, D. (1996). Expression and action of parathyroid hormone-related peptide in human cervical epithelial cells. *The American Journal of Physiology*, *271*(1 Pt. 1), C164–C171. https://doi.org/10.1152/ajpcell.1996.271.1.C164.

Kritmetapak, K., & Pongchaiyakul, C. (2019). Parathyroid hormone measurement in chronic kidney disease: from basics to clinical implications. *International Journal of Nephrology, 2019*, 5496710. https://doi.org/10.1155/2019/5496710.

Kukreja, S. C., Shevrin, D. H., Wimbiscus, S. A., Ebeling, P. R., Danks, J. A., Rodda, C. P., et al. (1988). Antibodies to parathyroid hormone-related protein lower serum calcium in athymic mouse models of malignancy-associated hypercalcemia due to human tumors. *The Journal of Clinical Investigation, 82*(5), 1798–1802. https://doi.org/10.1172/JCI113794.

Kushnir, M. M., Rockwood, A. L., Strathmann, F. G., Frank, E. L., Straseski, J. A., & Meikle, A. W. (2016). LC-MS/MS measurement of parathyroid hormone-related peptide. *Clinical Chemistry, 62*(1), 218–226. https://doi.org/10.1373/clinchem.2015.244012.

Lafferty, F. W. (1966). Pseudohyperparathyroidism. *Medicine, 45*(3), 247–260. https://doi.org/10.1097/00005792-196605000-00004.

Lee, C., Gardella, T. J., Abou-Samra, A. B., Nussbaum, S. R., Segre, G. V., Potts, J. T., Jr., et al. (1994). Role of the extracellular regions of the parathyroid hormone (PTH)/PTH-related peptide receptor in hormone binding. *Endocrinology, 135*(4), 1488–1495. https://doi.org/10.1210/endo.135.4.7523099.

Li, J., Camirand, A., Zakikhani, M., Sellin, K., Guo, Y., Luan, X. R., et al. (2022). Parathyroid hormone-related protein inhibition blocks triple-negative breast cancer expansion in bone through epithelial to mesenchymal transition reversal. *JBMR Plus, 6*(6), e10587. https://doi.org/10.1002/jbm4.10587.

Li, J., Karaplis, A. C., Huang, D. C., Siegel, P. M., Camirand, A., Yang, X. F., et al. (2011). PTHrP drives breast tumor initiation, progression, and metastasis in mice and is a potential therapy target. *The Journal of Clinical Investigation, 121*(12), 4655–4669. https://doi.org/10.1172/JCI46134.

Ling, P. J., A'Hern, R. P., & Hardy, J. R. (1995). Analysis of survival following treatment of tumour-induced hypercalcaemia with intravenous pamidronate (APD). *British Journal of Cancer, 72*(1), 206–209. https://doi.org/10.1038/bjc.1995.304.

Lins, L. E. (1978). Reversible renal failure caused by hypercalcemia. A retrospective study. *Acta Medica Scandinavica, 203*(4), 309–314. https://doi.org/10.1111/j.0954-6820.1978.tb14879.x.

Liu, B., Amizuka, N., Goltzman, D., & Rabbani, S. A. (1995). Inhibition of processing of parathyroid hormone-related peptide by anti-sense furin: Effect in vitro and in vivo on rat Leydig (H-500) tumor cells. *International Journal of Cancer, 63*(2), 276–281. https://doi.org/10.1002/ijc.2910630222.

Liu, B., Goltzman, D., & Rabbani, S. A. (1993). Regulation of parathyroid hormone-related peptide production in vitro by the rat hypercalcemic Leydig cell tumor H-500. *Endocrinology, 132*(4), 1658–1664. https://doi.org/10.1210/endo.132.4.8462465.

Liu, B., Goltzman, D., & Rabbani, S. A. (1995). Processing of pro-PTHRP by the prohormone convertase, furin: Effect on biological activity. *The American Journal of Physiology, 268*(5 Pt. 1), E832–E838. https://doi.org/10.1152/ajpendo.1995.268.5.E832.

Lu, C. M., Burton, W. D., Fitzgerald, R. L., Deftos, L. J., Buchholz, B. A., Vogel, J. S., et al. (2002). Mass spectrometric immunoassay for parathyroid hormone-related protein. *Analytical Chemistry, 74*(21), 5507–5512. https://doi.org/10.1021/ac020182a.

Luco, A., Li, J., Ochietti, B., Camirand, A., Karaplis, A., & Kremer, R. (2013). Parathyroid hormone-related peptide (PTHrP) blockade inhibits the development of bone metastasis and potentiates the effect of zoledronic acid in vitro and in vivo in a mouse model of breast tumor progression. In *Proceedings, annual meeting, American Society for Bone and Mineral Research, Baltimore, MD.*

Maioli, E., & Fortino, V. (2004). The complexity of parathyroid hormone-related protein signalling. *Cellular and Molecular Life Sciences: CMLS*, *61*(3), 257–262. https://doi.org/10.1007/s00018-003-3233-2.

Mak, I. W., Cowan, R. W., Turcotte, R. E., Singh, G., & Ghert, M. (2011). PTHrP induces autocrine/paracrine proliferation of bone tumor cells through inhibition of apoptosis. *PLoS One*, *6*(5), e19975. https://doi.org/10.1371/journal.pone.0019975.

Mangin, M., Ikeda, K., Dreyer, B. E., & Broadus, A. E. (1989). Isolation and characterization of the human parathyroid hormone-like peptide gene. *Proceedings of the National Academy of Sciences of the United States of America*, *86*(7), 2408–2412. https://doi.org/10.1073/pnas.86.7.2408.

Mannstadt, M., Jüppner, H., & Gardella, T. J. (1999). Receptors for PTH and PTHrP: Their biological importance and functional properties. *The American Journal of Physiology*, *277*(5), F665–F675. https://doi.org/10.1152/ajprenal.1999.277.5.F665.

Massfelder, T., Lang, H., Schordan, E., Lindner, V., Rothhut, S., Welsch, S., et al. (2004). Parathyroid hormone-related protein is an essential growth factor for human clear cell renal carcinoma and a target for the von Hippel-Lindau tumor suppressor gene. *Cancer Research*, *64*(1), 180–188. https://doi.org/10.1158/0008-5472.can-03-1968.

Mawer, E. B., Hayes, M. E., Heys, S. E., Davies, M., White, A., Stewart, M. F., et al. (1994). Constitutive synthesis of 1,25-dihydroxyvitamin D3 by a human small cell lung cancer cell line. *The Journal of Clinical Endocrinology and Metabolism*, *79*(2), 554–560. https://doi.org/10.1210/jcem.79.2.8045976.

Miao, D., Su, H., He, B., Gao, J., Xia, Q., Zhu, M., et al. (2008). Severe growth retardation and early lethality in mice lacking the nuclear localization sequence and C-terminus of PTH-related protein. *Proceedings of the National Academy of Sciences of the United States of America*, *105*(51), 20309–20314. https://doi.org/10.1073/pnas.0805690105.

Miki, T., Yano, S., Hanibuchi, M., Kanematsu, T., Muguruma, H., & Sone, S. (2004). Parathyroid hormone-related protein (PTHrP) is responsible for production of bone metastasis, but not visceral metastasis, by human small cell lung cancer SBC-5 cells in natural killer cell-depleted SCID mice. *International Journal of Cancer*, *108*(4), 511–515. https://doi.org/10.1002/ijc.11586.

Moseley, J. M., Kubota, M., Diefenbach-Jagger, H., Wettenhall, R. E., Kemp, B. E., Suva, L. J., et al. (1987). Parathyroid hormone-related protein purified from a human lung cancer cell line. *Proceedings of the National Academy of Sciences of the United States of America*, *84*(14), 5048–5052. https://doi.org/10.1073/pnas.84.14.5048.

Mula, R. V., Bhatia, V., & Falzon, M. (2010). PTHrP promotes colon cancer cell migration and invasion in an integrin α6β4-dependent manner through activation of Rac1. *Cancer Letters*, *298*(1), 119–127. https://doi.org/10.1016/j.canlet.2010.06.009.

Mundy, G. R. (2002). Metastasis to bone: Causes, consequences and therapeutic opportunities. *Nature Reviews. Cancer*, *2*(8), 584–593. https://doi.org/10.1038/nrc867.

Nagai, M., Kyakumoto, S., & Sato, N. (2000). Cancer cells responsible for humoral hypercalcemia express mRNA encoding a secreted form of ODF/TRANCE that induces osteoclast formation. *Biochemical and Biophysical Research Communications*, *269*(2), 532–536. https://doi.org/10.1006/bbrc.2000.2314.

Nguyen, M., He, B., & Karaplis, A. (2001). Nuclear forms of parathyroid hormone-related peptide are translated from non-AUG start sites downstream from the initiator methionine. *Endocrinology*, *142*(2), 694–703. https://doi.org/10.1210/endo.142.2.7944.

Nishi, S. P., Barbagelata, N. A., Atar, S., Birnbaum, Y., & Tuero, E. (2006). Hypercalcemia-induced ST-segment elevation mimicking acute myocardial infarction. *Journal of Electrocardiology*, *39*(3), 298–300. https://doi.org/10.1016/j.jelectrocard.2005.10.015.

Nishihara, M., Ito, M., Tomioka, T., Ohtsuru, A., Taguchi, T., & Kanematsu, T. (1999). Clinicopathological implications of parathyroid hormone-related protein in human colorectal tumours. *The Journal of Pathology*, *187*(2), 217–222. https://doi.org/10.1002/(SICI)1096-9896(199901)187:2<217::AID-PATH210>3.0.CO;2-0.

Nissenson, R. A., Abbott, S. R., Teitelbaum, A. P., Clark, O. H., & Arnaud, C. D. (1981). Endogenous biologically active human parathyroid hormone: measurement by a guanyl nucleotide-amplified renal adenylate cyclase assay. *The Journal of Clinical Endocrinology and Metabolism*, *52*(5), 840–846. https://doi.org/10.1210/jcem-52-5-840.

Nussbaum, S. R., Gaz, R. D., & Arnold, A. (1990). Hypercalcemia and ectopic secretion of parathyroid hormone by an ovarian carcinoma with rearrangement of the gene for parathyroid hormone. *The New England Journal of Medicine*, *323*(19), 1324–1328. https://doi.org/10.1056/NEJM199011083231907.

Onuma, E., Azuma, Y., Saito, H., Tsunenari, T., Watanabe, T., Hirabayashi, M., et al. (2005). Increased renal calcium reabsorption by parathyroid hormone-related protein is a causative factor in the development of humoral hypercalcemia of malignancy refractory to osteoclastic bone resorption inhibitors. *Clinical Cancer Research: An Official Journal of the American Association for Cancer Research*, *11*(11), 4198–4203. https://doi.org/10.1158/1078-0432.CCR-04-2531.

Orloff, J. J., Ganz, M. B., Nathanson, M. H., Moyer, M. S., Kats, Y., Mitnick, M., et al. (1996). A midregion parathyroid hormone-related peptide mobilizes cytosolic calcium and stimulates formation of inositol trisphosphate in a squamous carcinoma cell line. *Endocrinology*, *137*(12), 5376–5385. https://doi.org/10.1210/endo.137.12.8940360.

Orloff, J. J., & Stewart, A. F. (1995). The carboxy-terminus of parathyroid hormone—Inert or invaluable? *Endocrinology*, *136*(11), 4729–4731. https://doi.org/10.1210/endo.136.11.7588199.

Otieno, B. A., Krause, C. E., Jones, A. L., Kremer, R. B., & Rusling, J. F. (2016). Cancer diagnostics via ultrasensitive multiplexed detection of parathyroid hormone-related peptides with a microfluidic immunoarray. *Analytical Chemistry*, *88*(18), 9269–9275. https://doi.org/10.1021/acs.analchem.6b02637.

Paget, S. (1989). The distribution of secondary growths in cancer of the breast. *Cancer Metastasis Reviews*, *8*(2), 98–101.

Pearse, R. N., Sordillo, E. M., Yaccoby, S., Wong, B. R., Liau, D. F., Colman, N., et al. (2001). Multiple myeloma disrupts the TRANCE/osteoprotegerin cytokine axis to trigger bone destruction and promote tumor progression. *Proceedings of the National Academy of Sciences of the United States of America*, *98*(20), 11581–11586. https://doi.org/10.1073/pnas.201394498.

Pecherstorfer, M., Schilling, T., Blind, E., Zimmer-Roth, I., Baumgartner, G., Ziegler, R., et al. (1994). Parathyroid hormone-related protein and life expectancy in hypercalcemic cancer patients. *The Journal of Clinical Endocrinology and Metabolism*, *78*(5), 1268–1270. https://doi.org/10.1210/jcem.78.5.8175989.

Pelosof, L. C., & Gerber, D. E. (2010). Paraneoplastic syndromes: An approach to diagnosis and treatment. *Mayo Clinic Proceedings*, *85*(9), 838–854. https://doi.org/10.4065/mcp.2010.0099.

Pitarresi, J. R., Norgard, R. J., Chiarella, A. M., Suzuki, K., Bakir, B., Sahu, V., et al. (2021). PTHrP drives pancreatic cancer growth and metastasis and reveals a new therapeutic vulnerability. *Cancer Discovery*, *11*(7), 1774–1791. https://doi.org/10.1158/2159-8290.CD-20-1098.

Powell, D., Singer, F. R., Murray, T. M., Minkin, C., & Potts, J. T., Jr. (1973). Nonparathyroid humoral hypercalcemia in patients with neoplastic diseases. *The New England Journal of Medicine*, *289*(4), 176–181. https://doi.org/10.1056/NEJM197307262890403.

Powell, G. J., Southby, J., Danks, J. A., Stillwell, R. G., Hayman, J. A., Henderson, M. A., et al. (1991). Localization of parathyroid hormone-related protein in breast cancer metastases: Increased incidence in bone compared with other sites. *Cancer Research*, *51*(11), 3059–3061.

Rabbani, S. A., Gladu, J., Harakidas, P., Jamison, B., & Goltzman, D. (1999). Over-production of parathyroid hormone-related peptide results in increased osteolytic skeletal metastasis by prostate cancer cells in vivo. *International Journal of Cancer*, *80*(2), 257–264. https://doi.org/10.1002/(sici)1097-0215(19990118)80:2<257::aid-ijc15>3.0.co;2-3.

Rabbani, S. A., Gladu, J., Liu, B., & Goltzman, D. (1995). Regulation in vivo of the growth of Leydig cell tumors by antisense ribonucleic acid for parathyroid hormone-related peptide. *Endocrinology*, *136*(12), 5416–5422. https://doi.org/10.1210/endo.136.12.7588290.

Rankin, W., Grill, V., & Martin, T. J. (1997). Parathyroid hormone-related protein and hypercalcemia. *Cancer*, *80*(8 Suppl), 1564–1571. https://doi.org/10.1002/(sici)1097-0142(19971015)80:8+<1564::aid-cncr6>3.3.co;2-h.

Ratcliffe, W. A., Norbury, S., Stott, R. A., Heath, D. A., & Ratcliffe, J. G. (1991). Immunoreactivity of plasma parathyrin-related peptide: three region-specific radioimmunoassays and a two-site immunoradiometric assay compared. *Clinical Chemistry*, *37*(10 Pt 1), 1781–1787.

Rizzoli, R., Aubert, M. L., Sappino, A. P., & Bonjour, J. P. (1994). Cyclic AMP increases the release of parathyroid hormone-related protein from a lung-cancer cell line. *International Journal of Cancer*, *56*(3), 422–426. https://doi.org/10.1002/ijc.2910560323.

Rizzoli, R., & Bonjour, J. P. (1989). High extracellular calcium increases the production of a parathyroid hormone-like activity by cultured Leydig tumor cells associated with humoral hypercalcemia. *Journal of Bone and Mineral Research: the Official Journal of the American Society for Bone and Mineral Research*, *4*(6), 839–844. https://doi.org/10.1002/jbmr.5650040608.

Rizzoli, R., Pache, J. C., Didierjean, L., Bürger, A., & Bonjour, J. P. (1994). A thymoma as a cause of true ectopic hyperparathyroidism. *The Journal of Clinical Endocrinology and Metabolism*, *79*(3), 912–915. https://doi.org/10.1210/jcem.79.3.8077382.

Rizzoli, R., Thiébaud, D., Bundred, N., Pecherstorfer, M., Herrmann, Z., Huss, H. J., et al. (1999). Serum parathyroid hormone-related protein levels and response to bisphosphonate treatment in hypercalcemia of malignancy. *The Journal of Clinical Endocrinology and Metabolism*, *84*(10), 3545–3550. https://doi.org/10.1210/jcem.84.10.6026.

Rodan, S. B., Insogna, K. L., Vignery, A. M., Stewart, A. F., Broadus, A. E., D'Souza, S. M., et al. (1983). Factors associated with humoral hypercalcemia of malignancy stimulate adenylate cyclase in osteoblastic cells. *The Journal of Clinical Investigation*, *72*(4), 1511–1515. https://doi.org/10.1172/JCI111108.

Rodan, S. B., Wesolowski, G., Ianacone, J., Thiede, M. A., & Rodan, G. A. (1989). Production of parathyroid hormone-like peptide in a human osteosarcoma cell line: Stimulation by phorbol esters and epidermal growth factor. *The Journal of Endocrinology*, *122*(1), 219–227. https://doi.org/10.1677/joe.0.1220219.

Rong, J., Wang, S., Ding, Q., Yun, M., Zheng, Z., & Ye, S. (2013). Comparison of 18 FDG PET-CT and bone scintigraphy for detection of bone metastases in breast cancer patients. A meta-analysis. *Surgical Oncology*, *22*(2), 86–91. https://doi.org/10.1016/j.suronc.2013.01.002.

Roodman, G. D. (1997). Mechanisms of bone lesions in multiple myeloma and lymphoma. *Cancer*, *80*(8 Suppl), 1557–1563. https://doi.org/10.1002/(sici)1097-0142(19971015)80:8+<1557::aid-cncr5>3.3.co;2-k.

Roodman, G. D. (2004). Mechanisms of bone metastasis. *The New England Journal of Medicine*, *350*(16), 1655–1664. https://doi.org/10.1056/NEJMra030831.

Roodman, G. D. (2010). Pathogenesis of myeloma bone disease. *Journal of Cellular Biochemistry*, *109*(2), 283–291. https://doi.org/10.1002/jcb.22403.

Rouleau, M. F., Mitchell, J., & Goltzman, D. (1990). Characterization of the major parathyroid hormone target cell in the endosteal metaphysis of rat long bones. *Journal of Bone and Mineral Research: the Official Journal of the American Society for Bone and Mineral Research*, *5*(10), 1043–1053. https://doi.org/10.1002/jbmr.5650051008.

Schweitzer, D. H., Hamdy, N. A., Frölich, M., Zwinderman, A. H., & Papapoulos, S. E. (1994). Malignancy-associated hypercalcaemia: Resolution of controversies over vitamin D metabolism by a pathophysiological approach to the syndrome. *Clinical Endocrinology*, *41*(2), 251–256. https://doi.org/10.1111/j.1365-2265.1994.tb02538.x.

Sebag, M., Henderson, J., Goltzman, D., & Kremer, R. (1994). Regulation of parathyroid hormone-related peptide production in normal human mammary epithelial cells in vitro. *The American Journal of Physiology*, *267*(3 Pt 1), C723–C730. https://doi.org/10.1152/ajpcell.1994.267.3.C723.

Seitz, P.K., Zhang, R.W., Simmons, D.J., & Cooper, C.W. (1995). Effects of C-terminal parathyroid hormone-related peptide on osteoblasts. Mineral and Electrolyte Metabolism, 21(1-3), 180–183.

Sepulveda, V. A. T., Shen, X., & Falzon, M. (2002). Intracrine PTHrP protects against serum starvation-induced apoptosis and regulates the cell cycle in MCF-7 breast cancer cells. *Endocrinology*, *143*(2), 596–606. https://doi.org/10.1210/endo.143.2.8645.

Shell, J., Keutgen, X. M., Millo, C., Nilubol, N., Patel, D., Sadowski, S., et al. (2018). 68-Gallium DOTATATE scanning in symptomatic patients with negative anatomic imaging but suspected neuroendocrine tumor. *International Journal of Endocrine oncology*, *5*(1), IJE04. https://doi.org/10.2217/ije-2017-0005.

Shen, X., & Falzon, M. (2003). PTH-related protein modulates PC-3 prostate cancer cell adhesion and integrin subunit profile. *Molecular and Cellular Endocrinology*, *199*(1-2), 165–177. https://doi.org/10.1016/s0303-7207(02)00287-3.

Simpson, E. L., Mundy, G. R., D'Souza, S. M., Ibbotson, K. J., Bockman, R., & Jacobs, J. W. (1983). Absence of parathyroid hormone messenger RNA in nonparathyroid tumors associated with hypercalcemia. *The New England Journal of Medicine*, *309*(6), 325–330. https://doi.org/10.1056/NEJM198308113090601.

Smith, J. D., Wilson, S., & Schneider, H. G. (2018). Misclassification of calcium status based on albumin-adjusted calcium: studies in a tertiary hospital setting. *Clinical Chemistry*, *64*(12), 1713–1722. https://doi.org/10.1373/clinchem.2018.291377.

Southby, J., Kissin, M. W., Danks, J. A., Hayman, J. A., Moseley, J. M., Henderson, M. A., et al. (1990). Immunohistochemical localization of parathyroid hormone-related protein in human breast cancer. *Cancer Research*, *50*(23), 7710–7716.

Sterling, J. A., Oyajobi, B. O., Grubbs, B., Padalecki, S. S., Munoz, S. A., Gupta, A., et al. (2006). The hedgehog signaling molecule Gli2 induces parathyroid hormone-related peptide expression and osteolysis in metastatic human breast cancer cells. *Cancer Research*, *66*(15), 7548–7553. https://doi.org/10.1158/0008-5472.CAN-06-0452.

Stewart, A. F. (2005). Clinical practice. Hypercalcemia associated with cancer. *The New England Journal of Medicine*, *352*(4), 373–379. https://doi.org/10.1056/NEJMcp042806.

Stewart, A. F., Horst, R., Deftos, L. J., Cadman, E. C., Lang, R., & Broadus, A. E. (1980). Biochemical evaluation of patients with cancer-associated hypercalcemia: evidence for humoral and nonhumoral groups. *The New England Journal of Medicine*, *303*(24), 1377–1383. https://doi.org/10.1056/NEJM198012113032401.

Stewart, A. F., Vignery, A., Silverglate, A., Ravin, N. D., LiVolsi, V., Broadus, A. E., et al. (1982). Quantitative bone histomorphometry in humoral hypercalcemia of malignancy: Uncoupling of bone cell activity. *The Journal of Clinical Endocrinology and Metabolism*, *55*(2), 219–227. https://doi.org/10.1210/jcem-55-2-219.

Strewler, G. J., Budayr, A. A., Clark, O. H., & Nissenson, R. A. (1993). Production of parathyroid hormone by a malignant nonparathyroid tumor in a hypercalcemic patient. *The Journal of Clinical Endocrinology and Metabolism*, *76*(5), 1373–1375. https://doi.org/10.1210/jcem.76.5.7684395.

Strewler, G. J., Stern, P. H., Jacobs, J. W., Eveloff, J., Klein, R. F., Leung, S. C., et al. (1987). Parathyroid hormonelike protein from human renal carcinoma cells. Structural and functional homology with parathyroid hormone. *The Journal of Clinical Investigation, 80*(6), 1803–1807. https://doi.org/10.1172/JCI113275.

Sullivan, F. J. (1996). Palliative radiotherapy for lung cancer. In H. I. Pass, J. B. Mitchell, D. H. Johnson, & A. T. Turrisi (Eds.), *Lung cancer: Principles and practice* (pp. 775–789). Philadelphia: Lippincott-Raven.

Suva, L. J., Winslow, G. A., Wettenhall, R. E., Hammonds, R. G., Moseley, J. M., Diefenbach-Jagger, H., et al. (1987). A parathyroid hormone-related protein implicated in malignant hypercalcemia: Cloning and expression. *Science (New York, N.Y.), 237*(4817), 893–896. https://doi.org/10.1126/science.3616618.

Terpos, E., Christoulas, D., & Gavriatopoulou, M. (2018). Biology and treatment of myeloma related bone disease. *Metabolism, Clinical and Experimental, 80*, 80–90. https://doi.org/10.1016/j.metabol.2017.11.012.

Tisdale, M. J. (2002). Cachexia in cancer patients. *Nature Reviews. Cancer, 2*(11), 862–871. https://doi.org/10.1038/nrc927.

Truong, N. U., de Edwardes, M. D. B., Papavasiliou, V., Goltzman, D., & Kremer, R. (2003). Parathyroid hormone-related peptide and survival of patients with cancer and hypercalcemia. *The American Journal of Medicine, 115*(2), 115–121. https://doi.org/10.1016/s0002-9343(03)00310-3.

Vargas, S. J., Gillespie, M. T., Powell, G. J., Southby, J., Danks, J. A., Moseley, J. M., et al. (1992). Localization of parathyroid hormone-related protein mRNA expression in breast cancer and metastatic lesions by in situ hybridization. *Journal of Bone and Mineral Research: the Official Journal of the American Society for Bone and Mineral Research*, 7(8), 971–979. https://doi.org/10.1002/jbmr.5650070814.

Walls, J., Ratcliffe, W. A., Howell, A., & Bundred, N. J. (1994). Response to intravenous bisphosphonate therapy in hypercalcaemic patients with and without bone metastases: The role of parathyroid hormone-related protein. *British Journal of Cancer, 70*(1), 169–172. https://doi.org/10.1038/bjc.1994.270.

Waning, D. L., & Guise, T. A. (2014). Molecular mechanisms of bone metastasis and associated muscle weakness. *Clinical Cancer Research: An Official Journal of the American Association for Cancer Research, 20*(12), 3071–3077. https://doi.org/10.1158/1078-0432.CCR-13-1590.

Washam, C. L., Byrum, S. D., Leitzel, K., Ali, S. M., Tackett, A. J., Gaddy, D., et al. (2013). Identification of PTHrP(12-48) as a plasma biomarker associated with breast cancer bone metastasis. *Cancer Epidemiology, Biomarkers & Prevention: A Publication of the American Association for Cancer Research, cosponsored by the American Society of Preventive Oncology, 22*(5), 972–983. https://doi.org/10.1158/1055-9965.EPI-12-1318-T.

Weinman, E. J., Hall, R. A., Friedman, P. A., Liu-Chen, L. Y., & Shenolikar, S. (2006). The association of NHERF adaptor proteins with g protein-coupled receptors and receptor tyrosine kinases. *Annual Review of Physiology, 68*, 491–505. https://doi.org/10.1146/annurev.physiol.68.040104.131050.

Wimalawansa, S. J. (1994). Significance of plasma PTH-rp in patients with hypercalcemia of malignancy treated with bisphosphonate. *Cancer, 73*(8), 2223–2230. https://doi.org/10.1002/1097-0142(19940415)73:8<2223::aid-cncr2820730831>3.0.co;2-c.

Wu, T. L., Vasavada, R. C., Yang, K., Massfelder, T., Ganz, M., Abbas, S. K., et al. (1996). Structural and physiologic characterization of the mid-region secretory species of parathyroid hormone-related protein. *The Journal of Biological Chemistry, 271*(40), 24371–24381. https://doi.org/10.1074/jbc.271.40.24371.

Wysolmerski, J. J. (2012). Parathyroid hormone-related protein: An update. *The Journal of Clinical Endocrinology and Metabolism, 97*(9), 2947–2956. https://doi.org/10.1210/jc.2012-2142.

Yamada, T., Muguruma, H., Yano, S., Ikuta, K., Ogino, H., Kakiuchi, S., et al. (2009). Intensification therapy with anti-parathyroid hormone-related protein antibody plus zoledronic acid for bone metastases of small cell lung cancer cells in severe combined immunodeficient mice. *Molecular Cancer Therapeutics*, *8*(1), 119–126. https://doi.org/10.1158/1535-7163.MCT-08-0874.

Yamato, H., Nagai, Y., Inoue, D., Ohnishi, Y., Ueyama, Y., Ohno, H., et al. (1995). In vivo evidence for progressive activation of parathyroid hormone-related peptide gene transcription with tumor growth and stimulation of osteoblastic bone formation at an early stage of humoral hypercalcemia of cancer. *Journal of Bone and Mineral Research: the Official Journal of the American Society for Bone and Mineral Research*, *10*(1), 36–44. https://doi.org/10.1002/jbmr.5650100108.

Yang, D., Singh, R., Divieti, P., Guo, J., Bouxsein, M. L., & Bringhurst, F. R. (2007). Contributions of parathyroid hormone (PTH)/PTH-related peptide receptor signaling pathways to the anabolic effect of PTH on bone. *Bone*, *40*(6), 1453–1461. https://doi.org/10.1016/j.bone.2007.02.001.

Yin, J. J., Selander, K., Chirgwin, J. M., Dallas, M., Grubbs, B. G., Wieser, R., et al. (1999). TGF-beta signaling blockade inhibits PTHrP secretion by breast cancer cells and bone metastases development. *The Journal of Clinical Investigation*, *103*(2), 197–206. https://doi.org/10.1172/JCI3523.

Yoshida, A., Nakamura, Y., Shimizu, A., Harada, M., Kameda, Y., Nagano, A., et al. (2000). Significance of the parathyroid hormone-related protein expression in breast carcinoma. *Breast Cancer (Tokyo, Japan)*, 7(3), 215–220. https://doi.org/10.1007/BF02967463.

Yoshimoto, K., Yamasaki, R., Sakai, H., Tezuka, U., Takahashi, M., Iizuka, M., et al. (1989). Ectopic production of parathyroid hormone by small cell lung cancer in a patient with hypercalcemia. *The Journal of Clinical Endocrinology and Metabolism*, *68*(5), 976–981. https://doi.org/10.1210/jcem-68-5-976.

Zhang, H. W., Ding, J., Jin, J. L., Guo, J., Liu, J. N., Karaplis, A., et al. (2010). Defects in mesenchymal stem cell self-renewal and cell fate determination lead to an osteopenic phenotype in Bmi-1 null mice. *Journal of Bone and Mineral Research: the Official Journal of the American Society for Bone and Mineral Research*, *25*(3), 640–652. https://doi.org/10.1359/jbmr.090812.

Zhang, X. H., Wang, Q., Gerald, W., Hudis, C. A., Norton, L., Smid, M., et al. (2009). Latent bone metastasis in breast cancer tied to Src-dependent survival signals. *Cancer Cell*, *16*(1), 67–78. https://doi.org/10.1016/j.ccr.2009.05.017.

Zondek, H., Petrow, H., & Siebert, W. (1923). Die bedeutung der calcium-bestimmung im blute fur die diagnose der nierrenin-siffizientz. *Z. Clin. Med.*, *99*, 129–132.

Further reading

Chilco, P. J., Gerardi, J. M., Kaczmarczyk, S. J., Chu, S., Leopold, V., & Zajac, J. D. (1993). Calcitonin increases transcription of parathyroid hormone-related protein via cAMP. *Molecular and Cellular Endocrinology*, *94*(1), 1–7. https://doi.org/10.1016/0303-7207(93)90045-l.

El Abdaimi, K., Papavasiliou, V., Rabbani, S. A., Rhim, J. S., Goltzman, D., & Kremer, R. (1999). Reversal of hypercalcemia with the vitamin D analogue EB1089 in a human model of squamous cancer. *Cancer Research*, *59*(14), 3325–3328.

Firkin, F., Seymour, J. F., Watson, A. M., Grill, V., & Martin, T. J. (1996). Parathyroid hormone-related protein in hypercalcaemia associated with haematological malignancy. *British Journal of Haematology*, *94*(3), 486–492. https://doi.org/10.1046/j.1365-2141.1996.d01-1819.x.

Haq, M., Kremer, R., Goltzman, D., & Rabbani, S. A. (1993). A vitamin D analogue (EB1089) inhibits parathyroid hormone-related peptide production and prevents the development of malignancy-associated hypercalcemia in vivo. *The Journal of Clinical Investigation, 91*(6), 2416–2422. https://doi.org/10.1172/JCI116475.

Jiang, M., Chen, G., Lu, N., Zhang, Y., Jin, S., Karaplis, A., et al. (2015). Deficiency of the parathyroid hormone-related peptide nuclear localization and carboxyl terminal sequences leads to premature skin ageing partially mediated by the upregulation of p27. *Experimental Dermatology, 24*(11), 847–852. https://doi.org/10.1111/exd.12789.

Jüppner, H., Schipani, E., Bringhurst, F. R., McClure, I., Keutmann, H. T., Potts, J. T., Jr., et al. (1994). The extracellular amino-terminal region of the parathyroid hormone (PTH)/PTH-related peptide receptor determines the binding affinity for carboxyl-terminal fragments of PTH-(1-34). *Endocrinology, 134*(2), 879–884. https://doi.org/10.1210/endo.134.2.8299582.

Ralston, S. H., Cowan, R. A., Robertson, A. G., Gardner, M. D., & Boyle, I. T. (1984). Circulating vitamin D metabolites and hypercalcaemia of malignancy. *Acta Endocrinologica, 106*(4), 556–563. https://doi.org/10.1530/acta.0.1060556.

Seymour, J. F., & Gagel, R. F. (1993). Calcitriol: The major humoral mediator of hypercalcemia in Hodgkin's disease and non-Hodgkin's lymphomas. *Blood, 82*(5), 1383–1394.

Toribio, R. E., Brown, H. A., Novince, C. M., Marlow, B., Hernon, K., Lanigan, L. G., et al. (2010). The midregion, nuclear localization sequence, and C terminus of PTHrP regulate skeletal development, hematopoiesis, and survival in mice. *FASEB Journal: Official Publication of the Federation of American Societies for Experimental Biology, 24*(6), 1947–1957. https://doi.org/10.1096/fj.09-147033.

Yu, J., Papavasiliou, V., Rhim, J., Goltzman, D., & Kremer, R. (1995). Vitamin D analogs: New therapeutic agents for the treatment of squamous cancer and its associated hypercalcemia. *Anti-Cancer Drugs, 6*(1), 101–108.

CHAPTER SEVEN

Parathyroid hormone-related protein (PTHrP)-dependent modulation of gene expression signatures in cancer cells

Claudio Luparello* and Mariangela Librizzi
Dipartimento di Scienze e Tecnologie Biologiche, Chimiche e Farmaceutiche (STEBICEF), Università di Palermo, Palermo, Italia
*Corresponding author: e-mail address: claudio.luparello@unipa.it

Contents

Abstract

PTHrP is encoded by *PTHLH* gene which can generate by alternative promoter usage and splicing mechanisms at least three mature peptides of 139, 141 and 173 amino acids with distinct carboxy terminus. PTHrP may undergo proteolytic processing into smaller bioactive forms, comprising an amino terminus peptide, which is the mediator of the "classical" PTH-like effect, as well as midregion and carboxy terminus peptides that act as multifaceted critical regulator of proliferation, differentiation and apoptosis

Vitamins and Hormones, Volume 120
ISSN 0083-6729
https://doi.org/10.1016/bs.vh.2022.03.003

via the reprogramming of gene expression in normal and neoplastic cells. Moreover, a nuclear/nucleolar localization signal sequence is present in the [87–107] domain allowing PTHrP nuclear import and "intracrine" effect additional to the autocrine/paracrine one. Within the large number of data available in the literature on PTHrP bioactivities, the goal of this chapter is to pick up selected studies that report the detection of molecular signatures of cancer cell exposure to PTHrP, either as full-length protein or discrete peptides, demonstrated by individual gene or whole genome expression profiling, briefly recapitulating the biological implications associated with the specific gene activation or silencing.

1. Parathyroid hormone-related protein (PTHrP): A brief insight into its structure and function

PTHrP is encoded by the *PTHLH* gene located on the short arm of chromosome 12, which extends on more than 15 kb of genomic DNA with its nine exons, and produces multiple mRNA variants through the utilization of three different promoters and alternative splicing mechanisms that differentiate their 3′ ends encompassing both coding and untranslated regions. As summarized in Table 1, the resulting protein products are three isoforms of 139, 141, and 173 amino acids, the latter being human-specific, with distinct carboxy-terminus (Southby, Murphy, Martin, & Gillespie, 1996; Southby, O'Keeffe, Martin, & Gillespie, 1995). As reported in seminal papers of the early 90s (e.g., Soifer et al. (1992) and Yang et al. (1994)), PTHrP isoforms are polyhormones and therefore, in addition to their secretion as

Table 1 PTHrP mRNA variants produced by alternative splicing at the 5′ and 3′ ends.

Promoter	Exons included at the 5′ end	Ever-present exons	Exons included at the 3′ end → protein isoform
P1	I–II–III	V–VI	VII → 139 aa, VIII → 173aa, IX → 141 aa
	I–III		
	I		
P2	III	V–VI	VII → 139 aa, VIII → 173aa, IX → 141 aa
P3	IV	V–VI	VII → 139 aa, VIII → 173aa, IX → 141 aa

From Longo, A., Librizzi, M., Naselli, F., Caradonna, F., Tobiasch, E., & Luparello, C. (2013). PTHrP in differentiating human mesenchymal stem cells: transcript isoform expression, promoter methylation, and protein accumulation. *Biochimie, 95*, 1888–1896. https://doi.org/10.1016/j.biochi.2013.06.014.

full-length proteins, they are also subjected to post-translational processing by diverse classes of endoproteases into smaller secretory forms. One of them, the amino-terminus [1–34] fragment, presents homology with PTH and can activate the G protein-linked receptor PTH1R thereby triggering PTH-like activities (Thompson, Seedor, Fisher, Rosenblatt, & Rodan, 1988). In addition, midregion [38–94] and [67–86] peptides have been identified influencing transplacental calcium transport and the phenotypic expressions of tumoral and immortalized breast epithelial cells, including their growth and invasive behavior (Luparello, Birch, Gallagher, & Burtis, 1997; Luparello, Burtis, Raue, Birch, & Gallagher, 1995; Luparello, Librizzi, Asaro, Cruciata, & Caradonna, 2019; Luparello et al., 2001; Luparello, Sirchia, & Lo Sasso, 2008; Sirchia & Luparello, 2007; Sirchia et al., 2007; Strid, Care, Jansson, & Powell, 2002), whereas the carboxy-terminus [107–139] fragment (osteostatin) was proven to restrain osteoclast-mediated bone resorption (Fenton, Martin, & Nicholson, 1994). The receptors recognizing the non-N terminal domains of PTHrP and mediating their biological effects are still unknown. These and other less-studied PTHrP cleavage products and their functional roles were recently and comprehensively reviewed by Frieling and Lynch (2019). Noteworthy, a sequence mapped in the [87–107] domain of the protein possesses a Lys/Arg-rich bipartite sequence homologous to the nuclear/nucleolar localization signal (NLS) of SV40 large tumor antigen (Fig. 1). This may allow PTHrP nuclear import *via* NLS

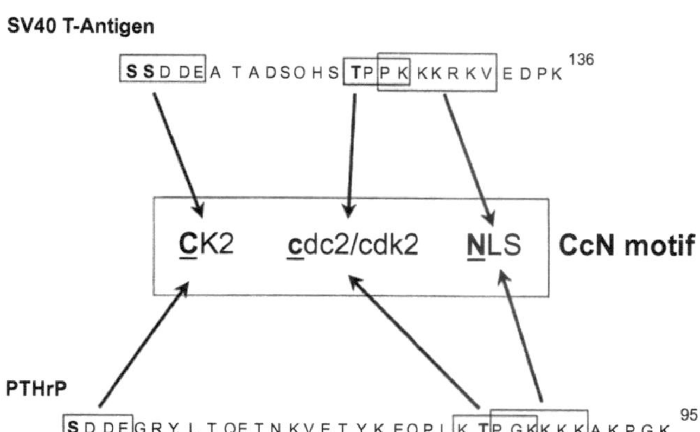

Fig. 1 Depiction of the nuclear targeting regulatory modules of the CcN motifs in SV40 large tumor antigen and PTHrP, comprising the NLS and the CK2- and cdc2/cdk2 sites. *From Lam, M. H., Thomas, R. J., Martin, T. J., Gillespie, M. T., & Jans, D. A. (2000). Nuclear and nucleolar localization of parathyroid hormone-related protein.* Immunology and Cell Biology, *78, 395–402. https://doi.org/10.1046/j.1440-1711.2000.00919.x.*

binding to an importin β_1/RanGDP complex, thus following an "intracrine" route supplementing the autocrine/paracrine counterpart. In this case, the endogenous PTHrP, translated from alternative start sites which circumvent the amino-terminus signal sequence, is diverted from its secretory pathway through the endoplasmic reticulum and re-directed to the cytoplasm for subsequent intake through the nuclear envelope (Lam, Briggs, et al., 1999; Lam et al., 1997; Lam et al., 2002; Lam, Thomas, Martin, Gillespie, & Jans, 2000; Massfelder et al., 1997).

Evidence collected over the past 30 years has highlighted that PTHrP, originally identified only as the mediator of the humoral hypercalcemia of malignancy syndrome (HHM; e.g., Suva et al., 1987), does not play roles limited to systemic calcium homeostasis as PTH, but is a multifaceted critical regulator of proliferation, differentiation and apoptosis, acting in a substantial way on the reprogramming of gene expression in normal and neoplastic cells. Within the large number of data available in the literature on PTHrP bioactivities, the goal of this chapter is to pick up selected studies that report the detection of molecular signatures of cancer cell exposure to PTHrP, either as full-length protein or discrete peptides, demonstrated by individual gene or whole genome expression profiling. Where available, the molecular, biochemical and/or physiological aspects associated with the specific gene activation or silencing in the target cell models under examination will be also briefly recapitulated.

2. Bone cancer cells

2.1 Osteosarcoma cells

It is well-known that about one-third of cancer patients exhibits elevated levels of circulating PTHrP and undergoes HHM onset, whose collateral skeletal complications represented by enhanced tumor cell survival and osteolytic bone metastatization are the most responsible of poor prognosis (Clines & Guise, 2005). In order to check the possible interference of PTHrP with programmed death of cancer cells, the molecular mechanisms underlying the anti-apoptotic effect exerted by amino- and carboxy-terminus PTHrP fragments on osteosarcoma (OS) Saos-2 cells were studied by Gagiannis et al. (2009) *via* transfection assays with both *BAX*- and *CD95*-luciferase reporter plasmids, the latter chosen for the p53-responsive element contained within its promoter. Both peptides were proven to restrain the transcriptional activation function of p53 in chemotherapeutic agent-treated cells, as confirmed by the increase of anti-apoptotic Bcl-2 and Bcl-xl and the decrease of pro-apoptotic Bax at the transcriptional and protein level upon treatment. In addition, downregulation of the pro-apoptotic BH3-only molecule PUMA, a direct transcriptional target of p53, by N- and C-terminal PTHrP was also observed. These data, corroborated also by the evidence of PTHrP fragment-mediated caspase inactivation and counteraction of drug-induced dissipation of mitochondrial transmembrane potential, gave a molecular explanation to the clinically-relevant chemoresistance conferred by PTHrP to cancer cells.

On the other hand, almost the totality of OS is characterized by null *TP53*alleles, coding for p53 protein, due to mutations and genomic rearrangements. Walia et al. (2016) examined the molecular effects of the upregulation of the axis PTHrP-cAMP-CREB1, which results in the phosphorylation and activation of the latter transcription factor, in p53-deficient cells derived from mouse osteoblastic OS models which, differently from normal osteoblasts, are dependent on continuous activation of this signalization for their survival and growth (Walkley, Walia, Ho, & Martin, 2014). Therefore, a PTHrP-specific gene signature array was defined and compared with a previously-published panel obtained from differentiating osteoblasts exposed to PTHrP [1–36] and [1–141] peptides (Allan et al., 2008), demonstrating an upregulation of about 70% of the tested genes in the OS experimental model. In particular, among CREB1 targets, the strongest molecular signatures were represented by *SOX9*, *NR4A2*, *NR4A3* and *TCF7* coding for the transcription factors SRY-box transcription factor 9, nuclear receptor subfamily 4 group A members 2 and -3, and T cell-specific transcription factor 7, respectively, the latter shown to promote oncogenic signaling (Wang et al., 2016). Other notable signatures were *AREG*, the amphiregulin gene, and *DUSP1* the dual specificity phosphatase 1 gene, whose products are linked to OS metastatization and poor prognosis (Liu, Tsao, & Hou, 2015; Lopes et al., 2017), the protooncogene *FOS* known to contribute to the development of OS through the upregulation of *FGFR1*, the fibroblast growth factor receptor 1 gene (Weekes et al., 2016), and *UGDH* coding for UDP-glucose 6-dehydrogenase implicated in the accumulation of glycosaminoglycans in the extracellular matrix (ECM) instrumental in the promotion of multiple pathways related to neoplastic growth, invasion and migration (Oyinlade et al., 2018). Hence, these data identified the PTHrP-cAMP-CREB1 axis as an eligible pathway for therapeutic intervention against OS.

2.2 Giant bone tumor cells

Giant cell tumor of the bone (GCT), a.k.a osteoclastoma, is an aggressive cancer histotype with a tendency to generate metastases. The cellular elements involved in the development of GCT are the osteoclast-like multinucleated giant cells, responsible of the GCT-associated osteolysis, and the stromal mesenchymal stem-like cells which are considered the true neoplastic cytotype capable to induce giant cell formation (Sobti, Agrawal, Agarwala, & Agarwal, 2016).

Cowan, Singh, and Ghert (2012) showed that stromal GCT cells expressed *PTHLH* and *PTH1R* genes, the latter being expressed also by the giant cells and their precursors, and demonstrated that exposure of GCT stromal cells to amino-terminus PTHrP fragment determined the upregulation of *RANKL*, coding for the receptor activator of nuclear factor κB ligand, resulting in an enhanced recruitment of monocytes to form the multinucleated cytotype. This was a first evidence of the central role played by PTHrP in GCT pathogenesis, following the earliest indications given several years before by the immunohistochemical and *in situ* hybridization data of Nakashima et al. (2003).

Subsequent larger-scale works by Mak, Turcotte, and Ghert (2012, 2013) were focused on cDNA microarray and proteomic analyses aimed to define a "reverse" PTHrP interactome, i.e., a global regulatory network map of relevant gene ontologies, processes, and pathways obtained starting from GCT stromal cells exposed to anti-PTHrP [1–34] antiserum to neutralize the effect of endogenous PTHrP on the stimulation of cell survival and growth. From the specific study of the apoptosis and cell-cycle linkages in the overall gene expression patterns, they found that the switch-off of PTHrP signalization contributed to cell cycle impairment and apoptosis promotion. In particular, the results showed an increase of the transcripts of the extrinsic apoptotic adaptor tumor necrosis factor (TNF)-related apoptosis-inducing ligand (TRAIL), the mitochondrial pro-apoptotic cytochrome C, the ubiquitin-conjugating enzyme E2I and the anti-proliferative Cip/Kip p21 factor. Conversely, members of the apoptosis-suppressing JAK/STAT signalization pathway were downregulated as well as the expression of *CCNE1* and *CDK2*, coding for cyclin E and cyclin dependent-kinase 2, and their upstream positive regulator *E2F*, the E2F transcription factor 1 gene, and *SKP*, the S-phase kinase-associated protein 2 gene whose product is responsible of the timed destruction of p21 (Wang, Nacusi, Sheaff, & Liu, 2005). Also, most of the integrin subunits, whose involvement in the promotion of cell proliferation, motility and invasion is widely acknowledged, were downregulated by anti-PTHrP treatment of GCT stromal cells, thereby further supporting the anti-GCT potential of PTHrP neutralization as a strategy against the development of the primary tumor.

3. Breast cancer cells

3.1 8701-BC cell line

The physiopathology of the mammary gland and PTHrP expression and signalization are highly intertwined. In fact, PTHrP is fundamental for mammary gland development and Ca^{++} transfer into milk during lactation (Hiremath & Wysolmerski, 2013; Uemura, Yasui, Yoneda, Irahara, & Aono, 1997). Also, it was proven to act as the mediator of the breast cancer-associated humoral hypercalcemia, to induce the bone resorption instrumental for cancer cell metastatization, and to intervene in the control of cell proliferative and invasive behavior itself (Luparello et al., 1997, 1995, 2001; Martin & Johnson, 2019).

Studies performed on 8701-BC tumor cells exposed to the *in vitro* invasion-promoting PTHrP [67–86] fragment demonstrated the upregulation of *HSP90*, *HSBP1* and *HSF1* and *-2*, coding for the 90 kDa-heat shock proteins α and -β, the heat shock factor-binding protein 1 and the heat shock factors 1 and -2, respectively. Interestingly, the increase of hsbp1 protein, undermining the negative regulation of activated hsf1 trimers on the promoter of *PLAU*, the urokinase-plasminogen activator gene, stimulated the expression and release of the extracellular lytic enzyme, thus substantiating the PTHrP [67–86]-mediated induction of cell invasiveness (Luparello et al., 1995; Luparello, Sirchia, & Pupello, 2003).

3.2 MDA-MB231 cell line

The MDA-MB231 cell line obtained from a highly-metastatic triple-negative breast cancer has been the object of several investigations on the molecular signatures associated to treatment with PTHrP. Integrin gene expression was examined in full-length PTHrH-overexpressing stable MDA-MB231 transfectants by Shen and Falzon (2006) and Anderson, Grabowska, and Watson (2007). The first found the selective upregulation of *ITGA6* and *-B4*, coding for the integrin subunits α6 and β4, which, in turn, enhanced drastically the survival rate and migratory behavior of tumor cells, also in light of the acknowledged integrin-mediated PTHrP action on the activation of the apoptosis-suppressing phosphatidylinositol 3-kinase (PI3K)/Akt pathway (Bhatia, Mula, Weigel, & Falzon, 2009). On their part, Anderson et al. (2007) demonstrated a significant upregulation of also *ITGA5* and *-B1*, the genes of the α_5 and β_1 subunits, in addition to *ITGA6* and *-B4*, that could be abrogated by either *PTHLH* silencing or NLS-mutant *PTHLH* expression. This underlined the need of PTHrP nuclear uptake to perform this function that resulted in an increased adhesivity of cells to ECM protein substrates. In this study, the transcriptional factor-like mechanism of action by PTHrP acting upon the *ITGA5* promoter region was also demonstrated through reporter construct and luciferase assays.

PTHrP [38–94] was described as the actual physiologic mid-region secretory form of the protein in earliest papers of the 90s (Soifer et al., 1992; Wu et al., 1996; Yang et al., 1994). It is known that MDA-MB231 cell growth is inhibited both *in vitro* and *in vivo* by treatment with mid-region PTHrP [38–94]-amide which, on the other hand, includes the bipartite NLS (Luparello et al., 2001). Sirchia et al. (2007) produced microscopical evidence of the import of the peptide into the cell nuclei, both *in situ* and isolated in tube, and of a direct binding of this peptide to metaphase chromosome spreads of MDA-MB231 cells. Of note, as shown in Fig. 2, PTHrP [38–94] fragment was found capable to bind also metaphase

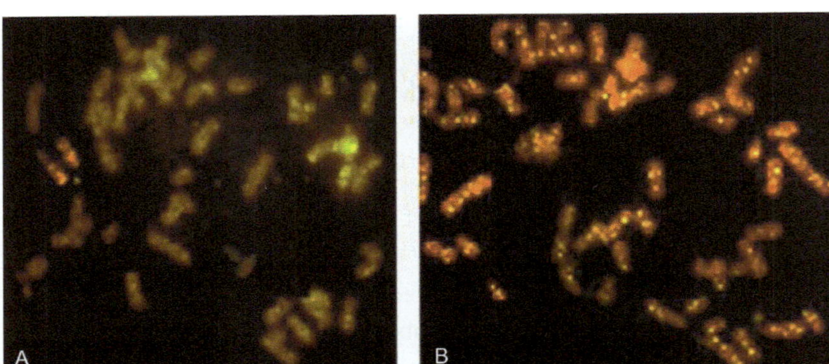

Fig. 2 Chromosomal decorations of: (A) a representative HB2 metaphase treated with 2 μg of mid-region PTHrP, and (B) a representative MDA-MB231 metaphase treated with 1 μg of mid-region PTHrP. *From Luparello, C., Librizzi, M., Asaro, D., Cruciata, I., & Caradonna, F. (2019). Mid-region parathyroid hormone-related protein is a genome-wide chromatin-binding factor that promotes growth and differentiation of HB2 epithelial cells from the human breast.* BioFactors (Oxford, England), 45, *279–288. https://doi.org/10.1002/biof.1484.*

chromosome spreads of non-neoplastic HB2 breast epithelial cells, although with a different microscopical "decoration" pattern, thus further substantiating the DNA-binding factor role played by the mid-region peptide (Luparello et al., 2019). From a molecular perspective, as shown in Table 2 band-shift assays proved the ability of the fragment to bind DNA sequences representing dimers or trimers of the 20-mer module 5′-AGTAGAATTCTAATATCTC-3′ in both orientations, a nucleotide stretch which can be found in diverse human genomic and mitochondrial DNA segments. DNase footprinting analyses demonstrated in more detail the existence of two critical sites in the module, i.e., $G^1A^2G^3T^4$ and $C^{18}T^{19}C^{20}$, for sequence recognition by PTHrP [38–94]-amide which appeared to be context-sensitive. Some gene signatures related to the 20-mer sequence were found down-regulated following exposure of MDA-MB231 cells to PTHrP [38–94]-amide. These included *COX6C*, coding for the mitochondrial cytochrome C oxidase subunit 6C, *OXR1*, coding for the reactive oxygen species (ROS)-protective oxidation resistance 1 factor (Yang et al., 2014), *FZD6*, the gene of the motility and invasion-promoting frizzled class receptor 6 (Corda et al., 2017), *HOX7/MSX1* whose product is the tumorigenesis-involved Msh homeobox 1 protein (Yue et al., 2018), and *TMCO4*, coding for transmembrane and coiled-coil domain-containing protein 4 whose implication in biological processes is still unknown. In addition, the transcriptional switch-on of the Bcl-xS product of *BCL2L1*, and of Bcl2-associated agonist of cell death, proteolipid protein-2, receptor-interacting serine/threonine-protein kinase 1, coded by *BAD*, *PLP2*, and *RIPK1*, respectively, and of multiple caspase members (i.e., the products of *CASP2*, *-5*, *-6*, *-7* and *-8*), all involved in the onset of apoptosis (Danial, 2008; Darding & Meier, 2012; Longo, Librizzi, & Luparello, 2013; Stevens & Oltean, 2019), was also reported (Luparello et al., 2008; Sirchia & Luparello, 2007). The mid-region PTHrP-responsive genes identified by these cumulative results substantiated the cytotoxic effect exerted by the peptide on this breast cancer cell line.

Dittmer et al. (2006) suppressed *PTHLH* expression in the MDA-MB231 cell line through RNA interference and studied the "reverse" PTHrP-associated gene signature pattern by microarray experiments, validating selected expression data also by RT-PCR and Western blotting. Among the about 200 genes whose expression levels were modified, they focused their analysis on the down-regulation of *ITGA6* confirming Shen and Falzon's data, *SERPINE1*, coding for the motility-inducing plasminogen activator inhibitor-1, and the oncopromoter *KISS-1*, and

Table 2 The "whole-genome PCR" strategy was employed to determine the specific DNA sequences recognized by mid-region PTHrP. The complexes between PTHrP and MDA-MB231 cell DNA digests, were isolated by binding to antibody-coated magnetic beads and the PTHrP-binding DNA fragments were enriched by PCR amplification and cloned in plasmid vectors. The table lists the sequences of the inserts that were obtained from 7 different recombinant colonies of transformed JM109 competent cells.

A	**Sequences of the DNA fragments bound to mid-region PTHrP**
Insert 1	5′-GAGTAGAATTCTAATATCTCGAGTAGAATT CTAATATCTCGAGATATTAGAATTCTACTC-3′
Insert 2	5′-GAGTAGAATTCTAATATCTCGAGATATTAG AATTCTACTC-3′
Insert 3	5′-GAGTAGAATTCTAATATCTCGAGATATTAG AATTCTACTC-3′
Insert 4	5′-GAGTAGAATTCTAATATCTCGAGATATTAG AATTCTACTC-3′
Insert 5	5′-GAGTAGAATTCTAATATCTCGAGATATTAG AATTCTACTC-3′
Insert 6	5′-GAGTAGAATTCTAATATCTCGAGATATTAG AATTCTACTC-3′
Insert 7	5′-GAGATATTAGAATTCTACTCGAGATATTAG AATTCTACTCGAGATATTAGAATTCTACTC-3′

B	Results of computer-assisted sequence comparison analysis		
Accession number	Denomination	Nucleotides	Identity
HSU25123	Human 16S ribosomal RNA homolog mRNA, mitochondrial gene encoding mitochondrial rRNA, partial sequence	1–20	20/20

AP005355.2	*Homo sapiens* genomic DNA, chromosome 8q23, clone: KB1189A8, complete sequence	91,930–91,947	18/18
AC107394.3	*Homo sapiens* BAC clone RP11–617114 from 4, complete sequence	123,222–123,205	18/18
AC113397.2	*Homo sapiens* chromosome 5 clone RP11-433B15, complete sequence	116,684–116,289	16/16
AL445985.10	Human DNA sequence from clone RP11-45B20 on chromosome 13, complete sequence	161,792–161,807	16/16
AL390791.15	Human DNA sequence from clone RP11-204A14 on chromosome 9, complete sequence	19,467–19,452	16/16
AC074101.5	*Homo sapiens* BAC clone RP11–646119 from 2, complete sequence	109,047–109,042	16/16
AF064857	*Homo sapiens* BAC derived from chromosome 21q22.3, complete sequence containing PEP19 (PCP4) gene	89,332–89,347	16/16
AL359472.13	Human DNA sequence from clone RP11-365 J7 on chromosome 13, complete sequence	24,161–24,176	16/16

From Sirchia, R., Priulla, M., Sciandrello, G., Caradonna, F., Barbata, G., & Luparello, C. (2007b). Mid-region parathyroid hormone-related protein (PTHrP) binds chromatin of MDA-MB231 breast cancer cells and isolated oligonucleotides "*in vitro*". *Breast Cancer Research and Treatment*, *105*, 105–116. https://doi.org/10.1007/s10549-006-9437-z

the concurrent upregulation of *PLAU*, *CTNNB1*, i.e., the gene of β-catenin, and five cell cycle-regulating genes, i.e., *CDC2*, *CDC25B* and *GRCC8* coding for cyclin-dependent kinase 1, cell division cycle 25B and trigger of mitotic entry 1, which are CDC2 direct and indirect activators, *CKS2* and *FBXO7* coding for the CDC2 substrates CDC28 protein kinase regulatory subunit 2 and F-box protein 7. Interestingly, PTHrP itself is a substrate for CDC2 which, by phosphorylation of the aminoacidic residue Thr^{85} close to the NLS, is able to inhibit the nuclear import of the protein, confining it to the cytosol (Lam, House, et al., 1999); therefore, the under-expression of the CDC2 system in *PTHLH*-overexpressing cells was interpreted as a way to guarantee its nuclear activity. In the same paper, the signatures were also associated to MDA-MB231 cell exposure to either PTHrP [1–34]- or [67–86] peptides in order to discriminate the involvement of PTH1R-dependent and -independent mechanisms of transcriptional control. The results unveiled a complex and, in some case, interlinked array of molecular responses since *CDC2* expression appeared to be regulated by amino-terminus PTHrP, *SERPINE1* and *CTNNB1* by mid-region PTHrP, *CDC25B* and *KISS-1* by both, and *PLAU* by neither, thereby suggesting the existence of at least three different pathways of gene regulation.

Immunohistochemical studies by Ibaragi et al. (2010) and Isowa et al. (2010) indicated a strong colocalization of PTHrP and both the extracellular lytic enzyme matrix metallopeptidase 13 (MMP13), and the angiogenetic and osteoclast-stimulating vascular endothelial growth factor (VEGF), respectively, in MDA-MB231 cells metastasizing to the bone marrow after intracardiac injection in nude mice. Assuming that *MMP13* and *VEGF* expression, instrumental in bone destruction for osteolytic metastasis, might be the result of a PTHrP-dependent autocrine stimulatory loop, they exposed cultured MDA-MB231 cells to PTHrP [1–34] and confirmed the related increase in both gene expression. In the case of *VEGF*, a promoter assay gave also evidence of a direct transcriptional effect by PTHrP. These upregulations were triggered by the activation of protein kinase C (PKCα) and extracellular signal-regulated kinase (ERK1/2) signalization, additionally involving the p38 MAPK contribution for *VEGF* signature, thus highlighting the opportunity of a therapeutic targeting of these pathways to counteract osteodegradation and malignant cell ingrowth.

3.3 MCF-7 cell line

In light of the negative correlation between *PTHLH* and *BMP6* expression, the latter being the gene of bone morphogenetic-6 protein, observed in breast cancer specimens, Mi et al. (2011) confirmed this finding also *in vitro* using MCF-7 breast tumor cells. In particular, the downregulation of *BMP6* in cultured cells was obtained by both transient transfection of *PTHLH* expression plasmid and exposure to PTHrP [1–40], but not [107–139] peptide, thereby demonstrating the implication of PTH1R in the observed downregulation. Being BMP-6 a mitogenic factor for this cell line, the PTHrP-induced reduction of its biosynthesis resulted in cell growth inhibition that appeared to be mediated by the PKA signaling pathway.

Substantial changes in gene expression pattern of *PTHLH*-overexpressing MCF-7 cells were more recently shown by Johnson et al. (2018) by sequence alignment and RNAseq bioinformatic analysis. Their data demonstrated the upregulation of tumor-inducing genes, such as *AREG* (Willmarth & Ethier, 2008) and *NRP1*, coding for neuropilin 1 (Arpel et al., 2016), and the downregulation of dormancy genes, such as *LIFR*, *SOCS3*, *AMOT*, *P4HA1*, *HIST1H2BK* and *SELENBP1*, whose products are LIF receptor subunit alpha, suppressor of cytokine signaling 3 factor, angiomotin, prolyl 4-hydroxylase subunit alpha-1, histone H2B type 1-K and selenium-binding protein 1, respectively, which enabled cells to exit quiescence and become proliferative (Kim et al., 2012). Moreover, enriched PTHrP-downstream pathways were brought to light, the more significant being those linked to calcium signaling, whose association with tumorigenesis is widely acknowledged (O'Grady & Morgan, 2019), cytokine–cytokine receptor interaction, chemokine signaling and inflammatory mediator regulation of transient receptor potential channels (Fig. 3). All these substantial effects on gene reprogramming appeared to be non-PTHR1-mediated actions, therefore unrelated to cAMP/PKA pathway activation, but the outcome of the signalization by domains different from amino-terminus PTHrP.

3.4 Animal studies

An *in vivo* approach to the analysis of PTHrP-associated molecular signatures in an animal cancer model was taken by Fleming et al. (2009) by conditional deletion of *PTHLH* in mouse mammary tumor virus (MMTV)-*neu* transgenic mice which determined a higher incidence of tumor development. Global gene expression analysis demonstrated the PTHrP ablation-dependent differential expression of 1283 out of 22,000 genes, which could be grouped significantly into the following functional Gene ontology (GO)-Biological process categories: ossification and bone remodeling, apoptosis, cell growth and lipid metabolism, and chemokine signaling and

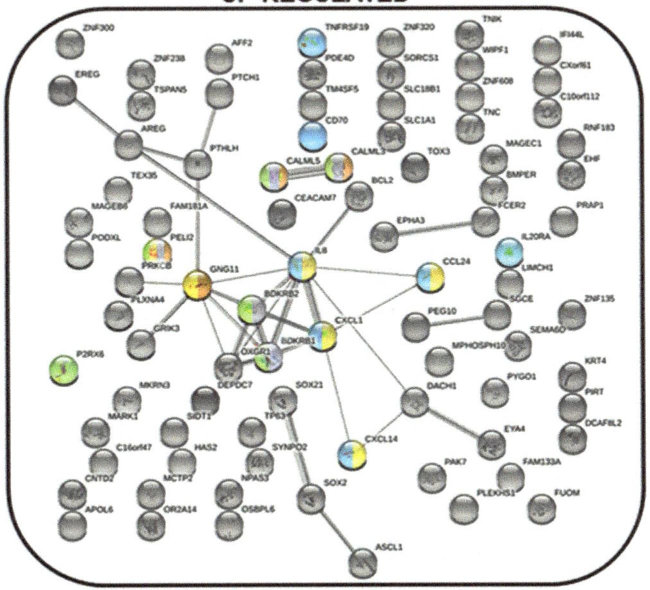

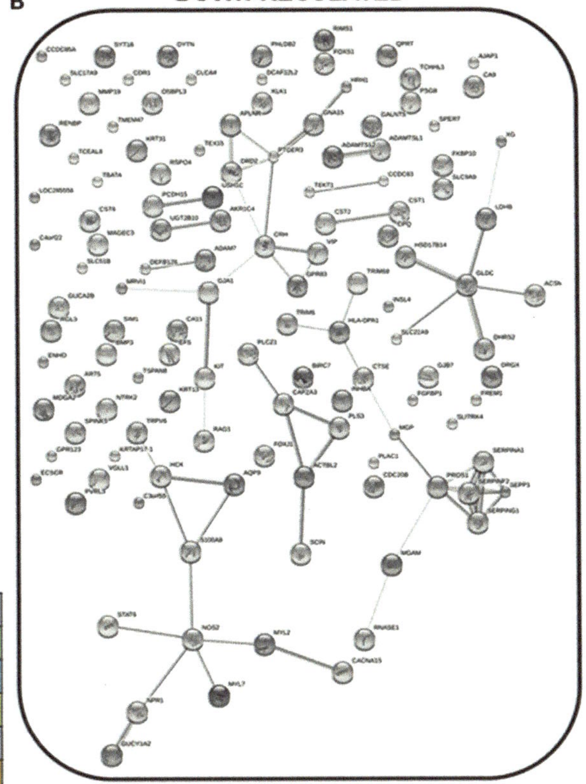

C

KEGG BIOLOGICAL PATHWAY (UP-REGULATED)	Gene count	False discovery rate
Calcium signaling pathway	6	0.0081
Cytokine-cytokine receptor interaction	7	0.0081
Chemokine signaling pathway	6	0.0081
Inflammatory mediator regulation of TRP channels	5	0.0081
Circadian entrainment	4	0.0324

Fig. 3 See figure legend on opposite page.

chemotaxis, the latter related to the signaling of tumor to immune cells. Among the highest ranking genes found, they have focused on *IGFBP4*, coding for insulin growth factor binding protein-4 (a suppressor of insulin growth factor-I; Conover, 2008), and *OPN*, coding for osteopontin (Nakamura et al., 2008), thereby confirming the role of PTHrP in osteoblast-osteoclast signaling and the existing functional overlap for PTHrP signalization between mammary cancer and bone tissue remodeling.

4. Cancer cells of the digestive system

4.1 Colorectal cancer cells

The promoting role played by PTHrP in the development and malignant adenoma-carcinoma progression of colorectal tumor was brought to light for the first time at the end of the 90s by Nishihara et al. (1999). For a better understanding of the molecular mechanisms of PTHrP stimulation of carcinogenesis, the associated molecular signatures were investigated, initially focusing on the transcriptional levels of integrin subunits. In the footstep of the data obtained on MDA-MB231 breast cancer cells already reported in this chapter, Shen, Rychahou, Evers, and Falzon (2007) showed the direct correlation between *PTHLH* and *ITGA6* and *-B4* expression also in LoVo colorectal adenocarcinoma cells *via* siRNA transfection experiments. *PTHLH* downregulation and the consequent decrease of $\alpha 6\beta 4$ integrin exposure on cell surface determined the lowering of PI3K-mediated AKT phosphorylation, involved in colorectal carcinogenesis, and also of *in vitro* cell proliferation, migration, invasion and anchorage-dependent growth rates, the latter as a marker of malignant phenotype. On the other hand, the same integrin upregulation and activation of PI3K-AKT pathway were found in xenografts derived from *PTHLH*-overexpressing LoVo cell clones injected

Fig. 3 STRING network analysis of (A) the top 250 upregulated genes, and (B) the top 250 downregulated genes (both with $\log_2$ fold change <-1 and $P<0.05$) in *PTHLH*-overexpressing MCF-7 cells. Colors of each node in (A) correspond to the KEGG pathway indicated in panel (C). (C) KEGG biological pathways significantly enriched for in the STRING analysis of the top 250 upregulated genes. Colors correspond to the nodes in panel (A). There was no significantly-enriched KEGG biological pathway for the top 250 downregulated genes. *From Johnson, R. W., Sun, Y., Ho, P., Chan, A., Johnson, J. A., Pavlos, N. J., Sims, N. A., & Martin, T. J. (2018). Parathyroid Hormone-Related Protein Negatively Regulates Tumor Cell Dormancy Genes in a PTHR1/ Cyclic AMP-Independent Manner.* Frontiers in Endocrinology, 9, *241. https://doi.org/10.3389/fendo.2018.00241.*

in athymic nude mice, where the expansion of a larger tumor mass was observed, confirming the *in vitro* data and the link between PTHrP and colon cancer susceptibility. In the same period, Anderson et al. (2007) examined the PTHrP-associated integrin signatures in *PTHLH*-overexpressing HT29 colorectal adenocarcinoma cells in parallel with transfected MDA-MB231 breast cancer cells, as previously reported in this chapter. Their results showed the upregulation of *ITGA5* and *ITGB1* with evidence of the indispensability of an intact NLS domain for this activity, and of a direct transcriptional effect of PTHrP on *ITGA5* promoter, in analogy with what found in breast carcinoma cells. These cumulative data suggested that PTHrP was not able to homogeneously increase integrin subunit expression and postulated the existence of a still unknown selective regulatory mechanism. A subsequent article by Mula, Bhatia, and Falzon (2010) shed some light on other gene signatures associated to PTHrP- and integrin α6β4-linked enhancement of LoVo cell migratory and invasive ability. In fact, their data demonstrated the upregulation of *TIAM1*, encoding for a RAC1-specific guanine nucleotide exchange factor, in *PTHLH*-overexpressing LoVo cells *via* the enhanced integrin α6β4 expression and the switching-on of the PI3K/AKT pathway, ultimately activating the Rho GTPase Rac1 whose promoting effect on colon cancer cell growth and migration is well-known (e.g., Kotelevets & Chastre, 2020). Fig. 4 recapitulates the Rac1-involving mechanism of the regulation of cell migration and invasion by PTHrP proposed by Mula and collaborators.

More recently, Calvo, Martín, de Boland, and Gentili (2014) and Calvo, Carriere, Martín, Gigola, and Gentili (2019) have looked into the involvement of PTHrP [1–34] in both cell cycle regulation of colon cancer cells and their interaction with the microenvironmental endothelial cells, instrumental for the tumor-associated angiogenetic process. In the first case, consistent with the observed decrease of the percentage of PTHrP-exposed tumoral Caco-2 cells in G_0/G_1 in favor of those in S and G_2/M phases, they found the up-regulation of *CDK6*, coding for the proliferation-promoter cyclin-dependent kinase 6, and the down-regulation of *CDKN1B*, coding for the cell cycle inhibitor $p27^{Kip1}$, as a result of a cooperative ERK1/2- and AKT-mediated transcriptional regulation. In the second case, the study was performed on both cultured Caco-2 and HCT116 cells, and xenografts of the latter cell line in nude mice. The data obtained showed that the amino-terminus peptide administered *in vivo* induced in the neoplastic cells an increased level of transcription of *VEGF*, which, in turn, determined a proliferative status of the endothelial cells and a considerable vessel neoformation. The same effect was observed also in both PTHrP-treated

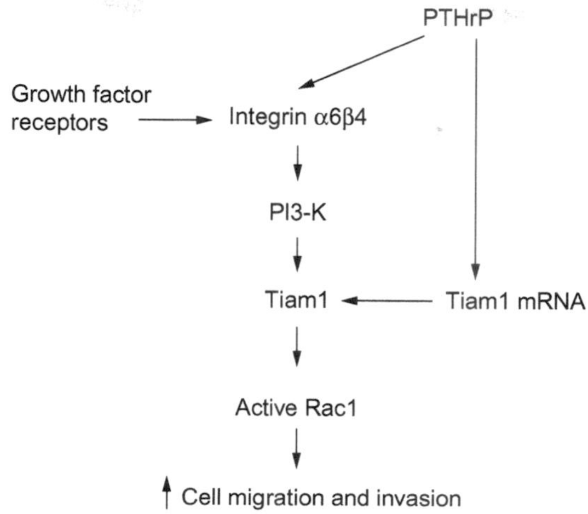

Fig. 4 Rac1-involving mechanism of the regulation of cell migration and invasion by PTHrP proposed by Mula and collaborators featuring (i) the PTHrP-stimulated increase of cell surface levels of integrin α6β4, (ii) the synergistic growth factor receptors/integrin α6β4 signalization activating the PI3K pathway, (iii) the resulting activation of the GTPase Rac1 through the Rac1-specific guanine nucleotide exchange factor Tiam1, (iv) the parallel PTHrP-mediated upregulation of Tiam1 mRNA levels *via* a transcriptional/post-transcriptional pathway, and (v) Rac1 activation-promoted increase of cell migration and invasion. *From Mula, R. V., Bhatia, V., & Falzon, M. (2010). PTHrP promotes colon cancer cell migration and invasion in an integrin α6β4-dependent manner through activation of Rac1.* Cancer Letters, 298, *119–127. https://doi.org/10.1016/j.canlet.2010.06.009.*

cell lines *in vitro* and, further, the upregulation of two other signatures associated to angiogenesis, i.e., *HIF1A* and *MMP9* (Deryugina & Quigley, 2015; Krock, Skuli, & Simon, 2011), encoding for the subunit α of the transcription factor HIF-1 (hypoxia-inducible factor 1) and matrix metalloproteinase-9, respectively, was also proven. These mRNA level increases, likely mediated by the ERK1/2 and PI3K/AKT signaling pathways also in this case, were corroborated by Western blot analyses and cell motility assays and confirmed that the amino-terminus domain of PTHrP mediates the communication between colon cancer and endothelial cells by triggering the release of angiogenic factors.

4.2 Liver cancer cells

PTHLH expression is known to be low in non-tumoral hepatitis/cirrhotic (HBV or HCV infection) liver tissue whereas being abundant in hepatocellular carcinoma. Thus, large scale studies by systems-theoretical analysis were performed by Huang et al. (2012) and Wang et al. (2012) identifying and comparing *PTHLH* feedback-mediated GO networks between the two model systems. Within the great assortment of molecular systems examined, in the tumoral tissue the following cell proliferation-associated gene signatures downstream of *PTHLH* were detected: *CST6*, *BUB1B*, *NTN1*, *PHLDA2* coding for cystatin E/M, BUB1 mitotic checkpoint serine/threonine kinase B, netrin1 and pleckstrin homology-like domain family A member 2 proteins, respectively. The products of these activated genes were proven to promote the progression of hepatocellular carcinoma or other neoplasms of the digestive tract (Han et al., 2019; Ma, Lou, & Jiang, 2020; Vigneswaran, Wu, & Zacharias, 2003; Zhuang, Yang, & Meng, 2018). Moreover, also the following downstream receptor signaling- and cell adhesion-associated signatures were put in evidence: *GNG10*, *GNAZ*, *PTHR2*, *RFC4*, *UBE2C*, *NRXN3*, *BAP1*, *PVRL2*, *TROAP* and *VCAN*, coding for the subunits γ_{10} of the G protein and α_1 of the G_z protein, parathyroid hormone-2 receptor, the subunit 4 of replication factor C, ubiquitin-conjugating enzyme E2 C, neurexin 3, BRCA1-associated protein 1, nectin cell adhesion molecule 2, trophinin-associated protein and versican, respectively. Literature data indicate that *RCF4*, *UBE2C* *TROAP* and *VCAN* overexpression is linked to the development of hepatocellular carcinoma (Hu et al., 2019; Naboulsi et al., 2016; Xiong et al., 2019; Yao et al., 2015). Thus, the cumulative results show the *PTHLH*-feedback stimulation of liver cancer growth with consequent poor prognosis and decreased survival rate.

4.3 Oral cancer cells

In order to clarify the molecular mechanisms of the development and progression of gingival squamous cell carcinoma, Chang et al. (2017) studied the molecular signatures associated to *PTHLP* overexpression in Ca9–22 tumor cells, which was proven to stimulate cell growth in an autocrine way, and compared the results obtained *in vitro* with those of a cohort of clinical patients. The bioinformatic data obtained through the Ingenuity Canonical Pathway analysis of microarrays from cultured cell preparations indicated that PTHrP mostly upregulated signatures belonging to the "Cyclins and cell cycle regulation" and "Estrogen-mediated S-phase entry" clusters. The comparison with the *in vivo* results provided more details on members of the groups critically enriched also in the patient cohort, i.e., *CCNA2, CCNE2* and *CDC25A* encoding for cyclin A2, -E2 and cell division cycle 25A factors. Since cyclins A2 and E2 are the rate-limiting molecules in G_2/M and G_1/S transitions, respectively, and cell division cycle 25A protein is a cell cycle regulator with peak expression in G_1, and in consideration that all of them have been considered prognostic factors for head and neck carcinogenesis (Lin, Zhang, & He, 2018; Nishioka et al., 2001; Zhao, Xie, & Miao, 2012), PTHrP appears to act as a key upstream activator regulating cell cycle progression thereby facilitating tumor growth.

5. Leukema/lymphoma adult T-cells

Adult T-cell leukemia/lymphoma, whose etiologic agent is the human T-lymphotropic virus type I (HTLV-I), is an aggressive malignancy of post-thymic pleomorphic activated T-lymphocytes, occurring in about 5% of the infected individuals after a long latency period. It is known that approximately 80% of patients develop the humoral hypercalcemia of malignancy syndrome and that PTHrP and its receptor are specifically upregulated during T-cell immortalization by HTLV-1, being also involved in the onset of the transformation process (Nadella et al., 2008). In order to identify the downstream targets of *PTHLH* responsible of the expression of the transformed phenotype, Shu et al. (2012) applied the *PTHLH* overexpression methodology to Jurkat cells, an *in vitro* model of $CD4^+$ T-cell leukemia, mainly focusing on gene signatures associated to the regulation of the transcription process and apoptosis. In particular, they showed the downregulation of *CD40LG, IFNG* and *ID1* encoding for CD40 ligand, interferon-γ and inhibitor of DNA binding-1 helix-loop-helix protein, respectively, the upregulation of *RANKL*, and also *DAPK1, VDR, PAX6, STAT4* and, at a lesser level, *TNFA* encoding for death-associated protein kinase-1, vitamin D receptor, the transcription factors paired box-6 and signal transducer and activator of transcription-4, and tumor necrosis factor α, respectively. Their findings were consistent with the contextually-reported significant decrease of tumor development *in vivo* by *PTHLP*-overexpressing cells injected in nude mice. In fact, the switching-on of CD40 ligand and interferon-γ pathways, the former being downstream to inhibitor of DNA binding-1 signalization whereas the latter being downregulated by vitamin D receptor activation, is necessary to block T-cell apoptosis (Chen & Liu, 2009; Laman, Claassen, & Noelle, 2017). Moreover, death-associated protein kinase-1 is known to enhance tumor necrosis factor α-induced apoptosis, and both are upregulated by receptor activator of nuclear factor κB ligand signalization (Lai & Chen, 2014). On the other hand, the anti-cancer activity of *PAX6* and *STAT4* upregulation was only hypothesized.

6. Prostate cancer cells

Studies on integrin gene signatures associated to *PTHLH* overexpression in androgen-independent PC-3 prostate adenocarcinoma cells were performed by Shen and Falzon (2003a) who showed the up-regulation of *ITGA1*, encoding for the α1 subunit, *-A5*, *-A6* and *-B4*, an effect abolished by the deletion of either multibasic clusters of the NLS. This demonstrated the requirement of the β importin-mediated nuclear translocation which was also confirmed by the lack of activity on integrin mRNA levels by cell exposure to NLS-devoid amino-terminus PTHrP. The enhanced surface exposure of the α1 integrin subunit was correlated to the observed increase of cell adhesion to type I collagen, fibronectin and laminin (Shen & Falzon, 2003b), whereas those of the α5 and α6 subunits to apoptosis resistance and bone metastasis potential of cancer cells (De Toni-Costes et al., 2010; King et al., 2008). Subsequent work was focused on PTHrP regulation of the pro-invasive integrin α6β4 and demonstrated that in *PTHLH*-overexpressing PC-3 tumor cells the activities of both *ITGA6* and *–B4* promoters were elevated. In light of the observed *PTHLH*-associated upregulation of NF-κB mRNA levels, the involvement of the putative NF-κB-binding sites within *ITGA6* and *-B4* promoters in the PI3K/AKT-dependent transcriptional upregulation was examined but a positive result was restricted only to the latter one, thereby suggesting the existence of an alternative pathway, such as the CREB/AP1 one, for the activation of *ITGA6* (Bhatia, Mula, & Falzon, 2013; Bhatia et al., 2009).

Ongkeko et al. (2014) examined the epithelial-mesenchymal transition (EMT)-associated signatures in PTHrP [1–141]- and PTHrP [1–173]-overexpressing DU-145 prostate tumor cells and found the upregulation of the mesenchymal markers vimentin and the transcription factor snail, encoded by *VIM* and *SNAI1*, and the concurrent downregulation of *CDH1*, encoding for E-cadherin, more prominent in case of the expression of the longer PTHrP form. This result was confirmed by reversal through the knock-down of *PTHLH* in PC-3 tumor cells and suggested that PTHrP may act through EMT to promote an aggressive and metastatic phenotype in prostate cancer. In the context of cancer cell–ECM interactions,

Park and McCauley (2012) demonstrated that the decrease of TNF-α mRNA levels is another *PTHLH*-associated signature that protects PC-3 and LNCaP cells from anoikis, i.e., substrate detachment-dependent apoptosis, thus promoting the formation of metastatic lesions in the bone which are instead prevented in *PTHLP*-knockdown cells. Interestingly, *TNF* downregulation was not induced in NLS-defective PTHrP-expressing LNCaP cells suggesting the existence of a still unclear intracrine mechanism of transcriptional control mediated by nuclear translocation of PTHrP.

As reported by Tsigelny, Burton, Sharikov, Hastings, and Deftos (2005), data from microarray and bioinformatics analyses demonstrated that transfection of DU-145 cells with PTHrP [1–173]- and amino terminus-devoid PTHrP [33–173] expression plasmids determined the modulation of the expression of a total of 245 annotated genes. In particular, differently to the amino-terminus lacking form, PTHrP [1–173] upregulated the expression of genes coding for aldo-keto reductase enzymes, cell adhesion and cytoskeleton-related proteins and growth and/or apoptosis factors, and downregulated the expression of genes involved in differentiation or development. On the other hand, PTHrP [33–173] was active in increasing the expression level of genes for MHC class I antigens, nuclear transcription factors, and proteins related to cell–cell interactions, proteasome function and arachidonic acid metabolism, while downregulating those coding for cytoskeleton- and cell-matrix interaction-related proteins. Tables 3 and 4 summarize the representative molecular signatures in the two experimental conditions. The cumulative results of this work showed that in prostate tumor cells PTHrP can play non-overlapping biological roles ascribable to its different domains and that multiple mechanisms based upon modulations of protein processing and changes in secretory pattern, receptor expression, and nuclear localization can account for the diversified effect of the PTHrP fragments in prostate cancer pathology.

In light of Tsigelny and coworkers' results indicating the overexpression of C1, C2, and C3 members of the aldo-keto reductase family as the most striking change induced by PTHrP [1–173] and of the evidence that this could play a role in androgen response, e.g., by negatively affecting 5α-dihydrotestosterone levels, interestingly DaSilva, Gioeli, Weber, and Parsons (2009) found an androgen-receptor stabilizing effect by PTHrP [1–34] signalization in LNCaP cells. This allowed cell growth under low androgen conditions and progression of hormone-refractory disease, and was upstream to the downregulation of the androgen-dependent *DKK1* gene,

Table 3 Genes regulated by PTHrP [1–173] but unaffected by PTHrP [33–173].

Gene	Unigene no	Region	Expression with PTHrP 1–173 transfection[a](%)
Upregulated to >200% of control			
Transforming growth factor, beta 2	Hs. 169300	1q41	265
Interferon, gamma-inducible protein 16	370873	1q23	264
ALL1-fused gene from chromosome 1q	75823	1q21	271
Interferon-induced protein 44	82316	1p31	344
Caspase 10, apoptosis-related cysteine protease	5353	2q33	255
Zinc finger protein 288	436987	3q13	281
Cysteine-rich with EGF-like domains 1	9383	3p25	253
I factor (complement)	312485	4q25	494
Rab coupling protein	96125	8p11	292
Syndecan 2	1501	8q22	253
Aldo-keto reductase family 1, member C1	306098	10p15	991
Aldo-keto reductase family 1, member C2	201967	10p15	1328
Aldo-keto reductase family 1, member C3	78183	10pl5	785
Hypothetical protein FLJ12616	287537	10q26	260
Nebulette	5025	10pl2	277
p8 protein (candidate of metastasis 1)	418692	16p11	263
Junction plakoglobin	2340	17q21	298
Downregulated to <40% of control			
Tumor necrosis factor receptor superfamily, member 25	299558	1p36	20
Pre-B-cell leukemia transcription factor 1	40822	1q23	23
Kinesin family member 5C	6641	2q23	39
Butyrylcholinesterase	422857	3q26	21
Somatostatin	12409	3q28	31
Hypothetical protein FLJ23235	211501	4p14	34

Table 3 Genes regulated by PTHrP [1–173] but unaffected by PTHrP [33–173].—cont'd

Gene	Unigene no	Region	Expression with PTHrP 1–173 transfection[a](%)
Hypothetical protein FLJ23548	22895	4q26	39
Hypothetical protein DKFZp434N1235	149030	4q28	12
Potassium intermediate/small conductance			
Calcium-activated channel, subfamily N, member 2	98280	5q22	16
Lysyl oxidase	102267	5q23	32
Diphtheria toxin receptor	799	5q23	36
Intestinal cell kinase	417022	6p12	37
Connexin 43	74471	6q21	29
Connective tissue growth factor	410037	6q23	33
Adipose differentiation-related protein	3416	9p22	34
Myotubularin-related protein 2	181326	11q22	38
Interleukin 23, alpha subunit p 19	98309	12ql3	33
High mobility group AT-hook 1	57301	6p21	36
Heat shock 105kD	36927	13q12	36
Glycine amidinotransferase	75335	15q15	39
Ubiquitin-like 5	386532	19p13	16
Melanoma antigen, family A, 8	37109	Xq28	39

[a]Expression is presented as a % of the expression by the vector-transfected control cells.
From Tsigelny, I., Burton, D. W., Sharikov, Y., Hastings, R. H., & Deftos, L. J. (2005). Coherent expression chromosome cluster analysis reveals differential regulatory functions of amino-terminal and distal parathyroid hormone-related protein domains in prostate carcinoma. *Journal of Biomedicine & Biotechnology*, 2005(4), 353–363. https://doi.org/10.1155/JBB.2005.353

coding for the Wnt inhibitor Dickkopf-related protein 1, thereby linking PTHrP to Wnt pathway and allowing for the development of Wnt-stimulated osteoblastic bone lesions of prostate cancer (Zhang et al., 2014).

Further androgen receptor-associated signatures, i.e., *PSA*, *TMPRSS2* and *FKBP5*, coding for prostate-specific antigen, transmembrane serine protease 2 and FKBP prolyl isomerase 5, respectively, appeared to be upregulated in prostate tumor cells exposed to PTHrP produced by

Table 4 Representative genes regulated by PTHrP [33–173].

Gene	Unigene no		Expression with PTHrP 33–173 transfection[a](%)
Upregulated genes			
Class 1 major histocompatibility complex proteins			
HLA-A	Hs. 181244	6p21	224
HLA-B	Hs. 77961	6p21	200
HLA-C	Hs. 274485	6p21	201
HLA-F	Hs. 411958	6p21	180
HLA-G	Hs. 512152	6p21	242
Proteasome-related proteins			
Proteasome (prosome, macropain) subunit, beta type, 8	Hs. 180062	6p21	358
Proteasome (prosome, macropain) subunit, beta type, 9	Hs. 381081	6p21	318
Ring finger protein 1, RING1	Hs. 202430	6p21	249
Proteins mediating cell–cell interactions			
E-cadherin, CDH1	Hs. 194657	16q22	334
P-cadherin, CDH3	Hs. 191842	16q22	229
Arachidonic acid pathway proteins			
Glutathione peroxidase 1, GPX1	Hs. 76686	3p21	200
Glutathione peroxidase 3, GPX3	Hs. 386793	5q23	718
Nuclear proteins			
Basic leucine nuclear zipper factor Jem-1, BLZF1	Hs. 444697	1q24	221
Kruppel-like factor 3, KLF3	Hs. 145754	4p14	197
Zinc finger protein 305, ZNF305	Hs. 134816	6p22	229
Downregulated genes			
Cell-Matrix/Cytoskeleton-related proteins			
Sterile alpha motif domain containing 4, SAMD4	Hs. 98259	14q22	22
Dual-specificity tyrosine-(Y)-phosphorylation-regulated kinase 4	Hs. 439530	12p13	33

Table 4 Representative genes regulated by PTHrP [33–173].—cont'd

Gene	Unigene no		Expression with PTHrP 33–173 transfection[a](%)
Myosin light polypeptide kinase, MYLK	Hs. 386078	3q21	10
Transmembrane 4 superfamily member 2, TM4SF2	Hs. 439586	Xp11	34
Lysyl oxidase-like 2, LOXL2	Hs. 83354	8p21	37
Dicer1, Dcr-1 homolog, DICER1	Hs. 87889	14q32	34

[a]Expression is presented as a % of the expression by vector-transfected control cells.
From Tsigelny, I., Burton, D. W., Sharikov, Y., Hastings, R. H., & Deftos, L. J. (2005). Coherent expression chromosome cluster analysis reveals differential regulatory functions of amino-terminal and distal parathyroid hormone-related protein domains in prostate carcinoma. *Journal of Biomedicine & Biotechnology*, 2005(4), 353–363. https://doi.org/10.1155/JBB.2005.353

cocultured neuroendocrine prostate cancer cells, a minor population in the heterogeneous neoplastic bulk. This transcriptional reprogramming was linked to increased docetaxel resistance of the neighboring target tumor cells, as also evidenced *in vivo* by mouse xenograft experiments (Cui et al., 2016).

7. Conclusions

Gene expression profiling has identified molecular signatures associated to full-length PTHrP or its distinct domains in different cancer models *in vitro* and *in vivo* and correlated to the modulation of the growth and metastatic functions of the specific cytotypes. In this review, we have examined a number of distinct sets of molecular signatures which are the result of large scale-bioinformatic investigations and data mining or, conversely, have been unveiled by lesser scale studies. The latter ones are summarized in Table 5 and grouped according to specific *Homo sapiens*' GO-Biological Process terms for instant synoptic analysis by the reader. These molecular markers identify important cellular responses and quantify the activation of transduction pathways, thus expanding the knowledge on PTHrP transcriptional activity and signaling mechanism, and, although rigorous assessment of the clinical relevance of PTHrP-associated gene signatures requires further studies, may even predict clinical outcomes, thereby being potentially routed to the development of targeted therapeutic treatments.

Table 5 Clustering of PTHrP-associated gene expression signatures according to selected *Homo sapiens'* GO-Biological Process terms.

GO-Biological process term	Gene	Up↑ Down↓	Cancer type	References
Apoptotic signaling pathway	*CASP5*	↑	Breast cancer	Luparello et al. (2008)
	DAPK1	↑	Leukemia/ lymphoma	Shu et al. (2012)
BMP signaling pathway	*BMP6*	↓	Breast cancer	Mi et al. (2011)
Canonical Wnt signaling pathway	*DKK1*	↓	Prostate cancer	DaSilva et al. (2009)
Cell adhesion	*ITGA1*	↑	Prostate cancer	Shen and Falzon (2003a)
	ITGA2	↑	Breast cancer	Shen and Falzon (2006)
	ITGA5	↑	Breast cancer	Anderson et al. (2007)
		↑	Colorectal cancer	Anderson et al. (2007)
		↑	Prostate cancer	Shen and Falzon (2003a)
	ITGA6	↑	Colorectal cancer	Shen et al. (2007)
		↑	Prostate cancer	Shen and Falzon (2003a)
	ITGB1	↑	Breast cancer	Anderson et al. (2007)
		↑	Colorectal cancer	Anderson et al. (2007)
	ITGB4	↑	Breast cancer	Shen and Falzon (2006)
		↑	Colorectal cancer	Shen et al. (2007)
		↑	Prostate cancer	Shen and Falzon (2003a)
Cellular response to heat	*HSBP1*	↑	Breast cancer	Luparello et al. (2003)

Table 5 Clustering of PTHrP-associated gene expression signatures according to selected *Homo sapiens'* GO-Biological Process terms.—cont'd

GO-Biological process term	Gene	Up↑ Down↓	Cancer type	References
Cellular response to stress	*FKBP5*	↑	Prostate cancer	Cui et al. (2016)
Cytokine-mediated signaling pathway	*PLP2*	↑	Breast cancer	Sirchia and Luparello (2007)
Epithelial to mesenchymal transition	*SNAI1*	↑	Prostate cancer	Shen and Falzon (2003a)
Execution phase of apoptosis	*CASP2*	↑	Breast cancer	Luparello et al. (2008)
	CASP6	↑	Breast cancer	Luparello et al. (2008)
	CASP7	↑	Breast cancer	Luparello et al. (2008)
	CASP8	↑	Breast cancer	Luparello et al. (2008)
Immune response	*RANKL*	↑	GCT	Cowan et al. (2012)
		↑	Leukemia/ lymphoma	Shu et al. (2012)
	CD40LG	↓	Leukemia/ lymphoma	Shu et al. (2012)
Intermediate filament organization	*VIM*	↑	Prostate cancer	Ongkeko et al. (2014)
Negative regulation of cell migration	*CDH1*	↓	Prostate cancer	Shen and Falzon (2003a)
Negative regulation of DNA-binding	*ID1*	↓	Leukemia/ lymphoma	Shu et al. (2012)
Positive regulation of angiogenesis	*VEGF*	↑	Breast cancer	Isowa et al. (2010)
		↑	Colorectal cancer	Calvo et al. (2019)
	HIF1A	↑	Colorectal cancer	Calvo et al. (2019)

Continued

Table 5 Clustering of PTHrP-associated gene expression signatures according to selected *Homo sapiens'* GO-Biological Process terms.—cont'd

GO-Biological process term	Gene	Up↑ Down↓	Cancer type	References
Positive regulation of apoptotic process	*BAX*	↓	Osteosarcoma	Gagiannis et al. (2009)
	CD95	↓	Osteosarcoma	Gagiannis et al. (2009)
	PUMA	↓	Osteosarcoma	Gagiannis et al. (2009)
	BAD	↑	Breast cancer	Luparello et al. (2008)
	BCL2L1	↑	Breast cancer	Luparello et al. (2008)
	RIPK1	↑	Breast cancer	Luparello et al. (2008)
	TNF	↓	Prostate cancer	Park and McCauley (2012)
Positive regulation of gene expression	*PAX6*	↑	Leukemia/ lymphoma	Shu et al. (2012)
Positive regulation of GTPase activity	*TIAM1*	↑	Colorectal cancer	Mula et al. (2010)
Positive regulation of transcription by RNA	*HSF1*	↑	Breast cancer	Luparello et al. (2003)
	HSF2	↑	Breast cancer	Luparello et al. (2003)
Polymerase II	*HOX7/ MSX1*	↓	Breast cancer	Sirchia and Luparello (2007)
Protein folding	*HSP90*	↑	Breast cancer	Luparello et al. (2003)
Proteolysis	*MMP9*	↑	Colorectal cancer	Calvo et al. (2019)
	MMP13	↑	Breast cancer	Ibaragi et al. (2010)
	PSA	↑	Prostate cancer	Cui et al. (2016)
	TMPRSS2	↑	Prostate cancer	Cui et al. (2016)

Table 5 Clustering of PTHrP-associated gene expression signatures according to selected *Homo sapiens'* GO-Biological Process terms.—cont'd

GO-Biological process term	Gene	Up↑ Down↓	Cancer type	References
Proton transmembrane transport	*COX6C*	↓	Breast cancer	Sirchia and Luparello (2007)
Receptor signaling pathway	*IFNG*	↓	Leukemia/ lymphoma	Shu et al. (2012)
via JAK-STAT	*STAT4*	↑	Leukemia/ lymphoma	Shu et al. (2012)
Regulation of cell cycle	*CDK6*	↑	Colorectal cancer	Calvo et al. (2014)
Regulation of cell cycle G1/S transition	*CDKN1B*	↓	Colorectal cancer	Calvo et al. (2014)
Response to oxidative stress	*OXR1*	↓	Breast cancer	Sirchia and Luparello (2007)
Vitamin D receptor signaling pathway	*VDR*	↑	Leukemia/ lymphoma	Shu et al. (2012)
Wnt signaling pathway	*FZD6*	↓	Breast cancer	Sirchia and Luparello (2007)

References

Allan, E. H., Häusler, K. D., Wei, T., Gooi, J. H., Quinn, J. M., Crimeen-Irwin, B., et al. (2008). EphrinB2 regulation by PTH and PTHrP revealed by molecular profiling in differentiating osteoblasts. *Journal of Bone and Mineral Research: the Official Journal of the American Society for Bone and Mineral Research, 23*, 1170–1181. https://doi.org/10.1359/jbmr.080324.

Anderson, J. A., Grabowska, A. M., & Watson, S. A. (2007). PTHrP increases transcriptional activity of the integrin subunit alpha5. *British Journal of Cancer, 96*, 1394–1403. https://doi.org/10.1038/sj.bjc.6603720.

Arpel, A., Gamper, C., Spenlé, C., Fernandez, A., Jacob, L., Baumlin, N., et al. (2016). Inhibition of primary breast tumor growth and metastasis using a neuropilin-1 transmembrane domain interfering peptide. *Oncotarget*, 7, 54723–54732. https://doi.org/10.18632/oncotarget.10101.

Bhatia, V., Mula, R. V., & Falzon, M. (2013). Parathyroid hormone-related protein regulates integrin α6 and β4 levels via transcriptional and post-translational pathways. *Experimental Cell Research, 319*, 1419–1430. https://doi.org/10.1016/j.yexcr.2013.03.003.

Bhatia, V., Mula, R. V., Weigel, N. L., & Falzon, M. (2009). Parathyroid hormone-related protein regulates cell survival pathways via integrin alpha6beta4-mediated activation of phosphatidylinositol 3-kinase/Akt signaling. *Molecular Cancer Research: MCR*, 7, 1119–1131. https://doi.org/10.1158/1541-7786.MCR-08-0568.

Calvo, N., Carriere, P., Martín, M. J., Gigola, G., & Gentili, C. (2019). PTHrP treatment of colon cancer cells promotes tumor associated-angiogenesis by the effect of VEGF. *Molecular and Cellular Endocrinology, 483*, 50–63. https://doi.org/10.1016/j.mce.2019.01.005.

Calvo, N., Martín, M. J., de Boland, A. R., & Gentili, C. (2014). Involvement of ERK1/2, p38 MAPK, and PI3K/Akt signaling pathways in the regulation of cell cycle progression by PTHrP in colon adenocarcinoma cells. *Biochemistry and Cell Biology-Biochimie et Biologie Cellulaire, 92*, 305–315. https://doi.org/10.1139/bcb-2013-0106.

Chang, W. M., Lin, Y. F., Su, C. Y., Peng, H. Y., Chang, Y. C., Hsiao, J. R., et al. (2017). Parathyroid hormone-like hormone is a poor prognosis marker of head and neck cancer and promotes cell growth via RUNX2 regulation. *Scientific Reports, 7*, 41131.

Chen, J., & Liu, X. (2009). The role of interferon gamma in regulation of CD4+ T-cells and its clinical implications. *Cellular Immunology, 254*, 85–90. https://doi.org/10.1016/j.cellimm.2008.09.001.

Clines, G. A., & Guise, T. A. (2005). Hypercalcaemia of malignancy and basic research on mechanisms responsible for osteolytic and osteoblastic metastasis to bone. *Endocrine-Related Cancer, 12*, 549–583. https://doi.org/10.1677/erc.1.00543.

Conover, C. A. (2008). Insulin-like growth factor-binding proteins and bone metabolism. *American journal of physiology. Endocrinology and Metabolism, 294*, E10–E14. https://doi.org/10.1152/ajpendo.00648.2007.

Corda, G., Sala, G., Lattanzio, R., Iezzi, M., Sallese, M., Fragassi, G., et al. (2017). Functional and prognostic significance of the genomic amplification of frizzled 6 (FZD6) in breast cancer. *The Journal of Pathology, 241*, 350–361. https://doi.org/10.1002/path.4841.

Cowan, R. W., Singh, G., & Ghert, M. (2012). PTHrP increases RANKL expression by stromal cells from giant cell tumor of bone. *Journal of Orthopaedic Research: Official Publication of the Orthopaedic Research Society, 30*, 877–884. https://doi.org/10.1002/jor.22020.

Cui, Y., Sun, Y., Hu, S., Luo, J., Li, L., Li, X., et al. (2016). Neuroendocrine prostate cancer (NEPCa) increased the neighboring PCa chemoresistance via altering the PTHrP/p38/Hsp27/androgen receptor (AR)/p21 signals. *Oncogene, 35*, 6065–6076. https://doi.org/10.1038/onc.2016.135.

Danial, N. N. (2008). BAD: Undertaker by night, candyman by day. *Oncogene, 27*(Suppl 1), S53–S70. https://doi.org/10.1038/onc.2009.44.

Darding, M., & Meier, P. (2012). IAPs: Guardians of RIPK1. *Cell Death and Differentiation, 19*, 58–66. https://doi.org/10.1038/cdd.2011.163.

DaSilva, J., Gioeli, D., Weber, M. J., & Parsons, S. J. (2009). The neuroendocrine-derived peptide parathyroid hormone-related protein promotes prostate cancer cell growth by stabilizing the androgen receptor. *Cancer Research, 69*, 7402–7411. https://doi.org/10.1158/0008-5472.CAN-08-4687.

De Toni-Costes, F., Despeaux, M., Bertrand, J., Bourogaa, E., Ysebaert, L., Payrastre, B., et al. (2010). A new alpha5beta1 integrin-dependent survival pathway through GSK3beta activation in leukemic cells. *PLoS One, 5*, e9807. https://doi.org/10.1371/journal.pone.0009807.

Deryugina, E. I., & Quigley, J. P. (2015). Tumor angiogenesis: MMP-mediated induction of intravasation- and metastasis-sustaining neovasculature. *Matrix Biology: Journal of the International Society for Matrix Biology, 44-46*, 94–112. https://doi.org/10.1016/j.matbio.2015.04.004.

Dittmer, A., Vetter, M., Schunke, D., Span, P. N., Sweep, F., Thomssen, C., et al. (2006). Parathyroid hormone-related protein regulates tumor-relevant genes in breast cancer cells. *The Journal of Biological Chemistry, 281*, 14563–14572. https://doi.org/10.1074/jbc.M510527200.

Fenton, A. J., Martin, T. J., & Nicholson, G. C. (1994). Carboxyl-terminal parathyroid hormone-related protein inhibits bone resorption by isolated chicken osteoclasts. *Journal of Bone and Mineral Research: the Official Journal of the American Society for Bone and Mineral Research*, *9*, 515–519. https://doi.org/10.1002/jbmr.5650090411.

Fleming, N. I., Trivett, M. K., George, J., Slavin, J. L., Murray, W. K., Moseley, J. M., et al. (2009). Parathyroid hormone-related protein protects against mammary tumor emergence and is associated with monocyte infiltration in ductal carcinoma in situ. *Cancer Research*, *69*, 7473–7479. https://doi.org/10.1158/0008-5472.CAN-09-0194.

Frieling, J. S., & Lynch, C. C. (2019). Proteolytic regulation of parathyroid hormone-related protein: Functional implications for skeletal malignancy. *International Journal of Molecular Sciences*, *20*, 2814. https://doi.org/10.3390/ijms20112814.

Gagiannis, S., Müller, M., Uhlemann, S., Koch, A., Melino, G., Krammer, P. H., et al. (2009). Parathyroid hormone-related protein confers chemoresistance by blocking apoptosis signaling via death receptors and mitochondria. *International Journal of Cancer*, *125*, 1551–1557. https://doi.org/10.1002/ijc.24471.

Han, P., Liu, J., Lei, Y., Lin, Z., Tian, D., & Yan, W. (2019). Netrin-1 promotes the collective cell migration of liver cancer cells in a 3D cell culture model. *Journal of Physiology and Biochemistry*, *75*, 489–498. https://doi.org/10.1007/s13105-019-00701-8.

Hiremath, M., & Wysolmerski, J. (2013). Parathyroid hormone-related protein specifies the mammary mesenchyme and regulates embryonic mammary development. *Journal of Mammary Gland Biology and Neoplasia*, *18*, 171–177. https://doi.org/10.1007/s10911-013-9283-7.

Hu, H., Xu, L., Chen, Y., Luo, S. J., Wu, Y. Z., Xu, S. H., et al. (2019). The upregulation of Trophinin-associated protein (TROAP) predicts a poor prognosis in hepatocellular carcinoma. *Journal of Cancer*, *10*, 957–967. https://doi.org/10.7150/jca.26666.

Huang, J., Wang, L., Jiang, M., Lin, H., Qi, L., & Diao, H. (2012). PTHLH coupling upstream negative regulation of fatty acid biosynthesis and Wnt receptor signal to downstream peptidase activity-induced apoptosis network in human hepatocellular carcinoma by systems-theoretical analysis. *Journal of Receptor and Signal Transduction Research*, *32*, 250–256. https://doi.org/10.3109/10799893.2012.700717.

Ibaragi, S., Shimo, T., Iwamoto, M., Hassan, N. M., Kodama, S., Isowa, S., et al. (2010). Parathyroid hormone-related peptide regulates matrix metalloproteinase-13 gene expression in bone metastatic breast cancer cells. *Anticancer Research*, *30*, 5029–5036.

Isowa, S., Shimo, T., Ibaragi, S., Kurio, N., Okui, T., Matsubara, K., et al. (2010). PTHrP regulates angiogenesis and bone resorption *via* VEGF expression. *Anticancer Research*, *30*, 2755–2767.

Johnson, R. W., Sun, Y., Ho, P., Chan, A., Johnson, J. A., Pavlos, N. J., et al. (2018). Parathyroid hormone-related protein negatively regulates tumor cell dormancy genes in a PTHR1/cyclic AMP-independent manner. *Frontiers in Endocrinology*, *9*, 241. https://doi.org/10.3389/fendo.2018.00241.

Kim, R. S., Avivar-Valderas, A., Estrada, Y., Bragado, P., Sosa, M. S., Aguirre-Ghiso, J. A., et al. (2012). Dormancy signatures and metastasis in estrogen receptor positive and negative breast cancer. *PLoS One*, *7*, e35569. https://doi.org/10.1371/journal.pone.0035569.

King, T. E., Pawar, S. C., Majuta, L., Sroka, I. C., Wynn, D., Demetriou, M. C., et al. (2008). The role of alpha 6 integrin in prostate cancer migration and bone pain in a novel xenograft model. *PLoS One*, *3*, e3535. https://doi.org/10.1371/journal.pone.0003535.

Kotelevets, L., & Chastre, E. (2020). Rac1 signaling: From intestinal homeostasis to colorectal cancer metastasis. *Cancers*, *12*, 665. https://doi.org/10.3390/cancers12030665.

Krock, B. L., Skuli, N., & Simon, M. C. (2011). Hypoxia-induced angiogenesis: Good and evil. *Genes & Cancer*, *2*, 1117–1133. https://doi.org/10.1177/1947601911423654.

Lai, M. Z., & Chen, R. H. (2014). Regulation of inflammation by DAPK. *Apoptosis: An International Journal on Programmed Cell Death*, *19*, 357–363. https://doi.org/https://doi.org/10.1007/s10495-013-0933-4.

Lam, M. H., Briggs, L. J., Hu, W., Martin, T. J., Gillespie, M. T., & Jans, D. A. (1999). Importin beta recognizes parathyroid hormone-related protein with high affinity and mediates its nuclear import in the absence of importin alpha. *The Journal of Biological Chemistry*, *274*, 7391–7398. https://doi.org/10.1074/jbc.274.11.7391.

Lam, M. H., House, C. M., Tiganis, T., Mitchelhill, K. I., Sarcevic, B., Cures, A., et al. (1999). Phosphorylation at the cyclin-dependent kinases site (Thr85) of parathyroid hormone-related protein negatively regulates its nuclear localization. *The Journal of Biological Chemistry*, *274*, 18559–18566. https://doi.org/10.1074/jbc.274.26.18559.

Lam, M. H., Olsen, S. L., Rankin, W. A., Ho, P. W., Martin, T. J., Gillespie, M. T., et al. (1997). PTHrP and cell division: Expression and localization of PTHrP in a keratinocyte cell line (HaCaT) during the cell cycle. *Journal of Cellular Physiology*, *173*, 433–446. https://doi.org/10.1002/(SICI)1097-4652(199712)173:3<433::AID-JCP16>3.0.CO;2-C.

Lam, M. H., Thomas, R. J., Loveland, K. L., Schilders, S., Gu, M., Martin, T. J., et al. (2002). Nuclear transport of parathyroid hormone (PTH)-related protein is dependent on microtubules. *Molecular Endocrinology (Baltimore, Md.)*, *16*, 390–401. https://doi.org/10.1210/mend.16.2.0775.

Lam, M. H., Thomas, R. J., Martin, T. J., Gillespie, M. T., & Jans, D. A. (2000). Nuclear and nucleolar localization of parathyroid hormone-related protein. *Immunology and Cell Biology*, *78*, 395–402. https://doi.org/10.1046/j.1440-1711.2000.00919.x.

Laman, J. D., Claassen, E., & Noelle, R. J. (2017). Functions of CD40 and its ligand, gp39 (CD40L). *Critical Reviews in Immunology*, *37*, 371–420. https://doi.org/10.1615/CritRevImmunol.v37.i2-6.100.

Lin, T., Zhang, B., & He, H. (2018). Identification of genes correlated with oral squamous cell carcinoma. *Journal of Cancer Research and Therapeutics*, *14*(Suppl), S675–S679. https://doi.org/10.4103/0973-1482.202893.

Liu, J. F., Tsao, Y. T., & Hou, C. H. (2015). Amphiregulin enhances intercellular adhesion molecule-1 expression and promotes tumor metastasis in human osteosarcoma. *Oncotarget*, *6*, 40880–40895. https://doi.org/10.18632/oncotarget.5679.

Longo, A., Librizzi, M., & Luparello, C. (2013). Effect of transfection with PLP2 antisense oligonucleotides on gene expression of cadmium-treated MDA-MB231 breast cancer cells. *Analytical and Bioanalytical Chemistry*, *405*, 1893–1901. https://doi.org/10.1007/s00216-012-6182-5.

Lopes, L., Tesser-Gamba, F., Petrilli, A. S., de Seixas Alves, M. T., Garcia-Filho, R. J., & Toledo, S. (2017). MAPK pathways regulation by DUSP1 in the development of osteosarcoma: Potential markers and therapeutic targets. *Molecular Carcinogenesis*, *56*, 1630–1641. https://doi.org/10.1002/mc.22619.

Luparello, C., Birch, M. A., Gallagher, J. A., & Burtis, W. J. (1997). Clonal heterogeneity of the growth and invasive response of a human breast carcinoma cell line to parathyroid hormone-related peptide fragments. *Carcinogenesis*, *18*, 23–29. https://doi.org/10.1093/carcin/18.1.23.

Luparello, C., Burtis, W. J., Raue, F., Birch, M. A., & Gallagher, J. A. (1995). Parathyroid hormone-related peptide and 8701-BC breast cancer cell growth and invasion in vitro: Evidence for growth-inhibiting and invasion-promoting effects. *Molecular and Cellular Endocrinology*, *111*, 225–232. https://doi.org/10.1016/0303-7207(95)03577-t.

Luparello, C., Librizzi, M., Asaro, D., Cruciata, I., & Caradonna, F. (2019). Mid-region parathyroid hormone-related protein is a genome-wide chromatin-binding factor that promotes growth and differentiation of HB2 epithelial cells from the human breast. *BioFactors (Oxford, England)*, *45*, 279–288. https://doi.org/10.1002/biof.1484.

Luparello, C., Romanotto, R., Tipa, A., Sirchia, R., Olmo, N., López de Silanes, I., et al. (2001). Midregion parathyroid hormone-related protein inhibits growth and invasion in vitro and tumorigenesis in vivo of human breast cancer cells. *Journal of Bone and Mineral Research: the Official Journal of the American Society for Bone and Mineral Research, 16*, 2173–2181. https://doi.org/10.1359/jbmr.2001.16.12.2173.

Luparello, C., Sirchia, R., & Lo Sasso, B. (2008). Midregion PTHrP regulates Rip1 and caspase expression in MDA-MB231 breast cancer cells. *Breast Cancer Research and Treatment, 111*, 461–474. https://doi.org/10.1007/s10549-007-9816-0.

Luparello, C., Sirchia, R., & Pupello, D. (2003). PTHrP [67-86] regulates the expression of stress proteins in breast cancer cells inducing modifications in urokinase-plasminogen activator and MMP-1 expression. *Journal of Cell Science, 116*, 2421–2430. https://doi.org/10.1242/jcs.00472.

Ma, Z., Lou, S., & Jiang, Z. (2020). PHLDA2 regulates EMT and autophagy in colorectal cancer via the PI3K/AKT signaling pathway. *Aging, 12*, 7985–8000. https://doi.org/10.18632/aging.103117.

Mak, I. W., Turcotte, R. E., & Ghert, M. (2012). Transcriptomic and proteomic analyses in bone tumor cells: Deciphering parathyroid hormone-related protein regulation of the cell cycle and apoptosis. *Journal of Bone and Mineral Research: the Official Journal of the American Society for Bone and Mineral Research, 27*, 1976–1991. https://doi.org/10.1002/jbmr.1638.

Mak, I. W., Turcotte, R. E., & Ghert, M. (2013). Parathyroid hormone-related protein (PTHrP) modulates adhesion, migration and invasion in bone tumor cells. *Bone, 55*, 198–207. https://doi.org/10.1016/j.bone.2013.02.020.

Martin, T. J., & Johnson, R. W. (2019). Multiple actions of parathyroid hormone-related protein in breast cancer bone metastasis. *British Journal of Pharmacology*. https://doi.org/10.1111/bph.14709. Advance online publication https://doi.org/10.1111/bph.14709.

Massfelder, T., Dann, P., Wu, T. L., Vasavada, R., Helwig, J. J., & Stewart, A. F. (1997). Opposing mitogenic and anti-mitogenic actions of parathyroid hormone-related protein in vascular smooth muscle cells: A critical role for nuclear targeting. *Proceedings of the National Academy of Sciences of the United States of America, 94*, 13630–13635. https://doi.org/10.1073/pnas.94.25.13630.

Mi, D., Zhang, M., Yan, J. D., Zhang, J., Wang, X., Wang, Q., et al. (2011). PTHrP inhibits BMP-6 expression through the PKA signaling pathway in breast cancer cells. *Journal of Cancer Research and Clinical Oncology, 137*, 295–303. https://doi.org/10.1007/s00432-010-0883-y.

Mula, R. V., Bhatia, V., & Falzon, M. (2010). PTHrP promotes colon cancer cell migration and invasion in an integrin α6β4-dependent manner through activation of Rac1. *Cancer Letters, 298*, 119–127. https://doi.org/10.1016/j.canlet.2010.06.009.

Naboulsi, W., Megger, D. A., Bracht, T., Kohl, M., Turewicz, M., Eisenacher, M., et al. (2016). Quantitative tissue proteomics analysis reveals Versican as potential biomarker for early-stage hepatocellular carcinoma. *Journal of Proteome Research, 15*, 38–47. https://doi.org/10.1021/acs.jproteome.5b00420.

Nadella, M. V., Shu, S. T., Dirksen, W. P., Thudi, N. K., Nadella, K. S., Fernandez, S. A., et al. (2008). Expression of parathyroid hormone-related protein during immortalization of human peripheral blood mononuclear cells by HTLV-1: Implications for transformation. *Retrovirology, 5*, 46. https://doi.org/10.1186/1742-4690-5-46.

Nakamura, H., Hiraga, T., Ninomiya, T., Hosoya, A., Fujisaki, N., Yoneda, T., et al. (2008). Involvement of cell-cell and cell-matrix interactions in bone destruction induced by metastatic MDA-MB-231 human breast cancer cells in nude mice. *Journal of Bone and Mineral Metabolism, 26*, 642–647. https://doi.org/10.1007/s00774-008-0857-1.

Nakashima, M., Nakayama, T., Ohtsuru, A., Fukada, E., Niino, D., Yamazumi, K., et al. (2003). Expression of parathyroid hormone (PTH)-related peptide (PthrP) and PTH/PTHrP receptor in osteoclast-like giant cells. *Pathology, Research and Practice, 199*, 85–92. https://doi.org/10.1078/0344-0338-00359.

Nishihara, M., Ito, M., Tomioka, T., Ohtsuru, A., Taguchi, T., & Kanematsu, T. (1999). Clinicopathological implications of parathyroid hormone-related protein in human colorectal tumours. *The Journal of Pathology, 187*, 217–222. https://doi.org/10.1002/(SICI)1096-9896(199901)187:2<217::AID-PATH210>3.0.CO;2-0.

Nishioka, K., Doki, Y., Shiozaki, H., Yamamoto, H., Tamura, S., Yasuda, T., et al. (2001). Clinical significance of CDC25A and CDC25B expression in squamous cell carcinomas of the oesophagus. *British Journal of Cancer, 85*, 412–421. https://doi.org/10.1054/bjoc.2001.1934.

O'Grady, S., & Morgan, M. P. (2019). Calcium transport and signalling in breast cancer: Functional and prognostic significance. *Seminars in Cancer Biology*. S1044-579X(19)30402-X. Advance online publication https://doi.org/10.1016/j.semcancer.2019.12.006.

Ongkeko, W. M., Burton, D., Kiang, A., Abhold, E., Kuo, S. Z., Rahimy, E., et al. (2014). Parathyroid hormone related-protein promotes epithelial-to-mesenchymal transition in prostate cancer. *PLoS One, 9*, e85803. https://doi.org/10.1371/journal.pone.0085803.

Oyinlade, O., Wei, S., Lal, B., Laterra, J., Zhu, H., Goodwin, C. R., et al. (2018). Targeting UDP-α-D-glucose 6-dehydrogenase inhibits glioblastoma growth and migration. *Oncogene, 37*, 2615–2629. https://doi.org/10.1038/s41388-018-0138-y.

Park, S. I., & McCauley, L. K. (2012). Nuclear localization of parathyroid hormone-related peptide confers resistance to anoikis in prostate cancer cells. *Endocrine-Related Cancer, 19*, 243–254. https://doi.org/10.1530/ERC-11-0278.

Shen, X., & Falzon, M. (2003a). Parathyroid hormone-related protein upregulates integrin expression via an intracrine pathway in PC-3 prostate cancer cells. *Regulatory Peptides, 113*, 17–29. https://doi.org/10.1016/s0167-0115(02)00293-8.

Shen, X., & Falzon, M. (2003b). PTH-related protein modulates PC-3 prostate cancer cell adhesion and integrin subunit profile. *Molecular and Cellular Endocrinology, 199*, 165–177. https://doi.org/10.1016/s0303-7207(02)00287-3.

Shen, X., & Falzon, M. (2006). PTH-related protein upregulates integrin alpha6beta4 expression and activates Akt in breast cancer cells. *Experimental Cell Research, 312*, 3822–3834. https://doi.org/10.1016/j.yexcr.2006.08.011.

Shen, X., Rychahou, P. G., Evers, B. M., & Falzon, M. (2007). PTHrP increases xenograft growth and promotes integrin alpha6beta4 expression and Akt activation in colon cancer. *Cancer Letters, 258*, 241–252. https://doi.org/10.1016/j.canlet.2007.09.010.

Shu, S. T., Dirksen, W. P., Lanigan, L. G., Martin, C. K., Thudi, N. K., Werbeck, J. L., et al. (2012). Effects of parathyroid hormone-related protein and macrophage inflammatory protein-1α in Jurkat T-cells on tumor formation in vivo and expression of apoptosis regulatory genes in vitro. *Leukemia & Lymphoma, 53*(4), 688–698. https://doi.org/10.3109/10428194.2011.626883.

Sirchia, R., & Luparello, C. (2007). Mid-region parathyroid hormone-related protein (PTHrP) and gene expression of MDA-MB231 breast cancer cells. *Biological Chemistry, 388*, 457–465. https://doi.org/10.1515/BC.2007.059.

Sirchia, R., Priulla, M., Sciandrello, G., Caradonna, F., Barbata, G., & Luparello, C. (2007). Mid-region parathyroid hormone-related protein (PTHrP) binds chromatin of MDA-MB231 breast cancer cells and isolated oligonucleotides "in vitro". *Breast Cancer Research and Treatment, 105*, 105–116. https://doi.org/10.1007/s10549-006-9437-z.

Sobti, A., Agrawal, P., Agarwala, S., & Agarwal, M. (2016). Giant cell tumor of bone - an overview. *The Archives of Bone and Joint Surgery, 4*, 2–9.

Soifer, N. E., Dee, K. E., Insogna, K. L., Burtis, W. J., Matovcik, L. M., Wu, T. L., et al. (1992). Parathyroid hormone-related protein. Evidence for secretion of a novel mid-region fragment by three different cell types. *The Journal of Biological Chemistry, 267*, 18236–18243.

Southby, J., Murphy, L. M., Martin, T. J., & Gillespie, M. T. (1996). Cell-specific and regulator-induced promoter usage and messenger ribonucleic acid splicing for parathyroid hormone-related protein. *Endocrinology, 137*, 1349–1357. https://doi.org/10.1210/endo.137.4.8625910.

Southby, J., O'Keeffe, L. M., Martin, T. J., & Gillespie, M. T. (1995). Alternative promoter usage and mRNA splicing pathways for parathyroid hormone-related protein in normal tissues and tumours. *British Journal of Cancer, 72*, 702–707. https://doi.org/10.1038/bjc.1995.397.

Stevens, M., & Oltean, S. (2019). Modulation of the apoptosis gene Bcl-x function through alternative splicing. *Frontiers in Genetics, 10*, 804. https://doi.org/10.3389/fgene.2019.00804.

Strid, H., Care, A., Jansson, T., & Powell, T. (2002). Parathyroid hormone-related peptide (38-94) amide stimulates ATP-dependent calcium transport in the basal plasma membrane of the human syncytiotrophoblast. *The Journal of Endocrinology, 175*, 517–524. https://doi.org/10.1677/joe.0.1750517.

Suva, L. J., Winslow, G. A., Wettenhall, R. E., Hammonds, R. G., Moseley, J. M., Diefenbach-Jagger, H., et al. (1987). A parathyroid hormone-related protein implicated in malignant hypercalcemia: Cloning and expression. *Science (New York, N.Y.), 237*, 893–896. https://doi.org/10.1126/science.3616618.

Thompson, D. D., Seedor, J. G., Fisher, J. E., Rosenblatt, M., & Rodan, G. A. (1988). Direct action of the parathyroid hormone-like human hypercalcemic factor on bone. *Proceedings of the National Academy of Sciences of the United States of America, 85*, 5673–5677. https://doi.org/10.1073/pnas.85.15.5673.

Tsigelny, I., Burton, D. W., Sharikov, Y., Hastings, R. H., & Deftos, L. J. (2005). Coherent expression chromosome cluster analysis reveals differential regulatory functions of amino-terminal and distal parathyroid hormone-related protein domains in prostate carcinoma. *Journal of Biomedicine & Biotechnology, 2005*, 353–363. https://doi.org/10.1155/JBB.2005.353.

Uemura, H., Yasui, T., Yoneda, N., Irahara, M., & Aono, T. (1997). Measurement of N- and C-terminal-region fragments of parathyroid hormone-related peptide in milk from lactating women and investigation of the relationship of their concentrations to calcium in milk. *The Journal of Endocrinology, 153*, 445–451. https://doi.org/10.1677/joe.0.1530445.

Vigneswaran, N., Wu, J., & Zacharias, W. (2003). Upregulation of cystatin M during the progression of oropharyngeal squamous cell carcinoma from primary tumor to metastasis. *Oral Oncology, 39*, 559–568. https://doi.org/10.1016/s1368-8375(03)00038-1.

Walia, M. K., Ho, P. M., Taylor, S., Ng, A. J., Gupte, A., Chalk, A. M., et al. (2016). Activation of PTHrP-cAMP-CREB1 signaling following p53 loss is essential for osteosarcoma initiation and maintenance. *eLife, 5*, e13446. https://doi.org/10.7554/eLife.13446.

Walkley, C. R., Walia, M. K., Ho, P. W., & Martin, T. J. (2014). PTHrP, its receptor, and protein kinase a activation in osteosarcoma. *Molecular & Cellular Oncology, 1*, e965624. https://doi.org/10.4161/23723548.2014.965624.

Wang, L., Huang, J., Jiang, M., Lin, H., Qi, L., & Diao, H. (2012). Activated PTHLH coupling feedback phosphoinositide to G-protein receptor signal-induced cell adhesion network in human hepatocellular carcinoma by systems-theoretic analysis. *TheScientificWorldJOURNAL, 2012*, 428979. https://doi.org/10.1100/2012/428979.

Wang, W., Nacusi, L., Sheaff, R. J., & Liu, X. (2005). Ubiquitination of p21Cip1/WAF1 by SCFSkp2: Substrate requirement and ubiquitination site selection. *Biochemistry, 44*, 14553–14564. https://doi.org/10.1021/bi051071j.

Wang, Y., Zhang, S., Xu, Y., Zhang, Y., Guan, H., Li, X., et al. (2016). Upregulation of miR-192 inhibits cell growth and invasion and induces cell apoptosis by targeting TCF7 in human osteosarcoma. *Tumour Biology: The Journal of the International Society for Oncodevelopmental Biology and Medicine, 37*, 15211–15220. https://doi.org/10.1007/s13277-016-5417-z.

Weekes, D., Kashima, T. G., Zandueta, C., Perurena, N., Thomas, D. P., Sunters, A., et al. (2016). Regulation of osteosarcoma cell lung metastasis by the c-Fos/AP-1 target FGFR1. *Oncogene, 35*, 2852–2861. https://doi.org/10.1038/onc.2015.344.

Willmarth, N. E., & Ethier, S. P. (2008). Amphiregulin as a novel target for breast cancer therapy. *Journal of Mammary Gland Biology and Neoplasia, 13*, 171–179. https://doi.org/10.1007/s10911-008-9081-9.

Wu, T. L., Vasavada, R. C., Yang, K., Massfelder, T., Ganz, M., Abbas, S. K., et al. (1996). Structural and physiologic characterization of the mid-region secretory species of parathyroid hormone-related protein. *The Journal of Biological Chemistry, 271*, 24371–24381. https://doi.org/10.1074/jbc.271.40.24371.

Xiong, Y., Lu, J., Fang, Q., Lu, Y., Xie, C., Wu, H., et al. (2019). UBE2C functions as a potential oncogene by enhancing cell proliferation, migration, invasion, and drug resistance in hepatocellular carcinoma cells. *Bioscience Reports, 39*. BSR20182384 https://doi.org/10.1042/BSR20182384.

Yang, K. H., dePapp, A. E., Soifer, N. E., Dreyer, B. E., Wu, T. L., Porter, S. E., et al. (1994). Parathyroid hormone-related protein: Evidence for isoform- and tissue-specific posttranslational processing. *Biochemistry, 33*, 7460–7469. https://doi.org/10.1021/bi00189a054.

Yang, M., Luna, L., Sørbø, J. G., Alseth, I., Johansen, R. F., Backe, P. H., et al. (2014). Human OXR1 maintains mitochondrial DNA integrity and counteracts hydrogen peroxide-induced oxidative stress by regulating antioxidant pathways involving p21. *Free Radical Biology & Medicine, 77*, 41–48. https://doi.org/10.1016/j.freeradbiomed.2014.09.003.

Yao, Z., Hu, K., Huang, H., Xu, S., Wang, Q., Zhang, P., et al. (2015). shRNA-mediated silencing of the RFC3 gene suppresses hepatocellular carcinoma cell proliferation. *International Journal of Molecular Medicine, 36*, 1393–1399. https://doi.org/10.3892/ijmm.2015.2350.

Yue, Y., Yuan, Y., Li, L., Fan, J., Li, C., Peng, W., et al. (2018). Homeobox protein MSX1 inhibits the growth and metastasis of breast cancer cells and is frequently silenced by promoter methylation. *International Journal of Molecular Medicine, 41*, 2986–2996. https://doi.org/10.3892/ijmm.2018.3468.

Zhang, H., Yu, C., Dai, J., Keller, J. M., Hua, A., Sottnik, J. L., et al. (2014). Parathyroid hormone-related protein inhibits DKK1 expression through c-Jun-mediated inhibition of β-catenin activation of the DKK1 promoter in prostate cancer. *Oncogene, 33*, 2464–2477. https://doi.org/10.1038/onc.2013.203.

Zhao, G. H., Xie, G. Q., & Miao, Y. Y. (2012). Expression and clinical significance of cyclin E2 in nasopharyngeal carcinoma. *Xi bao yu fen zi mian yi xue za zhi = Chinese Journal of Cellular and Molecular Immunology, 28*, 291–293.

Zhuang, L., Yang, Z., & Meng, Z. (2018). Upregulation of BUB1B, CCNB1, CDC7, CDC20, and MCM3 in tumor tissues predicted worse overall survival and disease-free survival in hepatocellular carcinoma patients. *BioMed Research International, 2018*, 7897346. https://doi.org/10.1155/2018/7897346.

CHAPTER EIGHT

Parathyroid hormone-related protein in breast cancer bone metastasis

Rachelle W. Johnson[a], Julie Rhoades[a,b], and T. John Martin[c,*]
[a]Department of Medicine, Division of Clinical Pharmacology, and Vanderbilt Center for Bone Biology, Vanderbilt University Medical Center, Nashville, TN, United States
[b]Department of Veterans Affairs, Tennessee Valley Healthcare System, Nashville, TN, United States
[c]St Vincent's Institute of Medical Research, University of Melbourne, St Vincent's Health, Melbourne, VIC, Australia
*Corresponding author: e-mail address: jmartin@svi.edu.au

Contents

Abstract

Parathyroid hormone-related protein (PTHrP) was discovered as the tumor product causing the humoral hypercalcemia of malignancy. Its structural similarity to the hormone, PTH, with 8 of the first 13 amino acids identical, was sufficient to explain the sharing by PTHrP and PTH of a common receptor, PTH1R, although the remainder of the sequences are unique. PTHrP has important roles in development of several organs, including breast and bone, and functions as a paracrine factor postnatally in these and other tissues. In addition to its hormonal role in cancer, PTHrP is produced by two thirds of primary breast cancers and 90% of bone metastases from breast cancer, leading to the concept that its production in bone by breast cancer cells promotes bone resorption, thus favoring tumor establishment and expansion, and an exit from tumor dormancy in bone through downregulation of leukemia inducing factor receptor (LIFR). Cancer production of PTHrP is increased by bone-derived growth factors, with particular attention paid to TGFβ, as well as by promoter-driven transcriptional effects, such as the hedgehog signaling factor, GLI2, and microenvironment effects including changes in underlying stiffness of substrates for cells. Although interest has been focused on

Vitamins and Hormones, Volume 120
ISSN 0083-6729
https://doi.org/10.1016/bs.vh.2022.04.006

PTHrP-induced bone resorption in bone metastasis, a mechanistically separate, protective effect against tumor progression has been proposed. Although there is conflicting mouse data, there are clinical studies suggesting that increased production of PTHrP by breast cancers confers upon them a less invasive phenotype, an effect distinct from the bone resorption-stimulating action that favors bone metastasis.

1. Introduction

Parathyroid hormone-related protein (PTHrP) was discovered through investigations into cancer-induced hypercalcemia in patients without bone metastases, the syndrome known as the humoral hypercalcemia of malignancy (HHM) (Suva et al., 1987). In 1941, when discussing a patient with renal carcinoma, a solitary metastasis and hypercalcemia, Fuller Albright suggested that some tumors might cause hypercalcemia by secreting PTH or something very like it. This gave rise in the next few decades to the concept of "ectopic PTH production" by cancers as the cause of nonmetastatic hypercalcemia. In subsequent years, a number of reports were published of patients with various cancers who seemed to fit this description (Gellhorn & Plimpton, 1956). Lafferty (Lafferty, 1966) reviewed the literature relating to 50 cases which fit at least one of three diagnostic criteria: (i) absence of bone metastasis or parathyroid adenoma, (ii) significant reduction of plasma calcium following tumor resection, and (iii) a positive PTH immunoassay of a tumor extract. 32 cases satisfied the first criterion, 20 the second and only 7 the third. This failure to demonstrate PTH immunoreactivity in tumor extracts might have been explained by the limited availability of assays; nevertheless, still it was accompanied by the conclusion that "*the synthesis of a PTH- like polypeptide by the malignant tumour itself appears to be the most likely mechanism for the hyperparathyroid like state.*"

2. Discovery and nature of PTHrP

Doubts about PTH as the cause of the cancer syndrome began to appear in the 1970s, when improved radioimmunoassays for PTH indicated that the immunoreactivity in tumor or plasma of patients with cancer differed from that of authentic PTH (Benson, Riggs, Pickard, & Arnaud, 1974; Melick, Martin, & Hicks, 1972; Riggs, Arnaud, Reynolds, & Smith, 1971; Roof, Carpenter, Fink, & Gordan, 1971), and with some assays PTH immunoreactivity could not be detected at all (Powell, Singer, Murray, Minkin, & Potts, 1973). Shortly afterwards, three clinical studies put beyond reasonable doubt the biochemical similarity between primary hyperparathyroidism and this syndrome of humoral hypercalcemia of malignancy (Kukreja, Shemerdiak, Lad, & Johnson, 1980; Rude et al., 1981; Stewart et al., 1980), by finding that patients with the HHM syndrome had hypercalcemia, hypophosphatemia and increased nephrogenous cyclic AMP, thus mimicking the biochemical features of primary hyperparathyroidism. Added to this was the finding that PTH mRNA could not be detected in any of 13 human and 3 animal tumors with the features of HHM (Simpson et al., 1983).

By that time, rapid, sensitive, robust biological assays of PTH had developed and extracts and culture supernatants of hypercalcemic animal and human tumors were found to contain PTH-like adenylate cyclase responses in osteoblast and kidney targets (Rodan et al., 1983; Stewart, Insogna, Goltzman, & Broadus, 1983). This paved the way for purification of PTHrP from a human lung cancer cell line (Moseley et al., 1987), a breast cancer (Burtis et al., 1987) and a renal cancer cell line (Strewler et al., 1987). Amino-acid sequencing (Suva et al., 1987) and cloning of its cDNA (Suva et al., 1987) and gene (Mangin, Ikeda, Dreyer, Milstone, & Broadus, 1988) showed 8 of the first 13 residues of PTHrP to be identical to those in PTH, with any remaining identities no more than expected by chance. The structural requirements for full biological activity of PTHrP were contained within the first 34 amino acids, as was known to be the case with PTH (Kemp et al., 1987). These findings explained the biochemical similarities between syndromes of PTH excess and nonmetastatic hypercalcemia in cancer.

3. Breast cancer and bone metastases before PTHrP

It had long been recognized that metastasis to bone occurs very commonly in patients with breast cancer. This was highlighted as long ago as 1890, when English surgeon Stephen Paget, in carrying out autopsies on several hundred women with breast cancer, observed the remarkable frequency of secondary growths in bone, especially at the ends of the femur and in the skull. He reached the conclusion that bone appears to be a preferred site of metastasis for breast cancer, rather than lung or liver. He concluded that "*in cancer of the breast the bones suffer in a special way, which cannot be explained by any theory of embolism alone. Some bones suffer more than others, the disease has its seats of election.*" This constituted the "seed and soil" hypothesis in which the "soil" of bone supports the survival and growth of breast cancer "seed," with the implication that bone has special properties which favor metastasis to that site. This concept continued to be validated by clinical observations and studies, and provided the stimulus to much research to the present day that seeks to identify the properties of the bone microenvironment that equip it so well to provide a favorable environment for tumor growth (Croucher, McDonald, & Martin, 2016).

An association between bone metastases in breast cancer and hypercalcemia had been long recognized (Milch & Changus, 1956), with an early suggested mechanism that the hypercalcemia was due to rapid localized bone disruption. Support for this direct tumor osteolysis hypothesis came from a study reporting hypercalcemia in patients with breast cancer metastatic to bone that progressed when treated with sex steroids (Woodard, Escher, & Farrow, 1954). Although such a local mechanism around tumor deposits seemed to imply excessive resorption due to osteoclasts, the relative paucity of osteoclasts in the region of bone loss was noted (Milch & Changus, 1956), and it was concluded that the process of malignant bone destruction was most likely due to direct tumor pressure or a physicochemical process due to the presence of tumor cells. If that were so it would be expected that the extent of lytic metastases would correlate with the severity of hypercalcemia. No such correlation was found in a clinical study of a variety of tumors including breast cancers (Ralston et al., 1984). In contrast, clinical studies in which renal handling of calcium and phosphate and nephrogenous cyclic AMP were examined, suggested that in up to 60% of hypercalcemic breast cancer patients, humoral-mediated factors were operative (Gallacher et al., 1990; Isales, Carcangiu, & Stewart, 1987). To what extent such humoral activity contributed to the hypercalcemia was not addressed. Overall, ample histological evidence was accumulated that human tumor breast cancer deposits in bone are surrounded by osteoclasts (Clohisy, Palkert, Ramnaraine, Pekurovsky, & Oursler, 1996; Mundy, 1987), a finding that supported the concept of peri-metastasis resorptive activity. Similar evidence was available from animal models of tumor osteolysis (Clohisy et al., 1996; Clohisy, Perkins, & Ramnaraine, 2000) and of metastasis to bone (Lelekakis et al., 1999).

Although these clinical studies of breast cancer were quite instructive, they received little attention at the time. Certainly, they provide in retrospect a link between breast cancer and PTHrP that came to be confirmed after the discovery of PTHrP. They added to other links between PTHrP and breast cancer, including the facts that a breast cancer was one of the sources of PTHrP that provided some protein sequence (Burtis et al., 1987), and that PTH-like immunoreactivity (nonparallel to standard PTH) was extracted from a breast cancer of a hypercalcemic patient (Melick et al., 1972).

4. Breast biology, breast cancer and PTHrP

After PTHrP was characterized, much evidence pointed to its involvement in breast biology. In parallel with studies in breast cancer, it was evident that PTHrP was involved in early breast development and in activation of the mature breast. PTHrP at high concentrations was found in the milk of several species, including that of humans (Budayr et al., 1989; Grill et al., 1992). Suckling in the rat resulted in increased urinary phosphate and cAMP (Yamamoto et al., 1992), and a venous - arterial gradient of plasma PTHrP was found in the goat mammary gland (Ratcliffe, 1992). Although it is still not clear what is the significance of such high concentrations of PTHrP in mammalian milk, its involvement in mammary development in addition to breast activation became apparent. Ablation of PTHrP in the mouse resulted in perinatal fatality because of skeletal deformities. When this phenotype in the mice was "rescued" by directing PTHrP production to cartilage with use of the collagen II promoter, the "rescued" mice failed to show early ductal development, thus providing evidence for a role for PTHrP in early mammary gland development (Wysolmerski et al., 1998; Wysolmerski, McCaughern-Carucci, Daifotis, Broadus, & Philbrick, 1995) and illustrating one of the important roles of PTHrP as a paracrine effector in normal breast biology.

In primary breast cancers removed at surgery, two thirds stained positively by immunohistochemistry for PTHrP (Southby et al., 1990), a finding confirmed by immunostaining in several clinical studies (Bouizar, Spyratos, & De vernejoul, 1999; Bundred et al., 1992; Kissin et al., 1993), as well as by using detection of mRNA by in situ hybridization (Vargas et al., 1992). These data were corroborated by a retrospective study that identified the gene for PTHrP (*PTHLH*) as a susceptibility locus for both estrogen receptor positive and negative breast cancers (Ghoussaini et al., 2012).

An important observation was made in a retrospective study of primary breast cancers and metastases in bone and soft tissue showing that 85% of bone metastases were positive for PTH compared to 18% of those in nonbone tissues (Powell et al., 1991). This led to the proposal that production of PTHrP as a bone resorbing factor might be a property of breast cancers that would favor their growth in bone, by allowing them to resorb a niche in which cancer cells could grow.

This suggestion was taken up in an experimental study in which the human breast cancer MDA-MB-231 cells were inoculated by intracardiac injection into nude nice and lytic bone deposits were analyzed after they developed. The mice had no elevated blood PTHrP levels but increased levels in marrow plasma and importantly, a neutralizing antibody against PTHrP largely prevented tumor establishment and growth (Guise et al., 1996), as did resorption inhibition with a bisphosphonate treatment (Yoneda et al., 1997). These and other similar experiments served to establish the concept that tumor production of a bone resorbing cytokine (in this case PTHrP) could facilitate cancer establishment in bone.

These findings, together with that of elevated circulating PTHrP levels in 70% of hypercalcemic patients with bone metastases in breast cancer (Grill et al., 1991), raised the question whether production of PTHrP in primary cancers might influence the development of metastases in bone. Several studies concluded that this was the case, but they were either retrospective, had limited numbers of patients or were carried out in selected groups with short follow-up, and/or in patients with advanced tumors (Bouizar et al., 1999; Bundred et al., 1992; Kissin et al., 1993).

In a long-term, prospective study of breast cancers obtained at surgery at a single center, positive immunostaining for PTHrP in the primary cancers was independently predictive of reduced metastases at all sites, including bone, in analyses carried out at 5 and 10 years (Henderson et al., 2006). A further study consistent with an action of PTHrP to confer a less invasive phenotype, was an analysis by immunostaining in two breast cancer patient groups that loss of PTHrP nuclear (but not cytoplasmic) staining was associated with an unfavorable prognosis (Tran et al., 2018). Other support for the concept came from a publicly available data base of independent transcript analysis of *Pthlh* ($n=3549$ patients) (Johnson et al., 2018; Lanczky et al., 2016), showing association between low *Pthlh* transcript levels and unfavorable prognosis, applying to both ER positive and ER negative breast cancer patients.

This surprising finding from the clinical studies suggested an action of PTHrP in breast cancer, probably early in tumor development, that was entirely distinct from its bone resorbing action in late cancer that could facilitate the invasion of bone (Guise et al., 1996; Powell et al., 1991). That proposition is argued however, and the question remains unanswered. Two studies in mouse models reached opposite conclusions. Ablation of the *Pthlh* gene in mammary epithelial cells of mice with cancers associated with MMTV-polyomavirus middle T antigen (PYMT) resulted in prolonged tumor latency and slowing of tumor growth and metastases (Li et al., 2011). The opposite result was obtained in a separate study in which *Pthlh* was ablated in mice with MMTV-*neu* induced carcinogenesis (Fleming et al., 2009).

A new dimension was added with the findings that overexpression of PTHrP in the human breast cancer cells, MCF7, conferred upon these cells the ability to colonize bone aggressively and induce osteolysis following intracardiac injection into nude mice (Thomas et al., 1999). A follow-up study many years later found that MCF7 cells remain relatively dormant in the bone marrow (Johnson et al., 2016); these findings thus suggested that overexpression of PTHrP may convert dormant tumor cells to an aggressive, osteolytic phenotype. This concept was further supported by the finding that PTHrP downregulates the pro-dormancy factor and tumor suppressor leukemia inhibitory factor receptor (LIFR) and other genes known to promote quiescence in breast cancer cells (Johnson et al., 2016), suggesting multiple potential mechanisms for PTHrP-mediated dormancy exit in bone. In further work it was shown that the effect of PTHrP on LIFR and other dormancy-promoting genes is the result of actions that do not require the cAMP/PKA signaling cascade induced by the N-terminal domain of PTHrP, in other words "noncanonical" actions of PTHrP are operating (Johnson et al., 2016). Collectively these data suggested that PTHrP regulates tumor dormancy through actions attributed to its other biological domains that do not act through PTH1R, such as its nuclear localization sequence and/or carboxy-terminal region. There is strong evidence for this potential role from previously published work demonstrating that PTHrP regulates cell proliferation. PTHrP increases the proliferative markers Ki67 and cyclin D1 in breast cancer (Phan & Croucher, 2020) and leads to an enrichment in cells in G2/M (Gujral, Burton, Terkeltaub, & Deftos, 2001); however, expression of the PTHrP nuclear localization sequence reduces cell cycle progression and proliferation (Gujral et al., 2001; Hastings et al., 2009). It is therefore possible that PTHrP regulates tumor dormancy through the NLS, but whether this mechanism requires the NLS residues alone or specifically the translocation of PTHrP out of the nucleus remains unknown.

5. Regulation of PTHrP in tumor-induced bone disease

Since PTHrP was established as a predictor of HHM and bone destruction, it became the focus of many groups to develop therapies for treatment of patients with tumor-induced bone diseases (TIBD). Early studies demonstrated in mouse models that inhibition of PTHrP with a neutralizing antibody reduced bone destruction (Guise et al., 1996), which led to a follow-up clinical trial. However, efforts at the time to test an inhibitory antibody in patients were halted. Despite few specifics regarding why the trials were discontinued, efforts were re-directed to target regulators of PTHrP, leading to decades of work to discover how PTHrP is regulated in bone-metastatic cancers.

5.1 TGF-β and the vicious cycle: Fig. 1 about here

Much of the focus on studying signaling pathways in breast cancer induced bone disease has been on the regulation of PTHrP by the transforming

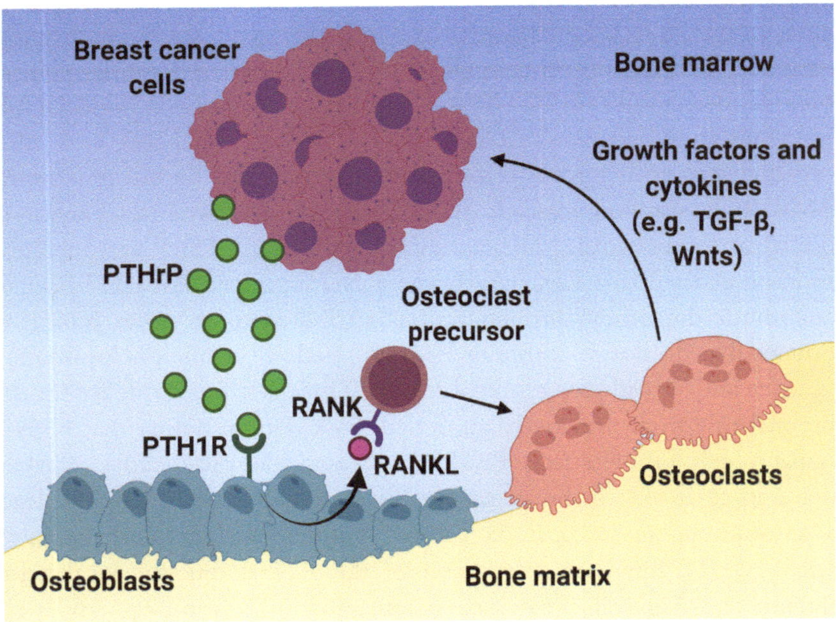

Fig. 1 PTHrP in the tumor-bone microenvironment. Bone-disseminated tumor cells secrete parathyroid hormone-related protein (PTHrP), which binds to the PTH receptor (PTH1R) on osteoblast lineage cells. PTHrP/PTH1R induces downstream signaling in osteoblasts, resulting in up-regulation of receptor activator of nuclear factor κB ligand (RANKL). RANKL expressed and secreted by osteoblasts binds to receptor activator of nuclear factor κB (RANK) on osteoclast precursor cells, which promotes osteoclast differentiation. This results in a localized increase in osteoclast-mediated bone resorption, which releases growth factors and cytokines that are stored within the bone matrix. As the bone is degraded, these matrix-derived factors fuel tumor growth in the microenvironment. Some of these factors, for example transforming growth factor β (TGF-β) and Wnts, can directly or indirectly stimulate PTHrP production by the tumor cells, thus further perpetuating tumor growth and bone destruction.

growth factor beta (TGF-β) signaling pathway (Fig. 1). Early studies by the Guise and Mundy labs demonstrated that tumoral expression of TGF-β type II receptor was required for tumors to produce PTHrP (Guise et al., 1996; Kakonen et al., 2002; Yin et al., 1999). In these and many other studies, it has been well established that loss of TGF-β response reduces tumor-induced bone destruction. Due to this, many groups have tested different TGF-β inhibitors in bone metastatic mouse models. While largely successful at reducing the tumor-induced osteolytic component, there have been mixed results in the effect of TGF-β inhibitors on tumor burden (Buenrostro et al., 2018; Mohammad et al., 2011. This suggests that while TGF-β is an important regulator of PTHrP, inhibition of TGF-β alone is not sufficient to eliminate tumor burden in the bone. While TGF-β inhibition reduces PTHrP, it does not eliminate PTHrP production in the cell, and it is therefore unclear whether the inability of TGF-β blockers to eliminate tumor burden is due to an incomplete blockade of PTHrP. TGF-β inhibitors have been used successfully for fibrotic diseases and some tumors, but the use of these inhibitors has not translated to clinical therapies for patients with bone metastatic disease (Lonning, Mannick, & McPherson, 2011). This is largely due to valid concerns that TGF-β plays disparate roles in early and metastatic tumors, which complicates the timing of therapy. In addition, TGF-β plays a critical role in many physiological processes, including immune responses (Bierie & Moses, 2009; Buenrostro et al., 2018; Lonning et al., 2011).

5.2 Transcriptional regulation of PTHrP expression by Gli2

The transcriptional regulation of the gene for PTHrP, *Pthlh*, is complex, with multiple promoter regions that control the production of multiple variants of PTHrP with different sizes and possible different functions. Thus, understanding the regulation of *Pthlh* has lagged behind our understanding of how PTHrP regulates hypercalcemia and bone metastatic disease. The *Pthlh* gene contains 3 separate promoters that span over 4 kb (Southby, O'Keeffe, Martin, & Gillespie, 1995). Promoter usage can vary between different physiological roles and tumors. Understanding how the promoters are regulated will be critical for understanding how to target PTHrP in bone metastatic cancers.

Multiple signaling molecules can bind to the *Pthlh* promoter. These include SMAD 2/3 (TGF-β), β-catenin (WNT), ETS, Sp1, and many others. Despite this, targeting these transcription factors in the pre-clinical setting has not been effective in blocking bone metastasis and bone metastatic progression, in part due to their important role in bone formation. For example, inhibition of WNT signaling impairs bone formation, leading to potential elevated fracture risk. Other signaling pathways have been implicated in the regulation of PTHrP in breast cancer induced bone disease, particularly the Hedgehog signaling transcription factor GLI2. The Mundy lab was the first to identify the Hedgehog transcription factor GLI2 as a driver of PTHrP expression in bone metastatic tumors (Sterling et al., 2006), following on the finding that Gli2 regulates PTHrP production in chrondrocytes. Subsequent studies have shown that GLI2 is regulated by both TGF-β and Wnt signaling (Johnson et al., 2011, 2014) and could potentially be a more bone specific target to reduce tumor-induced bone disease. GLI2 expression can be regulated by TGF-β to stimulate the transcription of PTHrP (Johnson et al., 2011), suggesting this may be an effective approach to regulate tumoral PTHrP production in both the primary tumor site and in bone metastatic disease. Due to the promise of GLI2 as a therapeutic target, therapeutic inhibitors of GLI2 and different delivery methods for these inhibitors have been identified, including nanoparticles, that may be useful for the treatment of TIBD in multiple bone metastatic models. It is likely that GLI2, a primarily developmental transcription factor, has other roles in mediating tumor metastasis to bone and bone destruction, beyond the regulation of PTHrP. Some areas of active investigation for GLI2 regulation in metastatic tumors include immune regulation and drug resistance; however, it is possible that these mechanisms may still link back to PTHrP, since PTHrP also has known roles in modulating immune cell activity and promoting cell survival (Park et al., 2013; Tovar Sepulveda, Shen, & Falzon, 2002; Yang et al., 2008; Younos et al., 2011).

6. Crosstalk with the microenvironment regulation of PTHrP

The primary tumor and bone microenvironments are well established regulators of PTHrP expression and other tumor-produced factors that contribute to bone metastatic disease. These interactions are multi-level including interactions with cells (osteoblast, osteoclasts, osteocytes, myeloid precursor/macrophages, fibroblasts), matrix (collagen, fibronectin), physical properties (rigidity, shear stress), cytokines/chemokines, and minerals (calcium). All of these have been shown to regulate tumor-produced PTHrP. For example, interactions with TGF-β1 producing macrophages can lead to increased PTHrP, more bone destruction, and tumor growth (Buenrostro et al., 2018; Buenrostro, Mulcrone, Owens, & Sterling, 2016; Buenrostro, Park, & Sterling, 2014; Danilin et al., 2012; Park et al., 2013).

The elevated rigidity of the bone microenvironment compared to soft tissue sites also upregulates PTHrP production by tumor cells through integrin signaling (Page et al., 2015; Ruppender et al., 2010). Given that there are fundamental differences in cellularity and stiffness within these different microenvironments, and that there are profoundly mixed reports on the role for PTHrP as a tumor promoter and prognostic factor, these features may be the deciding factor for whether PTHrP acts as a factor that promotes or inhibits tumorigenesis and progress in bone. It is therefore possible that the most effective way to ultimately target PTHrP will be through targeting of selected niches.

Funding information

Rachelle Johnson is supported by DoD Breakthrough Award W81XWH-18-1-0029 (RWJ). Julie Rhoades (Sterling) is supported by the Department of Veterans Affairs Merit Award 1I01BX001957. T. John Martin is supported by National Health and Medical Research Council (award No. 620203), and Victorian Government OIS Program.

References

Benson, R. C., Jr., Riggs, B. L., Pickard, B. M., & Arnaud, C. D. (1974). Immunoreactive forms of circulating parathyroid hormone in primary and ectopic hyperparathyroidism. *The Journal of Clinical Investigation*, *54*(1), 175–181. https://doi.org/10.1172/JCI107739.

Bierie, B., & Moses, H. L. (2009). Gain or loss of TGFbeta signaling in mammary carcinoma cells can promote metastasis. *Cell Cycle*, *8*(20), 3319–3327. https://doi.org/10.4161/cc.8.20.9727.

Bouizar, Z., Spyratos, F., & De vernejoul, M. C. (1999). The parathyroid hormone-related protein (PTHrP) gene: Use of downstream TATA promotor and PTHrP 1-139 coding pathways in primary breast cancers vary with the occurrence of bone metastasis. *Journal of Bone and Mineral Research*, *14*(3), 406–414. https://doi.org/10.1359/jbmr.1999.14.3.406.

Budayr, A. A., Halloran, B. P., King, J. C., Diep, D., Nissenson, R. A., & Strewler, G. J. (1989). High levels of a parathyroid hormone-like protein in milk. *Proceedings of the National Academy of Sciences of the United States of America*, *86*(18), 7183–7185. https://doi.org/10.1073/pnas.86.18.7183.

Buenrostro, D., Kwakwa, K. A., Putnam, N. E., Merkel, A. R., Johnson, J. R., Cassat, J. E., et al. (2018). Early TGF-beta inhibition in mice reduces the incidence of breast cancer induced bone disease in a myeloid dependent manner. *Bone*, *113*, 77–88. https://doi.org/10.1016/j.bone.2018.05.008.

Buenrostro, D., Mulcrone, P. L., Owens, P., & Sterling, J. A. (2016). The bone microenvironment: A fertile soil for tumor growth. *Current Osteoporosis Reports*, *14*(4), 151–158. https://doi.org/10.1007/s11914-016-0315-2.

Buenrostro, D., Park, S. I., & Sterling, J. A. (2014). Dissecting the role of bone marrow stromal cells on bone metastases. *BioMed Research International*, *2014*, 875305. https://doi.org/10.1155/2014/875305.

Bundred, N. J., Walker, R. A., Ratcliffe, W. A., Warwick, J., Morrison, J. M., & Ratcliffe, J. G. (1992). Parathyroid hormone related protein and skeletal morbidity in breast cancer. *European Journal of Cancer*, *28*(2–3), 690–692. https://doi.org/10.1016/s0959-8049(05)80127-3.

Burtis, W. J., Wu, T., Bunch, C., Wysolmerski, J. J., Insogna, K. L., Weir, E. C., et al. (1987). Identification of a novel 17, 000-dalton parathyroid hormone-like adenylate cyclase-stimulating protein from a tumor associated with humoral hypercalcemia of malignancy. *Journal of Biological Chemistry*, *262*(15), 7151–7156. Retrieved from https://www.ncbi.nlm.nih.gov/pubmed/3584110.

Clohisy, D. R., Palkert, D., Ramnaraine, M. L., Pekurovsky, I., & Oursler, M. J. (1996). Human breast cancer induces osteoclast activation and increases the number of osteoclasts at sites of tumor osteolysis. *Journal of Orthopaedic Research*, *14*(3), 396–402. https://doi.org/10.1002/jor.1100140309.

Clohisy, D. R., Perkins, S. L., & Ramnaraine, M. L. (2000). Review of cellular mechanisms of tumor osteolysis. *Clinical Orthopaedics and Related Research*, *373*, 104–114. https://doi.org/10.1097/00003086-200004000-00013.

Croucher, P. I., McDonald, M. M., & Martin, T. J. (2016). Bone metastasis: The importance of the neighbourhood. *Nature Reviews. Cancer*, *16*(6), 373–386. https://doi.org/10.1038/nrc.2016.44.

Danilin, S., Merkel, A. R., Johnson, J. R., Johnson, R. W., Edwards, J. R., & Sterling, J. A. (2012). Myeloid-derived suppressor cells expand during breast cancer progression and promote tumor-induced bone destruction. *Oncoimmunology*, *1*(9), 1484–1494. https://doi.org/10.4161/onci.21990.

Fleming, N. I., Trivett, M. K., George, J., Slavin, J. L., Murray, W. K., Moseley, J. M., et al. (2009). Parathyroid hormone-related protein protects against mammary tumor emergence and is associated with monocyte infiltration in ductal carcinoma in situ. *Cancer Research*, *69*(18), 7473–7479. https://doi.org/10.1158/0008-5472.CAN-09-0194.

Gallacher, S. J., Fraser, W. D., Patel, U., Logue, F. C., Soukop, M., Boyle, I. T., et al. (1990). Breast cancer-associated hypercalcaemia: A reassessment of renal calcium and phosphate handling. *Annals of Clinical Biochemistry*, *27*(Pt 6), 551–556. https://doi.org/10.1177/000456329002700605.

Gellhorn, A., & Plimpton, C. H. (1956). Hypercalcemia in malignant disease without evidence of bone destruction. *The American Journal of Medicine*, *21*(5), 750–759. Retrieved from https://www.ncbi.nlm.nih.gov/pubmed/13362306.

Ghoussaini, M., Fletcher, O., Michailidou, K., Turnbull, C., Schmidt, M. K., Dicks, E., et al. (2012). Genome-wide association analysis identifies three new breast cancer susceptibility loci. *Nature Genetics*, *44*(3), 312–318. https://doi.org/10.1038/ng.1049.

Grill, V., Hillary, J., Ho, P. M., Law, F. M., MacIsaac, R. J., MacIsaac, I. A., et al. (1992). Parathyroid hormone-related protein: A possible endocrine function in lactation. *Clinical Endocrinology*, *37*(5), 405–410. https://doi.org/10.1111/j.1365-2265.1992.tb02350.x.

Grill, V., Ho, P., Body, J. J., Johanson, N., Lee, S. C., Kukreja, S. C., et al. (1991). Parathyroid hormone-related protein: Elevated levels in both humoral hypercalcemia of malignancy and hypercalcemia complicating metastatic breast cancer. *The Journal of Clinical Endocrinology and Metabolism*, *73*(6), 1309–1315. https://doi.org/10.1210/jcem-73-6-1309.

Guise, T. A., Yin, J. J., Taylor, S. D., Kumagai, Y., Dallas, M., Boyce, B. F., et al. (1996). Evidence for a causal role of parathyroid hormone-related protein in the pathogenesis of human breast cancer-mediated osteolysis. *The Journal of Clinical Investigation*, *98*(7), 1544–1549. https://doi.org/10.1172/JCI118947.

Gujral, A., Burton, D. W., Terkeltaub, R., & Deftos, L. J. (2001). Parathyroid hormone-related protein induces interleukin 8 production by prostate cancer cells via a novel intracrine mechanism not mediated by its classical nuclear localization sequence. *Cancer Research*, *61*(5), 2282–2288. Retrieved from https://www.ncbi.nlm.nih.gov/pubmed/11280799.

Hastings, R. H., Montgrain, P. R., Quintana, R., Rascon, Y., Deftos, L. J., & Healy, E. (2009). Cell cycle actions of parathyroid hormone-related protein in non-small cell lung carcinoma. *American Journal of Physiology. Lung Cellular and Molecular Physiology*, *297*(4), L578–L585. https://doi.org/10.1152/ajplung.90560.2008.

Henderson, M. A., Danks, J. A., Slavin, J. L., Byrnes, G. B., Choong, P. F., Spillane, J. B., et al. (2006). Parathyroid hormone-related protein localization in breast cancers predict improved prognosis. *Cancer Research*, *66*(4), 2250–2256. https://doi.org/10.1158/0008-5472.CAN-05-2814.

Isales, C., Carcangiu, M. L., & Stewart, A. F. (1987). Hypercalcemia in breast cancer. Reassessment of the mechanism. *The American Journal of Medicine*, *82*(6), 1143–1147. https://doi.org/10.1016/0002-9343(87)90216-6.

Johnson, R. W., Finger, E. C., Olcina, M. M., Vilalta, M., Aguilera, T., Miao, Y., et al. (2016). Induction of LIFR confers a dormancy phenotype in breast cancer cells disseminated to the bone marrow. *Nature Cell Biology*, *18*(10), 1078–1089. https://doi.org/10.1038/ncb3408.

Johnson, R. W., Merkel, A. R., Page, J. M., Ruppender, N. S., Guelcher, S. A., & Sterling, J. A. (2014). Wnt signaling induces gene expression of factors associated with bone destruction in lung and breast cancer. *Clinical & Experimental Metastasis*, *31*(8), 945–959. https://doi.org/10.1007/s10585-014-9682-1.

Johnson, R. W., Nguyen, M. P., Padalecki, S. S., Grubbs, B. G., Merkel, A. R., Oyajobi, B. O., et al. (2011). TGF-beta promotion of Gli2-induced expression of parathyroid hormone-related protein, an important osteolytic factor in bone metastasis, is independent of canonical hedgehog signaling. *Cancer Research*, *71*(3), 822–831. https://doi.org/10.1158/0008-5472.CAN-10-2993.

Johnson, R. W., Sun, Y., Ho, P. W. M., Chan, A. S. M., Johnson, J. A., Pavlos, N. J., et al. (2018). Parathyroid hormone-related protein negatively regulates tumor cell dormancy genes in a PTHR1/cyclic AMP-independent manner. *Frontiers in Endocrinology (Lausanne)*, *9*, 241. https://doi.org/10.3389/fendo.2018.00241.

Kakonen, S. M., Selander, K. S., Chirgwin, J. M., Yin, J. J., Burns, S., Rankin, W. A., et al. (2002). Transforming growth factor-beta stimulates parathyroid hormone-related protein and osteolytic metastases via Smad and mitogen-activated protein kinase signaling pathways. *The Journal of Biological Chemistry*, *277*(27), 24571–24578. https://doi.org/10.1074/jbc.M202561200.

Kemp, B. E., Moseley, J. M., Rodda, C. P., Ebeling, P. R., Wettenhall, R. E., Stapleton, D., et al. (1987). Parathyroid hormone-related protein of malignancy: Active synthetic fragments. *Science*, *238*(4833), 1568–1570. https://doi.org/10.1126/science.3685995.

Kissin, M. W., Henderson, M. A., Danks, J. A., Hayman, J. A., Bennett, R. C., & Martin, T. J. (1993). Parathyroid hormone related protein in breast cancers of widely varying prognosis. *European Journal of Surgical Oncology*, *19*(2), 134–142. Retrieved from https://www.ncbi.nlm.nih.gov/pubmed/8491317.

Kukreja, S. C., Shemerdiak, W. P., Lad, T. E., & Johnson, P. A. (1980). Elevated nephrogenous cyclic AMP with normal serum parathyroid hormone levels in patients with lung cancer. *The Journal of Clinical Endocrinology and Metabolism*, *51*(1), 167–169. https://doi.org/10.1210/jcem-51-1-167.

Lafferty, F. W. (1966). Pseudohyperparathyroidism. *Medicine (Baltimore)*, *45*(3), 247–260. https://doi.org/10.1097/00005792-196605000-00004.

Lanczky, A., Nagy, A., Bottai, G., Munkacsy, G., Szabo, A., Santarpia, L., et al. (2016). miRpower: A web-tool to validate survival-associated miRNAs utilizing expression data from 2178 breast cancer patients. *Breast Cancer Research and Treatment*, *160*(3), 439–446. https://doi.org/10.1007/s10549-016-4013-7.

Lelekakis, M., Moseley, J. M., Martin, T. J., Hards, D., Williams, E., Ho, P., et al. (1999). A novel orthotopic model of breast cancer metastasis to bone. *Clinical & Experimental Metastasis*, *17*(2), 163–170. https://doi.org/10.1023/a:1006689719505.

Li, J., Karaplis, A. C., Huang, D. C., Siegel, P. M., Camirand, A., Yang, X. F., et al. (2011). PTHrP drives breast tumor initiation, progression, and metastasis in mice and is a potential therapy target. *The Journal of Clinical Investigation*, *121*(12), 4655–4669. https://doi.org/10.1172/JCI46134.

Lonning, S., Mannick, J., & McPherson, J. M. (2011). Antibody targeting of TGF-beta in cancer patients. *Current Pharmaceutical Biotechnology*, *12*(12), 2176–2189. https://doi.org/10.2174/138920111798808392.

Mangin, M., Ikeda, K., Dreyer, B. E., Milstone, L., & Broadus, A. E. (1988). Two distinct tumor-derived, parathyroid hormone-like peptides result from alternative ribonucleic acid splicing. *Molecular Endocrinology*, *2*(11), 1049–1055. https://doi.org/10.1210/mend-2-11-1049.

Melick, R. A., Martin, T. J., & Hicks, J. D. (1972). Parathyroid hormone production and malignancy. *British Medical Journal*, *2*(5807), 204–205. https://doi.org/10.1136/bmj.2.5807.204.

Milch, R. A., & Changus, G. W. (1956). Response of bone to tumor invasion. *Cancer*, *9*(2), 340–351. https://doi.org/10.1002/1097-0142(195603/04)9:2<340::aid-cncr2820090221->3.0.co;2-c.

Mohammad, K. S., Javelaud, D., Fournier, P. G., Niewolna, M., McKenna, C. R., Peng, X. H., et al. (2011). TGF-beta-RI kinase inhibitor SD-208 reduces the development and progression of melanoma bone metastases. *Cancer Research*, *71*(1), 175–184. https://doi.org/10.1158/0008-5472.CAN-10-2651.

Moseley, J. M., Kubota, M., Diefenbach-Jagger, H., Wettenhall, R. E., Kemp, B. E., Suva, L. J., et al. (1987). Parathyroid hormone-related protein purified from a human lung cancer cell line. *Proceedings of the National Academy of Sciences of the United States of America*, *84*(14), 5048–5052. https://doi.org/10.1073/pnas.84.14.5048.

Mundy, G. R. (1987). Tumor products and the hypercalcemia of malignancy. *Revue Médicale de Bruxelles*, *8*(6–7), 353–357. Retrieved from https://www.ncbi.nlm.nih.gov/pubmed/3671927.

Page, J. M., Merkel, A. R., Ruppender, N. S., Guo, R., Dadwal, U. C., Cannonier, S., et al. (2015). Matrix rigidity regulates the transition of tumor cells to a bone-destructive phenotype through integrin beta3 and TGF-beta receptor type II. *Biomaterials*, *64*, 33–44. https://doi.org/10.1016/j.biomaterials.2015.06.026.

Park, S. I., Lee, C., Sadler, W. D., Koh, A. J., Jones, J., Seo, J. W., et al. (2013). Parathyroid hormone-related protein drives a CD11b + Gr1 + cell-mediated positive feedback loop to support prostate cancer growth. *Cancer Research*, *73*(22), 6574–6583. https://doi.org/10.1158/0008-5472.CAN-12-4692.

Phan, T. G., & Croucher, P. I. (2020). The dormant cancer cell life cycle. *Nature Reviews. Cancer*, *20*(7), 398–411. https://doi.org/10.1038/s41568-020-0263-0.

Powell, D., Singer, F. R., Murray, T. M., Minkin, C., & Potts, J. T., Jr. (1973). Nonparathyroid humoral hypercalcemia in patients with neoplastic diseases. *The New England Journal of Medicine*, *289*(4), 176–181. https://doi.org/10.1056/NEJM197307262890403.

Powell, G. J., Southby, J., Danks, J. A., Stillwell, R. G., Hayman, J. A., Henderson, M. A., et al. (1991). Localization of parathyroid hormone-related protein in breast cancer metastases: Increased incidence in bone compared with other sites. *Cancer Research*, *51*(11), 3059–3061. Retrieved from https://www.ncbi.nlm.nih.gov/pubmed/2032246.

Ralston, S. H., Fogelman, I., Gardner, M. D., Dryburgh, F. J., Cowan, R. A., & Boyle, I. T. (1984). Hypercalcaemia of malignancy: Evidence for a nonparathyroid humoral agent with an effect on renal tubular handling of calcium. *Clinical Science (London, England)*, *66*(2), 187–191. https://doi.org/10.1042/cs0660187.

Ratcliffe, W. A. (1992). Role of parathyroid hormone-related protein in lactation. *Clinical Endocrinology*, *37*(5), 402–404. https://doi.org/10.1111/j.1365-2265.1992.tb02349.x.

Riggs, B. L., Arnaud, C. D., Reynolds, J. C., & Smith, L. H. (1971). Immunologic differentiation of primary hyperparathyroidism from hyperparathyroidism due to nonparathyroid cancer. *The Journal of Clinical Investigation*, *50*(10), 2079–2083. https://doi.org/10.1172/JCI106701.

Rodan, S. B., Insogna, K. L., Vignery, A. M., Stewart, A. F., Broadus, A. E., D'Souza, S. M., et al. (1983). Factors associated with humoral hypercalcemia of malignancy stimulate adenylate cyclase in osteoblastic cells. *The Journal of Clinical Investigation*, *72*(4), 1511–1515. https://doi.org/10.1172/JCI111108.

Roof, B. S., Carpenter, B., Fink, D. J., & Gordan, G. S. (1971). Some thoughts on the nature of ectopic parathyroid hormones. *The American Journal of Medicine*, *50*(5), 686–691. https://doi.org/10.1016/0002-9343(71)90124-0.

Rude, R. K., Sharp, C. F., Jr., Fredericks, R. S., Oldham, S. B., Elbaum, N., Link, J., et al. (1981). Urinary and nephrogenous adenosine 3′,5′-monophosphate in the hypercalcemia of malignancy. *The Journal of Clinical Endocrinology and Metabolism*, *52*(4), 765–771. https://doi.org/10.1210/jcem-52-4-765.

Ruppender, N. S., Merkel, A. R., Martin, T. J., Mundy, G. R., Sterling, J. A., & Guelcher, S. A. (2010). Matrix rigidity induces osteolytic gene expression of metastatic breast cancer cells. *PLoS One*, *5*(11), e15451. https://doi.org/10.1371/journal.pone.0015451.

Simpson, E. L., Mundy, G. R., D'Souza, S. M., Ibbotson, K. J., Bockman, R., & Jacobs, J. W. (1983). Absence of parathyroid hormone messenger RNA in nonparathyroid tumors associated with hypercalcemia. *The New England Journal of Medicine*, *309*(6), 325–330. https://doi.org/10.1056/NEJM198308113090601.

Southby, J., Kissin, M. W., Danks, J. A., Hayman, J. A., Moseley, J. M., Henderson, M. A., et al. (1990). Immunohistochemical localization of parathyroid hormone-related protein in human breast cancer. *Cancer Research*, *50*(23), 7710–7716. Retrieved from https://www.ncbi.nlm.nih.gov/pubmed/2253214.

Southby, J., O'Keeffe, L. M., Martin, T. J., & Gillespie, M. T. (1995). Alternative promoter usage and mRNA splicing pathways for parathyroid hormone-related protein in normal tissues and tumours. *British Journal of Cancer*, *72*(3), 702–707. https://doi.org/10.1038/bjc.1995.397.

Sterling, J. A., Oyajobi, B. O., Grubbs, B., Padalecki, S. S., Munoz, S. A., Gupta, A., et al. (2006). The hedgehog signaling molecule Gli2 induces parathyroid hormone-related peptide expression and osteolysis in metastatic human breast cancer cells. *Cancer Research*, *66*(15), 7548–7553. https://doi.org/10.1158/0008-5472.CAN-06-0452.

Stewart, A. F., Horst, R., Deftos, L. J., Cadman, E. C., Lang, R., & Broadus, A. E. (1980). Biochemical evaluation of patients with cancer-associated hypercalcemia: Evidence for humoral and nonhumoral groups. *The New England Journal of Medicine*, *303*(24), 1377–1383. https://doi.org/10.1056/NEJM198012113032401.

Stewart, A. F., Insogna, K. L., Goltzman, D., & Broadus, A. E. (1983). Identification of adenylate cyclase-stimulating activity and cytochemical glucose-6-phosphate dehydrogenase-stimulating activity in extracts of tumors from patients with humoral hypercalcemia of malignancy. *Proceedings of the National Academy of Sciences of the United States of America*, *80*(5), 1454–1458. https://doi.org/10.1073/pnas.80.5.1454.

Strewler, G. J., Stern, P. H., Jacobs, J. W., Eveloff, J., Klein, R. F., Leung, S. C., et al. (1987). Parathyroid hormonelike protein from human renal carcinoma cells. Structural and functional homology with parathyroid hormone. *The Journal of Clinical Investigation*, *80*(6), 1803–1807. https://doi.org/10.1172/JCI113275.

Suva, L. J., Winslow, G. A., Wettenhall, R. E., Hammonds, R. G., Moseley, J. M., Diefenbach-Jagger, H., et al. (1987). A parathyroid hormone-related protein implicated in malignant hypercalcemia: Cloning and expression. *Science*, *237*(4817), 893–896. https://doi.org/10.1126/science.3616618.

Thomas, R. J., Guise, T. A., Yin, J. J., Elliott, J., Horwood, N. J., Martin, T. J., et al. (1999). Breast cancer cells interact with osteoblasts to support osteoclast formation. *Endocrinology*, *140*(10), 4451–4458. https://doi.org/10.1210/endo.140.10.7037.

Tovar Sepulveda, V. A., Shen, X., & Falzon, M. (2002). Intracrine PTHrP protects against serum starvation-induced apoptosis and regulates the cell cycle in MCF-7 breast cancer cells. *Endocrinology*, *143*(2), 596–606. https://doi.org/10.1210/endo.143.2.8645.

Tran, T. H., Utama, F. E., Sato, T., Peck, A. R., Langenheim, J. F., Udhane, S. S., et al. (2018). Loss of nuclear localized parathyroid hormone-related protein in primary breast cancer predicts poor clinical outcome and correlates with suppressed stat 5 signaling. *Clinical Cancer Research*, *24*(24), 6355–6366. https://doi.org/10.1158/1078-0432.CCR-17-3280.

Vargas, S. J., Gillespie, M. T., Powell, G. J., Southby, J., Danks, J. A., Moseley, J. M., et al. (1992). Localization of parathyroid hormone-related protein mRNA expression in breast cancer and metastatic lesions by in situ hybridization. *Journal of Bone and Mineral Research*, *7*(8), 971–979. https://doi.org/10.1002/jbmr.5650070814.

Woodard, H. Q., Escher, G. C., & Farrow, J. H. (1954). Changes in the blood chemistry of patients with disseminated carcinoma of the breast during endocrine therapy. *Cancer*, 7(4), 744–757. https://doi.org/10.1002/1097-0142(195407)7:4<744::aid-cncr2820070414>3.0.co;2-0.

Wysolmerski, J. J., McCaughern-Carucci, J. F., Daifotis, A. G., Broadus, A. E., & Philbrick, W. M. (1995). Overexpression of parathyroid hormone-related protein or parathyroid hormone in transgenic mice impairs branching morphogenesis during mammary gland development. *Development*, *121*(11), 3539–3547. Retrieved from https://www.ncbi.nlm.nih.gov/pubmed/8582268.

Wysolmerski, J. J., Philbrick, W. M., Dunbar, M. E., Lanske, B., Kronenberg, H., & Broadus, A. E. (1998). Rescue of the parathyroid hormone-related protein knockout mouse demonstrates that parathyroid hormone-related protein is essential for mammary gland development. *Development*, *125*(7), 1285–1294. Retrieved from https://www.ncbi.nlm.nih.gov/pubmed/9477327.

Yamamoto, M., Fisher, J. E., Thiede, M. A., Caulfield, M. P., Rosenblatt, M., & Duong, L. T. (1992). Concentrations of parathyroid hormone-related protein in rat milk change with duration of lactation and interval from previous suckling, but not with milk calcium. *Endocrinology*, *130*(2), 741–747. https://doi.org/10.1210/endo.130.2.1733722.

Yang, L., Huang, J., Ren, X., Gorska, A. E., Chytil, A., Aakre, M., et al. (2008). Abrogation of TGF beta signaling in mammary carcinomas recruits Gr-1+CD11b+ myeloid cells that promote metastasis. *Cancer Cell*, *13*(1), 23–35. https://doi.org/10.1016/j.ccr.2007.12.004.

Yin, J. J., Selander, K., Chirgwin, J. M., Dallas, M., Grubbs, B. G., Wieser, R., et al. (1999). TGF-beta signaling blockade inhibits PTHrP secretion by breast cancer cells and bone metastases development. *The Journal of Clinical Investigation*, *103*(2), 197–206. https://doi.org/10.1172/JCI3523.

Yoneda, T., Sasaki, A., Dunstan, C., Williams, P. J., Bauss, F., De Clerck, Y. A., et al. (1997). Inhibition of osteolytic bone metastasis of breast cancer by combined treatment with the bisphosphonate ibandronate and tissue inhibitor of the matrix metalloproteinase-2. *The Journal of Clinical Investigation*, *99*(10), 2509–2517. https://doi.org/10.1172/JCI119435.

Younos, I., Donkor, M., Hoke, T., Dafferner, A., Samson, H., Westphal, S., et al. (2011). Tumor- and organ-dependent infiltration by myeloid-derived suppressor cells. *International Immunopharmacology*, *11*(7), 816–826. https://doi.org/10.1016/j.intimp.2011.02.021.

CHAPTER NINE

Stimulation of osteoclast formation and bone resorption by glucocorticoids: Synergistic interactions with the calcium regulating hormones parathyroid hormone and $1,25(OH)_2$-vitamin D3

Petra Henning[a], H. Herschel Conaway[b], and Ulf H. Lerner[a,*]
[a]Department of Internal Medicine and Clinical Nutrition, Institute of Medicine, Sahlgrenska Osteoporosis Centre and Centre for Bone and Arthritis Research at the Sahlgrenska Academy, University of Gothenburg, Gothenburg, Sweden
[b]Department of Physiology and Cell Biology, University of Arkansas for Medical Sciences, Little Rock, Arkansas, United States
*Corresponding author: e-mail address: ulf.lerner@gu.se

Contents

Abstract

Osteoporosis is a significant health problem, with skeletal fractures increasing morbidity and mortality. Excess glucocorticoids (GC) represents the leading cause of secondary osteoporosis. The first phase of glucocorticoid-induced osteoporosis is increased bone resorption. In this Chapter, *in vitro* studies of the direct glucocorticoid receptor (GR) mediated cellular effects of GC on osteoclasts to affect bone resorption and indirect effects on osteoblast lineage cells to increase the RANKL/OPG ratio and stimulate

Vitamins and Hormones, Volume 120
ISSN 0083-6729
https://doi.org/10.1016/bs.vh.2022.04.005

osteoclastogenesis and bone resorption are reviewed in detail, together with detailed descriptions of *in vivo* effects of GC in different portions of the skeleton in research animals and humans. Brief sections are devoted to contrasting functions of GC in osteonecrosis, vitamin D formation, *in vitro* and *in vivo* bone resorptive actions dependent on vitamin D receptor and vitamin D toxicity, as well as the molecular basis of GR action. Included are also more detailed assessments of the interactions of GC with the major calcium regulating hormones, $1,25(OH)_2$-vitamin D3 and parathyroid hormone, describing the *in vitro* increases in RANKL/OPG ratios, osteoclastogenesis and synergistic bone resorption that occurs when GC is combined with either $1,25(OH)_2$-vitamin D3 or parathyroid hormone. Additionally, a molecular basic for the synergistic interaction of GC with $1,25(OH)_2$-vitamin D3 is provided along with a suggested molecular basic for the interaction between GC and parathyroid hormone.

Abbreviations

ACTH	adrenocorticotropic hormone
AP-1	activation protein-1
CRE	cyclic AMP response element
CREB	cyclic AMP response element-binding protein
CSF-1	colony-stimulating factor-1
CTX	crosslinked C-telopeptide of type I collagen
DBD	DNA-binding domain
DCR	distal control region
FGF-23	fibroblast growth factor-23
GC	glucocorticoid
GM-CSF	granulocyte-macrophage colony-stimulating factor
GR	glucocorticoid receptor
GRE	glucocorticoid response element
11ß-HSD2	11ß-hydroxysteroid dehydrogenase type 2
IGF-1	insulin-like growth factor-1
ITAM	immunoreceptor tyrosine-based activation motif
LDB	ligand-binding domain
MAPK	mitogen-activated protein kinase
M-CSF	macrophage colony-stimulating factor
NFATC1	nuclear factor of activated T-cells, cytoplasmic 1
NF-κB	nuclear factor-κB
OPG	osteoprotegerin
RANK	receptor activator nuclear factor-κB
RANKL	receptor activator nuclear factor-κB ligand
TRAF-6	tumor necrosis factor receptor-associated factor 6

1. Introduction

Glucocorticoids (GC) are necessary for life and glucocorticoid receptors (GR) are found in most tissues and organs of the body. GR is a hormone-dependent transcription factor that regulates GC responsive genes. It is estimated that GC responsive genes may make up 3–10% of the human genome and GC are thought to play important roles in numerous cellular and organ processes (Nicolaides, Chrousos, & Kino, 2000).

GC have immunosuppressive and anti-inflammatory properties and there is widespread clinical use of GC in organ transplantation and diseases such as asthma, rheumatoid arthritis, multiple sclerosis, systemic lupus erythematosus and cancer (Canalis, Mazziotti, Giustina, & Bilezikian, 2007).

Excess levels of GC due to treatment or resulting endogenously (*e.g.*, pituitary tumor secreting ACTH (adrenocorticotropic hormone), adrenal tumor secreting cortisol, cancerous lesion secreting ACTH-like peptides) can result in abnormalities such as diabetes, central obesity, thinning of the skin, bruising, purple stria, loss of muscle mass, muscle weakness, osteoporosis, osteonecrosis, depression, and high blood pressure (Nieman, 2018; Soffer, Iannaccone, & Gabrilove, 1961) A rapidly developing abnormality caused by excess GC is skeletal fragility. GC therapy is the most common cause of secondary osteoporosis and the leading cause of non-traumatic osteonecrosis, with over 30% of patients experiencing an osteoporotic fracture and over 9% developing osteonecrosis (Weinstein, 2012).

Fracture incidence in osteoporosis correlates with increasing dosage and duration of GC treatment (Steinbuch, Youket, & Cohen, 2004). Increased vertebral and hip fractures have been reported with as little as 2.5 to 7.5 mg/day of prednisolone, while increases of greater than 15-fold in vertebral fractures and 7-fold in hip fractures have been reported for continuous use of 10 mg/day prednisone for 90 days (Steinbuch et al., 2004; Van Staa et al., 2003). Decreased bone quality also appears to occur following GC therapy for fractures occur at a higher bone mineral density than normally expected in osteoporotic subjects (Steinbuch et al., 2004). This decrease in bone strength may be the result of decreased skeletal vascularity and hydration of bone following glucocorticoid therapy (Nyman, Reyes, & Wang, 2005; Weinstein, 2010; Weinstein et al., 2010).

GC-induced osteoporosis is suggested to occur in two primary phases (Canalis et al., 2007). An initial, rapid phase of bone loss is due primarily to increased bone resorption. GC have been shown to enhance the expressions of the osteoclast stimulatory cytokines M-CSF and RANKL, while decreasing the expression of the osteoclast inhibitory decoy receptor OPG (Hofbauer et al., 1999; Rubin et al., 1998; Swanson, Lorentzon, Conaway, & Lerner, 2006). These effects promote an increased RANKL/OPG ratio causing increased formation, differentiation and function of osteoclasts. The active mode of mature osteoclast resorption is also thought to be increased by GC therapy (Soe & Delaisse, 2010). The rapid phase of osteoclast mediated bone loss is followed by a slower phase characterized by decreased bone formation. GC decrease the precursor pool of osteoblasts (Canalis, 1983, 1984; Ito, Suzuki, Kato, Takahashi, & Takagi, 2007; Pereira, Delany, & Canalis, 2002), decrease osteoblast differentiation and function (Canalis, 2005; Ohnaka, Tanabe, Kawate, Nawata, & Takayanagi, 2005; Smith & Frenkel, 2005) and increase osteoblast and osteocyte apoptosis (Liu et al., 2004; O'Brien et al., 2004). Initially, bone turnover is increased, but with chronic GC exposure, osteoclastic resorption slows and bone turnover is low because of decreased osteoblast lineage cells supporting osteoclastogenesis.

Osteonecrosis (avascular necrosis) is a degenerative bone disease associated with inadequate blood flow (Mankin, 1992; Murphey et al., 2014). It can occur in numerous bones, but often develops in the epiphysis of weight bearing long bones (Murphey et al., 2014). The femoral head is particularly susceptible, but osteonecrosis is also seen frequently in the humeral head, knee and talus (Murphey et al., 2014). The incidence of GC-induced osteonecrosis is related to dosage and duration of treatment and may occur in up to 40% of patients receiving long term GC therapy (Weinstein, 2012); however, even low dosages for short periods of time (*e.g.*, one methyl-prednisone treatment pack) may precipitate the disease (Dilisio, 2014).

A primary reason for the collapse of bone in osteonecrosis is increased bone resorption caused by increased numbers of osteoclasts, but it is doubtful that GC play a significant role in GC-induced osteonecrosis by increasing the RANKL/OPG ratio and stimulating osteoclastogenesis (Weinstein, 2007). Rather, current evidence favors increased osteoclastogenesis in GC-induced osteonecrosis being linked to the necrosis that occurs. In a recent study, data suggest the large increase of osteoclasts in osteonecrosis is due to necrotic osteocytes releasing DAMPs (damage-associated molecular pattern) that stimulate the C-type leptin receptor Mincle on osteoclasts. Mincle is then suggested to stimulate osteoclastogenesis *via* ITAM-mediated induction of calcium signaling and oxidative phosphorylation (Andreev et al., 2020).

Skeletal effects of GC in osteoporosis are believed to be due to direct effects of the steroids on bone cells (Conaway, Grigorie, & Lerner, 1996; Lowe, Gray, & Reid, 1992; Reid, Katz, Ibbertson, & Gray, 1986), but GCs have systemic effects that can affect bone as well. High concentrations of GC will decrease sex hormones (Besemer, Pereira, & Smit, 2011; Norbury, Herndon, Branski, Chinkes, & Jeschke, 2008), growth hormone and IGF-1 (Mazziotti et al., 2016), oppose calcium absorption (Klein, Arnaud, Gallagher, Deluca, & Riggs, 1977) and increase urinary calcium excretion (Laake, 1960). Increased bone resorption due to synergistic interactions of GC with vitamin D and parathyroid hormone have also been observed (Conaway, Grigorie, & Lerner, 1997) and GC have additionally been shown to alter the secretory dynamics of parathyroid hormone (Bonadonna et al., 2005) and increase the sensitivity of parathyroid hormone (Canalis, Bilezikian, Angeli, & Giustina, 2004). Previously, a primary theory for GC-induced osteoporosis was that decreased calcium absorption and increased calcium excretion decreased plasma calcium causing secondary hyperparathyroidism (Lukert & Adams, 1976; Raisz, 1980). Most measurements of serum parathyroid hormone have not been supportive of this idea (Paz-Pacheco, Fuleihan, & LeBoff, 1995) and there is not good agreement when densitometric and histomorphometric comparisons of parathyroid hormone alone and GC-induced osteoporotic effects on bone are evaluated (Rubin & Bilezikian, 2002). Thus, there has been only sparce support for this theory in recent years (Kaji, Yamauchi, Chihara, & Sugimoto, 2008).

Vitamin D3 is a fat-soluble vitamin obtained from animal-based foods in the diet and formed in the skin following UV-irradiation of 7-dehydrocholesterol (Bouillon et al., 2019; DeLuca, 2016). Vitamin D3 is converted to 25-OH-vitamin D3 by cytochrome P-450 enzymes (CYP2RI and CYP27A1) in the liver (Guo, Strugnell, Back, & Jones, 1993; Zhu, Ochalek, Kaufmann, Jones, & Deluca, 2013). 25-OH-vitamin D3 is the main circulating form of vitamin D3. It is hydroxylated by another cytochrome P-450 enzyme (CYP27B1) in the kidney (Murayama et al., 1999) to form 1,25$(OH)_2$-vitamin D3, the hormonally active form of vitamin D3. 1,25 $(OH)_2$-vitamin D3 stimulates calcium absorption in the gastrointestinal tract, increases renal reabsorption of calcium and stimulates bone resorption (DeLuca, 2016). In addition to these classical functions, numerous additional positive effects have been suggested (DeLuca, 2016). Unfortunately, high levels of vitamin D can result in vitamin D toxicity.

Vitamin D toxicity does not occur following exposure to sunlight (Holick, 2015). Vitamin D toxicity is characterized by hypercalcemia and usually develops following excessive intake of pharmacological preparations of vitamin D (Jones, 2008; Marcinowska-Suchowierska, Kupisz-Urbanska, Lukaszkiewicz, Pludowski, & Jones, 2018). Vitamin D toxicity can also develop endogenously from elevated production of 1,25$(OH)_2$-vitamin D3 in granulomatous disorders and in some lymphomas, as well as in idiopathic infantile hypercalcemia because of decreased 1,25$(OH)_2$-vitamin D3 metabolism (Tebben, Singh, & Kumar, 2016). The possibility of excessive production of 25(OH)-vitamin D3 and 1,25 $(OH)_2$-vitamin D3 have also been suggested in congenital disorders, such as Williams-Beuren

Syndrome (Marcinowska-Suchowierska et al., 2018). In endogenous vitamin D toxicity, hypercalcemia is due to increased 1,25$(OH)_2$-vitamin-D3, while in exogenous vitamin D toxicity, hypercalcemia is suggested to be mediated by high concentration of 25(OH)-vitamin D3 binding vitamin D receptor (VDR) (Marcinowska-Suchowierska et al., 2018).

Stimulation of *in vitro* bone resorption by 1,25$(OH)_2$-vitamin D3 was reported by Raisz et al. in 1972 using fetal rat long bones (Raisz, Trummel, Holick, & DeLuca, 1972). It is now well established that *in vitro* stimulation of osteoclastogenesis and bone resorption by 1,25$(OH)_2$-vitamin D3 is VDR dependent and due to an increased RANKL/OPG ratio in osteoblast-lineage cells (Palmqvist et al., 2006; Suda et al., 1999; Swanson et al., 2006). Recent studies have also established the *in vivo* requirement for VDR in mice (Mori et al., 2020). Administration of an increased dosage of 1,25 $(OH)_2$-vitamin D3 (5 μg/kg body weight/day) to nine-week-old male mice for 4 days increased the numbers of osteoclasts in bone and the concentrations of calcium, bone matrix derived crosslinked telopeptide of type 1 collagen (CTX-1) and fibroblast growth factor 23 (FGF23) in serum. These effects were blocked in the mice by an antibody to RANKL and with osteoblast-specific deletion of VDR, suggesting that *in vivo* effects of high dosage 1,25$(OH)_2$- vitamin D3 are also RANKL dependent and mediated by VDR in osteoblast-lineage cells.

As mentioned above, bone loss induced by GCs has been attributed to an initial phase of increased bone resorption and a delayed phase of decreased bone formation. In the present chapter, we focus on pathogenetic mechanisms involved in formation of osteoclasts by GC. Special focus will be on cellular and molecular interactions between GCs and the calcium regulating hormones 1,25$(OH)_2$–vitamin D3 and parathyroid hormone. For discussion on mechanisms involved in decreased bone formation, the readers are referred to recent reviews (Ahmad et al., 2019; Frenkel, White, & Tuckermann, 2015).

2. Pathogenetic mechanisms causing increased bone resorption in glucocorticoid-induced osteoporosis

2.1 Studies in humans

It is well recognized that GC treatment of humans decreases bone mass and increases fracture risk and that the risk is positively associated with duration of treatment and dose (Compston, 2018; van Staa, Leufkens, & Cooper, 2002). Bone is lost at both cortical and trabecular compartments, but more severely in trabecular bone (Dalle Carbonare et al., 2001; van Staa et al., 2002; Wetzsteon et al., 2009). In agreement with the latter finding, the highest incidence of GC-induced fractures occurs in the spine, in which a five-fold increase has been observed in patients taking prednisone at a dosage ≥5 mg/daily (van Staa et al., 2002). The risk of hip fracture was also increased by this dosage of GC treatment (van Staa et al., 2002). Furthermore, in a recent study it was additionally shown that patients with rheumatic arthritis have an increased risk of vertebral fractures with low doses of prednisolone daily (≤7.5 mg) (Abtahi et al., 2021). The fact that fracture risk can be noted after only 3 months of treatment with 5 mg/d of prednisolone in older women and men indicates that early onset increased bone resorption has an important role (van Staa et al., 2002). This is further suggested by the observations that vertebral fracture risk is decreased by bisphosphonate treatment (Allen, Yeung, Vandermeer, & Homik, 2016; Amiche et al., 2016) and that alendronate significantly decreases hip fracture risk in elderly women and men on prednisolone treatment (Axelsson, Nilsson, & Lorentzon, 2017). The observations that methylprednisolone treatment (15 mg/kg) for 10 days results in increased serum levels of CTX and an increased ratio of urinary calcium/creatinine demonstrate that GCs increase bone resorption early after initiation of therapy (Dovio et al., 2004). Further supporting the role of rapidly increased bone resorption is the finding of enhanced amounts of the resorption marker CTX in serum 96 h after initiation of prednisolone treatment (10–50 mg/d) (Brabnikova Maresova, Pavelka, & Stepan, 2013).

2.2 Studies in mice

In mice, most attention has been paid to effects by GC on bone formation. However, like findings in humans, an 80% increased concentration of the bone resorption marker CTX was found in serum after 7 and 28 days in Swiss Webster mice administered prednisolone (5 mg/kg) using slow-release pellets by *s.c.* implantation (Yao et al., 2008). It was also found that several genes associated with increased osteoclastogenesis such as *Csf1* and *Csf1r* mRNA (encoding M-CSF/CSF-1 and the receptor for M-CSF/CSF-1, respectively), important for proliferation and survival of osteoclast progenitors, and *Nfatc1* mRNA, the master transcription factor of osteoclastogenesis, were upregulated by GC treatment after 7 and 28 days. Furthermore, genes encoding receptor components in the ITAM (immunoreceptor tyrosine-based activation motif) pathway, known to be a crucial co-signaling pathway involved in osteoclastogenesis (Koga et al., 2004; Mocsai et al., 2004), such as *Fcrg*, *Oscar*, *Trem2* and *Dap12* mRNA were upregulated. Weinstein et al. treated Swiss Webster mice with prednisolone (2.1 mg/kg per day) using slow release *s.c.* pellets for 27 days and reported that this treatment decreased trabecular bone in vertebra due to increased formation of trabecular osteoclasts (Weinstein et al., 2002). This group later found that the identical treatment resulted in decreased cortical thickness due to enhanced numbers of endocortical osteoclasts in the femur, an effect associated with increased mRNA expression of *Calcr* (encoding calcitonin receptor, a hallmark of mature osteoclasts) (Piemontese et al., 2015). In a subsequent study, these investigators confirmed these findings and also reported that cortical area was decreased and cortical porosity enhanced in the femur and that these observations were associated with increased mRNA expression of *Acp5* (encoding TRAP), *Calcr* and *Ctsk* (encoding cathepsin K) in the cortical bone (Piemontese, Xiong, Fujiwara, Thostenson, & O'Brien, 2016). Similarly, Henneicke et al. have reported that treatment of mice with corticosterone using slow release pellets for 28 days resulted in increased serum levels of the resorption marker TRAP5b (released from osteoclasts), a response associated with increased numbers of endocortical osteoclasts in tibia giving rise to increased endocortical circumference and enhanced bone marrow area (Henneicke et al., 2011). Interestingly, no effects on cortical thickness were found, most likely because numbers of pericortical osteoclasts were decreased and pericortical circumference was increased. These effects were not seen in mice overexpressing the cortisol degrading enzyme 11ß-HSD2 in Col2.3 expressing cells, indicating that effects on osteoclasts were mediated indirectly through osteoblasts. These data indicate GC act differentially on osteoclastogenesis at different compartments in the cortical skeleton, similar to retinoids which increase pericortical and decrease endocortical numbers of osteoclasts (Lind et al., 2011; Lionikaite et al., 2018). In the studies by Piemontese et al. and Hennicke et al., trabecular bone mass in either the femur, tibia or vertebra was not affected by prednisolone or corticosterone (Henneicke et al., 2011; Piemontese et al., 2015, 2016). In other studies, however, prednisolone treatment (slow release pellet, 2.1. mg/kg per day for 28 days) in Swiss Webster mice, or treatment with similar pellets (12.4 mg/kg per day for 14 days) of mice on a FVB/N background, resulted in decreased trabecular bone mass in vertebra (Rauch et al., 2010; Weinstein et al., 2002; Weinstein, Jilka, Parfitt, & Manolagas, 1998). Thus, GC seem to be able to decrease both cortical and trabecular bone in mice, similar to observations in humans (van Staa et al., 2002).

Prednisolone treatment by slow-releasing pellets (2.1 mg/kg) in mice expressing the GC inactivating enzyme 11ß-HSD2 under control of the osteocalcin promoter results in decreased trabecular bone mass similar to that seen in wild type mice during 28 days of treatment, although the mice were resistant to decreased osteoblast and osteocyte apoptosis induced by prednisolone (O'Brien et al., 2004). This observation suggests that the loss of bone was due to effects on bone resorption and osteoclasts rather than on effects mediated by osteoblasts.

2.3 Effects by GCs in bone organ cultures

In organ cultured bone, early studies by Stern et al. found that hydrocortisone and dexamethasone at 1 μM inhibited calcium release stimulated by parathyroid hormone from fetal rat limb bones (Stern, 1969). However, testosterone and cholesterol also inhibited parathyroid hormone-induced calcium release, indicating a steroid non-specific effect. Also using fetal rat limb bones, Raisz et al. found that cortisol (1 μM) inhibited calcium release stimulated by either retinol, prostaglandin E_2, 25(OH)-vitamin D3 or parathyroid hormone (Raisz, Trummel, Wener, & Simmons, 1972). Lower concentrations of cortisol potentiated the transient inhibitory effect by calcitonin. This is most likely explained by subsequent findings showing that cortisol prevents the down-regulation of calcitonin receptors (Wada et al., 2001). An inhibitory effect on calcium release by cortisol or dexamethasone has also been observed in neonatal mouse calvarial bones stimulated by IL-1α or IL-1ß, probably due to inhibition of prostaglandin biosynthesis induced by these cytokines (Ishikawa et al., 1990; Marusic & Raisz, 1991). The first evidence that GC can stimulate bone resorption *in vitro* was provided by Reid et al. using neonatal mouse calvarial bones and demonstrating that 0.1 and 1 μM hydrocortisone caused a pronged stimulation of calcium release (Reid et al., 1986). It was also found that hydrocortisone caused an additive stimulation of calcium release induced by parathyroid hormone. The effects were inhibited by a bisphosphonate, indicating that enhanced calcium release was osteoclast-dependent. Later studies by the same group found that stimulation by hydrocortisone could only be seen in serum-free medium (Lowe et al., 1992). This explains why Ishikawa et al. did not find stimulation of calcium release by dexamethasone since the neonatal mouse calvarial bones in that study were incubated in the presence of 5% horse serum (Ishikawa et al., 1990). Using fetal rat parietal bones incubated in serum-free medium, Gronowicz et al. observed that 1 and 10 nM corticosterone or dexamethasone caused a transient increased release of calcium, associated with an increased number of osteoclasts (Gronowicz, McCarthy, & Raisz, 1990).

Using neonatal mouse calvarial bones incubated in medium with 0.1% albumin, we have reported that hydrocortisone and dexamethasone cause prolonged increased calcium release and bone matrix degradation with half maximal stimulation observed at 0.3 and 0.08 μM, respectively (Conaway et al., 1996). The fact that the effect was associated with increased lysosomal enzyme release and inhibited by calcitonin and bisphosphonate indicate that the GC had stimulated osteoclast formation and/or activity (Conaway et al., 1996). Subsequent studies found that GC stimulation of neonatal mouse calvarial bone resorption was associated with increased mRNA expression of *Acp5*, *Ctsk* and *Calcr*, indicating that increased bone resorption is due to activation of osteoclastogenesis by GCs. (Swanson et al., 2006). In agreement with observations by Raisz et al. (Raisz, Trummel, Wener, et al., 1972), both hydrocortisone and dexamethasone prevented the escape from calcitonin-induced inhibition of parathyroid hormone-stimulated calcium release (Conaway et al., 1996). The stimulatory effect by GCs seemed to be dependent on mitotic activity since both hydroxyurea and mitomycin C inhibited calcium release induced by hydrocortisone and dexamethasone. However, the data did not reveal which cells in the calvarial periosteum were targets for the GC-induced stimulation of proliferation. Stimulation of calcium release by hydrocortisone and dexamethasone was competitively inhibited by the GC-receptor antagonists progesterone and RU486, indicating an effect dependent on GR.

2.4 Direct effects of GCs on osteoclasts and osteoclast progenitors

Using osteoclasts disaggregated from neonatal rat long bones, Tobias & Chambers found that hydrocortisone (0.1–10 μM) and dexamethasone (0.001–1 μM) decreased the resorptive cavities formed by mature osteoclasts when the cells were incubated on dentine slices, an effect associated with decreased survival of the osteoclasts (Tobias & Chambers, 1989). This effect was inhibited by progesterone, indicating that the GCs caused their effect through GRs. Similar inhibition of mature osteoclast survival has also been observed by Dempster et al. (Dempster, Moonga, Stein, Horbert, & Antakly, 1997) Using neonatal rat osteoclasts incubated on bone slices, these authors found that GCs decreased the resorbed area, an effect associated with decreased numbers of osteoclasts and increased apoptosis. Treatment of RANKL-stimulated mouse bone marrow macrophages with dexamethasone (0.001–1 μM) has been found to result in decreased spreading of mature osteoclasts and decreased resorptive activity, due to disruption of M-CSF-induced organization of the cytoskeleton and actin ring (Kim et al., 2006). The effect has been attributed to inhibition of calpain-6, a non-proteolytic member of the calpain family, localized to the sealing zone and important for actin ring formation used by osteoclasts to form the Howship's lacuna, in which bone resorption takes place (Hong et al., 2011).

In contrast to these findings, GCs have been found to cause no effect on apoptosis (Sivagurunathan, Muir, Brennan, Seale, & Mason, 2005) or to decrease apoptosis (Weinstein et al., 2002). In the latter studies, dexamethasone (0.001–0.1 μM) decreased apoptosis in osteoclasts formed in mouse bone marrow cultures stimulated with M-CSF/RANKL. Dexamethasone also prevented apoptosis of these cells treated with the pro-apoptotic bisphosphonate alendronate due to inhibition of alendronate-induced activity of caspase-9 and -3.

When M-CSF/RANKL-stimulated purified mouse bone marrow macrophages were incubated on plastic with dexamethasone (0.1 μM), hydrocortisone (1 μM) or prednisolone (1 μM), we found that the GCs stimulated formation of giant osteoclasts at later stages of the 7 days cultures, whereas those incubated without GC disappeared at the end of the cultures (Conaway, Henning, Lie, Tuckermann, & Lerner, 2016). The difference was due to inhibition of apoptosis of the mature osteoclasts by GCs. Similarly, treating crude bone marrow cells on plastic dishes with either 1,25$(OH)_2$-vitamin D3 or RANKL, together with dexamethasone or hydrocortisone, resulted in formation of long lived giant osteoclasts. The formation of the oversized osteoclasts was not associated with any effect on the mRNA expression of osteoclast differentiation markers such *Acp5*, *Ctsk*, *c-Fos* or *Nfatc1*, nor with expression of *Dcstamp*, which has been suggested to be important for multinucleation. However, the formation of long-lived mature osteoclasts in cell cultures treated with dexamethasone was associated with decreased formation of DNA-fragments as markers of apoptosis. Similar to Kim et al., (2006), we found that dexamethasone at earlier stages of osteoclast formation decreased spreading of the osteoclasts, but this was followed by formation of the giant osteoclasts as described above. Importantly, no such giant osteoclasts were observed, nor did GC affect osteoclast apoptosis, when the cultures were performed on bone slices.

It is unclear why osteoclast morphology is so different and why osteoclasts are more susceptible to apoptosis when cells are incubated on plastic dishes compared to bone slices. It is likely, however, that recognition of molecules expressed by bone tissue changes the morphology and behavior of osteoclasts, making them more similar to those present on bone *in vivo*. These observations demonstrate the importance of performing experiments not only on plastic dishes, but also on bone when studying effects by different hormones, cytokines and other compounds on osteoclast formation.

Evidence that GCs may exert both stimulatory and inhibitory effects on osteoclast formation dependent on differentiation stage are the observations using RANKL- and M-CSF-stimulated human peripheral blood monocytes in 21-day cultures showing that dexamethasone (0.01 μM), when added during the first 14 days of culture, stimulated formation of osteoclasts, whereas when added during days 14–21 decreased osteoclast number (Hirayama, Sabokbar, & Athanasou, 2002). These findings suggest that GCs might increase proliferation and/or differentiation of osteoclast progenitors during the first 14 days, but inhibit osteoclasts when mature osteoclasts are formed during the last 7 days. An additional explanation why both stimulatory and inhibitory effects can be observed is the finding that dexamethasone at 0.01 μM potentiates M-CSF- and RANKL-stimulated formation of the resorbed area on whale dentine by human peripheral blood monocytes, whereas dexamethasone at 0.1 μM inhibits (Sivagurunathan et al., 2005). A time- and concentration dependent effect by dexamethasone has also been observed using mouse bone marrow macrophages stimulated by M-CSF/RANKL together with TGF-ß (Takuma et al., 2003).

As discussed above, there is evidence from mouse bone marrow macrophage cultures (Conaway et al., 2016) that the response of osteoclasts to GC might be different if cells are incubated on bone compared to incubations on plastic dishes. Interesting observations have been made when human $CD14^+$ monocytes have been incubated on bone disks together with M-CSF/RANKL in the presence of prednisolone (1.6 μM) (Soe & Delaisse, 2010). When the cells were incubated in the absence of prednisolone, many round excavations were formed, but in the presence of prednisolone the excavations were fewer and more trenchlike. In addition, whereas demineralized bone matrix collagen fibers remained at the bottom of the round excavations, no such matrix persisted in the trenchlike excavations. It was suggested that the GC changed the behavior of the osteoclasts from intermittent to continuous resorption and maintain the osteoclasts in the active mode of the resorption cycle.

We have made similar findings using M-CSF/RANKL-stimulated mouse bone marrow macrophages (Conaway et al., 2016). As discussed above, when cells were incubated on plastic dishes in the presence of dexamethasone, osteoclasts became oversized and remained in culture for long periods of time due to decreased apoptosis, whereas osteoclasts incubated in the absence of dexamethasone became apoptotic and disappeared. No such effects were seen when cells were incubated on bone disks; the morphology of the osteoclasts was similar with or without dexamethasone and the numbers of osteoclasts were similar at different time points. The mRNA expression of osteoclastic and osteoclastogenic genes such as *Acp5*, *Ctsk*, *Nfatc1* and *c-Fos* were not affected by dexamethasone, nor was the mRNA expression of the receptors (*Tnfrsf11a*, *Csf1r*) for RANKL or M-CSF affected. However, the activity of the mature osteoclasts was clearly enhanced resulting in increased numbers of resorption excavations and enhanced release of the bone matrix degradation fragment CTX. These effects were obtained by dexamethasone $\geq 0.01\ \mu M$ as well as by hydrocortisone and prednisolone. Interestingly, the GCs had to be present during the entire culture period; when added once the mature osteoclasts were formed, no such stimulation was seen. This finding indicates that GCs prime the mononucleated osteoclast progenitor cells resulting in more active mature osteoclasts. The stimulatory effect by dexamethasone was decreased by the GR antagonist RU486 (mifepristone) and when bone marrow macrophages from cells deficient of GR in myeloid cells (macrophages isolated from $GR^{LysMCre}$ mice) were employed. These findings, together with the finding that dexamethasone needs to be present during early phases of osteoclastogenesis, demonstrate that the effect is dependent on GR present in osteoclast progenitor cells. However, it remains to be shown which genes are regulated by GR to make the mature osteoclasts more active. The effect seems to be dependent on fully functional GR since dexamethasone could not stimulate osteoclast activity when cells from mice (GR^{dim}) with a lack-of-function mutation in the domain of GR important for homodimerization were used (see below).

2.5 Indirect effects by GCs on osteoclast formation through RANKL and OPG

Increased bone loss due to enhanced bone resorption is more often associated with increased formation of osteoclasts than with increased activity, such as observed in GC-treated mice (Henneicke et al., 2011; Piemontese et al., 2015, 2016; Weinstein et al., 2002). Several of the hormones and cytokines known to stimulate bone resorption increase osteoclast formation indirectly through osteoblasts, as suggested initially by Rodan & Martin (1981). The interaction is due to release of RANKL from the osteoblasts activating its cognate receptor RANK on osteoclast progenitor cells (Hsu et al., 1999; Kong et al., 1999). Downstream signaling involves activation of TRAF-6, MAPK and several osteoclastogenic transcription factors such as NF-kB, c-Fos-containing AP-1 and NFATc1, the latter considered being the master regulator of osteoclastogenesis (Okamoto et al., 2017). Subsequently, genes important for differentiation, fusion and bone resorbing activity are induced in the osteoclast progenitor cells such as *Acp5*, *Ctsk*, *Dcstamp*. Osteoclastogenesis is also regulated by the soluble, decoy receptor OPG, which acts by inhibiting the binding of RANKL to RANK (Simonet et al., 1997). OPG belongs to the same receptor superfamily as RANK and is expressed by several cell types, including osteoblasts and osteocytes. Osteoclastogenesis is dependent on the relative amounts of RANKL and OPG, as well as the expression of RANK. Osteoclastogenesis is also critically dependent on M-CSF, expressed by several cell types and important for the proliferation and survival of the osteoclast progenitor cells. Hormones and cytokines which stimulate osteoclast formation indirectly through osteoblasts do so by increasing the ratio RANKL/OPG. There is evidence that both late osteoblasts/osteocytes (Nakashima et al., 2011; Xiong et al., 2018) and bone forming osteoblasts (Fumoto, Takeshita, Ito, & Ikeda, 2014) produce RANKL. There is also ample evidence that GCs can stimulate osteoclast formation by regulating the ratio of RANKL/OPG.

The first indication that GC may affect the RANKL/OPG ratio was provided by Vidal et al. showing that hydrocortisone ($\geq$10 nM) transiently decreases *Opg* mRNA expression (encoded by the gene *Tnfrsf11b*) in primary human osteoblasts (Vidal, Brandstrom, Jonsson, & Ohlsson, 1998). A similar decrease of *Opg* expression was also observed in human bone marrow stromal cells and in the human osteoblastic cell line MG-63. Decreases of *Opg* mRNA and OPG protein by dexamethasone ($\geq$1 nM) have also been reported in immortalized human bone marrow stromal and fetal osteoblastic cell lines as well as in primary human osteoblasts and bone marrow stromal cells (Hofbauer et al., 1999). These authors also reported that dexamethasone increases the expression of *Rankl* mRNA (encoded by the gene *Tnfsf11*) in immortalized human osteoblastic cells, which was the first observation showing that GC can increase RANKL. A decrease of OPG protein by dexamethasone has additionally been reported in human osteoblastic cells by Brandstrom, Bjorkman, & Ljunggren (2001). A similar decrease of *Opg* mRNA expression and increase of *Rankl* mRNA expression has been observed by Sivagurunathan et al. using primary human osteoblasts treated with dexamethasone at and above 0.01 nM (Sivagurunathan et al., 2005). Additionally, Humphrey et al. (Humphrey, Williams, Davie, & Marshall, 2006) have reported that dexamethasone ($\geq$1 nM) and prednisolone ($\geq$ 10 nM) decrease the production of OPG protein in human fetal osteoblasts and MG-63 cells and increase the ratio *Rankl/Opg* mRNA in these cells. It was also found that so called dissociated GCs, such as deflazacort, thought at this time to have less side effects, were less potent compared to dexamethasone and prednisolone (Humphrey et al., 2006).

Bone resorption and osteoclast differentiation induced by GCs in neonatal mouse calvarial bones are associated with increased expression of *Rankl* mRNA and RANKL protein (Conaway et al., 2011; Swanson et al., 2006). In contrast to observations in isolated cells, dexamethasone increased *Opg* mRNA and had no effect on OPG protein in calvarial bones. The fact that addition of OPG to dexamethasone-stimulated calvaria inhibited bone resorption demonstrates that the effect of GC is dependent on an increased RANKL/OPG ratio. Interestingly, dexamethasone also increased *Rank* mRNA expression (encoded by the gene *Tnfrsf11a*) indicating that GC-induced RANK expression might contribute to increased osteoclast formation and bone resorption. Stimulation of *Rankl* and *Opg* mRNA expression by dexamethasone can also be observed in calvarial bones from GR^{dim} mice with a point mutation rendering the GR less prone to dimerize (Conaway et al., 2011).

The observations in bone cell cultures showing that GCs decrease OPG mRNA and protein expression are in agreement with findings *in vivo* showing a decrease in serum OPG in patients with chronic glomerulonephritis treated with GCs for 2 or 4 weeks (Sasaki et al., 2001, 2002). The decrease of serum OPG after 2 weeks of treatment with GC, and the reduced bone mineral density after 6 months, were associated with an increase of serum TRAP indicating that decreased OPG resulted in increased osteoclast formation (Sasaki et al., 2002). A decrease of serum OPG by GCs has also been observed in subsequent studies (Brabnikova Maresova et al., 2013; Mohamed & Abdel-Latif, 2011).

The observations showing that GCs decrease OPG and increase RANKL in different osteoblastic cell lines explain why GCs increase bone resorption in organ cultured fetal rat long bones and neonatal mouse calvaria. The finding may also explain the increase of endocortical osteoclasts and decrease of cortical thickness observed in mice after GC treatment

(Piemontese et al., 2015, 2016). Although bone mass is lost in both cortical and trabecular bone, the loss in trabecular bone is more pronounced in patients treated with GCs (Dalle Carbonare et al., 2001; van Staa et al., 2002; Wetzsteon et al., 2009), which is reconciled with findings in mice showing an increased number of trabecular osteoclasts after GC treatment (Weinstein et al., 2002).

Udagawa et al. were the first to report that dexamethasone alone had no effect on osteoclast formation in co-cultures of mouse spleen cells with either the preadipose mouse cell line MC3T3-G2PA6 or the mouse stromal cell line ST2 (Udagawa et al., 1989). Later, we have shown in mouse bone marrow cultures, containing both mesenchymal stromal cells and hematopoetic cells including osteoclast progenitors, that dexamethasone (100 nM) does not affect the mRNA expression of osteoclastic genes (*Calcr, Acp5, Ctsk*) or osteoclastogenic transcription factors (*Nfatc1, c-Fos*), nor are any osteoclasts formed in bone marrow cell cultures stimulated by dexamethasone, although 1,25$(OH)_2$-vitamin D3 increased osteoclast formation and expression of these genes in the bone marrow cultures (Conaway, Henning, Lie, Tuckermann, & Lerner, 2019). Furthermore, when these cells were incubated on bone slices and treated with dexamethasone, no osteoclasts were formed and no effect on release of CTX could be observed, in contrast to the stimulatory effects by 1,25$(OH)_2$-vitamin D3. These negative findings were explained by the fact that dexamethasone did not affect the mRNA or protein expression of RANKL, in contrast to 1,25 $(OH)_2$-vitamin D3. Similar to observations in human bone marrow stromal cells (Hofbauer et al., 1999), dexamethasone decreased mRNA and protein expression of OPG, but this effect was apparently not enough to increase osteoclastogenesis.

3. Glucocorticoid receptor

Like other members of the nuclear receptor family, GR binds to DNA and regulates gene expression. The gene encoding GR, *NR3C1*, was cloned during the 1980's (Hollenberg et al., 1985) and consists of 9 exons that by alternative translation initiation and splicing can give rise to several isoforms of which GRα is the main isoform. GR is structurally built up by a N-terminal transactivating domain (AF1), followed by a DNA-binding domain (DBD) and a C-terminal ligand-binding domain (LBD) containing a second transactivation domain (AF2). GRβ is a shorter isoform that lacks part of the C-terminal LBD rendering it incapable of GC binding and it has been suggested to be an inhibitor of GRα (Oakley, Jewell, Yudt, Bofetiado, & Cidlowski, 1999). Although GRβ has been shown by immunohistochemistry to be expressed by bone cells (osteoblasts, osteocytes and osteoclasts), the relative importance *versus* GRα has not been studied in bone (Beavan, Horner, Bord, Ireland, & Compston, 2001; Hartmann et al., 2016; Moutsatsou, Kassi, & Papavassiliou, 2012). Different translational initiation isoforms of GR have been shown to differ in their regulation of osteoblast apoptosis when overexpressed in cultured human osteosarcoma U2OS cells but the *in vivo* expression and importance of the isoforms have not been explored (Gross, Oakley, Scoltock, Jewell, & Cidlowski, 2011; Lu, Collins, Grissom, & Cidlowski, 2007). In addition to the genomic effects, rapid non-genomic effects of GR have been described, but to our knowledge this has not been studied in bone cells (Panettieri et al., 2019).

In the absence of GC binding, GR is located in the cytoplasm as monomers bound by chaperone proteins. Upon GC binding, GR conformation is changed and the nuclear localization signal is exposed and the receptor is translocated to the nucleus through the nuclear pore complex and binds to DNA to regulate gene expression (Fig. 1). In the nucleus, GR can both increase (transactivate) and decrease (transrepress) gene transcription.

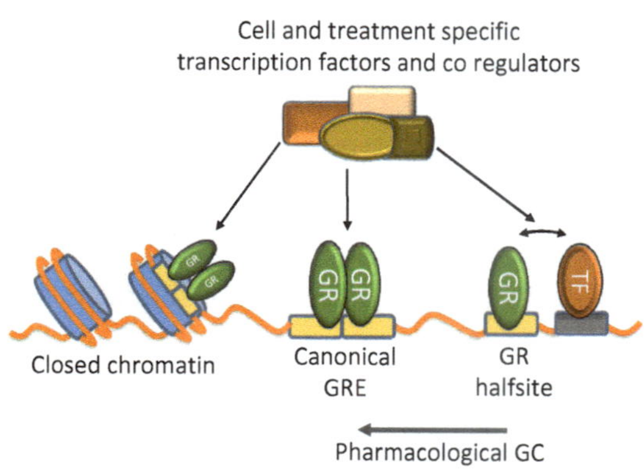

Fig. 1 Upon GC binding, GR translocates to the nucleus where it binds DNA and regulates gene expression. GR binds to canonical palindromic glucocorticoid response elements (GRE) and GR half sites in DNA. While the oligomeric status of GR bound to DNA has been challenged recently, most studies indicate that GR is in the form of homodimers when bound to canonical GRE and monomers when bound to half sites. The relative binding to canonical GRE increases with pharmacological GC treatment. Most binding to DNA by GR occurs in regions of pre-accessible chromatin, and the binding to GRE in closed chromatin seems to require a GR with retained dimerization functionality. The effects on gene expression are largely dependent on cell- and treatment-specific transcription factors (TF) and co-regulators.

Several different mechanisms of GR signaling have been described, and for a long time the dominant model for GR signaling was separated into GR either bound as homodimers directly to canonical glucocorticoid response elements (GREs) to activate gene transcription or as monomers bound to other transcriptions factors ("tethering") to transrepress genes. The anti-inflammatory effects by GC were largely attributed to monomeric GR tethering to NF-κB and AP-1 and decreasing the expression of inflammation activated genes (Heck et al., 1994; Ray & Prefontaine, 1994). With the development of more advanced microscopy and molecular biology techniques such as ChIP-seq, ChIP-exo, single cell sequencing and genome wide studies, the knowledge about GR signaling functions have expanded and changed (for review see (Vettorazzi, Nalbantoglu, Gebhardt, & Tuckermann, 2021)). Recent data suggest that tethering without direct DNA binding is not sufficient to regulate gene transcription *in vivo* (Escoter-Torres, Greulich, Quagliarini, Wierer, & Uhlenhaut, *2020*) and GR binding to DNA sequences within both AP-1 and NF-κB response elements have been reported and shown to be important for repression of inflammatory genes (Hudson et al., 2018; Weikum et al., 2017).

The canonical GRE is a 15 bp long sequence made up of two palindromic hexamer sequences separated by a 3 bp spacer (AGAACAnnn TGTTCA), to which GR binds as a homodimer in a head-to-head conformation. In addition to the canonical GRE and degenerated versions thereof, two other main classes of GR DNA binding sequences have been described; inverted-repeat GR binding sequences and GR half sites. The inverted-repeat GR binding sequence is believed to bind two monomers of GR in opposite direction on opposite sides of the DNA (review (Weikum, Knuesel, Ortlund, & Yamamoto, 2017)). GR half sites (AGAACA) have been shown to be occupied by both GRα and mutants with impaired dimerization function indicating that GR binding to GR half sites is in the form of monomers (Lim et al., 2015; Schiller, Chodankar, Watson, Stallcup, & Yamamoto, 2014). *In vitro* EMSA studies also show that GR can bind GR half sites as monomers (Watson et al., 2013). However, recent studies in cell culture using overexpression of GR- and GRmutant- GFP fusion proteins, indicate that monomer GR might not contribute largely to regulation of gene expression, and that GR is dimeric, or possibly even a tetramer, when bound to DNA also at GR half sites (Johnson, Paakinaho, Kim, Hager, & Presman, 2021; Presman et al., 2016).

At physiological GC levels, GR binding to half-sites and co-localization with tissue specific transcription factors are more prominent than binding to canonical palindromic sites in both liver and primary macrophages (Lim et al., 2015). Pharmacological treatment with GC increase the binding to palindromic sites at the expense of half-sites (Lim et al., 2015).

One of the most extensively studied GR mutants, GR^{dim} (mGR A465T, rGR A477T, hGR A458T), was originally designed to be devoid of dimerization capacity and used to study effects by monomeric GR (Reichardt et al., 1998). Although the DNA binding capacity is altered, with less binding to canonical palindromic GRE and retained binding to GR half sites (Lim et al., 2015; Watson et al., 2013), studies indicate that GR^{dim} can still form dimers when bound to DNA (Johnson et al., 2021; Presman et al., 2014). The differences in transcription regulation by wild type GR and GR^{dim} might be explained by impaired dimerization and altered DNA binding and differences in interaction with co-regulators and other transcription factors.

GR is ubiquitously expressed in most cell types, yet the response to GC varies dramatically, and it is obvious that not only DNA binding sequences specify gene expression regulation. Interaction with other transcription factors and co-regulators are crucial for GR mediated signaling and the effect of GC on gene expression is to a large extent cell- and gene-specific (Frank, Liu, & Ortlund, 2021; Lim et al., 2015; Rogatsky et al., 2003; Vettorazzi et al., 2021; Weikum, Knuesel, et al., 2017). Comparisons of genes regulated by GC using *in vitro* cultured human cell lines, A549 (lung carcinoma) and U2OS (osteosarcoma), show that the number of activated and repressed genes are largely different and that some genes are even differentially regulated in the two cell lines (Rogatsky et al., 2003; Schiller et al., 2014; Weikum, Knuesel, et al., 2017). The presence and binding of other transcriptions factors can facilitate the binding of GR to DNA by opening and remodeling chromatin, and GR can also facilitate the binding of other transcription factors (Biddie et al., 2011; John et al., 2011; Lee & Stallcup, 2017; Siersbaek et al., 2011). Of approximately 2.3-million identified 12-bp GRE in the mouse genome, only a small fraction (8236 and 3242 sites in mouse mammary adenocarcinoma and pituitary cells line, respectively) were found to be occupied by GR after hormone treatment and 95% of the GR bound sites were in regions of pre-accessible chromatin (John et al., 2011). Recently, it was shown, using the GR^{dim} mutant, that binding to closed chromatin required a fully functional dimerization of GR (Johnson et al., 2021). GR bound sites can be found both near (<100 bp) and far (>100 kb) from the transcription start site of regulated genes, indicating that GR can regulate gene transcription by both short and long range mechanisms (John et al., 2011; Reddy et al., 2009; Weikum, Knuesel, et al., 2017).

The effect on osteoclastogenesis by GC is largely attributed to increased expression of RANKL. GR binding sites have been described both in sequences close to the Rankl transcription start site (TSS) and in long distance enhancer regions located up to −76kb upstream from the mouse Rankl TSS (Kim, Yamazaki, Zella, Shevde, & Pike, 2006; Kitazawa & Kitazawa, 2002; Kitazawa, Kitazawa, & Maeda, 1999; Kitazawa, Kajimoto, Kondo, & Kitazawa, 2003; Lovsin & Marc, 2021; Nerenz, Martowicz, & Pike, 2008). These are described in more detail below together with possible interactions with signaling through vitamin D and parathyroid hormone receptors.

4. Interaction between glucocorticoids and calcium regulating hormones

Most cells in the body express GR and treatment with GCs affects the expression of a wide variety of genes, some of which may affect bone cells and/or interact with GCs on formation and activity of bone cells. In addition, patients treated with GCs suffer from pathological conditions, which may produce compounds able to affect the skeleton by themselves and/or interact with GCs to enhance bone resorption. This may further enhance the complexity of interactions between GCs and systemic factors affecting the skeleton. So far, most attention has been paid to the interactions between GCs and the calcium regulating hormones parathyroid hormone and 1,25 $(OH)_2$-vitamin D3.

The first evidence that GC can stimulate osteoclast formation was the finding by Udagawa et al. (1989) showing that, although dexamethasone itself had no effect on osteoclast formation in co-cultures of mouse spleen cells with either the preadipose mouse cell line MC3T3-G2PA6 or the mouse stromal cell line ST2, dexamethasone potentiated the stimulatory effect by 1,25$(OH)_2$-vitamin D3. The TRAP$^+$ multinucleated cells formed were able to resorb dentine and expressed calcitonin receptors demonstrating that they were true osteoclasts. A similar observation has been made in cultures of crude mouse bone marrow cells containing osteoclast progenitors and mesenchymal stromal cells, where dexamethasone potentiated 1,25$(OH)_2$-vitamin D3-stimulated osteoclast formation (Shuto, Kukita, Hirata, Jimi, & Koga, 1994). This effect was attributed to an inhibitory effect by dexamethasone on the production of GM-CSF (granulocyte-macrophage colony-stimulating factor).

Similar to osteoclast formation in bone marrow cell cultures, GCs synergistically potentiate osteoclast differentiation and bone resorption in organ cultured calvarial bones induced by 1,25$(OH)_2$-vitamin D3 (Conaway et al., 1997; Swanson et al., 2006). Bone resorption induced by parathyroid hormone was also found to be synergistically potentiated by GCs in the calvarial bones (Conaway et al., 1997). Effects were observed at and above 10 nM of dexamethasone and 30 nM of hydrocortisone, and at and above 0.01 nM parathyroid hormone or 1,25$(OH)_2$-vitamin D3. These effects were inhibited by the GR antagonist RU486. The interaction between dexamethasone and 1,25$(OH)_2$-vitamin D3 on bone resorption in mouse calvaria is explained by the finding that dexamethasone synergistically potentiates 1,25$(OH)_2$-vitamin D3-induced upregulation of RANKL mRNA and protein expression in the calvarial bones (Swanson et al., 2006). In contrast, dexamethasone did not affect the 1,25$(OH)_2$-vitamin D3-induced decrease of *Opg* mRNA or increase of vitamin D receptor mRNA expression.

In a recent study, we have investigated in more detail the interactions in mouse bone marrow cell cultures between GCs and the two calcium regulating hormones, parathyroid hormone and 1,25$(OH)_2$-vitamin D3 (Conaway et al., 2019). Confirming the studies by Udagawa et al. (1989) and Shuto et al. (1994), we observed that dexamethasone (100 nM) synergistically potentiated formation of osteoclasts induced by 1,25 $(OH)_2$-vitamin D3. This was not seen during the early phases of mature osteoclast formation, but at later stages, 6 days after initiation of the cultures. At this phase, osteoclasts incubated with 1,25$(OH)_2$-vitamin D3 in the absence of dexamethasone start to disappear due to apoptosis, whereas osteoclasts formed after stimulation with both 1,25$(OH)_2$-vitamin D3 and dexamethasone appear as giant, oversized osteoclasts (Fig. 2), similar to osteoclasts formed in mouse bone marrow macrophage cultures when stimulated by RANKL and GCs (Conaway et al., 2016). Due to the fusion of osteoclasts to abnormally giant osteoclasts, counting of the number of osteoclasts is not a relevant measure of osteoclastogenesis. The fact that intracellular enzyme activity of TRAP is also synergistically enhanced by dexamethasone suggests that the increased number is associated with increased osteoclast differentiation (Conaway et al., 2019). Similar observations of oversized osteoclasts at the end of the culture were also made when mouse bone marrow cells were incubated with 1,25$(OH)_2$-vitamin D3 and hydrocortisone (1 μM). Formation of the abnormally giant, pancake-like osteoclasts seems to be a

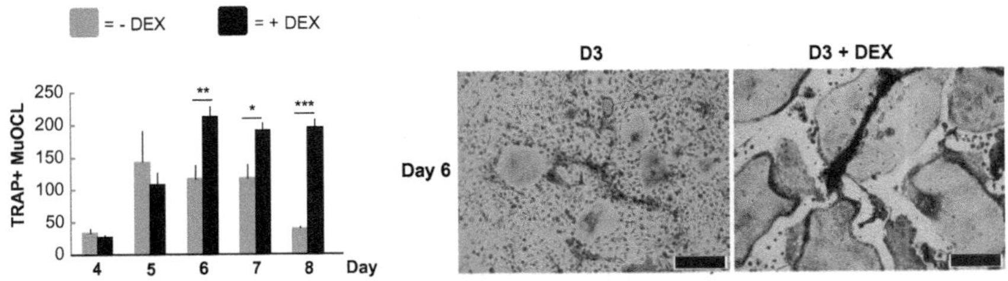

Fig. 2 To the left is shown the delayed potentiation by dexamethasone on osteoclast formation (TRAP$^+$MuOCL) in bone marrow cell cultures on plastic dishes stimulated by 1,25(OH)$_2$-vitamin D3 and its effect on prolonged survival of osteoclasts. To the right is shown the appearance of giant, oversized, widely spread osteoclasts caused by dexamethasone. Scale bar 200 µm. Values are means of 5–7 observations, and vertical bars represent SE. $^*P<0.05$, $^{**}P<0.01$, $^{***}P<0.001$. *Figure reproduced from Conaway, H. H., Henning, P., Lie, A., Tuckermann, J., & Lerner, U. H. (2019). Glucocorticoids employ the monomeric glucocorticoid receptor to potentiate vitamin D3 and parathyroid hormone-induced osteoclastogenesis.* FASEB J, 33*(12):14394–14409, 2019.*

phenomenon taking place when cells are incubated on plastic dishes, for these abnormally large-oversized osteoclasts were not observed when bone marrow cells were incubated on bone slices and treated with 1,25 (OH)$_2$-vitamin D3 and dexamethasone. Importantly, the synergistic potentiation of osteoclast formation was also seen on bone slices, but the large osteoclasts were not abnormally spread out (Fig. 3). The osteoclasts formed on bone were true bone resorbing osteoclasts since the number of resorption excavations and the release of the bone matrix degradation fragment CTX induced by 1,25(OH)$_2$-vitamin D3 were synergistically potentiated by dexamethasone (Fig. 4). Co-treatment with 1,25(OH)$_2$-vitamin D3 and dexamethasone (at and above 1 nM) caused synergistically enhanced mRNA expression of the osteoclastic genes *Calcr*, *Acp5* and *Ctsk*, a response which was more pronounced after 6 days than after 3 days, indicating that the synergistic interaction between activated vitamin D receptor and GR was a delayed phenomenon. Moreover, potentiation of the osteoclastogenic transcription factors *Nfatc1* and *c-Fos* was more pronounced at the later time points. These data indicate that formation of the large mature osteoclasts was not only dependent on fusion but involved increased osteoclast differentiation.

Since the bone marrow cell cultures contain both mesenchymal stromal cells and hematopoietic cells the possibility exists that the interaction between 1,25(OH)$_2$-vitamin D3 and GCs is due to effects in different cell types. Both stromal cells and osteoclast progenitors are well known to

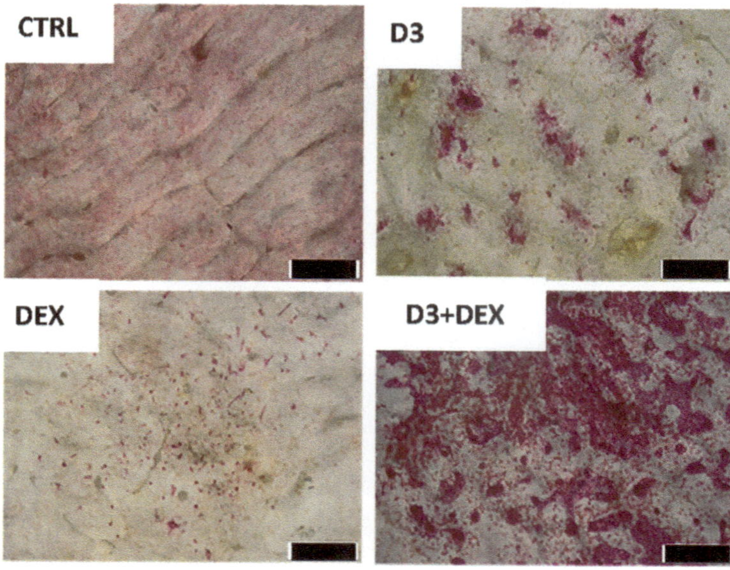

Fig. 3 Osteoclastogenesis in bone marrow cell cultures on bone slices stimulated by 1,25$(OH)_2$-vitamin D3 is potentiated by dexamethasone without formation of oversized, giant, widely spread out cells. Scale bar 200 μm. *Figure reproduced from Conaway, H. H., Henning, P., Lie, A., Tuckermann, J., & Lerner, U. H. (2019). Glucocorticoids employ the monomeric glucocorticoid receptor to potentiate vitamin D3 and parathyroid hormone-induced osteoclastogenesis.* FASEB J, 33*(12):14394–14409, 2019.*

express receptors for both hormones. In the case of osteoclast progenitors, activation of vitamin D receptors does not result in increased osteoclast formation. Activation in cell cultures of GR in mouse osteoclasts can prolong cell survival and increase bone resorbing activity without affecting osteoclast differentiation or number (Conaway et al., 2016; Weinstein et al., 2002). 1,25$(OH)_2$-vitamin D3 can stimulate osteoclast formation indirectly by increasing the RANKL/OPG ratio in bone marrow stromal cells and osteoblasts. As discussed above, there are also studies showing that GCs can increase the RANKL/OPG ratio in osteoblasts and in organ cultured bone. For these reasons, we assessed if the synergistic stimulation of osteoclast differentiation and formation caused by co-treatment with GCs and 1,25 $(OH)_2$-vitamin D3 was due to effects on RANKL and OPG in bone marrow cells. We showed that the mRNA and protein expression of RANKL was not affected by dexamethasone, but that the GC substantially potentiated the effect of 1,25$(OH)_2$-vitamin D3, an effect that was also seen at

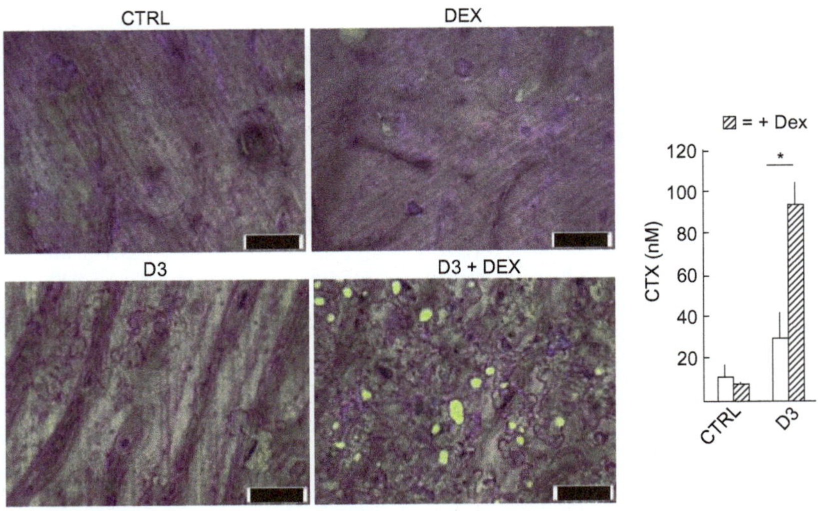

Fig. 4 To the left is shown that the amount of bone resorptive pits caused by bone marrow cells cultured on bone slices and stimulated by 1,25(OH)$_2$-vitamin D3 is potentiated by dexamethasone. To the right is shown that the release of the bone matrix degradation fragment CTX from the bone slices is synergistically enhanced by dexamethasone. Scale bar 100 μm. Values are means of 6 observations and vertical bars represent SE. $^*P<0.05$. *Figure reproduced from Conaway, H. H., Henning, P., Lie, A., Tuckermann, J., & Lerner, U. H. (2019). Glucocorticoids employ the monomeric glucocorticoid receptor to potentiate vitamin D3 and parathyroid hormone-induced osteoclastogenesis.* FASEB J, 33*(12):14394–14409, 2019.*

earlier phases of the cultures (day 3). The mRNA and protein expression of OPG was profoundly decreased both by dexamethasone and 1,25(OH)$_2$-vitamin D3. These findings show that co-treatment with 1,25(OH)$_2$-vitamin D3 and GC cause a large increase of the RANKL/OPG ratio in bone marrow cell cultures explaining the stimulatory effects on osteoclastogenesis. Further proof was the finding that both exogenously added OPG and soluble RANK abolished the effect by co-treatment without affecting mRNA expression of *Rankl*. Since the bone marrow cell culture is a crude system, the findings do not necessarily show that interactions occur in bone marrow stromal cells. However, we did show that the synergistic interactions on *Rankl* mRNA expression could be observed in purified mouse bone marrow stromal cells and in the bone marrow stromal cell line ST-2.

An additive or synergistic effect on *Rankl* mRNA production by dexamethasone together with 1,25(OH)$_2$-vitamin D3 in murine ST-2 cells and human MG63 and SaOS2 osteoblastic cells has also been reported by other groups and binding sites for GR and vitamin D receptor (VDR) have been identified both within 2 kb upstream the *Rankl* TSS and in multiple long distance enhancer regions (Kim et al., 2006; Kitazawa & Kitazawa, 2002; Kitazawa et al., 1999;).

Kitasawa et al. identified half sites for potential VDR and GR binding located −935 and – 640 bp upstream of the mouse *Rankl* TSS (Kitazawa et al., 1999). Functional studies focused primarily on the VDR and promoter-reporter studies and EMSA studies of nuclear proteins from ST2 cells confirmed the functionality of the VDR site in murine cells (Kitazawa & Kitazawa, 2002). *Rankl* mRNA induction was modest using dexamethasone alone and promoter-reporter studies were not performed with the combination of dexamethasone and 1,25$(OH)_2$-vitamin D3, limiting the possibility to evaluate the effect of mutation of the GRE half site (Kitazawa & Kitazawa, 2002). Soon after, the authors also verified a functional VDR site in human SaOS2 cells located −1584/−1570 upstream of the human *RANKL* TSS (Kitazawa et al., 2003). However, later studies by another group did not observe VDR binding to the −935 region in the mouse *Rankl* promoter by ChIP and no induction by 1,25$(OH)_2$-vitamin D3 was observed using a promoter-reporter containing a sequence spanning 2 kb upstream of the *Rankl* TSS (Fan et al., 2004). Recently, mice with deletion of the −510 to −1413 bp sequence up stream from the *Rankl* TSS were described (MacLeod et al., 2021). The deleted region includes the potential GR and VDR binding sites described above, but no reduction in basal levels of *Rankl* mRNA in bone were observed and mice had a normal bone phenotype, suggesting that physiological levels of GC and 1,25(OH)2-vitamin D3 did not affect RANKL through these sites. However, the importance of these sites for the regulation of RANKL by high or low levels of GC or 1,25 $(OH)_2$-vitamin D3 has not been investigated.

On the other hand, long distance enhancer regions upstream of the *Rankl* TSS were identified by Kim et al. using a combination of ChIP and DNA microarray (Kim et al., 2006). They identified 5 regions of VDR binding located −16 kb (mRLD1), −22 kb (mRLD2), −60 kb (mRLD3), −69 kb (mRLD4) and −76 kb (mRLD5) upstream of the mouse *Rankl* TSS and the most distal of these (mRLD5) conferred the 1,25$(OH)_2$-vitamin D3 response. The inducible activity of GR was also mapped to the mRLD5 locus. The authors could also show that when ST-2 cells were treated with dexamethasone alone, GR bound only to the mRLD5 region, but the combination of dexamethasone and 1,25$(OH)_2$-vitamin D3 mediated some binding to mRLD1 and mRLD3 in addition to the mRLD5 binding. Importantly, dexamethasone could also be demonstrated to enhance the

1,25(OH)$_2$-vitamin D3 initiated recruitment of RNA pol II to the mouse *Rankl* TSS. The −76 kb region has also been identified to be important for *Rankl* mRNA regulation by parathyroid hormone and gp130 cytokines (see further below) and this region has been termed the distal control region (DCR). Deletion of the DCR in mice results in decreased basal levels of *Rankl* mRNA, decreased numbers of osteoclasts in trabecular bone and increased bone mass (Fu, Manolagas, & O'Brien, 2006; Galli et al., 2008). The five distal enhancer regions were shown to be highly conserved in the human genome but at slightly different distances compared to the mouse genome and located −20 (hRLD1), −25 (hRLD2), −75 (hRLD3), −87 (hRLD4) and −96 kb (hRLD5) from the human *RANKL* TSS (Nerenz et al., 2008). VDR binding and transcription activating functions could be demonstrated at the most distal region corresponding to mRLD5, but, interestingly, a several-fold higher induction was observed at the most proximal region hRLD1. The transcriptional response at hRLD1 was also induced by dexamethasone and the combination of dexamethasone and 1,25(OH)$_2$-vitamin D3 showed an additive response (Nerenz et al., 2008) Since chromatin modifications can facilitate transcription factor binding, it is interesting to note that several reports have shown that 1,25(OH)$_2$-vitamin D3 alters H4 acetylation upstream of the *Rankl* TSS in both mouse and human cells (Fan et al., 2004; Kim et al., 2006; Nerenz et al., 2008).

Since parathyroid hormone also can enhance the RANKL/OPG ratio in bone marrow stromal cells and stimulates osteoclast formation, we have investigated if these effects can also be potentiated by GCs. In agreement with the findings in 1,25(OH)$_2$-vitamin D3-stimulated bone marrow cell cultures, dexamethasone synergistically potentiated the stimulatory effects by parathyroid hormone on osteoclast formation, expression of osteoclastic and osteoclastogenic genes (*Calcr, Acp5, Ctsk and Nfatc1*) and mRNA and protein expression of RANKL (Conaway et al., 2019). Similar to co-treatment with 1,25(OH)$_2$-vitamin D3 and GCs, and stimulation of pure bone marrow macrophages with RANKL and GCs, large, oversized osteoclasts were formed when experiments were performed on plastic dishes, further indicating that this abnormal response is due to activation of GR in osteoclasts. Importantly, no such abnormal osteoclasts were seen when cells were incubated on bone slices and co-treatment resulted in synergistically enhanced numbers of osteoclasts, pit excavations and release of CTX. The parathyroid hormone induced increase in *Rankl* mRNA levels is mediated *via* the parathyroid hormone receptor and downstream signaling by protein kinase A (PKA) and the cAMP response element (CRE)-binding protein (CREB) (Fu, Jilka, Manolagas, & O'Brien, 2002; Kondo, Guo, & Bringhurst, 2002; Lee & Lorenzo, 2002). CREB binding and *Rankl* induction have been mapped to the DCR region which binds GR, but the direct interaction between CREB and GR at this region has not been investigated (Fu et al., 2006; Kim, Yamazaki, Shevde, & Pike, 2007; Kim et al., 2006). Deletion of the DCR region blunts the parathyroid hormone induced increase in *Rankl* mRNA in bone marrow cell culture (Fu et al., 2006). CREB binding sites have also been reported in the more proximal part of the Rankl promoter, close to suggested GRE and VDR sites (Kitazawa & Kitazawa, 2002; Kitazawa et al., 1999; Park, Baek, Baek, & Kim, 2015). As mentioned above, mice where this proximal promoter region has been deleted have a normal bone phenotype (MacLeod et al., 2021). However, the *Rankl* mRNA response to parathyroid hormone was reduced in stromal cells isolated from these mice and the *in vivo* effects of either parathyroid hormone or GC in these mice have not been investigated.

The observation that interactions between dexamethasone and either 1,25(OH)$_2$-vitamin D3 or parathyroid hormone on osteoclast formation and RANKL protein were inhibited by the GR-antagonist RU486 and that no interaction was seen when bone marrow cells were employed from $GR^{Runx2Cre}$ mice, in which GR had been deleted in *Runx2*-expressing stromal cells, demonstrate the crucial role of GR for the effects by dexamethasone. The interaction between the hormones did not appear to be dependent on canonical GR signaling since osteoclast differentiation, osteoclast number and RANKL protein were also observed in cells from GR^{dim} mice. This was in contrast to the effect by GCs on bone resorbing activity of mature osteoclasts, which was not observed in cells from GR^{dim} mice (Conaway et al., 2016).

These findings show that GCs have the capacity to sensitize the RANKL-response to both 1,25(OH)$_2$-vitamin D3 and parathyroid hormone, which may explain the enhanced bone resorption seen during GC treatment. As discussed above, it is controversial if GC treatment results in hyperparathyroidism. However, it is important to point out that GCs can sensitize the RANKL-response at very low concentrations ($\geq$0.01 nM) of 1,25 (OH)$_2$-vitamin D3 and parathyroid hormone (Conaway et al., 1997), which are close to concentrations of these hormones found *in vivo* under normal conditions. Our findings demonstrate that one part of GC-induced increased resorption may be explained by effects of GR itself on transactivation of the RANKL promoter, but our findings indicate that it is more likely that it is the interactions between GR and other transcription factors which are crucial for increased transactivation of the RANKL promoter. Whether or not it might be interactions between GR and vitamin D receptor or transcription factors downstream of the parathyroid hormone receptor, which are important in patients treated with GCs, remains to be shown. Since many patients suffer from inflammatory disorders it should also be investigated if GR can interact with transcription factors activated by cytokines involved in inflammatory induced bone resorption.

5. Concluding remarks/summary

In GC-induced osteoporosis, an early, rapid phase of osteoclastic bone resorption due primarily to an increased RANKL/OPG ratio stimulated by GC is followed by a slower developing phase characterized by decreased osteoblastic bone formation. There are numerous studies, both in humans and animals, detailing the dosage and time dependence incidence of GC-induced osteoporosis. Both cortical and trabecular bone loss occur, with the greatest loss occurring in trabecular bone. The increase in fracture incidence is greatest in the spine, but also occurs in the hip as well.

GC have direct and indirect actions on osteoclasts mediated by GRs. GC act indirectly *via* osteoblasts/osteocytes to increase the RANKL/OPG ratio and stimulate osteoclast progenitors to mature, functional osteoclasts (Fig. 5A). Direct effects of GC on osteoclast activity have been mixed, some showing stimulation, others inhibition, possibly due to the multiple osteoclast preparations used for study and varying culture treatments. In general, lower dosage and early treatment of cultures with GC have been stimulatory and later treatment and higher dosages inhibitory. The ability of GC to decrease osteoclast apoptosis has also been suggested to be a direct effect, but this is observed on plastic dishes and has often been accompanied by formation of giant, pancake osteoclasts. When precursor cells are cultured on bone slices, apoptosis is not altered and giant, pancake osteoclasts do not develop. Another direct effect suggested for GC, observed in both mouse BMM and human cells, is the ability to prolong the active mode of osteoclast activity.

GC can not only function *via* the GR in bone cells to stimulate osteoclastogenesis, but the agents also have systemic (*e.g.*, GI tract, kidney, hormonal) actions that can affect bone. 1,25(OH)$_2$-vitamin D3 and parathyroid hormone are the two principal calcium regulating hormones that have been studied extensively with GC. Both 1,25(OH)$_2$-vitamin D3 and parathyroid hormone increase the RANKL/OPG ratio and stimulate

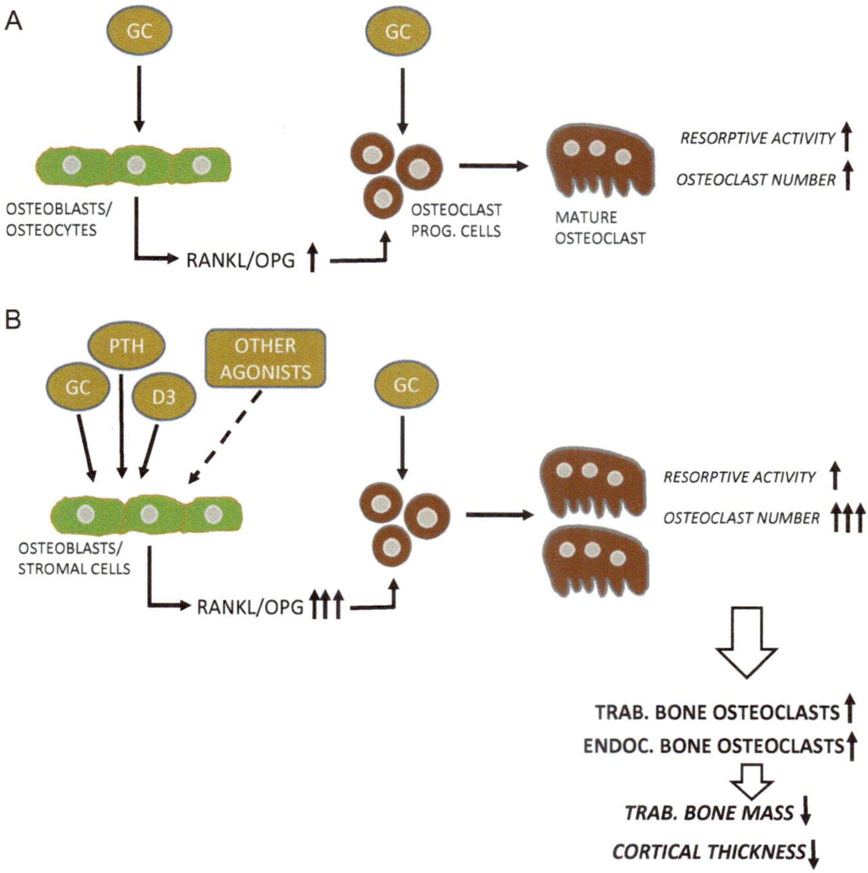

Fig. 5 Glucocorticoids decrease cortical and trabecular bone mass by increasing the numbers and activity of osteoclasts. (A) Glucocorticoids stimulate osteoclast formation by increasing the ratio of RANKL/OPG in osteoblasts/osteocytes. The resorptive activity of mature osteoclasts can be enhanced by glucocorticoids through activation of glucocorticoid receptors in osteoclast progenitor cells. (B) Stimulation of RANKL/OPG is particularly profound when glucocorticoids act in concert with osteoclast stimulators such as parathyroid hormone and 1,25$(OH)_2$-vitamin D3.

osteoclastogenesis and bone resorption. Addition of GC to either 1,25 $(OH)_2$ -vitamin D3 or parathyroid hormone bone cultures increases the RANKL/OPG ratio, causing synergistic stimulation of bone resorption (Fig. 5B). The synergism observed with parathyroid hormoneand GC *in vitro* suggests the possibility that this interaction might play a role in GC-induced osteoporosis.

Molecular studies offer an explanation for why synergistic interaction between GC and 1,25$(OH)_2$-vitamin D3 occurs. The mouse RANKL gene contains a distal control region located 76 kb upstream of the TSS that binds 1,25$(OH)_2$-vitamin D3 and GC. When both 1,25$(OH)_2$-vitamin D3 and GC are present, greater binding occurs and GC enhances 1,25$(OH)_2$ -vitamin D3 polymerase II recruitment. The −76 kb region of the RANKL gene also appears to be important for parathyroid hormone activation, but the details for interaction between GC and parathyroid hormone have not been determined.

References

Abtahi, S., Driessen, J. H. M., Burden, A. M., Souverein, P. C., van den Bergh, J. P., van Staa, T. P., et al. (2021). Low-dose oral glucocorticoid therapy and risk of osteoporotic fractures in patients with rheumatoid arthritis: A cohort study using the clinical practice research datalink. *Rheumatology (Oxford, England)*. https://doi.org/10.1093/rheumatology/keab548.

Ahmad, M., Hachemi, Y., Paxian, K., Mengele, F., Koenen, M., & Tuckermann, J. (2019). A Jack of all trades: Impact of glucocorticoids on cellular cross-talk in Osteoimmunology. *Frontiers in Immunology*, *10*, 2460. https://doi.org/10.3389/fimmu.2019.02460.

Allen, C. S., Yeung, J. H., Vandermeer, B., & Homik, J. (2016). Bisphosphonates for steroid-induced osteoporosis. *Cochrane Database of Systematic Reviews*, *10*, CD001347. https://doi.org/10.1002/14651858.CD001347.pub2.

Amiche, M. A., Albaum, J. M., Tadrous, M., Pechlivanoglou, P., Levesque, L. E., Adachi, J. D., et al. (2016). Efficacy of osteoporosis pharmacotherapies in preventing fracture among oral glucocorticoid users: A network meta-analysis. *Osteoporosis International*, *27*(6), 1989–1998. https://doi.org/10.1007/s00198-015-3476-4.

Andreev, D., Liu, M., Weidner, D., Kachler, K., Faas, M., Gruneboom, A., et al. (2020). Osteocyte necrosis triggers osteoclast-mediated bone loss through macrophage-inducible C-type lectin. *The Journal of Clinical Investigation*, *130*(9), 4811–4830. https://doi.org/10.1172/JCI134214.

Axelsson, K. F., Nilsson, A. G., & Lorentzon, M. (2017). Alendronate and hip fracture in patients using glucocorticoids-reply. *JAMA*, *318*(17), 1712. https://doi.org/10.1001/jama.2017.14295.

Beavan, S., Horner, A., Bord, S., Ireland, D., & Compston, J. (2001). Colocalization of glucocorticoid and mineralocorticoid receptors in human bone. *Journal of Bone and Mineral Research*, *16*(8), 1496–1504. https://doi.org/10.1359/jbmr.2001.16.8.1496.

Besemer, F., Pereira, A. M., & Smit, J. W. (2011). Alcohol-induced Cushing syndrome. Hypercortisolism caused by alcohol abuse. *The Netherlands Journal of Medicine*, *69*(7), 318–323.

Biddie, S. C., John, S., Sabo, P. J., Thurman, R. E., Johnson, T. A., Schiltz, R. L., et al. (2011). Transcription factor AP1 potentiates chromatin accessibility and glucocorticoid receptor binding. *Molecular Cell*, *43*(1), 145–155. https://doi.org/10.1016/j.molcel.2011.06.016.

Bonadonna, S., Burattin, A., Nuzzo, M., Bugari, G., Rosei, E. A., Valle, D., et al. (2005). Chronic glucocorticoid treatment alters spontaneous pulsatile parathyroid hormone secretory dynamics in human subjects. *European Journal of Endocrinology*, *152*(2), 199–205. https://doi.org/10.1530/eje.1.01841.

Bouillon, R., Marcocci, C., Carmeliet, G., Bikle, D., White, J. H., Dawson-Hughes, B., et al. (2019). Skeletal and Extraskeletal actions of vitamin D: Current evidence and outstanding questions. *Endocrine Reviews*, *40*(4), 1109–1151. https://doi.org/10.1210/er.2018-00126.

Brabnikova Maresova, K., Pavelka, K., & Stepan, J. J. (2013). Acute effects of glucocorticoids on serum markers of osteoclasts, osteoblasts, and osteocytes. *Calcified Tissue International, 92*(4), 354–361. https://doi.org/10.1007/s00223-012-9684-4.
Brandstrom, H., Bjorkman, T., & Ljunggren, O. (2001). Regulation of osteoprotegerin secretion from primary cultures of human bone marrow stromal cells. *Biochemical and Biophysical Research Communications, 280*(3), 831–835. https://doi.org/10.1006/bbrc.2000.4223.
Canalis, E. (1983). Effect of glucocorticoids on type I collagen synthesis, alkaline phosphatase activity, and deoxyribonucleic acid content in cultured rat calvariae. *Endocrinology, 112*(3), 931–939. https://doi.org/10.1210/endo-112-3-931.
Canalis, E. (1984). Effect of cortisol on periosteal and nonperiosteal collagen and DNA synthesis in cultured rat calvariae. *Calcified Tissue International, 36*(2), 158–166. https://doi.org/10.1007/BF02405312.
Canalis, E. (2005). Mechanisms of glucocorticoid action in bone. *Current Osteoporosis Reports, 3*(3), 98–102.
Canalis, E., Bilezikian, J. P., Angeli, A., & Giustina, A. (2004). Perspectives on glucocorticoid-induced osteoporosis. *Bone, 34*(4), 593–598. https://doi.org/10.1016/j.bone.2003.11.026.
Canalis, E., Mazziotti, G., Giustina, A., & Bilezikian, J. P. (2007). Glucocorticoid-induced osteoporosis: Pathophysiology and therapy. *Osteoporosis International, 18*(10), 1319–1328. https://doi.org/10.1007/s00198-007-0394-0.
Compston, J. (2018). Glucocorticoid-induced osteoporosis: An update. *Endocrine, 61*(1), 7–16. https://doi.org/10.1007/s12020-018-1588-2.
Conaway, H. H., Grigorie, D., & Lerner, U. H. (1996). Stimulation of neonatal mouse calvarial bone resorption by the glucocorticoids hydrocortisone and dexamethasone. *Journal of Bone and Mineral Research, 11*(10), 1419–1429. https://doi.org/10.1002/jbmr.5650111008.
Conaway, H. H., Grigorie, D., & Lerner, U. H. (1997). Differential effects of glucocorticoids on bone resorption in neonatal mouse calvariae stimulated by peptide and steroid-like hormones. *The Journal of Endocrinology, 155*(3), 513–521.
Conaway, H. H., Henning, P., Lie, A., Tuckermann, J., & Lerner, U. H. (2016). Activation of dimeric glucocorticoid receptors in osteoclast progenitors potentiates RANKL induced mature osteoclast bone resorbing activity. *Bone, 93*, 43–54. https://doi.org/10.1016/j.bone.2016.08.024.
Conaway, H. H., Henning, P., Lie, A., Tuckermann, J., & Lerner, U. H. (2019). Glucocorticoids employ the monomeric glucocorticoid receptor to potentiate vitamin D3 and parathyroid hormone-induced osteoclastogenesis. *The FASEB Journal, 33*(12), 14394–14409. https://doi.org/10.1096/fj.201802729RRR.
Conaway, H. H., Pirhayati, A., Persson, E., Pettersson, U., Svensson, O., Lindholm, C., et al. (2011). Retinoids stimulate periosteal bone resorption by enhancing the protein RANKL, a response inhibited by monomeric glucocorticoid receptor. *The Journal of Biological Chemistry, 286*(36), 31425–31436. https://doi.org/10.1074/jbc.M111.247734.
Dalle Carbonare, L., Arlot, M. E., Chavassieux, P. M., Roux, J. P., Portero, N. R., & Meunier, P. J. (2001). Comparison of trabecular bone microarchitecture and remodeling in glucocorticoid-induced and postmenopausal osteoporosis. *Journal of Bone and Mineral Research, 16*(1), 97–103. https://doi.org/10.1359/jbmr.2001.16.1.97.
DeLuca, H. F. (2016). Vitamin D: Historical overview. *Vitamins and Hormones, 100*, 1–20. https://doi.org/10.1016/bs.vh.2015.11.001.
Dempster, D. W., Moonga, B. S., Stein, L. S., Horbert, W. R., & Antakly, T. (1997). Glucocorticoids inhibit bone resorption by isolated rat osteoclasts by enhancing apoptosis. *The Journal of Endocrinology, 154*(3), 397–406.
Dilisio, M. F. (2014). Osteonecrosis following short-term, low-dose oral corticosteroids: A population-based study of 24 million patients. *Orthopedics, 37*(7), e631–e636. https://doi.org/10.3928/01477447-20140626-54.

Dovio, A., Perazzolo, L., Osella, G., Ventura, M., Termine, A., Milano, E., et al. (2004). Immediate fall of bone formation and transient increase of bone resorption in the course of high-dose, short-term glucocorticoid therapy in young patients with multiple sclerosis. *The Journal of Clinical Endocrinology and Metabolism, 89*(10), 4923–4928. https://doi.org/10.1210/jc.2004-0164.

Escoter-Torres, L., Greulich, F., Quagliarini, F., Wierer, M., & Uhlenhaut, N. H. (2020). Anti-inflammatory functions of the glucocorticoid receptor require DNA binding. *Nucleic Acids Research, 48*(15), 8393–8407. https://doi.org/10.1093/nar/gkaa565.

Fan, X., Roy, E. M., Murphy, T. C., Nanes, M. S., Kim, S., Pike, J. W., et al. (2004). Regulation of RANKL promoter activity is associated with histone remodeling in murine bone stromal cells. *Journal of Cellular Biochemistry, 93*(4), 807–818. https://doi.org/10.1002/jcb.20217.

Frank, F., Liu, X., & Ortlund, E. A. (2021). Glucocorticoid receptor condensates link DNA-dependent receptor dimerization and transcriptional transactivation. *Proceedings of the National Academy of Sciences of the United States of America, 118*(30). https://doi.org/10.1073/pnas.2024685118.

Frenkel, B., White, W., & Tuckermann, J. (2015). Glucocorticoid-induced osteoporosis. *Advances in Experimental Medicine and Biology, 872*, 179–215. https://doi.org/10.1007/978-1-4939-2895-8_8.

Fu, Q., Jilka, R. L., Manolagas, S. C., & O'Brien, C. A. (2002). Parathyroid hormone stimulates receptor activator of NFkappa B ligand and inhibits osteoprotegerin expression via protein kinase a activation of cAMP-response element-binding protein. *The Journal of Biological Chemistry, 277*(50), 48868–48875. https://doi.org/10.1074/jbc.M208494200.

Fu, Q., Manolagas, S. C., & O'Brien, C. A. (2006). Parathyroid hormone controls receptor activator of NF-kappaB ligand gene expression via a distant transcriptional enhancer. *Molecular and Cellular Biology, 26*(17), 6453–6468. https://doi.org/10.1128/MCB.00356-06.

Fumoto, T., Takeshita, S., Ito, M., & Ikeda, K. (2014). Physiological functions of osteoblast lineage and T cell-derived RANKL in bone homeostasis. *Journal of Bone and Mineral Research, 29*(4), 830–842. https://doi.org/10.1002/jbmr.2096.

Galli, C., Zella, L. A., Fretz, J. A., Fu, Q., Pike, J. W., Weinstein, R. S., et al. (2008). Targeted deletion of a distant transcriptional enhancer of the receptor activator of nuclear factor-kappaB ligand gene reduces bone remodeling and increases bone mass. *Endocrinology, 149*(1), 146–153. https://doi.org/10.1210/en.2007-0734.

Gronowicz, G., McCarthy, M. B., & Raisz, L. G. (1990). Glucocorticoids stimulate resorption in fetal rat parietal bones in vitro. *Journal of Bone and Mineral Research, 5*(12), 1223–1230. https://doi.org/10.1002/jbmr.5650051206.

Gross, K. L., Oakley, R. H., Scoltock, A. B., Jewell, C. M., & Cidlowski, J. A. (2011). Glucocorticoid receptor alpha isoform-selective regulation of antiapoptotic genes in osteosarcoma cells: A new mechanism for glucocorticoid resistance. *Molecular Endocrinology, 25*(7), 1087–1099. https://doi.org/10.1210/me.2010-0051.

Guo, Y. D., Strugnell, S., Back, D. W., & Jones, G. (1993). Transfected human liver cytochrome P-450 hydroxylates vitamin D analogs at different side-chain positions. *Proceedings of the National Academy of Sciences of the United States of America, 90*(18), 8668–8672. https://doi.org/10.1073/pnas.90.18.8668.

Hartmann, K., Koenen, M., Schauer, S., Wittig-Blaich, S., Ahmad, M., Baschant, U., et al. (2016). Molecular actions of glucocorticoids in cartilage and bone during health, disease, and steroid therapy. *Physiological Reviews, 96*(2), 409–447. https://doi.org/10.1152/physrev.00011.2015.

Heck, S., Kullmann, M., Gast, A., Ponta, H., Rahmsdorf, H. J., Herrlich, P., et al. (1994). A distinct modulating domain in glucocorticoid receptor monomers in the repression of activity of the transcription factor AP-1. *The EMBO Journal, 13*(17), 4087–4095.

Henneicke, H., Herrmann, M., Kalak, R., Brennan-Speranza, T. C., Heinevetter, U., Bertollo, N., et al. (2011). Corticosterone selectively targets endo-cortical surfaces by an osteoblast-dependent mechanism. *Bone, 49*(4), 733–742. https://doi.org/10.1016/j.bone.2011.06.013.

Hirayama, T., Sabokbar, A., & Athanasou, N. A. (2002). Effect of corticosteroids on human osteoclast formation and activity. *The Journal of Endocrinology, 175*(1), 155–163.

Hofbauer, L. C., Gori, F., Riggs, B. L., Lacey, D. L., Dunstan, C. R., Spelsberg, T. C., et al. (1999). Stimulation of osteoprotegerin ligand and inhibition of osteoprotegerin production by glucocorticoids in human osteoblastic lineage cells: Potential paracrine mechanisms of glucocorticoid-induced osteoporosis. *Endocrinology, 140*(10), 4382–4389. https://doi.org/10.1210/endo.140.10.7034.

Holick, M. F. (2015). Vitamin D is not as toxic as was once thought: A historical and an up-to-date perspective. *Mayo Clinic Proceedings, 90*(5), 561–564. https://doi.org/10.1016/j.mayocp.2015.03.015.

Hollenberg, S. M., Weinberger, C., Ong, E. S., Cerelli, G., Oro, A., Lebo, R., et al. (1985). Primary structure and expression of a functional human glucocorticoid receptor cDNA. *Nature, 318*(6047), 635–641. https://doi.org/10.1038/318635a0.

Hong, J. M., Teitelbaum, S. L., Kim, T. H., Ross, F. P., Kim, S. Y., & Kim, H. J. (2011). Calpain-6, a target molecule of glucocorticoids, regulates osteoclastic bone resorption via cytoskeletal organization and microtubule acetylation. *Journal of Bone and Mineral Research, 26*(3), 657–665. https://doi.org/10.1002/jbmr.241.

Hsu, H., Lacey, D. L., Dunstan, C. R., Solovyev, I., Colombero, A., Timms, E., et al. (1999). Tumor necrosis factor receptor family member RANK mediates osteoclast differentiation and activation induced by osteoprotegerin ligand. *Proceedings of the National Academy of Sciences of the United States of America, 96*(7), 3540–3545.

Hudson, W. H., Vera, I. M. S., Nwachukwu, J. C., Weikum, E. R., Herbst, A. G., Yang, Q., et al. (2018). Cryptic glucocorticoid receptor-binding sites pervade genomic NF-kappaB response elements. *Nature Communications, 9*(1), 1337. https://doi.org/10.1038/s41467-018-03780-1.

Humphrey, E. L., Williams, J. H., Davie, M. W., & Marshall, M. J. (2006). Effects of dissociated glucocorticoids on OPG and RANKL in osteoblastic cells. *Bone, 38*(5), 652–661. https://doi.org/10.1016/j.bone.2005.10.004.

Ishikawa, H., Tanaka, H., Iwato, K., Tanabe, O., Asaoku, H., Nobuyoshi, M., et al. (1990). Effect of glucocorticoids on the biologic activities of myeloma cells: Inhibition of interleukin-1 beta osteoclast activating factor-induced bone resorption. *Blood, 75*(3), 715–720.

Ito, S., Suzuki, N., Kato, S., Takahashi, T., & Takagi, M. (2007). Glucocorticoids induce the differentiation of a mesenchymal progenitor cell line, ROB-C26 into adipocytes and osteoblasts, but fail to induce terminal osteoblast differentiation. *Bone, 40*(1), 84–92. https://doi.org/10.1016/j.bone.2006.07.012.

John, S., Sabo, P. J., Thurman, R. E., Sung, M. H., Biddie, S. C., Johnson, T. A., et al. (2011). Chromatin accessibility pre-determines glucocorticoid receptor binding patterns. *Nature Genetics, 43*(3), 264–268. https://doi.org/10.1038/ng.759.

Johnson, T. A., Paakinaho, V., Kim, S., Hager, G. L., & Presman, D. M. (2021). Genome-wide binding potential and regulatory activity of the glucocorticoid receptor's monomeric and dimeric forms. *Nature Communications, 12*(1), 1987. https://doi.org/10.1038/s41467-021-22234-9.

Jones, G. (2008). Pharmacokinetics of vitamin D toxicity. *The American Journal of Clinical Nutrition, 88*(2), 582S–586S. https://doi.org/10.1093/ajcn/88.2.582S.

Kaji, H., Yamauchi, M., Chihara, K., & Sugimoto, T. (2008). Relationship between endogenous parathyroid hormone and bone metabolism/geometry in female patients treated with glucocorticoid. *Hormone and Metabolic Research, 40*(1), 60–65. https://doi.org/10.1055/s-2007-993170.

Kim, S., Yamazaki, M., Shevde, N. K., & Pike, J. W. (2007). Transcriptional control of receptor activator of nuclear factor-kappaB ligand by the protein kinase a activator forskolin and the transmembrane glycoprotein 130-activating cytokine, oncostatin M, is exerted through multiple distal enhancers. *Molecular Endocrinology*, *21*(1), 197–214. https://doi.org/10.1210/me.2006-0315.

Kim, S., Yamazaki, M., Zella, L. A., Shevde, N. K., & Pike, J. W. (2006). Activation of receptor activator of NF-kappaB ligand gene expression by 1,25-dihydroxyvitamin D3 is mediated through multiple long-range enhancers. *Molecular and Cellular Biology*, *26*(17), 6469–6486. https://doi.org/10.1128/MCB.00353-06.

Kim, H. J., Zhao, H., Kitaura, H., Bhattacharyya, S., Brewer, J. A., Muglia, L. J., et al. (2006). Glucocorticoids suppress bone formation via the osteoclast. *The Journal of Clinical Investigation*, *116*(8), 2152–2160. https://doi.org/10.1172/JCI28084.

Kitazawa, S., Kajimoto, K., Kondo, T., & Kitazawa, R. (2003). Vitamin D3 supports osteoclastogenesis via functional vitamin D response element of human RANKL gene promoter. *Journal of Cellular Biochemistry*, *89*(4), 771–777. https://doi.org/10.1002/jcb.10567.

Kitazawa, R., & Kitazawa, S. (2002). Vitamin D(3) augments osteoclastogenesis via vitamin D-responsive element of mouse RANKL gene promoter. *Biochemical and Biophysical Research Communications*, *290*(2), 650–655. https://doi.org/10.1006/bbrc.2001.6251.

Kitazawa, R., Kitazawa, S., & Maeda, S. (1999). Promoter structure of mouse RANKL/TRANCE/OPGL/ODF gene. *Biochimica et Biophysica Acta*, *1445*(1), 134–141.

Klein, R. G., Arnaud, S. B., Gallagher, J. C., Deluca, H. F., & Riggs, B. L. (1977). Intestinal calcium absorption in exogenous hypercortisonism. Role of 25-hydroxyvitamin D and corticosteroid dose. *The Journal of Clinical Investigation*, *60*(1), 253–259. https://doi.org/10.1172/JCI108762.

Koga, T., Inui, M., Inoue, K., Kim, S., Suematsu, A., Kobayashi, E., et al. (2004). Costimulatory signals mediated by the ITAM motif cooperate with RANKL for bone homeostasis. *Nature*, *428*(6984), 758–763. https://doi.org/10.1038/nature02444.

Kondo, H., Guo, J., & Bringhurst, F. R. (2002). Cyclic adenosine monophosphate/protein kinase a mediates parathyroid hormone/parathyroid hormone-related protein receptor regulation of osteoclastogenesis and expression of RANKL and osteoprotegerin mRNAs by marrow stromal cells. *Journal of Bone and Mineral Research*, *17*(9), 1667–1679. https://doi.org/10.1359/jbmr.2002.17.9.1667.

Kong, Y. Y., Yoshida, H., Sarosi, I., Tan, H. L., Timms, E., Capparelli, C., et al. (1999). OPGL is a key regulator of osteoclastogenesis, lymphocyte development and lymph-node organogenesis. *Nature*, *397*(6717), 315–323. https://doi.org/10.1038/16852.

Laake, H. (1960). The action of corticosteroids on the renal reabsorption of calcium. *Acta Endocrinologica*, *34*, 60–64. https://doi.org/10.1530/acta.0.xxxiv0060.

Lee, S. K., & Lorenzo, J. A. (2002). Regulation of receptor activator of nuclear factor-kappa B ligand and osteoprotegerin mRNA expression by parathyroid hormone is predominantly mediated by the protein kinase a pathway in murine bone marrow cultures. *Bone*, *31*(1), 252–259. https://doi.org/10.1016/s8756-3282(02)00804-9.

Lee, B. H., & Stallcup, M. R. (2017). Glucocorticoid receptor binding to chromatin is selectively controlled by the coregulator Hic-5 and chromatin remodeling enzymes. *The Journal of Biological Chemistry*, *292*(22), 9320–9334. https://doi.org/10.1074/jbc.M117.782607.

Lim, H. W., Uhlenhaut, N. H., Rauch, A., Weiner, J., Hubner, S., Hubner, N., et al. (2015). Genomic redistribution of GR monomers and dimers mediates transcriptional response to exogenous glucocorticoid in vivo. *Genome Research*, *25*(6), 836–844. https://doi.org/10.1101/gr.188581.114.

Lind, T., Lind, P. M., Jacobson, A., Hu, L., Sundqvist, A., Risteli, J., et al. (2011). High dietary intake of retinol leads to bone marrow hypoxia and diaphyseal endosteal mineralization in rats. *Bone*, *48*(3), 496–506. https://doi.org/10.1016/j.bone.2010.10.169.

Lionikaite, V., Gustafsson, K. L., Westerlund, A., Windahl, S. H., Koskela, A., Tuukkanen, J., et al. (2018). Clinically relevant doses of vitamin a decrease cortical bone mass in mice. *The Journal of Endocrinology*, *239*(3), 389–402. https://doi.org/10.1530/JOE-18-0316.

Liu, Y., Porta, A., Peng, X., Gengaro, K., Cunningham, E. B., Li, H., et al. (2004). Prevention of glucocorticoid-induced apoptosis in osteocytes and osteoblasts by calbindin-D28k. *Journal of Bone and Mineral Research*, *19*(3), 479–490. https://doi.org/10.1359/JBMR.0301242.

Lovsin, N., & Marc, J. (2021). Glucocorticoid receptor regulates TNFSF11 transcription by binding to glucocorticoid responsive element in TNFSF11 proximal promoter region. *International Journal of Molecular Sciences*, *22*(3). https://doi.org/10.3390/ijms22031054.

Lowe, C., Gray, D. H., & Reid, I. R. (1992). Serum blocks the osteolytic effect of cortisol in neonatal mouse calvaria. *Calcified Tissue International*, *50*(2), 189–192. https://doi.org/10.1007/BF00298799.

Lu, N. Z., Collins, J. B., Grissom, S. F., & Cidlowski, J. A. (2007). Selective regulation of bone cell apoptosis by translational isoforms of the glucocorticoid receptor. *Molecular and Cellular Biology*, *27*(20), 7143–7160. https://doi.org/10.1128/MCB.00253-07.

Lukert, B. P., & Adams, J. S. (1976). Calcium and phosphorus homeostasis in man. Effect of corticosteroids. *Archives of Internal Medicine*, *136*(11), 1249–1253.

MacLeod, R. S., Meyer, M. B., Xiong, J., Cawley, K. M., Liu, Y., Onal, M., et al. (2021). Deletion of a putative promoter-proximal Tnfsf11 regulatory region in mice does not alter bone mass or Tnfsf11 expression in vivo. *PLoS One*, *16*(5), e0250974. https://doi.org/10.1371/journal.pone.0250974.

Mankin, H. J. (1992). Nontraumatic necrosis of bone (osteonecrosis). *The New England Journal of Medicine*, *326*(22), 1473–1479. https://doi.org/10.1056/NEJM199205283262206.

Marcinowska-Suchowierska, E., Kupisz-Urbanska, M., Lukaszkiewicz, J., Pludowski, P., & Jones, G. (2018). Vitamin D toxicity-a clinical perspective. *Frontiers in Endocrinology*, *9*, 550. https://doi.org/10.3389/fendo.2018.00550.

Marusic, A., & Raisz, L. G. (1991). Cortisol modulates the actions of interleukin-1 alpha on bone formation, resorption, and prostaglandin production in cultured mouse parietal bones. *Endocrinology*, *129*(5), 2699–2706. https://doi.org/10.1210/endo-129-5-2699.

Mazziotti, G., Formenti, A. M., Adler, R. A., Bilezikian, J. P., Grossman, A., Sbardella, E., et al. (2016). Glucocorticoid-induced osteoporosis: Pathophysiological role of GH/IGF-I and PTH/VITAMIN D axes, treatment options and guidelines. *Endocrine*, *54*(3), 603–611. https://doi.org/10.1007/s12020-016-1146-8.

Mocsai, A., Humphrey, M. B., Van Ziffle, J. A., Hu, Y., Burghardt, A., Spusta, S. C., et al. (2004). The immunomodulatory adapter proteins DAP12 and fc receptor gamma-chain (FcRgamma) regulate development of functional osteoclasts through the Syk tyrosine kinase. *Proceedings of the National Academy of Sciences of the United States of America*, *101*(16), 6158–6163. https://doi.org/10.1073/pnas.0401602101.

Mohamed, G. B., & Abdel-Latif, E. A. (2011). Serum osteoprotegerin (OPG) in children with primary nephrotic syndrome. *Saudi Journal of Kidney Diseases and Transplantation*, *22*(5), 955–962.

Mori, T., Horibe, K., Koide, M., Uehara, S., Yamamoto, Y., Kato, S., et al. (2020). The vitamin D receptor in osteoblast-lineage cells is essential for the Proresorptive activity of 1alpha,25(OH)2D3 in vivo. *Endocrinology*, *161*(11). https://doi.org/10.1210/endocr/bqaa178.

Moutsatsou, P., Kassi, E., & Papavassiliou, A. G. (2012). Glucocorticoid receptor signaling in bone cells. *Trends in Molecular Medicine*, *18*(6), 348–359. https://doi.org/10.1016/j.molmed.2012.04.005.

Murayama, A., Takeyama, K., Kitanaka, S., Kodera, Y., Kawaguchi, Y., Hosoya, T., et al. (1999). Positive and negative regulations of the renal 25-hydroxyvitamin D3 1alpha-hydroxylase gene by parathyroid hormone, calcitonin, and 1alpha,25(OH)2D3 in intact animals. *Endocrinology*, *140*(5), 2224–2231. https://doi.org/10.1210/endo.140.5.6691.

Murphey, M. D., Foreman, K. L., Klassen-Fischer, M. K., Fox, M. G., Chung, E. M., & Kransdorf, M. J. (2014). From the radiologic pathology archives imaging of osteonecrosis: Radiologic-pathologic correlation. *Radiographics*, *34*(4), 1003–1028. https://doi.org/10.1148/rg.344140019.

Nakashima, T., Hayashi, M., Fukunaga, T., Kurata, K., Oh-Hora, M., Feng, J. Q., et al. (2011). Evidence for osteocyte regulation of bone homeostasis through RANKL expression. *Nature Medicine*, *17*(10), 1231–1234. https://doi.org/10.1038/nm.2452.

Nerenz, R. D., Martowicz, M. L., & Pike, J. W. (2008). An enhancer 20 kilobases upstream of the human receptor activator of nuclear factor-kappaB ligand gene mediates dominant activation by 1,25-dihydroxyvitamin D3. *Molecular Endocrinology*, *22*(5), 1044–1056. https://doi.org/10.1210/me.2007-0380.

Nicolaides, N. C., Chrousos, G., & Kino, T. (2000). Glucocorticoid Receptor. In K. R. Feingold, B. Anawalt, A. Boyce, G. Chrousos, W. W. de Herder, K. Dhatariya, ... D. P. Wilson (Eds.), *Endotext*. MA: South Dartmouth.

Nieman, L. K. (2018). Diagnosis of Cushing's syndrome in the modern era. *Endocrinology and Metabolism Clinics of North America*, *47*(2), 259–273. https://doi.org/10.1016/j.ecl.2018.02.001.

Norbury, W. B., Herndon, D. N., Branski, L. K., Chinkes, D. L., & Jeschke, M. G. (2008). Urinary cortisol and catecholamine excretion after burn injury in children. *The Journal of Clinical Endocrinology and Metabolism*, *93*(4), 1270–1275. https://doi.org/10.1210/jc.2006-2158.

Nyman, J. S., Reyes, M., & Wang, X. (2005). Effect of ultrastructural changes on the toughness of bone. *Micron*, *36*(7–8), 566–582. https://doi.org/10.1016/j.micron.2005.07.004.

Oakley, R. H., Jewell, C. M., Yudt, M. R., Bofetiado, D. M., & Cidlowski, J. A. (1999). The dominant negative activity of the human glucocorticoid receptor beta isoform. Specificity and mechanisms of action. *The Journal of Biological Chemistry*, *274*(39), 27857–27866. https://doi.org/10.1074/jbc.274.39.27857.

O'Brien, C. A., Jia, D., Plotkin, L. I., Bellido, T., Powers, C. C., Stewart, S. A., et al. (2004). Glucocorticoids act directly on osteoblasts and osteocytes to induce their apoptosis and reduce bone formation and strength. *Endocrinology*, *145*(4), 1835–1841. https://doi.org/10.1210/en.2003-0990.

Ohnaka, K., Tanabe, M., Kawate, H., Nawata, H., & Takayanagi, R. (2005). Glucocorticoid suppresses the canonical Wnt signal in cultured human osteoblasts. *Biochemical and Biophysical Research Communications*, *329*(1), 177–181. https://doi.org/10.1016/j.bbrc.2005.01.117.

Okamoto, K., Nakashima, T., Shinohara, M., Negishi-Koga, T., Komatsu, N., Terashima, A., et al. (2017). Osteoimmunology: The conceptual framework unifying the immune and skeletal systems. *Physiological Reviews*, *97*(4), 1295–1349. https://doi.org/10.1152/physrev.00036.2016.

Palmqvist, P., Lundberg, P., Persson, E., Johansson, A., Lundgren, I., Lie, A., et al. (2006). Inhibition of hormone and cytokine-stimulated osteoclastogenesis and bone resorption by interleukin-4 and interleukin-13 is associated with increased osteoprotegerin and decreased RANKL and RANK in a STAT6-dependent pathway. *The Journal of Biological Chemistry*, *281*(5), 2414–2429. https://doi.org/10.1074/jbc.M510160200.

Panettieri, R. A., Schaafsma, D., Amrani, Y., Koziol-White, C., Ostrom, R., & Tliba, O. (2019). Non-genomic effects of glucocorticoids: An updated view. *Trends in Pharmacological Sciences*, *40*(1), 38–49. https://doi.org/10.1016/j.tips.2018.11.002.

Park, H. J., Baek, K., Baek, J. H., & Kim, H. R. (2015). The cooperation of CREB and NFAT is required for PTHrP-induced RANKL expression in mouse osteoblastic cells. *Journal of Cellular Physiology*, *230*(3), 667–679. https://doi.org/10.1002/jcp.24790.

Paz-Pacheco, E., Fuleihan, G. E., & LeBoff, M. S. (1995). Intact parathyroid hormone levels are not elevated in glucocorticoid-treated subjects. *Journal of Bone and Mineral Research*, *10*(11), 1713–1718. https://doi.org/10.1002/jbmr.5650101114.

Pereira, R. C., Delany, A. M., & Canalis, E. (2002). Effects of cortisol and bone morphogenetic protein-2 on stromal cell differentiation: Correlation with CCAAT-enhancer binding protein expression. *Bone*, *30*(5), 685–691. https://doi.org/10.1016/s8756-3282(02)00687-7.

Piemontese, M., Onal, M., Xiong, J., Wang, Y., Almeida, M., Thostenson, J. D., et al. (2015). Suppression of autophagy in osteocytes does not modify the adverse effects of glucocorticoids on cortical bone. *Bone*, *75*, 18–26. https://doi.org/10.1016/j.bone.2015.02.005.

Piemontese, M., Xiong, J., Fujiwara, Y., Thostenson, J. D., & O'Brien, C. A. (2016). Cortical bone loss caused by glucocorticoid excess requires RANKL production by osteocytes and is associated with reduced OPG expression in mice. *American Journal of Physiology. Endocrinology and Metabolism*, *311*(3), E587–E593. https://doi.org/10.1152/ajpendo.00219.2016.

Presman, D. M., Ganguly, S., Schiltz, R. L., Johnson, T. A., Karpova, T. S., & Hager, G. L. (2016). DNA binding triggers tetramerization of the glucocorticoid receptor in live cells. *Proceedings of the National Academy of Sciences of the United States of America*, *113*(29), 8236–8241. https://doi.org/10.1073/pnas.1606774113.

Presman, D. M., Ogara, M. F., Stortz, M., Alvarez, L. D., Pooley, J. R., Schiltz, R. L., et al. (2014). Live cell imaging unveils multiple domain requirements for in vivo dimerization of the glucocorticoid receptor. *PLoS Biology*, *12*(3), e1001813. https://doi.org/10.1371/journal.pbio.1001813.

Raisz, L. G. (1980). Effect of corticosteroids on calcium metabolism. *Progress in Biochemical Pharmacology*, *17*, 212–219.

Raisz, L. G., Trummel, C. L., Holick, M. F., & DeLuca, H. F. (1972). 1,25-dihydroxycholecalciferol: A potent stimulator of bone resorption in tissue culture. *Science*, *175*(4023), 768–769. https://doi.org/10.1126/science.175.4023.768.

Raisz, L. G., Trummel, C. L., Wener, J. A., & Simmons, H. (1972). Effect of glucocorticoids on bone resorption in tissue culture. *Endocrinology*, *90*(4), 961–967. https://doi.org/10.1210/endo-90-4-961.

Rauch, A., Seitz, S., Baschant, U., Schilling, A. F., Illing, A., Stride, B., et al. (2010). Glucocorticoids suppress bone formation by attenuating osteoblast differentiation via the monomeric glucocorticoid receptor. *Cell Metabolism*, *11*(6), 517–531. https://doi.org/10.1016/j.cmet.2010.05.005. S1550-4131(10)00158-0 [pii].

Ray, A., & Prefontaine, K. E. (1994). Physical association and functional antagonism between the p65 subunit of transcription factor NF-kappa B and the glucocorticoid receptor. *Proceedings of the National Academy of Sciences of the United States of America*, *91*(2), 752–756. https://doi.org/10.1073/pnas.91.2.752.

Reddy, T. E., Pauli, F., Sprouse, R. O., Neff, N. F., Newberry, K. M., Garabedian, M. J., et al. (2009). Genomic determination of the glucocorticoid response reveals unexpected mechanisms of gene regulation. *Genome Research*, *19*(12), 2163–2171. https://doi.org/10.1101/gr.097022.109.

Reichardt, H. M., Kaestner, K. H., Tuckermann, J., Kretz, O., Wessely, O., Bock, R., et al. (1998). DNA binding of the glucocorticoid receptor is not essential for survival. *Cell*, *93*(4), 531–541.

Reid, I. R., Katz, J. M., Ibbertson, H. K., & Gray, D. H. (1986). The effects of hydrocortisone, parathyroid hormone and the bisphosphonate, APD, on bone resorption in neonatal mouse calvaria. *Calcified Tissue International*, *38*(1), 38–43. https://doi.org/10.1007/BF02556593.

Rodan, G. A., & Martin, T. J. (1981). Role of osteoblasts in hormonal control of bone resorption- -a hypothesis. *Calcified Tissue International*, *33*(4), 349–351.

Rogatsky, I., Wang, J. C., Derynck, M. K., Nonaka, D. F., Khodabakhsh, D. B., Haqq, C. M., et al. (2003). Target-specific utilization of transcriptional regulatory surfaces by the glucocorticoid receptor. *Proceedings of the National Academy of Sciences of the United States of America*, *100*(24), 13845–13850. https://doi.org/10.1073/pnas.2336092100.

Rubin, M. R., & Bilezikian, J. P. (2002). Clinical review 151: The role of parathyroid hormone in the pathogenesis of glucocorticoid-induced osteoporosis: A re-examination of the evidence. *The Journal of Clinical Endocrinology and Metabolism*, *87*(9), 4033–4041. https://doi.org/10.1210/jc.2002-012101.

Rubin, J., Biskobing, D. M., Jadhav, L., Fan, D., Nanes, M. S., Perkins, S., et al. (1998). Dexamethasone promotes expression of membrane-bound macrophage colony-stimulating factor in murine osteoblast-like cells. *Endocrinology*, *139*(3), 1006–1012. https://doi.org/10.1210/endo.139.3.5778.

Sasaki, N., Kusano, E., Ando, Y., Nemoto, J., Iimura, O., Ito, C., et al. (2002). Changes in osteoprotegerin and markers of bone metabolism during glucocorticoid treatment in patients with chronic glomerulonephritis. *Bone*, *30*(6), 853–858. https://doi.org/10.1016/s8756-3282(02)00742-1.

Sasaki, N., Kusano, E., Ando, Y., Yano, K., Tsuda, E., & Asano, Y. (2001). Glucocorticoid decreases circulating osteoprotegerin (OPG): Possible mechanism for glucocorticoid induced osteoporosis. *Nephrology, Dialysis, Transplantation*, *16*(3), 479–482. https://doi.org/10.1093/ndt/16.3.479.

Schiller, B. J., Chodankar, R., Watson, L. C., Stallcup, M. R., & Yamamoto, K. R. (2014). Glucocorticoid receptor binds half sites as a monomer and regulates specific target genes. *Genome Biology*, *15*(7), 418. https://doi.org/10.1186/s13059-014-0418-y.

Shuto, T., Kukita, T., Hirata, M., Jimi, E., & Koga, T. (1994). Dexamethasone stimulates osteoclast-like cell formation by inhibiting granulocyte-macrophage colony-stimulating factor production in mouse bone marrow cultures. *Endocrinology*, *134*(3), 1121–1126. https://doi.org/10.1210/endo.134.3.8119150.

Siersbaek, R., Nielsen, R., John, S., Sung, M. H., Baek, S., Loft, A., et al. (2011). Extensive chromatin remodelling and establishment of transcription factor 'hotspots' during early adipogenesis. *The EMBO Journal*, *30*(8), 1459–1472. https://doi.org/10.1038/emboj.2011.65.

Simonet, W. S., Lacey, D. L., Dunstan, C. R., Kelley, M., Chang, M. S., Luthy, R., et al. (1997). Osteoprotegerin: A novel secreted protein involved in the regulation of bone density. *Cell*, *89*(2), 309–319.

Sivagurunathan, S., Muir, M. M., Brennan, T. C., Seale, J. P., & Mason, R. S. (2005). Influence of glucocorticoids on human osteoclast generation and activity. *Journal of Bone and Mineral Research*, *20*(3), 390–398. https://doi.org/10.1359/JBMR.041233.

Smith, E., & Frenkel, B. (2005). Glucocorticoids inhibit the transcriptional activity of LEF/TCF in differentiating osteoblasts in a glycogen synthase kinase-3beta-dependent and -independent manner. *The Journal of Biological Chemistry*, *280*(3), 2388–2394. https://doi.org/10.1074/jbc.M406294200.

Soe, K., & Delaisse, J. M. (2010). Glucocorticoids maintain human osteoclasts in the active mode of their resorption cycle. *Journal of Bone and Mineral Research*, *25*(10), 2184–2192. https://doi.org/10.1002/jbmr.113.

Soffer, L. J., Iannaccone, A., & Gabrilove, J. L. (1961). Cushing's syndrome. A study of fifty patients. *The American Journal of Medicine*, *30*(1), 129–146.

Steinbuch, M., Youket, T. E., & Cohen, S. (2004). Oral glucocorticoid use is associated with an increased risk of fracture. *Osteoporosis International*, *15*(4), 323–328. https://doi.org/10.1007/s00198-003-1548-3.

Stern, P. H. (1969). Inhibition by steroids of parathyroid hormone-induced Ca45 release from embryonic rat bone in vitro. *The Journal of Pharmacology and Experimental Therapeutics*, *168*(2), 211–217.

Suda, T., Takahashi, N., Udagawa, N., Jimi, E., Gillespie, M. T., & Martin, T. J. (1999). Modulation of osteoclast differentiation and function by the new members of the tumor necrosis factor receptor and ligand families. *Endocrine Reviews*, *20*(3), 345–357. https://doi.org/10.1210/edrv.20.3.0367.

Swanson, C., Lorentzon, M., Conaway, H. H., & Lerner, U. H. (2006). Glucocorticoid regulation of osteoclast differentiation and expression of receptor activator of nuclear factor-kappaB (NF-kappaB) ligand, osteoprotegerin, and receptor activator of NF-kappaB in mouse calvarial bones. *Endocrinology*, *147*(7), 3613–3622. https://doi.org/10.1210/en.2005-0717.

Takuma, A., Kaneda, T., Sato, T., Ninomiya, S., Kumegawa, M., & Hakeda, Y. (2003). Dexamethasone enhances osteoclast formation synergistically with transforming growth factor-beta by stimulating the priming of osteoclast progenitors for differentiation into osteoclasts. *The Journal of Biological Chemistry*, *278*(45), 44667–44674. https://doi.org/10.1074/jbc.M300213200.

Tebben, P. J., Singh, R. J., & Kumar, R. (2016). Vitamin D-mediated hypercalcemia: Mechanisms, diagnosis, and treatment. *Endocrine Reviews*, *37*(5), 521–547. https://doi.org/10.1210/er.2016-1070.

Tobias, J., & Chambers, T. J. (1989). Glucocorticoids impair bone resorptive activity and viability of osteoclasts disaggregated from neonatal rat long bones. *Endocrinology*, *125*(3), 1290–1295. https://doi.org/10.1210/endo-125-3-1290.

Udagawa, N., Takahashi, N., Akatsu, T., Sasaki, T., Yamaguchi, A., Kodama, H., et al. (1989). The bone marrow-derived stromal cell lines MC3T3-G2/PA6 and ST2 support osteoclast-like cell differentiation in cocultures with mouse spleen cells. *Endocrinology*, *125*(4), 1805–1813. https://doi.org/10.1210/endo-125-4-1805.

Van Staa, T. P., Laan, R. F., Barton, I. P., Cohen, S., Reid, D. M., & Cooper, C. (2003). Bone density threshold and other predictors of vertebral fracture in patients receiving oral glucocorticoid therapy. *Arthritis and Rheumatism*, *48*(11), 3224–3229. https://doi.org/10.1002/art.11283.

van Staa, T. P., Leufkens, H. G., & Cooper, C. (2002). The epidemiology of corticosteroid-induced osteoporosis: A meta-analysis. *Osteoporosis International*, *13*(10), 777–787. https://doi.org/10.1007/s001980200108.

Vettorazzi, S., Nalbantoglu, D., Gebhardt, J. C. M., & Tuckermann, J. (2021). A guide to changing paradigms of glucocorticoid receptor function-a model system for genome regulation and physiology. *The FEBS Journal*. https://doi.org/10.1111/febs.16100.

Vidal, N. O., Brandstrom, H., Jonsson, K. B., & Ohlsson, C. (1998). Osteoprotegerin mRNA is expressed in primary human osteoblast-like cells: Down-regulation by glucocorticoids. *The Journal of Endocrinology*, *159*(1), 191–195.

Wada, S., Yasuda, S., Nagai, T., Maeda, T., Kitahama, S., Suda, S., et al. (2001). Regulation of calcitonin receptor by glucocorticoid in human osteoclast-like cells prepared in vitro using receptor activator of nuclear factor-kappaB ligand and macrophage colony-stimulating factor. *Endocrinology*, *142*(4), 1471–1478. https://doi.org/10.1210/endo.142.4.8085.

Watson, L. C., Kuchenbecker, K. M., Schiller, B. J., Gross, J. D., Pufall, M. A., & Yamamoto, K. R. (2013). The glucocorticoid receptor dimer interface allosterically transmits sequence-specific DNA signals. *Nature Structural & Molecular Biology*, *20*(7), 876–883. https://doi.org/10.1038/nsmb.2595.

Weikum, E. R., de Vera, I. M. S., Nwachukwu, J. C., Hudson, W. H., Nettles, K. W., Kojetin, D. J., et al. (2017). Tethering not required: The glucocorticoid receptor binds directly to activator protein-1 recognition motifs to repress inflammatory genes. *Nucleic Acids Research*, *45*(14), 8596–8608. https://doi.org/10.1093/nar/gkx509.

Weikum, E. R., Knuesel, M. T., Ortlund, E. A., & Yamamoto, K. R. (2017). Glucocorticoid receptor control of transcription: Precision and plasticity via allostery. *Nature Reviews. Molecular Cell Biology*, *18*(3), 159–174. https://doi.org/10.1038/nrm.2016.152.

Weinstein, R. S. (2007). Is long-term glucocorticoid therapy associated with a high prevalence of asymptomatic vertebral fractures? *Nature Clinical Practice. Endocrinology & Metabolism*, *3*(2), 86–87. https://doi.org/10.1038/ncpendmet0372.

Weinstein, R. S. (2010). Glucocorticoids, osteocytes, and skeletal fragility: The role of bone vascularity. *Bone, 46*(3), 564–570. https://doi.org/10.1016/j.bone.2009.06.030.

Weinstein, R. S. (2012). Glucocorticoid-induced osteoporosis and osteonecrosis. *Endocrinology and Metabolism Clinics of North America, 41*(3), 595–611. https://doi.org/10.1016/j.ecl.2012.04.004.

Weinstein, R. S., Chen, J. R., Powers, C. C., Stewart, S. A., Landes, R. D., Bellido, T., et al. (2002). Promotion of osteoclast survival and antagonism of bisphosphonate-induced osteoclast apoptosis by glucocorticoids. *The Journal of Clinical Investigation, 109*(8), 1041–1048. https://doi.org/10.1172/JCI14538.

Weinstein, R. S., Jilka, R. L., Parfitt, A. M., & Manolagas, S. C. (1998). Inhibition of osteoblastogenesis and promotion of apoptosis of osteoblasts and osteocytes by glucocorticoids. Potential mechanisms of their deleterious effects on bone. *The Journal of Clinical Investigation, 102*(2), 274–282. https://doi.org/10.1172/JCI2799.

Weinstein, R. S., Wan, C., Liu, Q., Wang, Y., Almeida, M., O'Brien, C. A., et al. (2010). Endogenous glucocorticoids decrease skeletal angiogenesis, vascularity, hydration, and strength in aged mice. *Aging Cell, 9*(2), 147–161. https://doi.org/10.1111/j.1474-9726.2009.00545.x.

Wetzsteon, R. J., Shults, J., Zemel, B. S., Gupta, P. U., Burnham, J. M., Herskovitz, R. M., et al. (2009). Divergent effects of glucocorticoids on cortical and trabecular compartment BMD in childhood nephrotic syndrome. *Journal of Bone and Mineral Research, 24*(3), 503–513. https://doi.org/10.1359/jbmr.081101.

Xiong, J., Cawley, K., Piemontese, M., Fujiwara, Y., Zhao, H., Goellner, J. J., et al. (2018). Soluble RANKL contributes to osteoclast formation in adult mice but not ovariectomy-induced bone loss. *Nature Communications, 9*(1), 2909. https://doi.org/10.1038/s41467-018-05244-y.

Yao, W., Cheng, Z., Busse, C., Pham, A., Nakamura, M. C., & Lane, N. E. (2008). Glucocorticoid excess in mice results in early activation of osteoclastogenesis and adipogenesis and prolonged suppression of osteogenesis: A longitudinal study of gene expression in bone tissue from glucocorticoid-treated mice. *Arthritis and Rheumatism, 58*(6), 1674–1686. https://doi.org/10.1002/art.23454.

Zhu, J. G., Ochalek, J. T., Kaufmann, M., Jones, G., & Deluca, H. F. (2013). CYP2R1 is a major, but not exclusive, contributor to 25-hydroxyvitamin D production in vivo. *Proceedings of the National Academy of Sciences of the United States of America, 110*(39), 15650–15655. https://doi.org/10.1073/pnas.1315006110.

CHAPTER TEN

Evolution of intra-operative parathyroid hormone and its application in parathyroid surgery

Nikita Machado[a] and Scott M. Wilhelm[b,*]
[a]Department of General Surgery, University Hospitals Conneaut, OH, Clinical Assistant Professor of Surgery, Case Western Reserve School of Medicine, Cleveland, OH, United States
[b]Department of General and Endocrine Surgery, University Hospitals Cleveland, OH, Professor of Surgery, Case Western Reserve School of Medicine, Cleveland, OH, United States
*Corresponding author: e-mail address: scott.wilhelm@uhhospitals.org

Contents

Abstract

The history and evolution of parathyroid hormone is a true testament to interdisciplinary collaboration among anatomists, biochemists and surgeons. The parathyroid glands were the last endocrine glands to be discovered in the mid-1800s. Over the next century, progress in the evaluation of primary hyperparathyroidism, the identification of parathyroid hormone (PTH) and its application for use in the field of parathyroid surgery have led to a significant improvement in surgical cure rates, accompanied by a shift toward minimally invasive surgical options. This chapter provides a historical lens through which we can view these relatively recent advancements, as well as the current role of parathyroid hormone, both with regards to pre-operative localization and intra-operative detection of abnormal glands. Furthermore, we discuss the importance of parathyroid hormone in the management of complex multiglandular disease and reoperative cases, as well as the significance of persistently elevated PTH levels post-parathyroidectomy.

Vitamins and Hormones, Volume 120
ISSN 0083-6729
https://doi.org/10.1016/bs.vh.2022.04.009

1. Discovery of the parathyroid gland

The parathyroid glands hold the honor of being the last endocrine glands to be discovered, owing in part to their unique anatomical position. While there was a clear lag in causal association, there is definitely historical proof of the effects of hyperparathyroidism, notably in "fossil skeletons" discovered in Illinois, North America in the 1930s (Denninger, 1931) as well as human remains found in the Dakleh Oasis, Egypt (Cook, Molto, & Anderson, 1988). Both showed signs of osteitis fibrosa cystica (von Recklinghausen disease).

The glands themselves evaded detection for centuries, due to a number of reasons, including anatomic variability, patients with greater than four glands, and those patients with parathyroid glands found either in the mediastinum or thymus. Credit for the first description of the parathyroid glands in animals goes to Sir Richard Owen, the Hunterian Professor and Conservator of the Museum of the Royal College of Surgeons of England (Modarai, Mrcs, Sawyer, Ellis, & Frcs, 2004). In the winter of 1849–1850, he took up the challenge of dissecting the carcass of the first Great Indian Rhinoceros purchased by the Zoological Society of London a decade before. His detailed account of this process (thought to be in 1862, but in reality in 1852) (Cave, 1953) made mention of a "small, compact yellow glandular body attached to the thyroid at the point where the vein emerged." His original preparation, which comprises a portion of the rhino's larynx and trachea, as well as the lateral lobe of the thyroid and parathyroid attached to it, remains at the Hunterian Museum to this day. The one-horned Indian Rhinoceros (*Rhinoceros Unicornis*) has now become the mascot and symbol of International Endocrine surgeons. Its likeness has been captured by artist David Mackay Harrison as a desktop sculpture for the International Association of Endocrine Surgeons (IAES)—see Fig. 1.

The next major breakthrough in this regard came at the hands of Ivan Viktor Sandstrom, a Swedish medical student at the University of Uppsala (Sandstorm, 1880). During his stint as a summer research assistant in 1887, he observed a new gland during the dissection of the neck of a dog. He was able to replicate these findings while studying the thyroid of a cat, rabbit, horse and ox. This was followed by his dissection of 50 human bodies to corroborate these findings. This lead to his publication "On a New Gland in Man and Fellow Animals." He was also the first to coin the term

Fig. 1 The one-horned Indian Rhinoceros (*Rhinoceros Unicornis*), the mascot and symbol of the IAES, by artist David Mackay Harrison.

"glandulae parathyroideae." Unfortunately, his recognition for the same did not come during his lifetime, despite the translation of his manuscript into German and its submission to a journal headed by none other than Virchow. The manuscript was denied on the basis of excessive length. Throughout the rest of his life, he suffered from psychotic episodes, alcohol and drug addiction and sadly committed suicide in 1889 (Johansson, 2015). Luckily, two abstracts in a German yearbook of 1880 remained, allowing him to receive all the credit he was due, albeit posthumously.

Despite the discovery of the anatomic glands, it would be another few decades before they were linked to a specific endocrine function. In Fig. 2, a young Dr. Wilhelm stands next to a plaque commemorating the work of Dr. Sandstrom in Uppsala, Sweden. Dr. Wilhelm was attending and presenting his research as a Fellow in Endocrine surgery at the International Association of Endocrine Surgeons meeting in Uppsala, Sweden in 2004. The plaque, seen in Fig. 3, roughly translates to "In Memory of Ivar Sandstrom, The Discoverer of the Parathyroid glands. This plaque was commissioned by the Faculty of Medicine, Uppsala."

The French physiologist Eugene Gley was the first person to identify the connection between removal of the parathyroid glands and tetany, a phenomenon that was all too common during the era of early thyroid surgery (Gley, 1891). He first reported these findings in 1891, after a series of experiments on rabbits and rats. He noticed that the animals developed tetany and died after removal of both thyroid and parathyroid glands. Interestingly, he noted comparable findings even with a selective parathyroid injury model. This finding was subsequently confirmed in 1896 by

Fig. 2 Dr. Wilhelm, next to a plaque commemorating the work of Dr. Sandstrom in Uppsala, Sweden.

Fig. 3 Plaque close up, translated to "In Memory of Ivar Sandstrom, The Discoverer of the Parathyroid glands. This plaque was commissioned by the Faculty of Medicine, Uppsala."

the Italian researchers Giulio Vassale and Francesco Generali, (Vassale & Generali, 1896) who were able to induce paralytic, convulsive symptoms in animals whose parathyroid glands had been removed, leaving behind the thyroid gland.

Almost two decades later, Jacob Erdheim, from the University of Vienna, repeated the study done by Gley, and confirmed the presence of tetany in rats whose parathyroid glands were destroyed by cautery (Erdheim, 1906). He expounded on Gley's findings and went on to describe features associated with chronic parathyroid insufficiency as well. A paper published in 1907 noted the enlargement of the parathyroid glands in conditions such as osteomalacia and osteitis fibrosa cystica (Erdheim, 1907). He correctly postulated that there was a connection between the parathyroid glands and bone disease. However, he erroneously believed that the bone disease was the primary pathology, and that the parathyroid enlargement was a compensatory mechanism.

This view was shared by the majority at the time, and it was not altered until other researchers began to postulate the primary role of parathyroid gland abnormalities causing these symptoms. A significant example of this was the suggestion in 1908 by MacCallum and Voegtlin that the neuromuscular symptoms thought to be due to thyroidectomies were actually due to removal of the parathyroid glands (MacCallum & Voegtlin, 1909). They believed that the parathyroid glands controlled calcium production and that removing them lead to a decrease in calcium levels in the blood, causing symptoms like tetany.

Furthermore, it was only in 1915 that Friedrich Schlagenhaufer, a professor of pathology in Vienna, postulated that enlarged parathyroid glands might actually be the cause of bone disease and not the result (Schlagenhaufer, 1915). He noted two patients with osteomalacia who were found to have single parathyroid tumors on autopsy.

2. History of parathyroid hormone

The initial forays into the discovery of parathyroid hormone were by Moussu in 1898. He was able to temporarily treat tetany in dogs who had undergone a parathyroidectomy, by a combination of subcutaneous and intravenous injections of water and glycerin extracts of horse parathyroid glands.

This was then extrapolated to humans in 1909 by Berkeley and Beebe, who were able to relieve hypocalcemia and tetany in humans by using a bovine parathyroid extract (Berkeley & Beebe, 1909).

Significant advancement in the disease process was made by James Bertram Collip, a Canadian biochemist. Initially named parathyrin, he was able to isolate a protein, in association with Leitch, and use it to treat tetany in a patient in 1925 (Collip, 1925). In his published paper, he did acknowledge that a similar effort had been made in 1923 by Adolph Hanson, a prior student of Harvey Cushing, working independently in Michigan (Hanson, 1923). However, Collip's methodology was more sophisticated and his study more detailed, leading him to garner most of the credit for this discovery.

Mention must be made of the researcher Fuller Albright who dedicated such a significant portion of his life to studying parathyroid disease. He discussed how hyperparathyroidism could be related to abnormalities in single, multiple glands or even related to parathyroid cancer. He authored a book with Reifenstein (Albright & Reifenstein, 1948) titled "The Parathyroid Glands and Metabolic Bone Disease" in 1948 which even made the distinction between primary and secondary hyperparathyroidism. Coincidentally, Albright had previously elected to study with Erdheim at his lab, which likely sparked his interest in the field, and he continued to remain in touch with him long after Albright was considered an expert in the field of parathyroid disease (Cope, 1966).

The isolate developed by Collip was further purified and refined by Rasmussen and Craig in 1959 (Rasmussen & Craig, 1959) It was at this time that they had a sufficiently pure isolate to definitively characterize the protein and analyze its structure. They noted at this time that parathyroid hormone is a complex polypeptide hormone.

3. Early parathyroid surgery

Significant advancements were made in the late 1800s-early 1900s with transplantation of the parathyroid glands as autografts in animals. A pupil of Billroth, by the name of Eiselberg, performed autografts in cats by transplanting half of the parathyroid and thyroid between the rectus fascia and peritoneum. Follow up extending to a month later demonstrated no signs of tetany. Even better, histological studies showed evidence of neovascularization (von Eiselsberg, 1892).

The first known surgical removal of an abnormal parathyroid gland was attributed to the Viennese surgeon Felix Mandl, who operated on a patient with osteitis fibrosa cystica in 1925. Unfortunately, that patient's disease process recurred, and he passed away 7 years later despite repeated attempts at surgery. The tumor described at the original operation was most likely a parathyroid carcinoma (Mandl, 1925).

Other initial forays into parathyroid surgery were all in a similar vein. Parathyroid surgery in the era before the routine use of parathyroid hormone was routinely based on gross morphology of the parathyroid glands, as well as the experience of the operating surgeon in identifying potentially abnormal tissue. Patients were often told that they might need repeat cervical explorations, and the surgeries themselves could last upwards of 15 h.

4. Evolution of intraoperative parathyroid hormone

The first group of immunoassays to detect hormones were developed in the mid-20th century and were centered around insulin, followed by glucagon and growth hormone. In 1963, Berson and Yalow extrapolated this to parathyroid hormone, and were able to create one of the first immunoassays derived from both bovine and human parathyroid extract (Berson, Yalow, Aurbach, & Potts, 1963). This new technology was adapted by surgeons to assist with immune-assay guided parathyroidectomies; however, the results weren't necessarily accurate due to the fact that the initial immunoassay captured both active and inactive components. Yalow went on to receive the Nobel Prize in Physiology & Medicine in 1977, for the work that she and Berson had done in developing radio-immunoassays to trace hormones within the human body.

In 1987, Nussbaum et al. were the first to document a highly sensitive, two side immunoradiometric assay to identify the levels of parathyroid hormone, then called parathyrin. They postulated that this test had a very high sensitivity for PTH levels (1 ng/mL), did not detect inactive compounds and had what was at that time considered a quick turnaround time of 24 h (Nussbaum et al., 1987). They subsequently modified this assay, a year later, with the help of a higher incubation temperature and the addition of a kinetic enhancer. This allowed them to significantly reduce their result time to around 15 min (Barczyński, Gołkowski, & Nawrot, 2015) While this test was the forerunner of modern day intraoperative PTH testing, they chose to obtain their results postoperatively, and other than sparking academic interest, this study did not alter the time-tested standard of bilateral neck exploration.

In 1990, a group of researchers in Paris, headed by Y. Chapuis, began studying the utility of 1–84 parathyroid hormone (PTH) levels for patients with ultrasound findings suggestive of parathyroid adenomas (Chapuis, Fulla, Icard, & Nonnemacher, 1990). They further evaluated PTH levels intraoperatively in combination with the measurement of urinary cAMP levels in a cohort of 45 patients (Chapuis et al., 1992). They were able to perform a focused operation in 42 out of the 45 patients, with a very small percentage requiring a bilateral neck exploration. These studies were forerunners for current surgical management.

Similar progress was being made across the Atlantic by George Irvin and his colleagues in Chicago. They had around two decades of experience with parathyroidectomies, however, continued to note a postoperative failure rate of around 7%, which they attributed to misdiagnosis as well as difficulty in identifying all the hypersecreting glands. In 1991, they published a series of 21 patients who had undergone parathyroidectomies with the addition of a quick-turnover intraoperative PTH (iPTH) assay (Irvin, Dembrow, & Prudhomme, 1991). This technique, a slight modification of the one introduced by Nussbaum, involved drawing whole blood samples to test for intact PTH levels 10 min after specimen removal. They found quantitative evidence that all hyperfunctioning parathyroid tissue had been removed in 19 out of the 21 patients and were able to continue exploration in the remaining patients to ablate parathyroid tissue. This was a significant milestone in the journey toward focused parathyroid surgery.

5. Dissemination of intraoperative parathyroid hormone

Over the course of the next decade, the use of IOPTH to assist with parathyroidectomies became more commonplace. As more institutions gained familiarity with the technique, timing and nuances of intraoperative PTH testing, studies emerged from multiple centers describing their experience with an increasing number of patients. One of the initial studies in this regard was by Irvin et al., detailing their experience with 89 patients, some of which included reoperative candidates with primary hyperparathyroidism (Boggs et al., 1996). They obtained PTH samples at a few crucial points (twice prior to excision of the abnormal gland, following that at 5- and 10-min intervals after excision). Further samples were taken if needed. The required goal was a decrease in serum PTH levels to less than 50% of the preoperative value. The operation was deemed successful if the patient had a serum calcium that returned to normal limits within 24 h after surgery, as well as at dedicated follow-up visits over the next year.

They found that serum calcium levels returned to normal in all but one of their patients. Significantly, IOPTH testing was extremely useful in patients who had equivocal sestamibi scans, parathyroid hyperplasia or reoperative fields.

A similar study came out of Duke University Medical Center. Published in 1999, this article by Garner et al. evaluated their experience using IOPTH with a cohort of 130 patients (Garner & Leight, 1999). The study found that IOPTH measurements allowed for proper identification and complete excision in 126/130 cases (96.2%). This was regardless of whether the pathology was a single adenoma or multiglandular hyperplasia.

As institutions became more familiar with the use of IOPTH, they developed various criteria to assess whether an adequate surgical resection had been performed. The studies varied with regards to the timing of lab draws, the degree of decrease in PTH levels and how many criteria were necessary to accurately predict the effectiveness of surgery. However, these criteria varied in sensitivity and specificity. The most common out of these were the Miami criteria, which required a drop in PTH level of $\geq$50% from the highest known PTH level (whether pre-incision or pre-excision of gland) at 10 min after excision of the abnormal gland had been completed. Further studies were performed to assess the effectiveness of the Miami criteria vs others in the field, and found the Miami criteria to be the most accurate (Carneiro et al., 2003). Interestingly, another study that evaluated the Miami criteria found an increased level of accuracy when the decrease in PTH level was combined with a final PTH level that was in the normal reference range (dubbed, "Dual Criteria"). Understandably, this was considered to be a more stringent, accurate set of criteria due to the fact that it would require an absolute return to normal values, as opposed to a relative decrease in PTH value from an abnormal pre-operative value (Chiu et al., 2006).

6. Use of parathyroid hormone testing in reoperative cases

Primary hyperparathyroidism remains one of the most common causes of hypercalcemia. We now know that the most common cause of primary hyperparathyroidism is a solitary parathyroid adenoma (80–85% of cases), followed by parathyroid hyperplasia, double parathyroid adenomas and finally parathyroid cancer (<1%) (Marcocci & Cetani, 2011). This variation in etiology is also displayed when it comes to pre-operative localization studies. For initial operations, the majority of lesions are identified by a combination of non-invasive studies, most commonly ultrasound and sestamibi scans (Johnson, Tublin, & Ogilvie, 2007). Ultrasounds are available at virtually all facilities and have demonstrated reasonable accuracy in diagnosis (studies have reported a sensitivity of 76% and positive predictive value of 93%) (Cheung, Wang, Farrokhyar, Roman, & Sosa, 2012). Understandably, smaller glands (<200 g), ectopic locations of parathyroid glands and operator experience all factor into the diagnostic accuracy of parathyroid ultrasounds. Another non-invasive pre-operative localization study that is commonly used is sestamibi scanning with ^{99m}Tc, which involves isotope uptake and washout to differentiate parathyroid tissue from thyroid tissue. When combined, USG and sestamibi scanning show up to 82% sensitivity and 92% positive predictive value for parathyroid adenomas (Hajioff, Iyngkaran, Panagamuwa, Hill, & Stearns, 2004; Prasannan, Davies, Bochner, Kollias, & Malycha, 2007). However, despite improvements in our current arsenal of pre-operative imaging, 10–20% of patients are still found to have equivocal or discordant imaging studies, that do not allow for accurate pre-operative localization (Treglia, Trimboli, Huellner, & Giovanella, 2018). Multiple factors can influence this disparity, including smaller gland size, concomitant thyroid disease (goiter, thyroid adenomas), parathyroid glands that are histologically composed of more chief cells than oxyntic cells, ectopic locations etc. (Mihai, Gleeson, Buley, Roskell, & Sadler, 2006). In these situations, we often need to turn to more advanced imaging studies, often times in combination with PTH testing. Imaging studies include 4D-CT, MRI, SPECT imaging. PTH sampling can often be invaluable in these cases, including methods like bilateral internal jugular venous sampling (BIJVS), selective venous sampling (SVS), or ultrasound guided parathyroid biopsy with PTH washout.

6.1 Radio-guided selective venous sampling

Reoperative parathyroid surgery is often fraught with a higher risk of complications, including recurrent laryngeal nerve injury and chronic hypoparathyroidism. Scar tissue from prior surgery alters normal anatomic planes, increasing the difficulty of dissection. It is imperative, in these cases, to attempt accurate pre-operative localization as much as possible, in order to guide surgical dissection.

This procedure works on the basis of a differential gradient in serum PTH levels based on anatomic location and proximity to the parathyroid pathology. The right femoral vein is usually cannulated, and a 4 or 5Fr sheath is advanced proximally. An initial sample is taken from the superior vena cava as a control. Samples can also be obtained from the azygos, thymic, internal mammary or superior intercostal veins. Ultimately, selective catheterization of the superior, middle and inferior thyroid veins is essential for accurate readings (England, Geer, & Deipolyi, 2018). Each sample is around 2–3 mL in volume.

A ratio of 1.5, 2.5 or 3 for selected veins as opposed to the peripheral control vein are associated with a positive predictive value of 72%, 75% and 83%, respectively (Ogilvie et al., 2006). Multiple studies have touted the sensitivity of selective venous sampling at around 83–91% for previously unidentified parathyroid glands (even more sensitive for hyperplasia than single adenomas) (Gough, 2006; Reidel et al., 2006).

6.2 Bilateral internal jugular venous sampling

While selective/super selective venous sampling has been shown to be useful when it comes to localizing parathyroid disease with equivocal/discordant imaging findings, there remains a small, but present risk of angiography related complications. These include allergic reactions, acute kidney injury related to contrast, local complications such as infection and hematomas, as well as vascular complications at the access site (pseudoaneurysms, arteriovenous fistulae as well as vascular thrombosis) (Image-Guided Interventions E-Book: Expert Radiology Series—Matthew A. Mauro, Kieran P.J. Murphy, Kenneth R. Thomson, Anthony C. Venbrux, Robert A. Morgan—Google Books, n.d.; England et al., 2018).

A study performed by Ito et al., at the University of Wisconsin, Madison sought to evaluate the utility of bilateral internal jugular venous sampling, using IOPTH technology (Ito, Sippel, Lederman, & Chen, 2007). The technique involved intraoperative access to both internal jugular veins (BIJ), with venous blood drawn from bilateral sides after induction of anesthesia but prior to incision. Lateralizing BIJ PTH levels were defined as >5% difference between samples from the left and right internal jugular veins. IOPTH was also measured, and adequate surgical excision considered to have been performed using the Miami criteria. They found BIJ PTH to be diagnostic in 62.9% of the cases of primary hyperparathyroidism. If the sestamibi scan and BIJ PTH studies localized to the same site, the likelihood of finding abnormal hyperfunctioning glands increased to 98.2%. Interestingly, in the setting of negative sestamibi scans, the BIJ PTH still managed to localize the disease in 57.8% of cases, finding bilateral hyperplasia in 11.1% of cases. This data shows that BIJ PTH localization is clearly of value in complex multiglandular disease, even more so when other pre-operative studies like sestamibi scans fail to localize the pathological gland. Unfortunately, BIJ PTH was not shown to be as useful in this study when it came to localization in secondary/tertiary hyperparathyroidism. A randomized controlled trial performed by Barczynski et al. (Barczynski, Konturek, Hubalewska-Dydejczyk, Cichon, & Nowak, 2009) corroborated these findings as well. Their experience showed that the use of BIJ PTH sampling in patients with negative sestamibi scans not only increased the rate of true positive lateralization of the disease (from 20.5% with ultrasound alone to 43.6%), but also decreased the rate of conversion from minimally invasive parathyroidectomy to bilateral neck exploration from 59.7% to 15%. Furthermore, the rate of bilateral neck explorations decreased by almost 33.3%, indicating that this lateralizing study certainly has its uses in more complicated cases of hyperparathyroidism.

6.3 USG guided parathyroid biopsy with PTH washout

The American Association of Endocrine Surgeons has recommended that an ultrasound guided fine needle aspiration(FNA) of parathyroid tissue with PTH washout be considered in patients who are surgical candidates for a parathyroidectomy (primarily reoperative) but have pre-operative localization studies that are discordant or equivocal (Wilhelm et al., 2016). A recent systematic review by Castellana et al. evaluated the experience of multiple institutions over the last 20 years with this technique (Castellana et al., 2019). In patients who have a suspicious lesion on ultrasound with discordant scintigraphy, an FNA of the suspected parathyroid tissue is performed using a 21- to 27-gauge needle, which is subsequently washed out with 0.5–1 cc of either normal saline or PTH-free serum. A positive result for this test had different cutoffs depending on the paper studied, but it varied from simply higher than the serum PTH level to twice the serum PTH level. The abnormal PTH level ranged from 40 to 103 pg/mL across the studies performed. In the personal experience of the senior author of this chapter, often times the PTH levels from FNA are in the thousands (data not published).

The pooled sensitivity for all studies was found to be 95%, with a positive predictive value of 97%. An important point to remember is that though an FNA-PTH that is positive indicates that the tissue of origin is parathyroid, a negative test does not necessarily exclude parathyroid tissue.

While not common, FNA-PTH can be associated with certain side effects. Biopsies can cause alteration in the tissue histology, especially following hemorrhage, abscess or inflammation. This can cause difficulty in subsequent surgery, include fibrosis or pseudo-invasion which may alter final pathology (Norman, Politz, & Browarsky, 2007). Another concern is that parathyroid tissue can disseminate, due to its ability to grow even when autotransplanted. A minor concern exists for parathyromatosis, whether benign or malignant (Kim et al., 2017).

7. Elevated parathyroid hormone levels postsurgery

A curative resection for primary hyperparathyroidism is defined as a normal calcium level 6 months postparathyroidectomy. Multiple studies, however, have demonstrated that a subset of these normocalcemic patients may still have an elevated parathyroid hormone level postoperatively. A number of mechanisms have been discussed to explain this phenomenon. One of the well described etiologies includes an alteration in the calcium sensing receptors of the remaining parathyroid glands, which can lead to a higher set point for PTH secretion. Patients may also have reduced peripheral sensitivity to PTH due to long standing disease. Other possible causes include a renal calcium leak, fall in the glomerular filtration rate (GFR), vitamin D deficiency or relative postoperative hypocalcemia causing a compensatory mechanism (de la Plaza Llamas et al., 2018).

An important point of distinction is to evaluate whether or not elevated PTH levels after surgery are linked with the risk of recurrent disease. Multiple studies have been performed in this regard, which found no difference in the rate of recurrent primary hyperparathyroidism between patients with or without elevated post-operative PTH in patients with initial normocalcemia (Carsello, Yen, & Wang, 2012; Yen, Wilson, Krzywda, & Sugg, 2006).

Another study by Caldwell et al. noted that almost a third of patients who had undergone a minimally invasive parathyroidectomy were found to have an elevated PTH level post-operatively, despite normal serum calcium levels (Caldwell, Laux, Clark, Kim, & Rubin, 2019). They found that a higher pre-operative PTH level and a lower serum calcium level were more likely to be associated with elevated PTH level postoperatively.

This lends credence to the fact that a simple elevation in postoperative PTH levels does not imply recurrent disease, but may be a temporary, dynamic biochemical process that is an independent clinical entity (Goldfarb, Gondek, Irvin, & Lew, 2011).

Regardless, it is important to stress the need for long-term follow-up in these patients, with clinical and laboratory surveillance, in order to evaluate for possible secondary causes of hyperparathyroidism.

8. Future directions

From the discovery of the parathyroid gland in the Indian Rhinoceros to the work of Dr. Sandstrom in characterizing the gland in humans along with the discovery of parathyroid hormone by Moussu, Berkeley and Beebe, we have come a long way in a relatively short time. Thanks to the ingenuity of hormonal researchers such as Berson, Yalow, and Nussbaum we were given a tool (the intraoperative PTH or IOPTH assay) to use during Parathyroid surgery to guide pioneering surgeons such as Dr. George Irvin and Dr. Chapuis to construct a new paradigm in parathyroidectomy. Their vision led to a movement from the traditional bilateral neck exploration to a minimally invasive parathyroidectomy which is now performed worldwide. From here, we have seen new operative approaches such as MIVAP (Miccoli, Berti, Materazzi, & Donatini, 2003) (Minimally invasive video assisted parathyroidectomy) championed by Dr. Miccoli, transaxillary "scarless" parathyroidectomy done laparoscopically and now robotically assisted. With each new surgical iteration designed to decrease incision size, postoperative pain and wound healing, we must be sure to carefully monitor outcomes, rates of surgical cure/failure and patient safety.

As to the hormone itself, parathyroid hormone and its use also continue to evolve. As mentioned above, historically, extracts of parathyroid hormone were successfully used to treat hypoparathroidism, hypocalcemia and tetany. The complication of permanent hypoparathyroidism can be as debilitating a as hyperparathyroidism to a patient and in many ways more difficult to treat. To this end, Natpara, a human recombinant 1–84 parathyroid hormone was created in 2013. It is a subcutaneously injected hormone that can replace naturally occurring parathyroid hormone in patients with hypoparathyroidism. It was approved for use by the FDA in 2015. But as with all new approvals, rigorous testing and long-term patient follow-up are critical. In 2019, the drug was recalled by its parent company and given a black box warning by the FDA for an increased risk of the development of

Osteosarcoma in patients using Natpara. At this time, there is only a compassionate use availability of this drug under strict FDA guidelines.

So, from four small glands that lie deep in the neck behind the thyroid gland, we are provided with a fascinating history of discovery, researcher perseverance, surgical ingenuity and innovation and a better appreciation for the Indian Rhinoceros, our modern day Unicorn.

The authors have no financial disclosures nor any conflicts of interest to report.

References

Albright, F., & Reifenstein, E. C. (1948). *The parathyroid glands and metabolic bone disease*. Baltimore: Williams & Wilkins.

Barczyński, M., Gołkowski, F., & Nawrot, I. (2015). The current status of intraoperative iPTH assay in surgery for primary hyperparathyroidism. *Gland Surgery*, *4*(1), 36–43. https://doi.org/10.3978/j.issn.2227-684X.2015.01.01.

Barczynski, M., Konturek, A., Hubalewska-Dydejczyk, A., Cichon, S., & Nowak, W. (2009). Utility of intraoperative bilateral internal jugular venous sampling with rapid parathyroid hormone testing in guiding patients with a negative sestamibi scan for minimally invasive parathyroidectomy-a randomized controlled trial. *Langenbeck's Archives of Surgery*, *394*(5), 827–835. https://doi.org/10.1007/s00423-009-0516-6.

Berkeley, Y. W. N., & Beebe, S. P. (1909). A contribution to the physiology and chemistry of the parathyroid glands. *J Med Re*, *20*, 149–157.

Berson, S. A., Yalow, R. S., Aurbach, G. D., & Potts, J. T. (1963). Immunoassay of bovine and human parathyroid hormone. *Proceedings of the National Academy of Sciences*, *49*(5), 613–617. https://doi.org/10.1073/pnas.49.5.613.

Boggs, J. E., Irvin, G. L., Molinari, A. S., Deriso, G. T., Shaha, A. R., Watson, C. G., et al. (1996). Intraoperative parathyroid hormone monitoring as an adjunct to parathyroidectomy. *Surgery*, *120*(6), 954–958. https://doi.org/10.1016/S0039-6060(96)80040-7.

Caldwell, M., Laux, J., Clark, M., Kim, L., & Rubin, J. (2019). Persistently elevated PTH after parathyroidectomy at one year: Experience in a tertiary referral center. *Journal of Clinical Endocrinology and Metabolism*, *104*(10), 4473–4480. https://doi.org/10.1210/jc.2019-00705.

Carneiro, D. M., Solorzano, C. C., Nader, M. C., Ramirez, M., Irvin, G. L., Udelsman, R., et al. (2003). Comparison of intraoperative iPTH assay (QPTH) criteria in guiding parathyroidectomy: Which criterion is the most accurate? *Surgery*, *134*(6), 973–979. https://doi.org/10.1016/j.surg.2003.06.001.

Carsello, C. B., Yen, T. W. F., & Wang, T. S. (2012). Persistent elevation in serum parathyroid hormone levels in normocalcemic patients after parathyroidectomy: Does it matter? *Surgery (United States)*, *152*(4), 575–583. https://doi.org/10.1016/j.surg.2012.07.005.

Castellana, M., Virili, C., Palermo, A., Giorgino, F., Giovanella, L., & Trimboli, P. (2019). Primary hyperparathyroidism with surgical indication and negative or equivocal scintigraphy: Safety and reliability of PTH washout. A systematic review and meta-analysis. *European Journal of Endocrinology*, *181*(3), 245–253. BioScientifica Ltd. https://doi.org/10.1530/EJE-19-0160.

Cave, A. (1953). In E. A. Underwood, & C. Singer (Eds.), *Vol. 2. Science medicine and history* (pp. 217–222). London: Oxford University Press.

Chapuis, Y., Fulla, Y., Icard, P., & Nonnemacher, L. (1990). Peroperative assay of active parathormone 1-84 in surgery of primary hyperparathyroidism. *Presse Medicale (Paris, France: 1983)*, *19*(31), 1461–1462. https://pubmed.ncbi.nlm.nih.gov/2146643/.

Chapuis, Y., Icard, P., Fulla, Y., Nonnenmacher, L., Bonnichon, P., Louvel, A., et al. (1992). Parathyroid adenomectomy under local anesthesia with intra-operative monitoring of UcAMP and/or 1-84 PTH. *World Journal of Surgery*, *16*(4), 570–575. https://doi.org/10.1007/BF02067323.

Cheung, K., Wang, T. S., Farrokhyar, F., Roman, S. A., & Sosa, J. A. (2012). A meta-analysis of preoperative localization techniques for patients with primary hyperparathyroidism. *Annals of Surgical Oncology*, *Vol. 19*(2), 577–583, Springer. https://doi.org/10.1245/s10434-011-1870-5.

Chiu, B., Sturgeon, C., Angelos, P., Talpos, G. B., Grant, C. S., Snyder, S., et al. (2006). Which intraoperative parathyroid hormone assay criterion best predicts operative success? A study of 352 consecutive patients. *Archives of Surgery*, *141*(5), 483–488. https://doi.org/10.1001/archsurg.141.5.483.

Collip, J. (1925). The extraction of parathyroid hormone which will prevent or control parathyroid tetany and which regulates the level of blood calcium. *The Journal of Biological Chemistry*, *63*, 395–438.

Cook, M., Molto, E., & Anderson, C. (1988). Possible case of hyperparathyroidism in a roman period skeleton from the Dakhleh oasis, Egypt, diagnosed using bone histomorphometry. *American Journal of Physical Anthropology*, *75*(1), 23–30. https://doi.org/10.1002/ajpa.1330750104.

Cope, O. (1966). The study of hyperparathyroidism at the Massachusetts General Hospital. *The New England Journal of Medicine*, *274*(21), 1174–1182. https://doi.org/10.1056/NEJM196605262742105.

de la Plaza Llamas, R., Ramia Ángel, J. M., Arteaga Peralta, V., García Amador, C., López Marcano, A. J., Medina Velasco, A. A., et al. (2018). Elevated parathyroid hormone levels after successful parathyroidectomy for primary hyperparathyroidism: A clinical review. *European Archives of Oto-Rhino-Laryngology*, *275*(3), 659–669. https://doi.org/10.1007/s00405-017-4836-9.

Denninger, H. (1931). *Osteitis fibrosa in a skeleton of a prehistoric American Indian*. American Medical Association.

England, R., Geer, E., & Deipolyi, A. (2018). Role of venous sampling in the diagnosis of endocrine disorders. *Journal of Clinical Medicine*, 7(5), 114. https://doi.org/10.3390/jcm7050114.

Erdheim, J. (1906). Tetania parathyreopriva. *Mitteilungen aus d. Grenzgebuten der Medizin und Chirurgie*, *16*, 632–744.

Erdheim, J. (1907). Ueber Epithelkorperbefunde bei Osteomalacie. *Sitzungsb. d. Akad. d. Wiss., Wien*, *116*, 311.

Garner, S. C., & Leight, G. S. (1999). Initial experience with intraoperative PTH determinations in the surgical management of 130 consecutive cases of primary hyperparathyroidism. *Surgery*, *126*(6), 1132–1138. https://doi.org/10.1067/msy.2099.101429.

Gley, E. (1891). Sur les fonctions du corps thyroide. *CR Soc Biol*, *43*, 841.

Goldfarb, M., Gondek, S., Irvin, G. L., & Lew, J. I. (2011). Normocalcemic parathormone elevation after successful parathyroidectomy: Long-term analysis of parathormone variations over 10 years. *Surgery*, *150*(6), 1076–1084. https://doi.org/10.1016/j.surg.2011.09.017.

Gough, I. (2006). Reoperative parathyroid surgery: The importance of ectopic location and multigland disease. *ANZ Journal of Surgery*, *76*(12), 1048–1050. https://doi.org/10.1111/j.1445-2197.2006.03931.x.

Hajioff, D., Iyngkaran, T., Panagamuwa, C., Hill, D., & Stearns, M. P. (2004). Preoperative localization of parathyroid adenomas: Ultrasonography, sestamibi scintigraphy, or both? *Clinical Otolaryngology and Allied Sciences*, *29*(5), 549–552. https://doi.org/10.1111/j.1365-2273.2004.00858.x.

Hanson, A. (1923). An elementary chemical study of the parathyroid glands of cattle. *Military Surgeon, 53*, 280–284.
Image-Guided Interventions E-Book: Expert Radiology Series - Matthew A. Mauro, Kieran P.J. Murphy, Kenneth R. Thomson, Anthony C. Venbrux, Robert A. Morgan - Google Books. (n.d.). Retrieved July 18, 2020, from https://books.google.com/books?hl=en&lr=&id=C4LWDwAAQBAJ&oi=fnd&pg=PP1&ots=Yc05jujGzo&sig=0notpAxXbLauQXc1HXDw8TdBKXM#v=onepage&q&f=false. n.d.
Irvin, G. L., Dembrow, V. D., & Prudhomme, D. L. (1991). Operative monitoring of parathyroid gland hyperfunction. *The American Journal of Surgery, 162*(4), 299–302. https://doi.org/10.1016/0002-9610(91)90135-Z.
Ito, F., Sippel, R., Lederman, J., & Chen, H. (2007). The utility of intraoperative bilateral internal jugular venous sampling with rapid parathyroid hormone testing. *Annals of Surgery, 245*(6), 959–963. https://doi.org/10.1097/01.sla.0000255578.11198.ff.
Johansson, H. (2015). The Uppsala anatomist ivar sandström and the parathyroid gland. *Upsala Journal of Medical Sciences, 120*(2), 72–77. https://doi.org/10.3109/03009734.2015.1027426.
Johnson, N. A., Tublin, M. E., & Ogilvie, J. B. (2007). Parathyroid imaging: Technique and role in the preoperative evaluation of primary hyperparathyroidism. *American Journal of Roentgenology, 188*(6), 1706–1715. https://doi.org/10.2214/AJR.06.0938.
Kim, J., Horowitz, G., Hong, M., Orsini, M., Asa, S. L., & Higgins, K. (2017). The dangers of parathyroid biopsy. *Journal of Otolaryngology - Head & Neck Surgery = Le Journal d'oto-Rhino-Laryngologie et de Chirurgie Cervico-Faciale, 46*(1), 4. https://doi.org/10.1186/s40463-016-0178-7.
MacCallum, W. G., & Voegtlin, C. (1909). On the relation of tetany to the parathyroid glands and to calcium metabolism. *Journal of Experimental Medicine, 11*(1), 118–151. https://doi.org/10.1084/jem.11.1.118.
Mandl, F. (1925). Therapeutic attempt for osteitis fibrosa generalisata via the excision of parathyroid tumours. *Wiener Klinische Wochenschrift, 38*, 13434.
Marcocci, C., & Cetani, F. (2011). Primary hyperparathyroidism. *New England Journal of Medicine, 365*(25), 2389–2397. https://doi.org/10.1056/NEJMcp1106636.
Miccoli, P., Berti, P., Materazzi, G., & Donatini, G. (2003). Minimally invasive video assisted parathyroidectomy (MIVAP). *European Journal of Surgical Oncology : The Journal of the European Society of Surgical Oncology and the British Association of Surgical Oncology, 29*(2), 188–190. https://doi.org/10.1053/EJSO.2002.1313.
Mihai, R., Gleeson, F., Buley, I. D., Roskell, D. E., & Sadler, G. P. (2006). Negative imaging studies for primary hyperparathyroidism are unavoidable: Correlation of sestamibi and high-resolution ultrasound scanning with histological analysis in 150 patients. *World Journal of Surgery, 30*(5), 697–704. https://doi.org/10.1007/s00268-005-0338-9.
Modarai, B., Mrcs, B., Sawyer, A., Ellis, M. H., & Frcs, C. (2004). The glands of Owen. *Journal of the Royal Society of Medicine, 97*.
Norman, J., Politz, D., & Browarsky, I. (2007). Diagnostic aspiration of parathyroid adenomas causes severe fibrosis complicating surgery and final histologic diagnosis. *Thyroid, 17*(12), 1251–1255. https://doi.org/10.1089/thy.2007.0081.
Nussbaum, S. R., Zahradnik, R. J., Lavigne, J. R., Brennan, G. L., Nozawa-Ung, K., Kim, L. Y., et al. (1987). Highly sensitive two-site immunoradiometric assay of parathyrin, and its clinical utility in evaluating patients with hypercalcemia. *Clinical Chemistry, 33*(8), 1364–1367. https://doi.org/10.1093/clinchem/33.8.1364.
Ogilvie, C. M., Brown, P. L., Matson, M., Dacie, J., Reznek, R. H., Britton, K., et al. (2006). Selective parathyroid venous sampling in patients with complicated hyperparathyroidism. *European Journal of Endocrinology, 155*(6), 813–821. https://doi.org/10.1530/eje.1.02304.
Prasannan, S., Davies, G., Bochner, M., Kollias, J., & Malycha, P. (2007). Minimally invasive parathyroidectomy using surgeon-performed ultrasound and sestamibi. *ANZ Journal of Surgery*, 77(9), 774–777. https://doi.org/10.1111/j.1445-2197.2007.04227.x.

Rasmussen, H., & Craig, L. (1959). Purification of parathormone by use of countercurrent distribution. *Journal of the American Chemical Society, 81*(5003), 28.

Reidel, M. A., Schilling, T., Graf, S., Hinz, U., Nawroth, P., Büchler, M. W., et al. (2006). Localization of hyperfunctioning parathyroid glands by selective venous sampling in reoperation for primary or secondary hyperparathyroidism. *Surgery, 140*(6), 907–913. https://doi.org/10.1016/j.surg.2006.06.037.

Sandstorm, I. V. (1880). *Omen ny Kortel hos mennisken och atskilige baggdjur Lakarefore rings.* Upsala: Lakarefore rings.

Schlagenhaufer, F. (1915). Zwei Fälle von Parathyreoideatumoren. Offizielles Protokoll der kk Gesellschaft der Aerzte in Wien Sitzung von. *Wien Klin Wchenschr, 28*(2), 1362.

Treglia, G., Trimboli, P., Huellner, M., & Giovanella, L. (2018). Imaging in primary hyperparathyroidism: Focus on the evidence-based diagnostic performance of different methods. *Minerva Endocrinologica, 43*(2), 133–143, Edizioni Minerva Medica. https://doi.org/10.23736/S0391-1977.17.02685-2.

Vassale, G., & Generali, F. (1896). Sugli effeti dell'estirpazione delle ghiandole paratiroide. *Riv Pat01 Nerv Ment, 1*, 95–99.

von Eiselsberg, A. (1892). Ü eber erfolgreiche Einheilung der Katzenshilddru¨se in die Bauchdecke und Auftreten von Tetanie nach deren Extirpation. *Wien KlinWochenschr, 5*, 81–85.

Wilhelm, S. M., Wang, T. S., Ruan, D. T., Lee, J. A., Asa, S. L., Duh, Q.-Y., et al. (2016). The American Association of Endocrine Surgeons Guidelines for definitive Management of Primary Hyperparathyroidism. *JAMA Surgery, 151*(10), 959. https://doi.org/10.1001/jamasurg.2016.2310.

Yen, T. W. F., Wilson, S. D., Krzywda, E. A., & Sugg, S. L. (2006). The role of parathyroid hormone measurements after surgery for primary hyperparathyroidism. *Surgery, 140*(4), 665–674. https://doi.org/10.1016/j.surg.2006.07.006.

CHAPTER ELEVEN

Gene expression profiles in parathyroid adenoma and normal parathyroid tissue

Sang Wan Kim*
Department of Internal Medicine, Seoul National University College of Medicine and Seoul Metropolitan Government Boramae Medical Center, Seoul, Republic of Korea
*Corresponding author: e-mail address: swkimmd@snu.ac.kr

Contents

Abstract

A parathyroid adenoma comprises 80–85% as a cause of primary hyperparathyroidism. The clonal origin of most parathyroid adenomas suggests a defect at the level of the gene controlling growth of the parathyroid cell or the expression of parathyroid hormone (PTH). Two genes, *MEN1* and *CCND1*, a tumor suppressor and a proto-oncogene respectively, have been solidly established as primary tumorigenic drivers in parathyroid adenomas. As well, germline and somatic mutation of other genes involved in cell cycle regulation or PTH regulation have been discovered in parathyroid adenomas. Moreover, comparative genomic studies between parathyroid adenomas and normal parathyroid tissues have suggested more complex genetic landscape. Microarray analysis have revealed differential expression profiles of genes involved in cell cycle regulation, growth factors, apoptotic pathway, or PTH synthesis or regulation pathway such as *CASR*, *GCM2* and *KL* (Klotho). Furthermore, recent next-generation sequencing analysis reconfirmed previous finding or revealed novel finding, suggesting signal peptidase complex subunit (SPCS2), ribosomal proteins (RPL23, RPL26, RPN1, RPS25),

Vitamins and Hormones, Volume 120
ISSN 0083-6729
https://doi.org/10.1016/bs.vh.2022.04.003

the endoplasmic reticulum membrane (SEC11C, SEC11A, SEC61G), Klotho, cyclin D1, β-catenin, VDR, CaSR and GCM2 may be important factors contributing to the parathyroid adenoma.

1. Introduction

Primary hyperparathyroidism (PHPT) is a common endocrine disorder characterized by dysregulation of parathyroid hormone (PTH) release. Inappropriately high PTH secretion leads to hypercalcemia that may cause diverse symptoms and complications including osteoporosis and nephrolithiasis. The only definitive treatment for PHPT is surgical. Patients with symptomatic PHPT should be advised to undergo surgery. The incidence of PHPT is estimated to be 3:1000 in the general population and most cases are sporadic, single-gland parathyroid adenoma (Brewer, Costa-Guda, & Arnold, 2019). However, 5–10% of the patients may be associated with syndromic and hereditary disease, like multiple endocrine neoplasia type 1 (MEN1), type 2A (MEN2A), type 4 (MEN4) and hyperparathyroidism jaw-tumor syndrome or the non-syndromic familial form of the disease, the familial isolated primary hyperparathyroidism.

Herein, I outline novel contributions and major discoveries that have been made in the field of parathyroid adenoma research over the last few years. I describe the genetic characteristics of sporadic forms of PHPT and the identification of somatic mutations that lead to excess PTH production. Differential gene expression profiles between parathyroid adenoma and normal parathyroid tissues are highlighted, as well as key signaling pathways and molecular mechanisms that may drive cell proliferation and constitutive PTH production in parathyroid adenomas. Additionally, the recent application of next-generation sequencing (NGS) methods and transcriptome profiling in parathyroid adenoma are described.

1.1 Synthesis of PTH

In humans, the gene encoding PTH resides on the short arm of chromosome 11. PTH gene expression is limited to the chief cells in the parathyroid gland. PTH is first synthesized as prepro-PTH, a 115 amino acid precursor peptide and later matures into its 84-amino acid form. Prepro-PTH contains an amino-terminal, 25 amino acid pre-sequence (the signal peptide) and a small pro-sequence of subsequent 6 amino acids, both of which direct the precursor molecule to the endoplasmic reticulum (ER) and the secretory pathway. Prepro-PTH crosses the ER membrane into the ER lumen.

The pre-sequence is removed during this process. Pro-PTH is transported to the Golgi network. During the transit through the Golgi apparatus, the pro-sequence is cleaved, and the 84-amino acid mature PTH is packaged into secretory granules.

1.2 Regulation of PTH secretion

The major physiological function of the parathyroid glands is to act as a "calciostat," sensing the prevailing blood ionized calcium level and adjusting the secretion of PTH accordingly. The relationship between ionized calcium and PTH secretion is a steep sigmoidal one, allowing significant changes in PTH secretion in response to very small changes in blood ionized calcium. There are variables modulating the overall secretory rate of normal parathyroid cells; changes in (1) the minute-to-minute secretion of preformed PTH, (2) the rate of PTH degradation, (3) the level of PTH gene expression, (4) parathyroid cellular proliferation, (5) parathyroid cellular apoptosis and (6) proportion of active and inactive secretory cells in parathyroid gland (Brown, 2015).

Extracellular calcium is "measured" at the surface of the parathyroid cell through a calcium-sensing receptor (CaSR) that is abundantly expressed at the plasma membrane of these cells (Brown et al., 1993). Increased levels of extracellular calcium suppress PTH secretion, whereas diminished levels increase PTH secretion. The CaSR is a class C G-protein coupled receptor (GPCR), and contains calcium-binding elements in its large extracellular domain. Binding of calcium (or calcimimetic agents such as cinacalcet) to the CaSR triggers activation of heterotrimeric G proteins comprising the alpha-subunits $G\alpha q$ and $G\alpha 11$, $G\alpha i$, and G12/13, which activate phospholipase C, inhibit adenylyl cyclase and stimulate Rho kinase, respectively (Luttrell & Lefkowitz, 2002; Swarthout, Doggett, Lemker, & Partridge, 2001). This results in an increase in intracellular free calcium and a decrease in 3′,5′-cyclic adenosine monophosphate (cAMP) levels in parathyroid cells.

2. Summary of germline and somatic genetics in sporadic parathyroid adenoma

Before I would focus on gene expression profile of parathyroid adenoma compared to normal parathyroid gland, I would briefly summarize germline and somatic genetics in sporadic parathyroid adenoma.

The large majority of PHPT cases are attributable to sporadic, single-gland parathyroid adenoma, in which two genes, *MEN1* and *CCND1*, a tumor suppressor and a proto-oncogene respectively, have

Table 1 Overview of general germline and somatic genetic alteration in parathyroid adenoma.

Gene	Established or hypothesized role	Primary mechanism	Clinical setting
MEN1	Tumor suppressor	Germline	MEN1[a]
		Somatic	Sporadic PA[b]
CCND1	Oncogene	Somatic	Sporadic PA[c]
CDKN1B(p27), *CDKN2C*(p18), *CDKN2B*(p15), *CDKN1A*(p21)	Tumor suppressor	Germline	MEN1 variant[a]
CDKN1B(p27)	Tumor suppressor	Somatic	Sporadic PA[c]
ZFX	Oncogene	Somatic	Sporadic PA[c]
EZH2	Oncogene	Somatic	Sporadic PA[c]
CTNNB1(β-catenin)	Oncogene	Somatic	Sporadic PA[c]
GCM2	Oncogene	Germline	FIHP[b]
CDC73(parafibromin)	Tumor suppressor	Germline	HPT-JT[a], FIHP[b]
		Somatic	Sporadic PA[c]
CASR	Tumor suppressor	Germline	FHH1[a], FIHP[d]
RET	Oncogene	Germline	MEN2A[a]

[a]Almost always present (81–100%).
[b]Sometimes present (11–40%).
[c]Rarely present (≤10%).
[d]Frequently present (41–80%).
MEN1: multiple endocrine neoplasia type 1; PA: parathyroid adenoma; FIHP: familial isolated hyperparathyroidism; HPT-JT: hyperparathyroidism-jaw tumor; FHH1: familial hypocalciuric hypercalcemia type 1; MEN2A: multiple endocrine neoplasia type 2A.
Adapted from Brewer, K., Costa-Guda, J., & Arnold, A. (2019). Molecular genetic insights into sporadic primary hyperparathyroidism. Endocrine-Related Cancer, 26(2), R53–R72; Juhlin, C. C., & Erickson, L. A. (2021). Genomics and Epigenomics in parathyroid neoplasia: From bench to surgical pathology practice. Endocrine Pathology, 32(1), 17–34.

been solidly established as primary tumorigenic drivers in parathyroid adenomas (Brewer et al., 2019). In addition, other genes, some involved in pathways leading to *CCND1* activation, *MEN1* inactivation, or PTH dysregulation characteristic of PHPT, have been and continue to be explored as candidate parathyroid tumor-driving genes (summarized in Table 1).

3. Comparative genomic expression

Gene expression studies have identified genes with a potential role in the pathogenesis and pathophysiology of parathyroid adenoma. Despite inter-study heterogeneity of gene expression data, which may be accounted for by use of different reference tissues (adjacent normal parathyroid tissue or, normal parathyroid gland) and different diagnostic criteria (Arvai et al., 2012; Chai et al., 2019; Forsberg et al., 2005; Haven et al., 2004; Hu et al., 2021; Morrison et al., 2004), such studies have proven valuable in the identification of genes and signaling pathways with a potential role in the pathogenesis and pathophysiology of parathyroid adenoma.

3.1 Genes involved in cell cycle regulation

Cyclin D1 is unique as the only cyclin clearly and *CCND1*, which encodes cyclin D1, is a solidly established driver of parathyroid neoplasia (Brewer et al., 2019). *CCND1* was first identified as a human oncogene via its involvement in parathyroid adenomas, of which a subset contains a pericentromeric inversion of chromosome 11. This rearrangement was found to result in juxtaposition of a specific region of chromosome 11q13 with PTH regulatory region (Arnold et al., 1989; Arnold, Staunton, Kim, Gaz, & Kronenberg, 1988). When parathyroid adenoma, and histologically normal parathyroid tissue were studied by comparative genomic hybridization, cyclin D1 overexpression are common in parathyroid adenomas (Hemmer et al., 2001). Cyclin D1 (and c-Jun) showed increased expression in adenomas vs normal parathyroids by microarray analysis and qRT-PCR (Forsberg et al., 2005). Collectively, these studies suggest an oncogenic role of these genes in parathyroid tumor development (Fig. 1).

Several mechanisms of abnormal cell cycle regulatory pathway have been described in sporadic parathyroid adenoma, including elevated expression of negative regulators of cell dependent kinase inhibitor (CDKI), inactivating mutations, or reduced expression of CDKI (Costa-Guda & Arnold, 2014).

CDKN1B, which encodes P27KIP1 (p27), is the most extensively studied of the CDKN1 family in parathyroid tumors, and, like other members of its family, p27 targets both CDK2 and CDK4/6. Confirmed somatic mutations of *CDNK1B* are rare, occurring in about 1% or fewer cases of

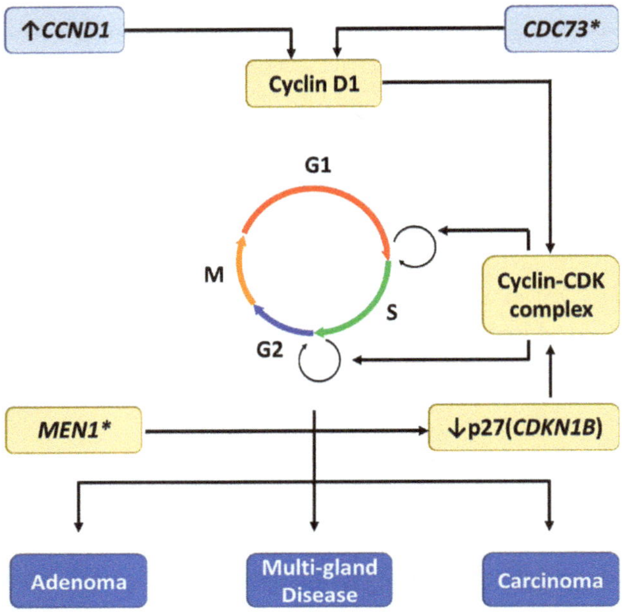

Fig. 1 Parathyroid tumorigenesis mechanisms via the cyclins pathway. Cyclin D1, encoded by *CCND1* gene, is upregulated in parathyroid adenomas. MEN1 functions as a classic tumor-suppressor gene and is frequently associated with somatic loss of heterozygosity. MEN1 gene inactivation results in a reduction of nuclear expression levels of p27, which stops progression of the cell cycle by inhibiting CDK complexes acting at multiple phases of the cell cycle, especially at G1-S and S-G2 checkpoints, and the loss of negative control of cell cycle progression. Inactivating somatic and germline mutations of CDC73 are frequently identified in patients with parathyroid carcinoma. *CDKN1B* gene, encoding the CDK inhibitor p27, has been reported to be downregulated in adenomas compared to normal tissues. *: Inactivating mutation.

sporadic parathyroid adenoma (Alvelos et al., 2013; Borsari et al., 2017; Costa-Guda, Marinoni, Molatore, Pellegata, & Arnold, 2011; Cromer et al., 2012; Newey et al., 2012).

The relative p27, p18 and p21 expression were significantly reduced in parathyroid adenomas compared to normal parathyroid gland biopsies (Buchwald, Akerstrom, & Westin, 2004). In addition, the high rate of Ki-67 expression also may indicate susceptibility of parathyroid adenomas to clonal proliferation(Arvai et al., 2012).

3.2 Genes involved in growth factors

Growth factors can be involved in cellular growth, proliferation, tissue repairing and homeostasis and any imbalance could result in tumorigenesis

(Segiet, Deska, Michalski, Gawrychowski, & Wojnicz, 2015). Angiogenesis is a crucial process for the emergence and survival of proliferative lesions, as it involves the neovascularization to ensure that oxygen and other necessary nutrients will be supplied to the tumor. Factors that are found to be overexpressed in PHPT include vascular endothelial growth factor (VEGF) (Segiet, Deska, et al., 2015), fibroblast growth factor (FGF) (Segiet et al., 2015), transforming growth factor beta (TGFβ) (Schuster & Krieglstein, 2002) and insulin growth factor-1 (IGF-1) (Segiet, Deska, et al., 2015). Genes affecting growth factor expression are found to be involved in sporadic PHPT.

3.3 Genes involved in apoptotic pathway

Apoptosis is a form of programmed cell death, which maintains the tissue homeostasis and the certain number of cells (Segiet et al., 2017). There are mechanisms involved in the pathogenesis of PHPT, which include an imbalance between apoptotic and anti-apoptotic factors, some of which are tumor necrosis factor-related apoptosis-inducing ligand (TRAIL), Fas receptor, the BCL family, the mouse double minute 2 homolog (MDM2) and tumor protein 53 (p53) (Segiet et al., 2017; Segiet, Deska, et al., 2015).

cDNA microarrays revealed elevated expression of both pro-apoptotic (*FAS* receptor, *TRAIL*, *CASP8*, and *CASP4*) and anti-apoptotic (*cIAP1*, *APOLLON*) genes in benign proliferative lesions compared to those in normal gland. A series of alterations in gene expression such as overexpression of apollon, p53 and suppression of p27, bcl-2, bax lead to uncontrolled cell proliferation (Arvai, et al. 2012). This study suggested that parathyroid adenomas contain cell populations with proliferative activity (clonal proliferation), but the weak immunoreactive expression of p53 combined with the relatively strong expression of bcl-2 might contribute to a slow glandular growth (Wang, Johansson, Kvasnicka, Farnebo, & Grimelius, 1996). The strong expression of bcl-2 may contribute to the characteristic slow progression of these tumors (Hadar et al., 2005).

3.4 Genes involved in PTH synthesis or regulation pathway

3.4.1 GCM2

GCM2, also known as GCMB, is a parathyroid-specific transcription factor known to be crucial for parathyroid gland development (Gunther et al., 2000), and germline inactivating mutations in the *GCM2* gene cause human hypoparathyroidism (Ding, Buckingham, & Levine, 2001). Some *GCM2* germline variants have been proposed as potential predisposition alleles in sporadic parathyroid tumors (D'Agruma et al., 2014; Guan et al., 2017; Marchiori et al., 2017). Other activating *GCM2* variants have been identified in sporadic parathyroid adenoma: V382M (Mannstadt, Holick, Zhao, & Juppner, 2011) and Y394S (Guan et al., 2017). GCM2 mRNA expression was upregulated in abnormal parathyroid glands of hyperparathyroidism and was downregulated in response to hypocalcemia(Kebebew et al., 2004). Most recently transcriptome analysis revealed GCM2 may be important factors contributing to the parathyroid adenoma gene expression profile (Hu, et al. 2021).

3.4.2 CaSR

Mutations in the CaSR gene (*CASR*) are associated with the hereditary syndromes familial hypocalciuric hypercalcemia (FHH) and neonatal severe hyperparathyroidism (NSHPT), the former of which is associated with hemizygous *CASR* inactivation and the latter with homozygous mutation (Hendy, D'Souza-Li, Yang, Canaff, & Cole, 2000; Pollak et al., 1993) (Fig. 2). In addition, *CASR* was suspected to play a driver role in sporadic parathyroid tumorigenesis due to the decreased sensitivity to serum calcium in patients with parathyroid adenomas.

The expression of the CaSR mRNA and protein was reduced in most cases of primary hyperparathyroidism, compared with strong expression normal parathyroid tissue (Gogusev et al., 1997). Recent RNA-sequencing analysis and immunohistochemical staining analysis also confirmed that CaSR expression was significantly decreased in parathyroid adenoma compared with normal parathyroid tissue samples (Chai et al., 2019; Hong et al., 2016).

In addition to inactivating mutation or reduced expression of CaSR, elevated expression of negative regulators of CaSR have been reported in parathyroid adenoma compared to normal gland (Balenga et al., 2017; Koh et al., 2011).

The regulator of G protein signaling 5 gene (*RGS5*) was identified as the highly expressed gene in parathyroid adenomas compared with matched-pair normal glands by microarray analysis (Koh et al., 2011). Transient expression of RGS5 in cells stably expressing CaSR resulted in dose-dependent abrogation of calcium-stimulated inositol trisphosphate production and ERK1/2 phosphorylation. Furthermore, RGS5-nullizygous mice display reduced plasma PTH levels, an outcome consistent with attenuated opposition to CaSR activity.

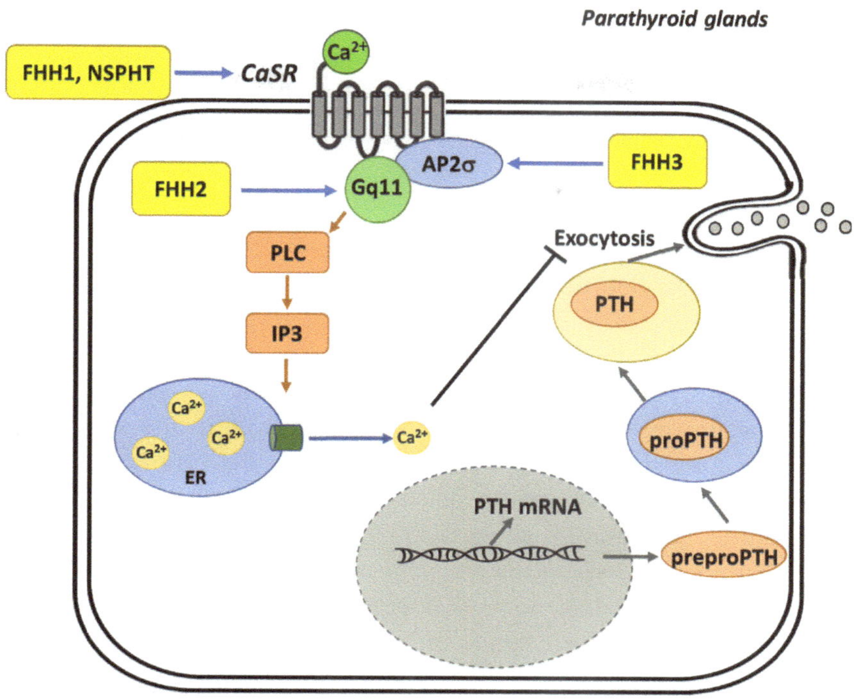

Fig. 2 PTH synthesis and secretion mediated by CaSR activation and familial primary hyperparathyroidism associated with CaSR and its downstream signaling. In the parathyroid gland, CaSR (calcium-sensing receptor) activation by Ca^{2+} or other compatible ligands, induces Gq/11-mediated activation of phospholipase C (PLC), increases inositol 1,4,5-trisphospate (IP3) and leads to intracellular Ca^{2+} mobilization, which acts as a secondary messenger of different cell-dependent actions, such as PTH synthesis regulation. FHH1; familial hypocalciuric hypercalcemia type 1, FHH2; familial hypocalciuric hypercalcemia type 2, FHH3; familial hypocalciuric hypercalcemia type 3, NSPHT; neonatal severe primary hyperparathyroidism.

The *GPR64/ADGRG2* gene, that encodes the orphan adhesion GPCR, has been reported as upregulated in parathyroid adenomas compared to normal parathyroid glands (Balenga et al., 2017). Gαs, cAMP, protein kinase A, and cAMP response element binding protein (CREB) were identified as the signaling cascade downstream of GPR64. The truncated constitutively active GPR64 physically interacts with CaSR and attenuates the CaSR-mediated intracellular Ca^{2+} signaling and cAMP suppression in HEK293 cells. This study indicates that GPR64 may be a physiologic regulator of PTH release that is dysregulated in parathyroid tumors.

Therefore, although impaired *CASR* expression or signaling may play a role in the dysregulation of calcium-PTH homeostasis central to the phenotype of PHPT (Farnebo et al., 1997; Imanishi et al., 2001; Varshney et al., 2013), the lack of clonally selected somatic inactivating genetic lesions at its locus strongly suggests that its dysfunction is generally a secondary effect rather than a cause of sporadic parathyroid tumorigenesis.

3.4.3 Klotho

The type I membrane protein Klotho encompasses 1012 amino acids with a large extracellular domain and a short cytoplasmic portion (Kuro-o et al., 1997). Klotho regulates serum calcium concentration directly at the level of the kidney and indirectly in parathyroid glands. At low serum calcium concentrations, Klotho stimulates calcium channel TRPV5 activity, which leads to an increased tubular reabsorption of calcium in the kidneys. In addition, Klotho may recruit Na^+/K^+-ATPase to the plasma membrane of parathyroid chief cells, generating a transmembrane voltage gradient that mediates PTH secretion(Imura et al., 2007). As deletion of Klotho functions in parathyroid gland hyperplasia, this gene may also contribute to the development of parathyroid adenoma (Fan et al., 2018). Klotho mRNA levels and protein expression were significantly decreased in parathyroid adenomas compared to normal tissues. In addition, Klotho mRNA levels were inversely correlated to serum calcium, and calcium dose-dependently decreased Klotho mRNA expression in normal parathyroid cells in vitro (Bjorklund, Krajisnik, Akerstrom, Westin, & Larsson, 2008). Recent RNA-sequencing analysis and immunohistochemical staining analysis confirmed that Klotho expression was downregulated in parathyroid adenoma compared with normal parathyroid tissue samples(Chai et al., 2019; Hong et al., 2016). By contrast, FGF23 a Klotho ligand, was not identified as a differentially expressed gene (DEG) in that study(Chai et al., 2019).

3.5 Next-generation sequencing analysis

Recent comparative transcriptome analyses reconfirmed previous finding or revealed novel finding. Chae et al. performed RNA-seq to compare the DEGs between 10 parathyroid adenomas and 5 normal parathyroid samples. They revealed that upregulated genes in parathyroid adenomas, such as *MED12, KMT5A, BMP2K,* and *ATAD*, which are implicated in cell proliferation and transcriptional regulation, supporting the notion that they potentially contribute to tumorigenesis (Chai et al., 2019). Furthermore, Protein-protein interaction (PPI) network analysis identified a densely connected sub-module, comprising eight hub molecules. These hub nodes,

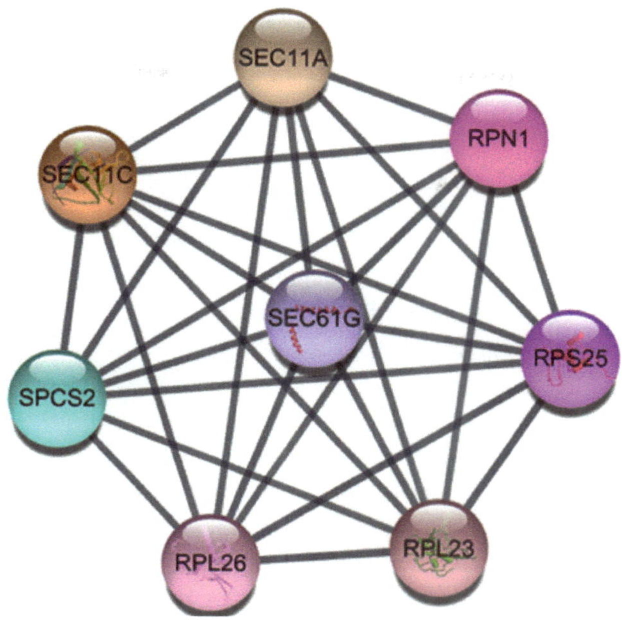

Fig. 3 The most significant protein-protein interaction sub-module detected by the MCODE Algorithm. Signal Peptidase Complex Subunit 2 (SPCS2), Ribosomal Protein L23 (RPL23), Ribosomal Protein L26 (RPL26), Ribophorin I (RPN1), SEC11 Homolog C, Signal Peptidase Complex Subunit (SEC11C), SEC11 Homolog A, Signal Peptidase Complex Subunit (SEC11A), Ribosomal Protein S25 (RPS25), and Sec61 Translocon Gamma Subunit (SEC61G). These hub molecules might play crucial roles in development of parathyroid adenoma. Of them, RPL23 has the highest score, indicating that it is the most strongly interconnected with its neighbor proteins.

identified by computational measurement of their degree of relationship to neighbor nodes, are considered to have pivotal roles in networks. Hub molecules identified in the same study were associated with signal peptidase complex subunit (SPCS2), ribosomal proteins (RPL23, RPL26, RPN1, RPS25), and the ER membrane (SEC11C, SEC11A, SEC61G) (Fig. 3). Prepro-PTH is synthesized as a 115-amino acid precursor and converted to pro-PTH upon cleavage of its signal peptide within the ER. This PPI analysis suggest the importance of the proteolytic process of signal peptide cleavage in the ER in the development of parathyroid adenoma.

Recent multiplatform NGS analysis combined with whole-exome sequencing and RNA-seq provides a more comprehensive view of the genetic landscape. They showed that *MEN1, EZH2, CASR, EZH1, ZFX, mTOR* and *FAT1* mutations were revealed in parathyroid adenoma. According to the RNA-seq data clustering analysis in the same study, cyclin

D1, β-catenin, VDR, CaSR and GCM2 may be important factors contributing to the parathyroid adenoma gene expression profile (Hu et al., 2021).

3.6 Differential microRNA expression

MicroRNAs (miRNA) are small non-coding, single-stranded RNAs, 19–25 nucleotides long, which act as a negative regulator through multiple mechanisms including decreased translation, increased degradation of the target messenger RNA (mRNA) or both. MiRNAs can inhibit the translation and stability of mRNA. Differentially expressed miRNAs have been reported in parathyroid adenomas and carcinomas, compared to normal parathyroid tissue (Verdelli, Forno, Vaira, & Corbetta, 2015). MiRNA-296 and miR-139 were downregulated in both parathyroid adenoma and carcinoma compared with normal parathyroid glands (Corbetta et al., 2010). Expression of miR-296 is involved in control of global mobility of epithelial cells via transcriptional repression of the plasticity in cellular polarity. Therefore, the downregulation of miR-296 induces an exaggerated cellular migration and invasiveness (Silva-Figueroa & Perrier, 2018). The C19MC is the largest human microRNA gene cluster discovered so far, located on chromosome 19q13.4. C19MC expression has been closely related to stem cell biology and tumorigenesis (Tsai, Kao, Chen, Chen, & Lin, 2009; Vaira et al., 2012). Interestingly, miR517C, a specific member of this miRNA group, is overexpressed in parathyroid carcinomas while its expression is decreased in parathyroid adenomas compared with normal parathyroid tissues (Vaira et al., 2012).

Acknowledgment

This work was supported by grants from the National Research Foundation of Korea (2017R1A2B2004708).

References

Alvelos, M. I., Vinagre, J., Fonseca, E., Barbosa, E., Teixeira-Gomes, J., Sobrinho-Simoes, M., et al. (2013). MEN1 intragenic deletions may represent the most prevalent somatic event in sporadic primary hyperparathyroidism. *European Journal of Endocrinology*, *168*(2), 119–128. https://doi.org/10.1530/EJE-12-0327.

Arnold, A., Kim, H. G., Gaz, R. D., Eddy, R. L., Fukushima, Y., Byers, M. G., et al. (1989). Molecular cloning and chromosomal mapping of DNA rearranged with the parathyroid hormone gene in a parathyroid adenoma. *The Journal of Clinical Investigation*, *83*(6), 2034–2040. https://doi.org/10.1172/JCI114114.

Arnold, A., Staunton, C. E., Kim, H. G., Gaz, R. D., & Kronenberg, H. M. (1988). Monoclonality and abnormal parathyroid hormone genes in parathyroid adenomas. *The New England Journal of Medicine*, *318*(11), 658–662. https://doi.org/10.1056/NEJM198803173181102.

Arvai, K., Nagy, K., Barti-Juhasz, H., Petak, I., Krenacs, T., Micsik, T., et al. (2012). Molecular profiling of parathyroid hyperplasia, adenoma and carcinoma. *Pathology Oncology Research*, *18*(3), 607–614. https://doi.org/10.1007/s12253-011-9483-7.

Balenga, N., Azimzadeh, P., Hogue, J. A., Staats, P. N., Shi, Y., Koh, J., et al. (2017). Orphan adhesion GPCR GPR64/ADGRG2 is overexpressed in parathyroid tumors and attenuates calcium-sensing receptor-mediated signaling. *Journal of Bone and Mineral Research*, *32*(3), 654–666. https://doi.org/10.1002/jbmr.3023.

Bjorklund, P., Krajisnik, T., Akerstrom, G., Westin, G., & Larsson, T. E. (2008). Type I membrane klotho expression is decreased and inversely correlated to serum calcium in primary hyperparathyroidism. *The Journal of Clinical Endocrinology and Metabolism*, *93*(10), 4152–4157. https://doi.org/10.1210/jc.2008-0564.

Borsari, S., Pardi, E., Pellegata, N. S., Lee, M., Saponaro, F., Torregrossa, L., et al. (2017). Loss of p27 expression is associated with MEN1 gene mutations in sporadic parathyroid adenomas. *Endocrine*, *55*(2), 386–397. https://doi.org/10.1007/s12020-016-0941-6.

Brewer, K., Costa-Guda, J., & Arnold, A. (2019). Molecular genetic insights into sporadic primary hyperparathyroidism. *Endocrine-Related Cancer*, *26*(2), R53–R72. https://doi.org/10.1530/ERC-18-0304.

Brown, E. M. (2015). Control of parathyroid hormone secretion by its key physiological regulators. In J. P. Bilezikian (Ed.), *The Parathyroids: Basic and clinical concepts* (3rd ed., pp. 101–118). San Diego: Academic Press.

Brown, E. M., Gamba, G., Riccardi, D., Lombardi, M., Butters, R., Kifor, O., et al. (1993). Cloning and characterization of an extracellular ca(2+)-sensing receptor from bovine parathyroid. *Nature*, *366*(6455), 575–580. https://doi.org/10.1038/366575a0.

Buchwald, P. C., Akerstrom, G., & Westin, G. (2004). Reduced p18INK4c, p21CIP1/WAF1 and p27KIP1 mRNA levels in tumours of primary and secondary hyperparathyroidism. *Clinical Endocrinology*, *60*(3), 389–393. https://doi.org/10.1111/j.1365-2265.2004.01995.x.

Chai, Y. J., Chae, H., Kim, K., Lee, H., Choi, S., Lee, K. E., et al. (2019). Comparative gene expression profiles in parathyroid adenoma and Normal parathyroid tissue. *Journal of Clinical Medicine*, *8*(3). https://doi.org/10.3390/jcm8030297.

Corbetta, S., Vaira, V., Guarnieri, V., Scillitani, A., Eller-Vainicher, C., Ferrero, S., et al. (2010). Differential expression of microRNAs in human parathyroid carcinomas compared with normal parathyroid tissue. *Endocrine-Related Cancer*, *17*(1), 135–146. https://doi.org/10.1677/ERC-09-0134.

Costa-Guda, J., & Arnold, A. (2014). Genetic and epigenetic changes in sporadic endocrine tumors: Parathyroid tumors. *Molecular and Cellular Endocrinology*, *386*(1–2), 46–54. https://doi.org/10.1016/j.mce.2013.09.005.

Costa-Guda, J., Marinoni, I., Molatore, S., Pellegata, N. S., & Arnold, A. (2011). Somatic mutation and germline sequence abnormalities in CDKN1B, encoding p27Kip1, in sporadic parathyroid adenomas. *The Journal of Clinical Endocrinology and Metabolism*, *96*(4), E701–E706. https://doi.org/10.1210/jc.2010-1338.

Cromer, M. K., Starker, L. F., Choi, M., Udelsman, R., Nelson-Williams, C., Lifton, R. P., et al. (2012). Identification of somatic mutations in parathyroid tumors using whole-exome sequencing. *The Journal of Clinical Endocrinology and Metabolism*, *97*(9), E1774–E1781. https://doi.org/10.1210/jc.2012-1743.

D'Agruma, L., Coco, M., Guarnieri, V., Battista, C., Canaff, L., Salcuni, A. S., et al. (2014). Increased prevalence of the GCM2 polymorphism, Y282D, in primary hyperparathyroidism: Analysis of three Italian cohorts. *The Journal of Clinical Endocrinology and Metabolism*, *99*(12), E2794–E2798. https://doi.org/10.1210/jc.2014-2857.

Ding, C., Buckingham, B., & Levine, M. A. (2001). Familial isolated hypoparathyroidism caused by a mutation in the gene for the transcription factor GCMB. *The Journal of Clinical Investigation*, *108*(8), 1215–1220. https://doi.org/10.1172/JCI13180.

Fan, Y., Liu, W., Bi, R., Densmore, M. J., Sato, T., Mannstadt, M., et al. (2018). Interrelated role of klotho and calcium-sensing receptor in parathyroid hormone synthesis and parathyroid hyperplasia. *Proceedings of the National Academy of Sciences of the United States of America*, *115*(16), E3749–E3758. https://doi.org/10.1073/pnas.1717754115.

Farnebo, F., Enberg, U., Grimelius, L., Backdahl, M., Schalling, M., Larsson, C., et al. (1997). Tumor-specific decreased expression of calcium sensing receptor messenger ribonucleic acid in sporadic primary hyperparathyroidism. *The Journal of Clinical Endocrinology and Metabolism*, *82*(10), 3481–3486. https://doi.org/10.1210/jcem.82.10.4300.

Forsberg, L., Bjorck, E., Hashemi, J., Zedenius, J., Hoog, A., Farnebo, L. O., et al. (2005). Distinction in gene expression profiles demonstrated in parathyroid adenomas by high-density oligoarray technology. *European Journal of Endocrinology*, *152*(3), 459–470. https://doi.org/10.1530/eje.1.01864.

Gogusev, J., Duchambon, P., Hory, B., Giovannini, M., Goureau, Y., Sarfati, E., et al. (1997). Depressed expression of calcium receptor in parathyroid gland tissue of patients with hyperparathyroidism. *Kidney International*, *51*(1), 328–336. https://doi.org/10.1038/ki.1997.41.

Guan, B., Welch, J. M., Vemulapalli, M., Li, Y., Ling, H., Kebebew, E., et al. (2017). Ethnicity of patients with germline GCM2-activating variants and primary hyperparathyroidism. *Journal of the Endocrine Society*, *1*(5), 488–499. https://doi.org/10.1210/js.2017-00043.

Gunther, T., Chen, Z. F., Kim, J., Priemel, M., Rueger, J. M., Amling, M., et al. (2000). Genetic ablation of parathyroid glands reveals another source of parathyroid hormone. *Nature*, *406*(6792), 199–203. https://doi.org/10.1038/35018111.

Hadar, T., Shvero, J., Yaniv, E., Ram, E., Shvili, I., & Koren, R. (2005). Expression of p53, Ki-67 and Bcl-2 in parathyroid adenoma and residual normal tissue. *Pathology & Oncology Research*, *11*(1), 45–49. https://doi.org/10.1007/BF03032405.

Haven, C. J., Howell, V. M., Eilers, P. H., Dunne, R., Takahashi, M., van Puijenbroek, M., et al. (2004). Gene expression of parathyroid tumors: Molecular subclassification and identification of the potential malignant phenotype. *Cancer Research*, *64*(20), 7405–7411. https://doi.org/10.1158/0008-5472.CAN-04-2063.

Hemmer, S., Wasenius, V. M., Haglund, C., Zhu, Y., Knuutila, S., Franssila, K., et al. (2001). Deletion of 11q23 and cyclin D1 overexpression are frequent aberrations in parathyroid adenomas. *The American Journal of Pathology*, *158*(4), 1355–1362. https://doi.org/10.1016/S0002-9440(10)64086-2.

Hendy, G. N., D'Souza-Li, L., Yang, B., Canaff, L., & Cole, D. E. (2000). Mutations of the calcium-sensing receptor (CASR) in familial hypocalciuric hypercalcemia, neonatal severe hyperparathyroidism, and autosomal dominant hypocalcemia. *Human Mutation*, *16*(4), 281–296. https://doi.org/10.1002/1098-1004(200010)16:4<281::AID-HUMU1>3.0.CO;2-A.

Hong, A. R., Kim, Y. A., Bae, J. H., Min, H. S., Kim, J. H., Shin, C. S., et al. (2016). A possible link between parathyroid hormone secretion and local regulation of GABA in human parathyroid adenomas. *The Journal of Clinical Endocrinology and Metabolism*, *101*(6), 2594–2601. https://doi.org/10.1210/jc.2015-4329.

Hu, Y., Zhang, X., Wang, O., Cui, M., Li, X., Wang, M., et al. (2021). Integrated whole-exome and transcriptome sequencing of sporadic parathyroid adenoma. *Frontiers in Endocrinology*, *12*, 631680. https://doi.org/10.3389/fendo.2021.631680.

Imanishi, Y., Hosokawa, Y., Yoshimoto, K., Schipani, E., Mallya, S., Papanikolaou, A., et al. (2001). Primary hyperparathyroidism caused by parathyroid-targeted overexpression of cyclin D1 in transgenic mice. *The Journal of Clinical Investigation*, *107*(9), 1093–1102. https://doi.org/10.1172/JCI10523.

Imura, A., Tsuji, Y., Murata, M., Maeda, R., Kubota, K., Iwano, A., et al. (2007). Alpha-klotho as a regulator of calcium homeostasis. *Science*, *316*(5831), 1615–1618. https://doi.org/10.1126/science.1135901.

Kebebew, E., Peng, M., Wong, M. G., Ginzinger, D., Duh, Q. Y., & Clark, O. H. (2004). GCMB gene, a master regulator of parathyroid gland development, expression, and regulation in hyperparathyroidism. *Surgery*, *136*(6), 1261–1266. https://doi.org/10.1016/j.surg.2004.06.056.

Koh, J., Dar, M., Untch, B. R., Dixit, D., Shi, Y., Yang, Z., et al. (2011). Regulator of G protein signaling 5 is highly expressed in parathyroid tumors and inhibits signaling by the calcium-sensing receptor. *Molecular Endocrinology*, *25*(5), 867–876. https://doi.org/10.1210/me.2010-0277.

Kuro-o, M., Matsumura, Y., Aizawa, H., Kawaguchi, H., Suga, T., Utsugi, T., et al. (1997). Mutation of the mouse klotho gene leads to a syndrome resembling ageing. *Nature*, *390*(6655), 45–51. https://doi.org/10.1038/36285.

Luttrell, L. M., & Lefkowitz, R. J. (2002). The role of beta-arrestins in the termination and transduction of G-protein-coupled receptor signals. Journal of Cell Science, 115(Pt 3), 455–465. Retrieved from https://www.ncbi.nlm.nih.gov/pubmed/11861753, DOI: https://doi.org/10.1242/jcs.115.3.455.

Mannstadt, M., Holick, E., Zhao, W., & Juppner, H. (2011). Mutational analysis of GCMB, a parathyroid-specific transcription factor, in parathyroid adenoma of primary hyperparathyroidism. *The Journal of Endocrinology*, *210*(2), 165–171. https://doi.org/10.1530/JOE-10-0247.

Marchiori, E., Pelizzo, M. R., Herten, M., Townsend, D. M., Rubello, D., & Boschin, I. M. (2017). Specifying the molecular pattern of sporadic parathyroid tumorigenesis-the Y282D variant of the GCM2 gene. *Biomedicine & Pharmacotherapy*, *92*, 843–848. https://doi.org/10.1016/j.biopha.2017.05.028.

Morrison, C., Farrar, W., Kneile, J., Williams, N., Liu-Stratton, Y., Bakaletz, A., et al. (2004). Molecular classification of parathyroid neoplasia by gene expression profiling. *The American Journal of Pathology*, *165*(2), 565–576. https://doi.org/10.1016/S0002-9440(10)63321-4.

Newey, P. J., Nesbit, M. A., Rimmer, A. J., Attar, M., Head, R. T., Christie, P. T., et al. (2012). Whole-exome sequencing studies of nonhereditary (sporadic) parathyroid adenomas. *The Journal of Clinical Endocrinology and Metabolism*, *97*(10), E1995–E2005. https://doi.org/10.1210/jc.2012-2303.

Pollak, M. R., Brown, E. M., Chou, Y. H., Hebert, S. C., Marx, S. J., Steinmann, B., et al. (1993). Mutations in the human ca(2+)-sensing receptor gene cause familial hypocalciuric hypercalcemia and neonatal severe hyperparathyroidism. *Cell*, *75*(7), 1297–1303. https://doi.org/10.1016/0092-8674(93)90617-y.

Schuster, N., & Krieglstein, K. (2002). Mechanisms of TGF-beta-mediated apoptosis. *Cell and Tissue Research*, *307*(1), 1–14. https://doi.org/10.1007/s00441-001-0479-6.

Segiet, O. A., Deska, M., Michalski, M., Gawrychowski, J., & Wojnicz, R. (2015). Molecular profiling in primary hyperparathyroidism. *Head & Neck*, *37*(2), 299–307. https://doi.org/10.1002/hed.23656.

Segiet, O. A., Deska, M., Mielanczyk, L., Brzozowa-Zasada, M., Bula, G., Gawrychowski, J., et al. (2017). Expression of TRAIL and Fas in primary hyperparathyroidism. *Journal of Investigative Surgery*, *30*(4), 242–246. https://doi.org/10.1080/08941939.2016.1236159.

Segiet, O. A., Michalski, M., Brzozowa-Zasada, M., Piecuch, A., Zaba, M., Helewski, K., et al. (2015). Angiogenesis in primary hyperparathyroidism. *Annals of Diagnostic Pathology*, *19*(2), 91–98. https://doi.org/10.1016/j.anndiagpath.2015.01.002.

Silva-Figueroa, A. M., & Perrier, N. D. (2018). Epigenetic processes in sporadic parathyroid neoplasms. *Molecular and Cellular Endocrinology*, *469*, 54–59. https://doi.org/10.1016/j.mce.2017.04.007.

Swarthout, J. T., Doggett, T. A., Lemker, J. L., & Partridge, N. C. (2001). Stimulation of extracellular signal-regulated kinases and proliferation in rat osteoblastic cells by parathyroid hormone is protein kinase C-dependent. *The Journal of Biological Chemistry*, *276*(10), 7586–7592. https://doi.org/10.1074/jbc.M007400200.

Tsai, K. W., Kao, H. W., Chen, H. C., Chen, S. J., & Lin, W. C. (2009). Epigenetic control of the expression of a primate-specific microRNA cluster in human cancer cells. *Epigenetics*, *4*(8), 587–592. https://doi.org/10.4161/epi.4.8.10230.

Vaira, V., Elli, F., Forno, I., Guarnieri, V., Verdelli, C., Ferrero, S., et al. (2012). The microRNA cluster C19MC is deregulated in parathyroid tumours. *Journal of Molecular Endocrinology*, *49*(2), 115–124. https://doi.org/10.1530/JME-11-0189.

Varshney, S., Bhadada, S. K., Saikia, U. N., Sachdeva, N., Behera, A., Arya, A. K., et al. (2013). Simultaneous expression analysis of vitamin D receptor, calcium-sensing receptor, cyclin D1, and PTH in symptomatic primary hyperparathyroidism in Asian Indians. *European Journal of Endocrinology*, *169*(1), 109–116. https://doi.org/10.1530/EJE-13-0085.

Verdelli, C., Forno, I., Vaira, V., & Corbetta, S. (2015). MicroRNA deregulation in parathyroid tumours suggests an embryonic signature. *Journal of Endocrinological Investigation*, *38*(4), 383–388. https://doi.org/10.1007/s40618-014-0234-y.

Wang, W., Johansson, H., Kvasnicka, T., Farnebo, L. O., & Grimelius, L. (1996). Detection of apoptotic cells and expression of Ki-67 antigen, Bcl-2, p53 oncoproteins in human parathyroid adenoma. *APMIS: Acta Pathologica, Microbiologica, et Immunologica Scandinavica*, *104*(11), 789–796. https://doi.org/10.1111/j.1699-0463.1996.tb04944.x.

CHAPTER TWELVE

Renal hyperparathyroidism

Aiji Yajima[a,b,c,*], Ken Tsuchiya[c], Makoto Kuro-o[d], Pablo Urena[e], Yoshihiro Tominaga[f], Manabu Okada[f], Toshihiro Ichimori[f], Toshihide Tomosugi[f], Takahisa Hiramitsu[f], Taro Murata[b], Masaki Nakamura[g], Masahiko Sasaki[b], Akemi Ito[h], and Kosaku Nitta[i]

[a]Department of Anatomy, Cell Biology and Physiology, Indiana University, School of Medicine, Indianapolis, IN, United States
[b]Department of Urology, Tokyo, Teishin Hospital, Tokyo, Japan
[c]Department Blood Purification, Kidney Center, Tokyo Women's Medical University, Tokyo, Japan
[d]Division of Anti-aging Medicine, Center for Molecular Medicine, Jichi Medical University, Shimotsuke, Japan
[e]Division of Nephrology, Clinique du Landy, Saint Ouen, France
[f]Department of Endocrine Surgery, Nagoya Daini Red Cross Hospital, Nagoya, Japan
[g]Department of Nephrology and Urology, NTT East Kanto Hospital, Tokyo, Japan
[h]Ito Bone Histomorphometry Institute, Niigata, Japan
[i]Department of Medicine, Kidney Center, Tokyo Women's Medical University, Tokyo, Japan
*Corresponding author: e-mail address: a-and-y@rj8.so-net.ne.jp

Contents

Vitamins and Hormones, Volume 120
ISSN 0083-6729
https://doi.org/10.1016/bs.vh.2022.04.010

Abstract

The number of the patients with chronic kidney disease is now increasing in the world. The pathophysiology of renal hyperparathyroidism is closely associated with Klotho-FGF-endocrine axes, which must be solved definitively as early as possible. It was revealed that the expression of *fgf23* is activated by calciprotein particles, which induces vascular ossification. And it is well known that phosphorus overload directly increases parathyroid hormone and hyperparathyroid bone disease develops in those subjects. On the other hand, low turnover bone disease is often recently. Both the patients with chronic kidney disease suffering from hyperparathyroid bone disease or low turnover bone disease are associated with increased fracture risk. Micropetrosis may be one of the causes of increased fracture risk in the subjects with low turnover bone disease. In this chapter, we now describe the diagnosis, pathophysiology and treatments of renal hyperparathyroidism.

1. Pathophysiology of renal hyperparathyroidism

1.1 Histology of parathyroid gland (Fig. 1)

The parathyroid gland has a fat-rich stromal component. The parenchymal cells are arranged in a cordlike pattern and are separated by stromal elements. The parenchymal cells include dark chief cells, clear chief cells, oxyphilic cells, and transitional oxyphilic cells. The chief cells are polyhedral with

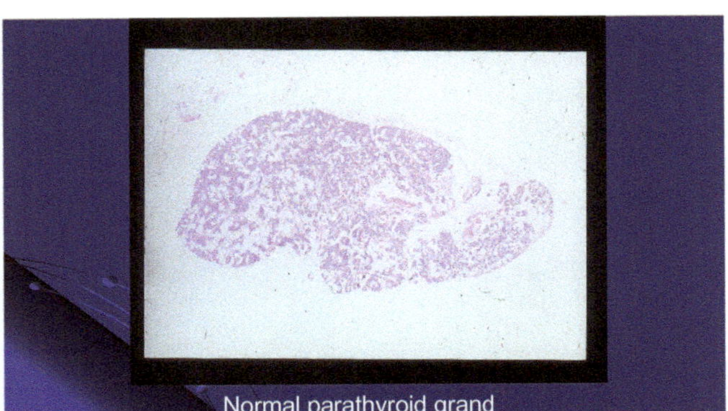

Fig. 1 Normal parathyroid gland has fat-rich stromal component. The gland size is 5 × 4 × 2 mm, and the maximal normal weight is about 60 mg.

centrally placed round nuclei with dense chromatin and inconspicuous nucleoli. The oxyphilic cells are larger than the chief cells and have abundant granular eosinophilic cytoplasm filled with mitochondria and nuclei. The nuclei are larger and more vesicular than those of chief cells. Oxyphilic nodules are commonly found in the parathyroid glands of older individuals. Transitional oxyphilic cells are smaller than mature eosinophilic cells and contra hyperplasia in primary hyperparathyroidism (PHPT). In the past, all four parathyroid glands were thought to be enlarged; however, in some cases fewer than four glands are involved.

Primary chief cell hyperplasia is characterized by an absolute increase in parenchymal cell mass, resulting from the proliferation of chief cells, oxyphilic cells, and transitional oxyphilic cells with unknown stimuli for parathyroid hormone (PTH) hypersecretion. Hyperplastic disease occurs in multiple endocrine neoplasia (MEN) type 1, MEN type 2, and other familial parathyroid diseases (Martin, Kayath, Vieira, & Nosē-Alberti, 1998; Navey-Many & Volovelsky, 2020; Tominaga et al., 1995).

1.2 Klotho-FGF-23-endocrine axes and calcium–phosphate homeostasis in renal hyperparathyroidism (Fig. 2) (Brown et al., 1993)

αKlotho is the obligate coreceptor for fibroblast growth factor 23 (FGF23), a phosphaturic hormone whose serum levels start increasing at an early stage

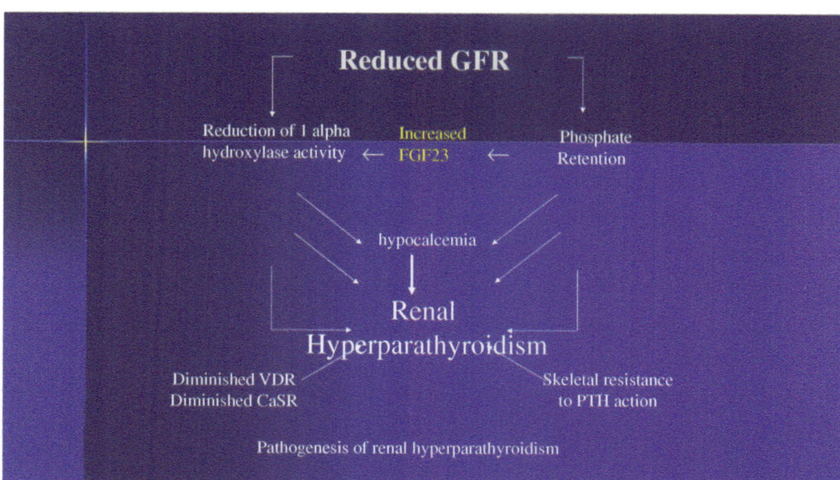

Fig. 2 As the pathophysiology of renal hyperparathyroidism, hypocalcemia, hyperphosphatemia and deficiency of biologically activated vitamin D are the causes of parathyroid hyperplasia.

during chronic kidney disease (CKD) progression. FGF23 acts on the proximal tubules in the kidney and suppresses Na-P cotransporters, which increases phosphate excretion in urine. FGF23 also suppresses 1 alpha hydroxylase, decreases serum active vitamin D levels, and reduces absorption of calcium and phosphate from the intestine. Thus, FGF23 functions as a counter-regulatory hormone against active vitamin D. Thus, an increase in FGF23 causes a decrease in biologically active vitamin D production, followed by a decrease in PTH secretion. FGF23 increases phosphate secretion per nephron and thus compensates the reduction of the number of nephrons during CKD progression to prevent accumulation of phosphate. Then, it is possible that low turnover osteomalacia develops in these patients. Thus, the uretic effect of phosphorus caused by FGF23 induces an increase of phosphorus secretion at first. However, further reduction of nephron number induces an increase of serum phosphorus levels, which increases serum PTH, thereby, forming CKD-MBD. The decrease in active vitamin D and increase in PTH can cause disturbed bone mineralization, potentially woven bone, microcracks, and fracture (Isakova et al., 2011; Kuro-o et al., 1997; Quarles, 2012).

The Klotho proteins, αKlotho and βKlotho, are essential components of endocrine FGF receptor complexes, as they are required for the high-affinity binding of FGF19, FGF21 and FGF23 to their cognate FGF receptors (FGFRs). Collectively, these proteins form a unique endocrine system that governs multiple metabolic processes in mammals.

FGF23 is secreted by osteocytes in response to phosphate intake and binds to αKlotho-FGFR complexes, which are expressed most abundantly in renal tubules, to regulate mineral metabolism. Growing evidence suggests that the FGF-Klotho endocrine system also has a crucial role in the pathophysiology of aging-related disorders, including diabetes, cancer, arteriosclerosis and chronic kidney disease. Therefore, targeting the FGF-Klotho endocrine axes might have therapeutic benefit in multiple systems; investigation of the crystal structures of FGF-Klotho-FGFR complexes is paving the way for the development of drugs that can regulate these axes (Kuro-o, 2019).

Recently, it was revealed that FGF23 synthesis is stimulated by both calciprotein particles (CPP) and sclerostin (Akiyama et al., 2020; Ito et al., 2021; Kuro-o, 2019).

By the way, Klotho is required for FGF23 to bind to FGFR with high affinity, and it is the binary complex of Klotho and FGFR that functions as the physiological receptor for FGF23.

Recently, it was revealed that klotho is closely associated with cancer. Further research is needed to clarify this matter (Onishi et al., 2020; Pako et al., 2020; Rubinstein et al., 2019; Shmulevich et al., 2020).

PTH acts on PTH receptors on the osteocyte and induces expression of Receptor Activator of NF-kappa B Ligand (RANKL) (Nakashima et al., 2011), which activates osteoclastic bone resorption and releases calcium and phosphorus from the bone to the blood. In addition, PTH also acts on the distal tubules in the kidney and induces the absorption of calcium from the tubules. Moreover, PTH acts on the proximal tubules and stimulates 1 alpha hydroxylase, which increases active vitamin D and serum calcium levels. A calcium sensing receptor (CaSR), located on the parathyroid cell membrane, regulates extracellular and intracellular calcium levels and controls the secretion of PTH (Komaba et al., 2010; Tominaga et al., 1992).

Moreover, FGF23, a phosphaturine, acts on Klotho and the FGF receptor complex on the parathyroid cell membrane and suppresses parathyroid cell proliferation and secretion of PTH. The mechanism of phosphate homeostasis has been briefly elucidated. FGF23 is expressed from the osteocyte and osteoblastic osteocyte by phosphate loading. FGF23 acts on the proximal tubules in the kidney and suppresses Na-P cotransporters, which increases phosphate excretion in urine. FGF23 also suppresses 1 alpha hydroxylase, decreases serum active vitamin D levels, and reduces absorption of calcium and phosphate from the intestine. Thus, FGF23 suppresses parathyroid function.

CPP caused by phosphate retention due to overdose of phosphate intake and decreased urinary phosphate excretion stimulates the secretion of FGF23 from osteoblastic osteocytes and osteoblasts. Increased FGF23 is closely associated with the progression of renal hyperparathyroidism (Fig. 2). In addition, decreased vitamin D directly enhances parathyroid cell proliferation and PTH secretion. Hypocalcemia induced by deficiency of active vitamin D and hyperphosphatemia stimulates the proliferation of parathyroid cells to synthesize PTH. Although all of the causes of renal hyperparathyroidism have not been clarified, it is obvious that hyperphosphatemia is one of the causes of renal hyperparathyroidism (Slatopolsky et al., 1996).

2. Causes of renal hyperparathyroidism

2.1 Hypocalcemia

In CKD patients, there is a reduction in the activity of the enzyme that converts vitamin D into calcitriol, which reduces calcitriol production and leads to reduced intestinal absorption of calcium (LaClaire et al., 2005). The reduction in intestinal calcium absorption caused by decreased calcitriol results from the decrease in the expression of intestinal proteins involved in calcium transport and which are regulated by calcitriol, such as calbindins and transient receptor potential vanilloid subfamily member 6 (TRPV6). In addition, calcium is no longer properly reabsorbed by the diseased kidneys, even if paradoxically a certain degree of hypocalciuria is usually observed. Therefore, this results in lowered plasma ionized calcium levels. Hypocalcemia then develops, leading to increased PTH secretion (Man & Jungers, 1998). In addition, there is a dysregulation of PTH secretion in CKD in response to serum calcium levels, to the point that a higher concentration of ionized calcium is required to suppress PTH secretion. Parathyroid cells do in fact become less sensitive to extracellular calcium's suppressive activity on PTH secretion, due to decreased parathyroid CaSR expression (Zitt et al., 2013). They are also less sensitive to calcitriol's inhibitory effect on PTH gene expression due to the decreased expression of the parathyroid's vitamin D receptor (VDR). And parathyroid cells are not affected by FGF23 downregulation as a result of decreased expression of FGF type 1 and alpha-klotho receptors.

2.2 Decreased circulating calcitriol levels

Calcitriol plasma levels decrease in parallel with CKD progression. This decline is due to thinning of the renal parenchyma, which produces calcitriol, but it is also secondary to at least four other mechanisms: a lack of substrate, i.e., natural vitamin D or cholecalciferol and its hydroxylated form, 25(OH)D, often found in CKD. The percentage of end-stage CKD patients with vitamin D insufficiency (25(OH)D between 15 and 30 ng/mL) and deficiency (25(OH)D $<$15 ng/mL) raises to 80–85% (Metzger et al., 2013). Moreover, hyperphosphatemia directly inhibit 1α-hydroxylase activity and calcitriol production in the kidney in these patients. High serum phosphorus levels stimulate FGF23 production by bone cells. Increased bone production of FGF23 protein, one of the most powerful inhibitors of renal 1α-hydroxylase and calcitriol production. FGF23 also stimulates vitamin D degradation by stimulating 24-hydroxylase (CYP24A1) activity; and, finally, the decreased expression and function of the PTH type 1 receptor in the kidney limits PTH's stimulatory activity on calcitriol (Torres & Broe, 2012). A decline in circulating calcitriol levels may promote the development of renal hyperparathyroidism by several mechanisms: lowered serum calcium, secondary to lowered intestinal absorption of calcium, leads to elevated PTH with a loss of calcitriol's negative feedback to parathyroid gland production. The response of parathyroid gland to vitamin D is decreased because the number of VDR is decreased in human and in the experimental model with CKD (Fukuda et al., 1993). And CaSR expression is decreased in parathyroid cells, which then become less sensitive to suppressive activity of parathyroid hyperplasia by calcium (Pearce et al., 1996).

2.3 Hyperphosphatemia

In patients with CKD, serum phosphates, which are essentially derived from food and drink (sodas, canned foods and processed foods containing rapidly absorbed inorganic phosphates), are not eliminated effectively by diseased kidneys. Hyperphosphatemia thus develops and plays a major role in renal hyperparathyroidism management (Galitzer, Ben-Dov, Silvet, & Naveh-Many, 2010; Shilo et al., 2017; Silver & Naveh-Manny, 2011; Torres et al., 2009). This phosphate retention is not apparent in the early stages of CKD, since PTH and FGF23 act to downregulate the proximal renal tubule's maximum phosphate transfer and its tubular reabsorption. However, it becomes perceptible at more advanced stages of CKD (3b to 5) when nephron loss reaches a significant threshold. In addition to its indirect activity, mediated by calcium and calcitriol, phosphate directly stimulates PTH production and secretion. Hyperphosphatemia appears to lead to a decreased number of CaSRs in the parathyroid glands. Phosphate ions can also bind to specific binding sites in the CaSR structure and reinforce its inactive form, thus making calcium's activating activity less effective (Geng et al., 2016). The glands thus become less sensitive to calcium's inhibitory action on PTH production, resulting in PTH hypersecretion (Kifor et al., 1996). Phosphate also supposedly has a direct effect on PTH production via a post-transcriptional effect, which involves the binding of *trans*-acting proteins or AUF1 to cis elements located in the 3′ untranslated region (3′UTR) of the messenger RNA of PTH. The interactions between these proteins and the PTH messenger RNA are orchestrated by the isomerase Pin1. In short, an increase in phosphate stabilizes PTH mRNA and promotes PTH production (Silver & Naveh-Manny, 2011). It has also been demonstrated that certain microRNAs, such as miR-148 and let-7, can play an important role in regulating PTH production in CKD. Levels of these microRNAs are elevated in CKD and inhibiting them prevents an increase in PTH secretion (Shilo et al., 2017).

2.4 Parathyroid cell proliferation and tumorigenesis

The characteristics of histopathology of parathyroid glands in patients with CKD can be summarized as follows: (1) asymmetric enlargement, (2) nodularity, and (3) increased oxyphilic and transitional oxyphilic cells. Under the same stimuli, parathyroid glands proliferate asymmetrically (Fig. 3); the reason for this has not yet been clarified. Hyperplasia of parathyroid glands can be classified into four types: diffuse hyperplasia, early nodularity in diffuse hyperplasia, nodular, and single nodular hyperplasia (Fig. 4) (Tominaga et al., 1996, 1995, 1992). The weight of the glands is related to the pattern of hyperplasia. The glandular weight of a normal gland is approximately 40 mg; if more than 60 mg, the gland is considered to be abnormal and hyperplastic. When the glandular weight is small, almost all glands show diffuse hyperplasia, whereas, when it is over 500 mg, more than 85% of glands show nodular hyperplasia (Fig. 5). We hypothesize that hyperplasia is a progressive process and that parathyroid glands progress from diffuse hyperplasia to early nodular, nodular, and finally single nodular hyperplasia. In a gland with single nodular hyperplasia, the nodule progresses remarkably, suppresses surrounding nodules, and becomes adenoma-like (Fig. 6) (Tominaga et al., 1996, 1995, 1992).

From our studies, we became aware that the incidence of graft-dependent recurrent HPT was more frequent when nodular hyperplastic tissue was transplanted than when diffuse hyperplasia was grafted. Evaluation

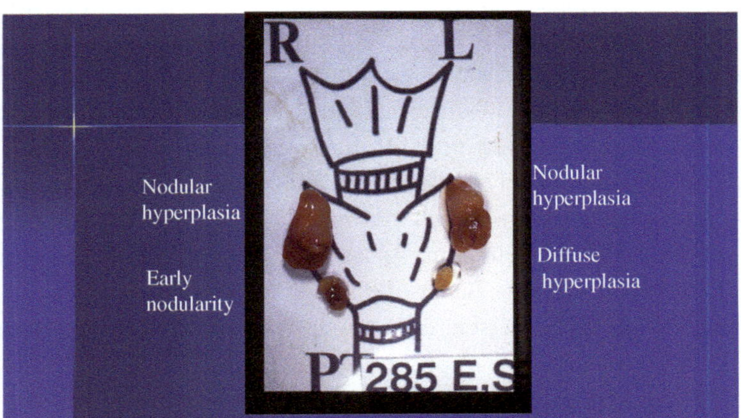

Fig. 3 Asymmetric enlargement of parathyroid glands, nodularity and oxyphilic are characteristics of renal hyperparathyroidism. Hyperplastic parathyroid gland is classified into the four types, including diffuse type, early nodularity in diffuse hyperplasia, nodular hyperplasia and single nodular hyperplasia.

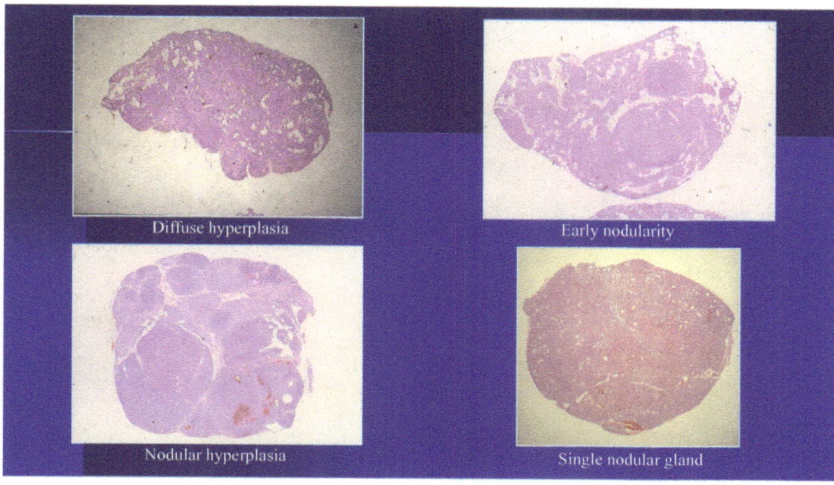

Fig. 4 (Upper left) diffuse type of parathyroid hyperplasia; (upper right) early nodularity in diffuse hyperplasia; (lower left) Nodular hyperplasia; (lower right) single nodular hyperplasia.

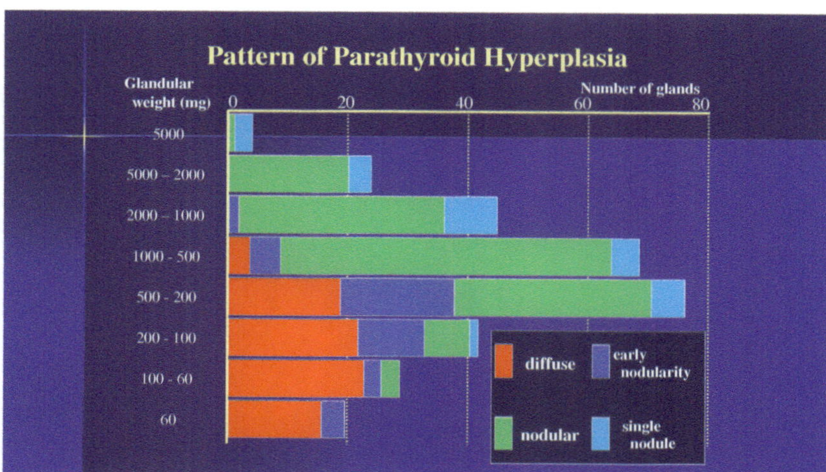

Fig. 5 The relationship between pattern of parathyroid hyperplasia and glandular weight of the resected each parathyroid gland in Nagoya Daini Red Cross Hospital, Nagoya, Japan. When the glandular weight is <200 mg, almost all glands show diffuse hyperplasia, and when it is more than 500 mg, >85% of the glands show nodular hyperplasia.

of the parathyroid cells by DNA analysis and Ki 67 histoimmunostaining showed that the cells in the nodular hyperplasia had a high growth potential (Figs. 7–10) (Tominaga et al., 1992). Histopathological examination revealed that the parathyroid cells in each nodule were homogeneous in hematoxylin

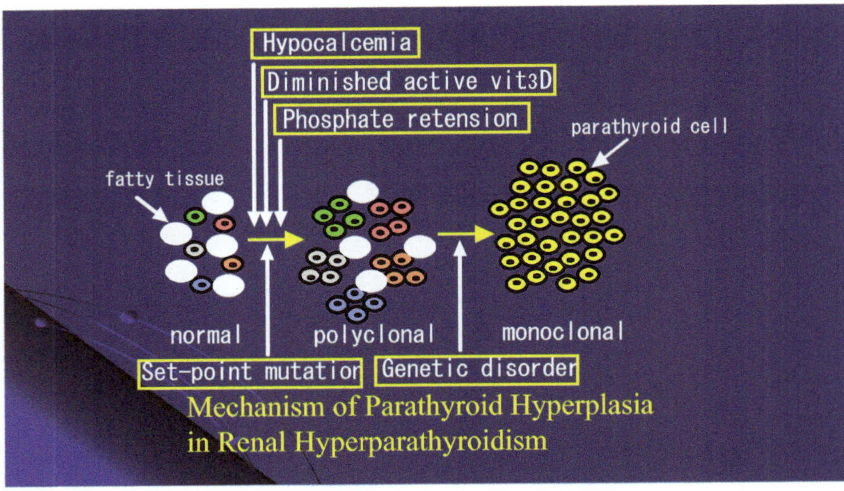

Fig. 6 Mechanism of parathyroid hyperplasia in renal hyperparathyroidism. In CKD subjects, hypocalcemia, hyperphosphatemia and diminished biologically activated vitamin D induces polyclonal hyperplasia of parathyroid glands, thereafter, monoclonal hyperplasia of parathyroid glands is induced by genetic disorder (Tominaga et al., 1996, 1995, 1992).

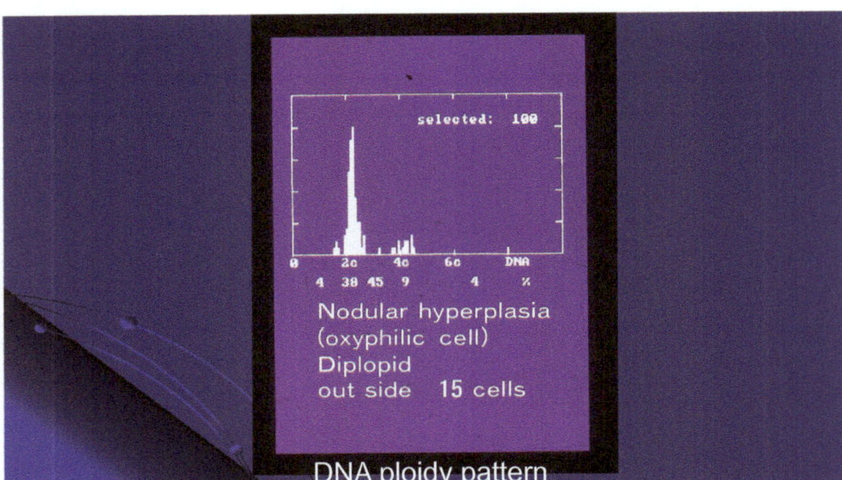

Fig. 7 Growth potential of nodular hyperplasia was analyzed by DNA analysis.

and eosin staining. Moreover, histoimmunostaining showed that the cells in each nodule proliferated monoclonally (Tominaga et al., 1996). Clonal analysis confirmed that the diffuse hyperplastic cells are polyclonal, whereas, each nodule proliferates monoclonally (Fig. 11). It has been speculated that the elucidating of each nodule is an independent process.

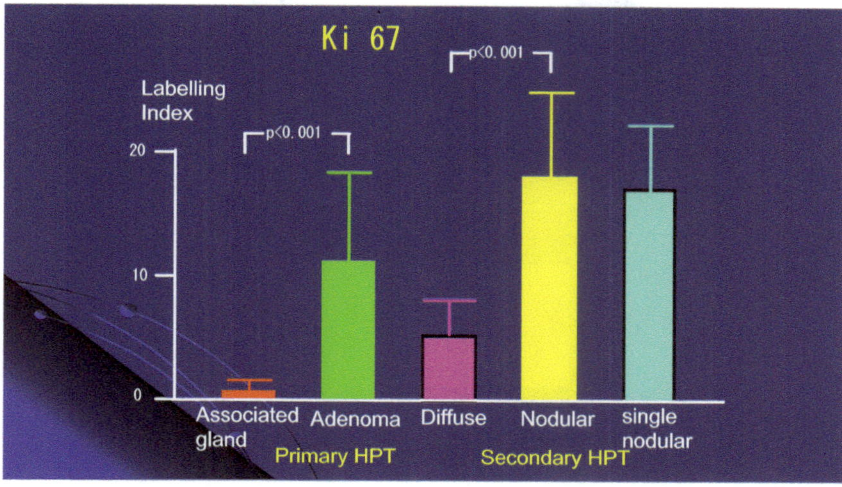

Fig. 8 Growth potential according to Ki67 staining-Ki staining is increased in nodular hyperplastic parathyroid tissue as compared to diffuse hyperplastic parathyroid tissue of the parathyroid glands.

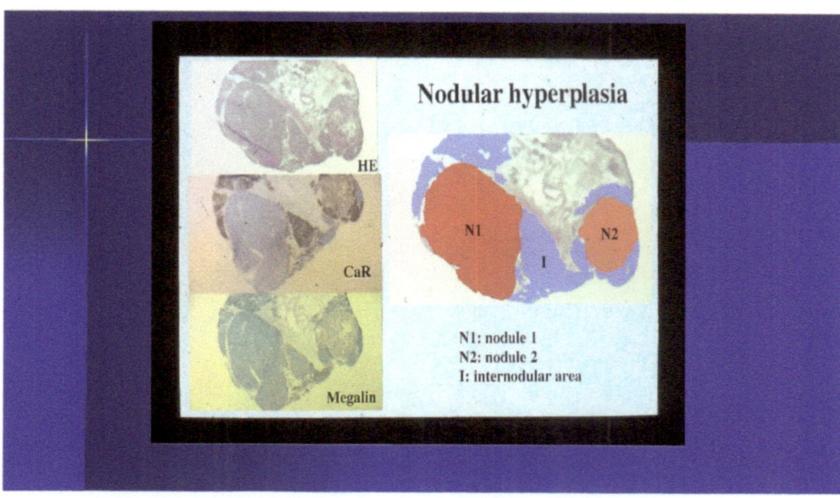

Fig. 9 Nodular hyperplasia. The cells in each nodule proliferate homogeneously.

The mechanism of tumorigenesis in each nodule has been evaluated. In PHPT, parathyroid adenoma proliferates monoclonally and sometimes as part of a genetic disorder (Figs. 9–13) (Fukuda et al., 1993; Tominaga et al., 1999). In cyclin D1 transgenic mouse, parathyroid adenomas and HPT develop progressively and such a mechanism occurs in approximately 25% of patients with parathyroid adenomas in PHPT (Imanishi et al., 2001).

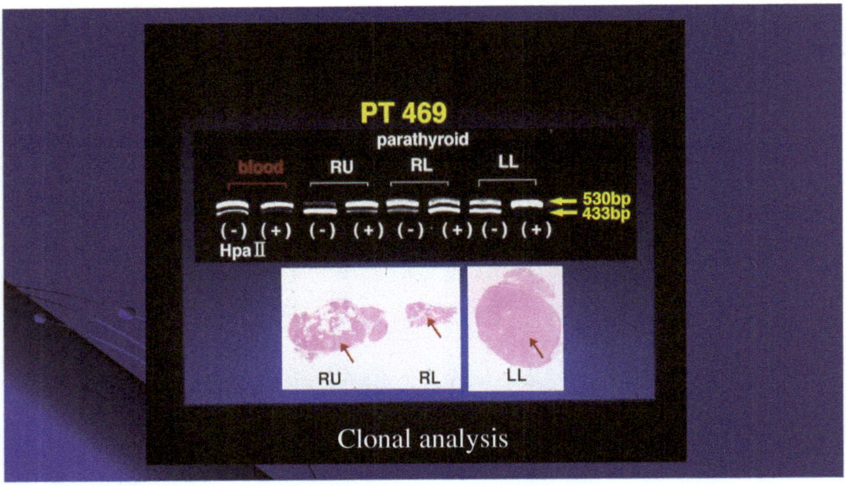

Fig. 10 Clonal analysis of the diffuse and nodular hyperplasia of parathyroid glands (Tominaga et al., 1996).

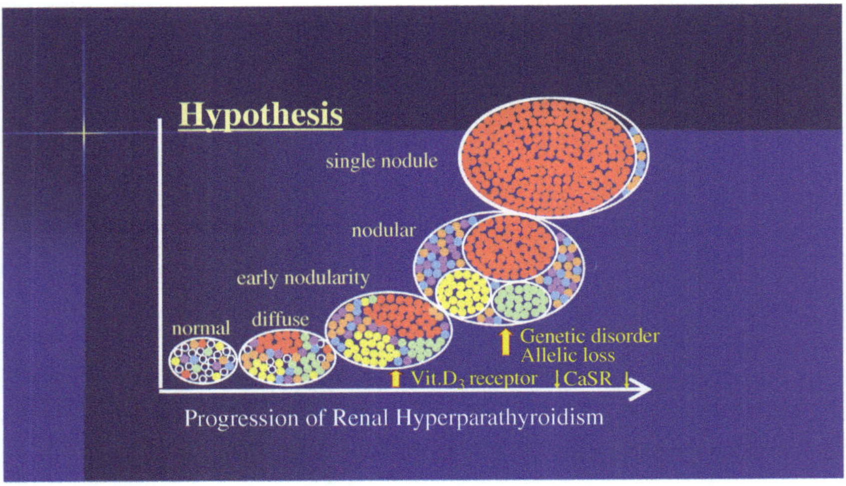

Fig. 11 Mechanism of parathyroid hyperplasia in renal hyperparathyroidism. Our hypothesis concerning the proliferative process of parathyroid hyperplasia in renal hyperparathyroidism. In chronic renal failure the parathyroid glands initially grow diffusely and polyclonally, and the cells in the nodules are subsequently transformed monoclonally (multiclonal), perhaps induced by some kinds of genetic abnormalities, such as the multiple endocrine neoplasia gene. Finally, one nodule with aggressive proliferative potential is markedly enlarged (monoclonal). Vit D_3, vitamin D_3.

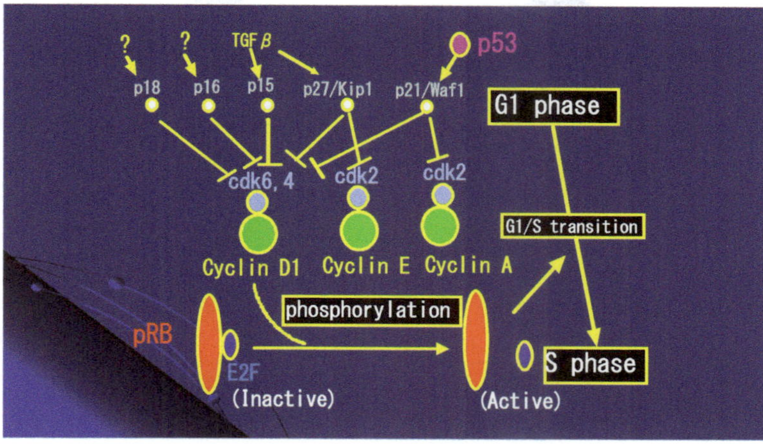

Fig. 12 Mechanism of Cyclin D1. Mechanism of tumorigenesis in hyperparathyroidism is not yet elucidated. In adenoma of primary hyperparathyroidism, the role of cyclin D1/PRAD has been gradually understood. Cyclin D1 converts cyclin dependent kinase four or six and pRB then transitions from the G1 phase to the G2 phase.

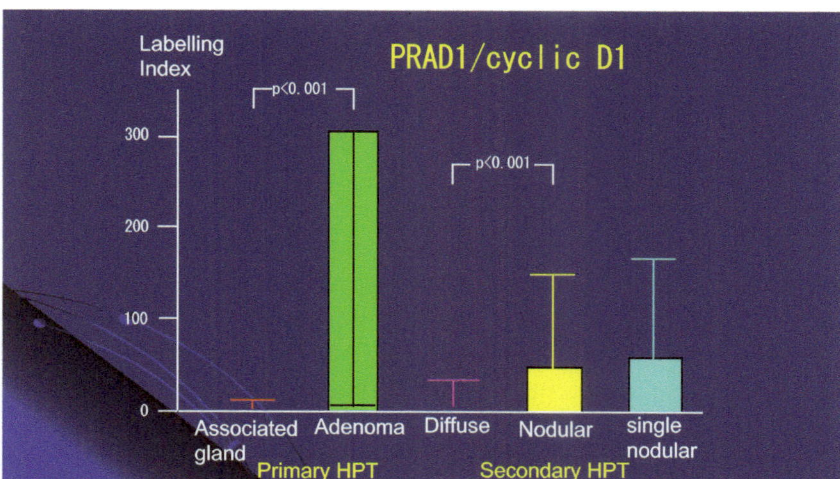

Fig. 13 Expression of Cyclin D1 in each specimen. Overexpression of cyclin D1 can be observed only in PHPT. In renal hyperparathyroidism, the expression could not be detected because other mechanisms of tumorigenesis are related to parathyroid cell growth.

Genetic defect in chromosome 11 encoding menin was detected in 25% of primary adenomas. However, in about half of the PHPT patients, the mechanism of tumorgenesis could not be elucidated.

The mechanism of tumorigenesis in renal hyperparathyroidism with monoclonal or single nodular hyperplasia is unclear; however, the mechanism might be different from PHPT. Recently, HRPT2, encoding a parafibromin tumor suppressor gene, has been reported. The HRPT2 gene abnormality is responsible for HPT jaw tumor syndrome, and approximately 35% of the members of HPT jaw tumor syndrome have parathyroid cancer. It has been confirmed that suppressed parafibromin induces parathyroid carcinoma. However, disorders of expressed parafibromin are not always present.

From our experience of parafibromin retaining positivity in lung metastatic nodules, we confirm parafibromin is not an important factor in the transformation to parathyroid carcinoma.

3. Medical treatment of renal hyperparathyroidism (Table 1)

Renal hyperparathyroidism is an important and frequent complication of CKD. Increased secretion of FGF23 from the bone due to decreased GFR and accumulation of P decreases the production of active vitamin D and serum Ca levels, resulting in increased production and secretion of PTH (Guitérrez, 2010). Various agents that consider these pathogenic mechanisms have been developed and applied in clinical practice. This section describes the medical treatment of renal hyperparathyroidism.

3.1 Management goals

The Japanese Society for Dialysis Therapy published clinical practice guidelines for CKD-MBD in 2013 (Fukagawa et al., 2013). According to these guidelines, it is recommended that the target serum levels of Ca and P are within 8.4–10.0mg/dL and 3.5–6.0mg/dL, respectively, and the serum PTH concentration is within 60–240 pg/mL when optimal serum levels of Ca and P have been achieved. In addition, aggressive treatment should be considered when there are obvious clinical symptoms of renal hyperparathyroidism or when the parathyroid glands are obviously swollen, even if the biochemical parameters are well controlled. Although evidence is scarce, in CKD stage 1–4, it is desirable to keep each parameter within the standard range for the general population. Medical treatments included diet, phosphorus adsorbents, vitamin D receptor activators (VDRAs), and calcimimetics.

3.2 Diet therapy and phosphorus restriction

Phosphorus restriction is important in the management of CKD-MBD. Phosphorus is mainly present in proteins. Protein-restricted diet is useful

Table 1 Medical treatment for renal hyperparathyroidism.

CKD stage	Non-dialysis CKD	Dialysis introduction period	Maintenance dialysis
Diet therapy	Phosphorus restriction (vitamin D intake) depending on the stages	Phosphorus restriction	Phosphorus restriction
Dialysis		Sufficient dialysis	Sufficient dialysis
Dialysate calcium concentration		2.75–3.00 mEq/L	2.5–3.0 mEq/L
Phosphorus adsorbents	Calcium carbonate	Sevelamer hydrochloride	Sevelamer hydrochloride
	Lanthanum carbonate	Calcium carbonate	Calcium carbonate
	Ferric citrate	Lanthanum carbonate	Lanthanum carbonate
		Bixalomer	Bixalomer
		Ferric citrate	Ferric citrate
		Sucroferric oxyhydroxide	Sucroferric oxyhydroxide
Vitamin D receptor activators (VDRAs)	Oral VDRAs	Oral VDRAs	Oral VDRAs
		Intravenous VRDAs	Intravenous VRDAs
Calcimimetics			Oral calcimimetics
			Intravenous calcimimetics

for phosphorus restriction, although phosphorus contained in proteins has a relatively low absorption rate. In addition, sarcopenia has been reported in CKD patients in recent years (Sabatino, Cuppari, Stenvinkel, Lindholm, & Avesani, 2021). Therefore, whether to prioritize phosphorus restriction or protein intake should be considered based on the individual condition. Inorganic phosphorus, present in drinking water and food additives has a high absorption rate, which is troublesome for phosphorus management (Kalantar-Zadeh et al., 2010). However, extraosseous ossification is significantly affected and increased by organic phosphorus intake. Since food additives are abundant in processed foods, it is recommended to cook and consume raw ingredients such as fish and meat before consumption. It is also recommended that foods with a low phosphorus/protein ratio should be eaten (Noori et al., 2010). It is necessary to avoid undernutrition due to radical protein restriction and practice efficient dietary content management.

3.3 Phosphorus adsorbents

At the advent of phosphorus adsorbent therapy, only calcium carbonate was available. Aluminum hydroxide has a good phosphorus adsorption capacity, but is contraindicated because it induces osteomalacia and encephalopathy (Hodsman & Steer, 1992; Petit, 1985). Ca-free phosphorus adsorbents have emerged as key drugs for phosphorus management. In the 2000s, sevelamer hydrochloride, lanthanum carbonate, bixalomer, ferric citrate, and sucroferric oxyhydroxide were developed. Calcium carbonate is cheap and has an excellent phosphorus adsorption capacity; however, it has been reported that long-term Ca load leads to vascular calcification and poor prognosis (Jamal et al., 2013). Sevelamer hydrochloride is reportedly more effective in suppressing vascular calcification as compared to calcium-containing phosphate binders (Block et al., 2005), but management with sevelamer hydrochloride alone is difficult due to the large tablets and constipation of the patients. As to phosphate binders developed after sevelamer hydrochloride, there are few side effects, but evidence of their calcification-suppressing effects is still insufficient. Iron-containing phosphorus absorbents have added value for iron supplementation for the patients with iron deficiency.

3.4 Vitamin D receptor activators

Decreased level of active vitamin D is significantly involved in the pathogenic mechanism of renal hyperparathyroidism, and VDRAs play an important role in the treatment and prevention of progression of renal hyperparathyroidism. VDRAs include oral (alfacalcidol, calcitriol, and falecalcitriol) and intravenous preparations (maxacalcitol and calcitriol). Although intravenous VDRAs are expensive, they are less prone to cause hypercalcemia and are superior to oral VDRAs in terms of patient adherence. However, both VDRAs have a limited impact on highly advanced renal hyperparathyroidism.

3.5 Drugs for osteoporosis

Renal hyperparathyroidism contributes to an increased risk of fractures in CKD patients (Danese et al., 2006; Jadoul et al., 2006). Although drugs such as bisphosphonates and receptor activator of nuclear factor kappa-B ligand (RANKL) inhibitors are effective treatments for bone loss and hypercalcemia (Bover et al., 2017), their effects are temporary, and they have the disadvantage of increasing PTH levels. Moreover, the use of bisphosphonates or RANKL inhibitors is not generally approved in dialysis patients.

3.6 Calcimimetics

Cinacalcet hydrochloride has been used as the first calcimimetic agent in Japan since 2008. Calcimimetics act on Ca-sensing receptors in the parathyroid gland and suppress PTH synthesis and secretion (Behets et al., 2015; Goodman, 2004; Yajima et al., 2017). The advent of calcimimetics has had the greatest impact on the medical treatment of renal hyperparathyroidism. In fact, the number of parathyroidectomy (PTX) in Japan has drastically dropped since 2008 (Fig. 14) (Tominaga et al., 2016). Calcimimetics lower not only the serum PTH levels but also the serum Ca and P levels by decreasing osteoclastic bone resorption. Furthermore, calcimimetics are expected to reduce the size of the parathyroid glands and vascular calcification (EVOLVE Trial Investigators et al., 2012; Raggi et al., 2011; Wheeler et al., 2014). However, it is often difficult to take cinacalcet hydrochloride

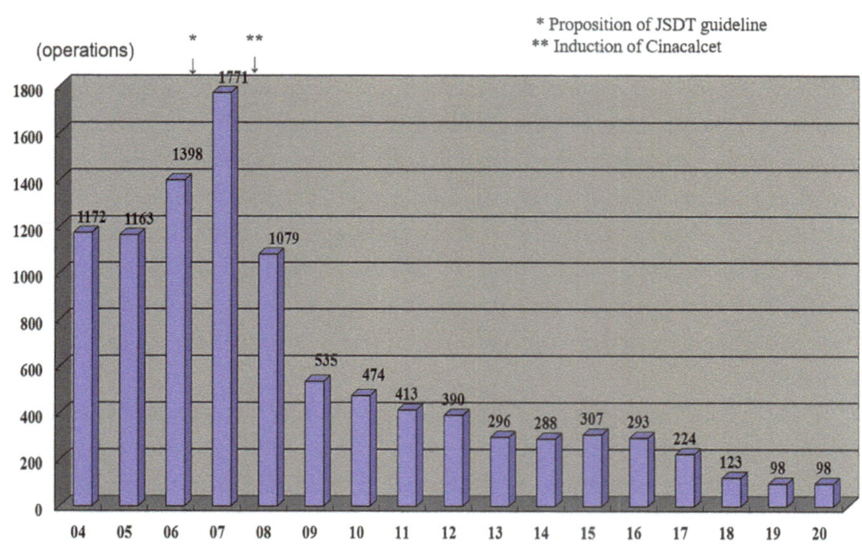

Fig. 14 Annual number of PTX for renal hyperparathyroidism in Japan (data from the Parathyroid Surgeons' Society of Japan).

due to the gastrointestinal side effects such as nausea. After cinacalcet, Evocalcet, an oral calcimimetic with fewer gastrointestinal symptoms, and Etelcalcetide, an intravenous calcimimetic, were developed. Although the number of PTX has decreased recently, there are still some patients with renal hyperparathyroidism who are resistant to calcimimetics. In such patients, PTX may be difficult because of the enlargement and/or adhesion of the parathyroid glands to the tissue around parathyroid glands due to these conservative therapies.

However, it is often difficult to take cinacalcet hydrochloride due to the gastrointestinal side effects such as nausea. Then, Evocalcet, an oral calcimimetic with fewer gastrointestinal symptoms (Kawata et al., 2018), and Etelcalcetide, an intravenous calcimimetic (Martin et al., 2014), were developed.

3.7 Parathyroidectomy

3.7.1 Surgical procedure

The surgical procedure for patients with renal hyperparathyroidism at our hospital is described. Sufficient dialysis is performed before surgery to eliminate hyperpotassemia.

After intubation using a nerve integrity monitor tube, an arterial line is reserved for intraoperative blood pressure measurement and blood sampling for intraoperative intact PTH monitoring. Moreover, in patients with hungry bone syndrome (total ALP >500 IU/L), catheterization of the central vein is performed for calcium replacement after the PTX.

A skin incision of approximately 5 cm is made according to the wrinkles of the neck, and a sufficient flap is made on the underside of the platysma muscle. The strap muscles are separated in the midline from the upper part of the thyroid cartilage to the sternal notch. The left thyroid lobe is exposed, and the inferior thyroid artery and recurrent laryngeal nerve (RLN) are identified. The cross point of the artery and nerve is called "the landmark." The RLN is detected by intraoperative neurophysiological monitoring, but does not generally need to be exposed, because adhesions around the nerve make it difficult to perform surgery when reoperation or thyroid surgery is required.

The thyroid capsule is carefully cut to avoid injury to the RLN, and the left superior parathyroid gland is exposed. The resected glands are placed in bottles containing sterile ice-cold saline. The left inferior parathyroid gland is resected. The thymic tongue is carefully removed in the cranial direction as much as possible to avoid injury to the RLN (Fig. 15).

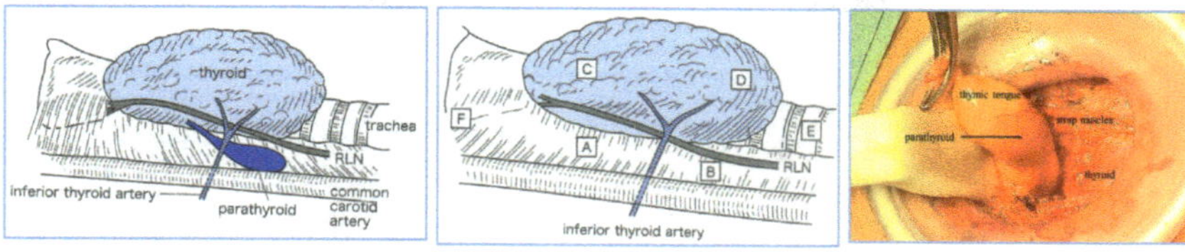

Fig. 15 Migration of the superior gland. The superior gland may sometimes fall and is often found in Area B. The cross point of the inferior thyroid artery and RLN is called "the landmark." Areas A, B, C and D are divided by the landmark. The superior glands sometimes fall and is often found in Area B. Removal of the thymic tongue is routinely performed because the supernumerary parathyroid glands or the lower parathyroid glands are often found in the thymic tongue.

The parathyroid glands have the same origin as the thymus, and originate from the endodermal cells of the third pharyngeal sac. Many descend with the thymus, separate from the thymus on the way, and are located on the caudal side of the thyroid gland. For this reason, excess glands in the thymus and the inferior glands are often found in the thymus (mostly at the top of the thymus). When eight parathyroid glands were removed at our department, as described above, five of these were present in the thymus. Additionally, the parathyroid glands are rich in variety, such as undescended and completely buried in the thyroid.

On the right side, the parathyroid gland and thyroid tongue are excised using the same procedure as on the left side, and 10 min after the excision, intact PTH measurements are taken. If the intact PTH level is reduced to one-third of that before removal, no further search of the supernumerary gland is performed. After confirming that the resected specimen is the parathyroid gland using rapid pathological diagnosis, a total of 90 mg is cut into small pieces in order from the gland closest to normal, and autotransplanted into the forearm muscle. If the RTx has been planned, the weight of the transplanted gland increases to 120 mg.

3.7.2 An algorithm to identify parathyroid glands

Identification of the parathyroid glands is crucial for a successful parathyroid operation. To achieve this, it is essential for parathyroid surgeons to be familiar with the anatomy of the parathyroid glands and their variations. When searching for the parathyroid gland, it is important to search in order from the most frequently present site. It is difficult to identify the parathyroid glands and avoid missing glands. Therefore, we propose an algorithm for finding all parathyroid glands (Hibi et al., 2002).

By dividing the area around the thyroid gland into Areas A–F by the RLN and the landmark, failure to find the parathyroid gland could be minimized. Area A is superior from the landmark and dorsal from the RLN; Area B: inferior from the landmark and dorsal from the RLN; Area C: superior from the landmark and ventral from the RLN; Area D: inferior from the landmark and ventral from the RLN; Area E: the paratracheal area including the thyrothymic ligament and thymic tongue; Area F: superior from the upper pole of the thyroid lobe (Fig. 15).

The superior parathyroid gland is most frequently located in area A. If the gland cannot be identified in Area A, then Area B should be searched. The superior gland may sometimes fall into the paraoesophageal area and may extend caudally toward the posterior mediastinum (Fig. 15).

Taping the inferior thyroid artery and searching area B may help identify the superior parathyroid gland. The hilus of the gland is located in the same position as the normal gland and can be easily ligated. Occasionally, a significantly enlarged superior parathyroid gland can be found in the posterior mediastinum, but since there is no inflow of blood vessels from the caudal side, the hilus will be reached when tracing to the cranial side. If the gland cannot be detected, Area C should be searched. Additionally, there may be a superior parathyroid gland around the Berry's ligament, where the RLN enters the larynx. To find this gland, the superior thyroid arteries and veins need to be treated, and the superior thyroid pole must be translocated.

Most frequently, the inferior parathyroid glands are located in Area D. If the inferior gland cannot be found, Area E should be searched. In our group, thymic tongue removal is a routine procedure (Fig. 16). If an undescended thymus can be detected, the thymic tissue must be carefully resected because the inferior gland can sometimes be located at the top of the undescended thymic tongue (Fig. 15).

The surrounding carotid artery is dissected to open the carotid sheath. If the parathyroid gland cannot be found, it may be an intrathyroidal parathyroid gland, in which case a partial thyroidectomy is required. Using this algorithm, it is possible to find four or more parathyroid glands in 96% of cases.

3.7.3 Intraoperative intact PTH monitoring

3.7.3.1 The necessity of intraoperative PTH monitoring

It is necessary to remove all the parathyroid glands (PTGs) while performing PTX for renal hyperparathyroidism because remnant PTGs can be stimulated in patients with end-stage renal disease and may cause persistent or

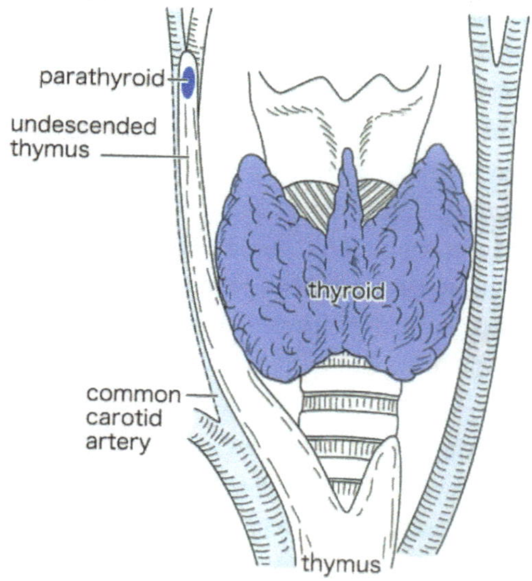

Fig. 16 Undescended parathyroid gland. The undescended thymus with parathyroid gland located on top of the undescended thymic tongue.

recurrent renal hyperparathyroidism (Keutman, Sauer, Hendy, O'Riodan, & Potts, 1978). Re-PTX in the neck for persistent or recurrent renal hyperparathyroidism can be a risk factor for recurrent laryngeal nerve injury due to the presence of adhesions. However, since the size and number of PTGs differ among individuals, intraoperative confirmation of complete PTG removal can be difficult. Intraoperative PTH monitoring helps increase the success rates of PTX. During intraoperative PTH monitoring, preoperative PTH level less than the cut-off value after removing the PTGs suggests complete removal of the PTGs.

3.7.4 Measurement of intact PTH

PTH is a single chain polypeptide hormone that consists of 84 amino acids (Segre, D'Amour, Hultman, & Potts, 1981). PTH is metabolized into multiple fragments in the liver and the kidneys. Whole PTH, which is described as 1–84 PTH, is a biologically active hormone. The 2nd generation intact PTH assay kit currently available helps measure the 1–84 PTH level using enzyme-linked immunosorbent assay (ELISA) with a one-step sandwich method. However, this assay may lack specificity due to cross-reactivity of the antibodies with 7–84 PTH. The intact PTH assay is suitable for intraoperative monitoring as the assay duration is only 15–20 min. The 7–84 PTH fragment is mainly excreted in the urine. The blood concentration of 7–84 PTH differs among individuals (Brossard et al., 1996). In patients with end-stage renal disease, the blood concentration of 7–84 PTH is higher than healthy individuals due to its poor via the kidneys (Carneiro, Solorzano, Nader, Ramirez, & Irvin, 2003; Tassone, Gianotti, Emmolo, Ghio, & Borreta, 2009). The pharmacological half-life of 7–84 PTH is several hours and is longer than 1–84 PTH, which has a pharmacological half-life of approximately 3–4 min. These factors imply that it is difficult to establish a cut-off value for predicting successful PTX for renal hyperparathyroidism, similar to the Miami criteria for PTX for primary hyperparathyroidism. In the Miami criteria, an intact PTH level 10 min after PTX, i.e., <50% of the preoperative intact PTH level, indicates successful PTX (complete removal of pathologic parathyroid glands). The sensitivity, specificity, and accuracy of the Miami criteria have been shown to be excellent at 97.6%, 93.3%, and 97.3%, respectively (Barczynski, Konturek, Hubalewska-Dydejczyk, Cichon, & Nowak, 2009; Irvin, Dembrow, & Prudhomme, 1993). This criterion was established using the Quick PTH assay kit (Nichols Institute Diagnostics, San Juan Capistrano, CA, USA), which makes it possible to measure 1–84 PTH concentration within 15 min (Matsuoka et al., 2007). Several studies have been performed to establish the intraoperative PTH criteria for successful PTX for renal hyperparathyroidism (Hiramitsu, Tominaga, Okada, Yamamoto, & Kobayashi, 2015; Ikeda, Kurihara, Morita, Miyabe, & Takami, 2007; Lokey et al., 2000). Unfortunately, the above Quick PTH assay kit is no longer available in the market. The assay kit currently available for intraoperative monitoring apart from the Quick PTH assay kit only helps determine intact PTH. Despite the difficulties mentioned, some studies have reported using intact PTH assay kits during PTX for renal hyperparathyroidism (Xu et al., 2021; Zhang et al., 2016).

Recently, a 3rd third-generation intact PTH assay kit that could measure 1–84 PTH alone within 20 min without the cross-reactivity of 7–84 PTH has become available in the market. This new kit is expected to offer more accurate intraoperative intact PTH monitoring, even in patients with end-stage renal disease.

3.7.5 Intraoperative intact PTH monitoring for renal hyperparathyroidism

The established timing for blood sampling and the cut-off values for intraoperating intact PTH monitoring for renal hyperparathyroidism vary various reports. One report mentioned a 70% decrease in intact PTH

10 min after total PTX, and transcervical thymectomy demonstrated excellent sensitivity (97.5%), specificity (52.2%), and accuracy (92.9%) (Zhang et al., 2016). Another report mentioned an 88.9% decrease in intact PTH levels 20 min after PTX (without thymectomy) with a sensitivity of 78.6% and a specificity of 88.5% (Xu et al., 2021). A recent study showed the superiority of the 3rd generation intact PTH assay kits over the 2nd generation intact PTH assay kit for intraoperative intact PTH monitoring. The study reported an 87.5% decrease in the 3rd generation intact PTH assay 20 min after total PTX, and transcervical thymectomy demonstrated a sensitivity of 84.0% and specificity of 94.7% (Miki, Sumitomo, Inoue, Kita, & Monden, 1996).

3.8 Parathyroid carcinoma

Parathyroid carcinoma arising in renal hyperparathyroidism is very rare, and its diagnosis and treatment are challenging (Keutmann, Niall, O'Riordan, & Potts, 1975). The involvement of surrounding tissues (e.g., thyroid gland, muscles, nerves, and esophagus) and the occurrence of cervical lymph node and distant metastases (lung, liver, and bone) histologically diagnosed as parathyroid gland tissues are indicators of parathyroid carcinoma. Histopathologic findings are evaluated based on the criteria of Castleman and Schantz as follows: the presence of a fibrous capsule or trabeculae, a trabecular or rosette-like cellular architecture, and nuclear mitoses and the invasion of the capsule or vasculature (Tominaga & Rundolph, 2021). Many cases are difficult to diagnose based on histopathologic findings alone. Unlike other endocrine tumors in which hormone production decreases with malignant transformation, parathyroid carcinoma causes severe elevation of serum calcium levels and renal and bone symptoms (Schantz & Castleman, 1973).

Surgical resection is considered the only effective treatment for parathyroid carcinoma. The most common site of invasion is the ipsilateral thyroid lobe (Koea & Shaw, 1999), and if parathyroid carcinoma is assumed, en bloc resection that includes the ipsilateral thyroid lobe should be considered, although complete resection is difficult in many cases (Hundahl, Fleming, Fremgen, & Menck, 1999). The reported 5- and 10-year survival rates for parathyroid carcinoma are 85% and 49%, respectively (Schantz & Castleman, 1973). Regarding prognosis, overall survival and recurrence-free survival times have been shown to be longer for patients who undergo extensive en bloc surgery than for those who undergo tumor resection alone (Koea & Shaw, 1999; Ricci et al., 2012). In the presence of lymph node metastasis, en bloc resection of the parathyroid carcinoma, as well as the ipsilateral thyroid lobe, is recommended. However, it should be noted that prophylactic lateral neck dissection has not been shown to improve survival and may only increase the occurrence of adverse events; thus, it is not recommended at this time (Machado & Wilhelm, 2019).

4. Renal bone disease

PTH is an important factor regulating bone microarchitecture, tissue quality and bone mass in CKD patients. In general, intact PTH and whole PTH (PTH (1–84)) are adopted to evaluate parathyroid function in these subjects (Goodman et al., 1994; Kazama et al., 2004; Machado & Wilhelm, 2019; Quarles, Lobaugh, & Murphy, 1992; Slatopolsky et al., 2000; Yajima et al., 2003).

Intact serum PTH was measured using an immunoradiometric assay (Allegro Intact PTH; Nichols Institute Diagnostics, San Juan Capistrano, CA, USA) or an electrochemiluminescence immunoassay (Elecsys PTH; Roche Diagnostics, GmbH, Mannheim, Germany), which represents a combined measurement of both PTH (1–84) and PTH (7–84). PTH (1–84) is measured by a new immunoradiometric (IRMA) assay (Scantibodies Labolatories, Santee, CA, USA) (Slatopolsky et al., 2000). PTH (7–84) is secreted from urine, and accumulates in patients with end-stage renal disease (Kazama et al., 2004; Slatopolsky et al., 2000). PTH (1–84) activates bone turnover, but PTH (7–84) has antagonistic effects on PTH (1–84) in bone (Slatopolsky et al., 2000). Skeletal resistance is more severe and common in CKD patients, who are treated with continuous ambulatory peritoneal dialysis (CAPD) (Tanno et al., 2003) and those suffering from diabetes (Bover et al., 2014). In addition, racial differences in skeletal resistance must be taken into consideration in these patients (Jorgetti, dos Reis, & Ott, 2014).

Increased multinucleated osteoclasts (Fig. 17), woven bone, fibrous tissue, increased mononuclear and cuboidal osteoblasts and increased bone formation rate are the characteristics of hyperparathyroid bone disease caused by renal hyperparathyroidism (Malluche & Faugere, 1990). Tartrate resistant acid phosphatase (TRAP) staining and Cathepsin K staining are used to detect osteoclasts clearly (Nakamura, Aoyama, Yamaguchi, & Sasano, 2021). Both bone turnover and osteocytic perilacunar/canalicular turnover are generally increased in patients with renal hyperparathyroidism (Malluche & Faugere, 1990; Parfitt, 1998; Yajima et al., 2018). By the way, TRAP staining for osteocytes yielded negative results in maintenance

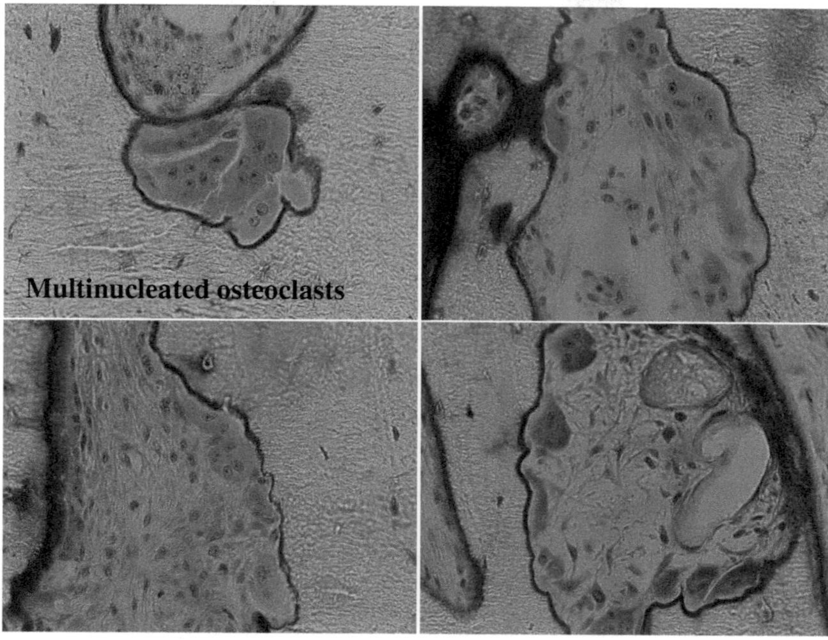

Fig. 17 Multinucleated osteoclasts are abundant in patients with renal hyperparathyroidism.

hemodialysis (HD) patients, probably because osteocytes had been exposed for an extended period of time to moderate or high PTH levels, resulting in downregulation of osteocytic TRAP expression in these patients. So bone resorption by osteocytes was determined according to the shape of the lacunar walls in these patients.

Cortical bone accounts for about 75% of total bone volume but only about 25% of bone surface in non-CKD adults (Parfitt, 1998). Fracture generally initiates from cortical bone, but there are few papers on cortical bone in the field of CKD (Jamal, Gilbert, Gordon, & Bauer, 2006; Schober et al., 1998; Yajima, Inaba, Ogawa, et al., 2007; Yajima, Inaba, Tominaga, & Ito, 2007; Yajima et al., 2013). Bone resorption by osteoclasts at the endocortical surface, intracortical area and periosteal surface contribute to reduced cortical bone area (Power et al., 2003; Yajima, Inaba, Ogawa, et al., 2007).

To avoid the severe cortical bone loss caused by cortical thinning and increased pore size, minimodeling on trabecular bone surfaces, which is defined by direct bone apposition to these surfaces without prior bone resorption, may partially compensate for cortical bone loss in patients with renal hyperparathyroidism (Yajima, Inaba, Tominaga, et al., 2007). Minimodeling is abundant when cortical thickness is small and intracortical resorption spaces are enlarged. Severe cortical thinning caused by increased bone resorption at the endocortical surface and increased cortical porosity caused by increased bone resorption at the intracortical resorption spaces are the characteristics of hyperparathyroid bone disease in long-term HD patients. In these patients, minimodeling area at the endocortical surface and that at the intracortical resorption spaces were also increased. Then, minimodeling at the endocortical surface reduces the rate of cortical thinning and expansion of marrow cavity is retarded by that at the intracortical resorption spaces (Yajima, Inaba, Tominaga, et al., 2007). Minimodeling is associated with the formation of only lamellar bone and is initiated at the quiescent surface, therefore, smooth cement lines are observed (Yajima, Inaba, Tominaga, et al., 2007).

4.1 Impact of treatments on hyperparathyroid bone disease

4.1.1 Parathyroidectomy

We treat hyperparathyroid bone disease caused by renal hyperparathyroidism with total parathyroidectomy followed by immediate autotransplantation of 150 mg of diffuse hyperplastic parathyroid tissue (parathyroidectomy), that may or may not be followed by 2.0 μg/day of alfacalcidol administration.

As a result, osteoclast surface (Oc.S/BS) on endocortical, intracortical and cancellous surfaces significantly decreased, whereas-active (Type-2) osteoblast surface with subsequent osteoid formation (Ob.S/BS) increases on cancellous surface 1 week after surgery (Yajima et al., 2003). Decreased Oc.S/BS and increased Ob.S/BS continued at least through 4 week postoperatively. Double labeling is predominant in patients with renal hyperparathyroidism, but single labeling was predominant in patients with severe hypoparathyroidism soon after total parathyroidectomy. Bone formation rate was significantly decreased after the surgery. As a result, it seems that low turnover osteomalacia developed in cancellous bone soon after parathyroidectomy (Malluche & Faugere, 1990; Yajima, Inaba, Tominaga, & Ito, 2008; Yajima et al., 2010, 2001, 2019).

Oc.S/BS decreased at endocortical surface and intracortical surface, but did not change at periosteal surface 1 week after parathyroidectomy. And Ob.S/BS did not increase at the three surfaces, including endocortical, intracortical and periosteal surfaces in cortical bone 1 week after parathyroidectomy. At 4 weeks after the surgery, Oc.S/BS decreased and Ob.S/BS increased at endocortical and intracortical surfaces 4 weeks after parathyroidectomy. But bone loss is irreversible around the periosteal surface, therefore, special attention should be paid to the cortical bone loss in dialysis patients. Therefore, it was revealed that improvement of high turnover bone disease in cortical bone lagged behind that in cancellous bone. It is likely that cortical bone loss is irreversible even after parathyroidectomy although it is possible that cortical bone loss can be reversed over a long period of observation (Wang et al., 2019; Yajima, Inaba, Ogawa, et al., 2007).

In the findings published by the papers (Yajima et al., 2010, 2019), increased osteocyte death was followed by increased mineralization by osteocytes if 2.0 μg/day of alfacalcidol is orally administered after parathyroidectomy (Fig. 18) and hypomineralized bone area was reduced in these patients (Marie & Glorieux, 1983). But mineralization by osteocytes was not well observed if vitamin D was not administered after the surgery. Vitamin D receptors on the osteocytes have been already found and reported (Lam et al., 2014), however, it has not been detected that vitamin D administration induces mineralization by osteocytes. Serum vitamin D levels of dialysis

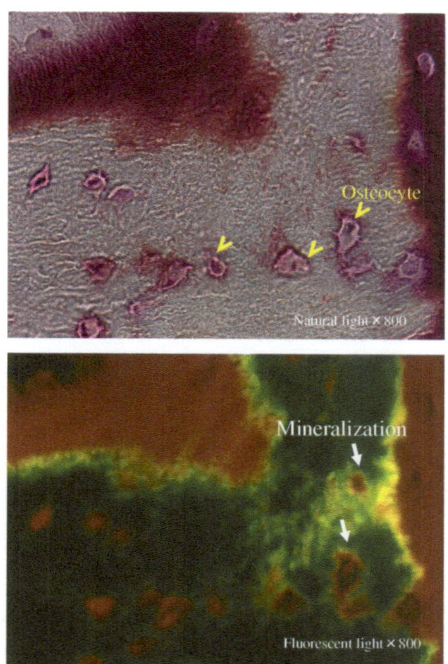

Fig. 18 The acute and marked mineralization on the osteocyte lacunar wall was demonstrated 1 week after parathyroidectomy in hemodialysis patients receiving 2.0 μg/day of alfacalcidol.

patients are extremely low and biologically active vitamin D was administered to the patients. As a result, it was revealed that mineralization was activated by osteocytes, but not osteoblasts.

4.2 Bone disease caused by long-term hypoparathyroidism

Fracture risk is high in both the patients with renal hyperparathyroidism and hypoparathyroiduism (Coco & Rush, 2000; Danese et al., 2006). Micropetrosis was defined and reported decades ago (Frost, 1960).

Micropetrosis is described as small regions of highly mineralized bone that develop when the osteocytes die, and the osteocyte perilacunar/canalicular system fills with mineral (Fig. 19) (Busse et al., 2010; Carpentier et al., 2012; Schmidt et al., 2018). The osteocyte lacunae are mineralized gradually after osteocyte death due to the reduction of serum PTH in these patients. Perilacunar areas also may become highly mineralized. Micropetrotic bone is predisposed to developing microcracks in these highly mineralized areas, and the combination of hypermineralization and microcracking are thought to be a cause of increased bone fragility (Frost, 1960; Yajima et al., 2021). Calcium, phosphorus and bone fluid stagnate in the canaliculi, which induces development of micropetrosis in CKD patients. We have reported on micropetrosis in maintenance hemodialysis patients (Yajima et al., 2021). Micropetrotic areas measured histomorphometrically were significantly greater in patients with long-term hypoparathyroidism as compared to those in patients with renal hyperparathyroidism. However, because most of the bone is not micropetrotic, the larger areas of normally mineralized bone will mask the presence of localized micropetrosis. Recent evidence showed microcrack accumulation in association with highly mineralized areas of bone devoid of osteocytes in non-CKD subjects. Therefore, special attention should be paid to avoid long-term hypoparathyroidism in hemodialysis patients.

4.3 Osteocytic perilacunar/canalicular turnover

In the past, the osteocyte was thought to be a relatively inert bone cell. However, osteocytic perilacunar/canalicular turnover, which consists of osteolysis followed by mineral deposited on lacunar walls, was demonstrated by normal light microscopy and backscattered electron microscopy in CKD subjects (Bonewald, 2010; Yajima et al., 2010, 2018, 2019). Osteoblasts initially convert to osteoblastic osteocytes and osteoid osteocytes, which subsequently mature. In some mature osteocytes, osteolysis and defective

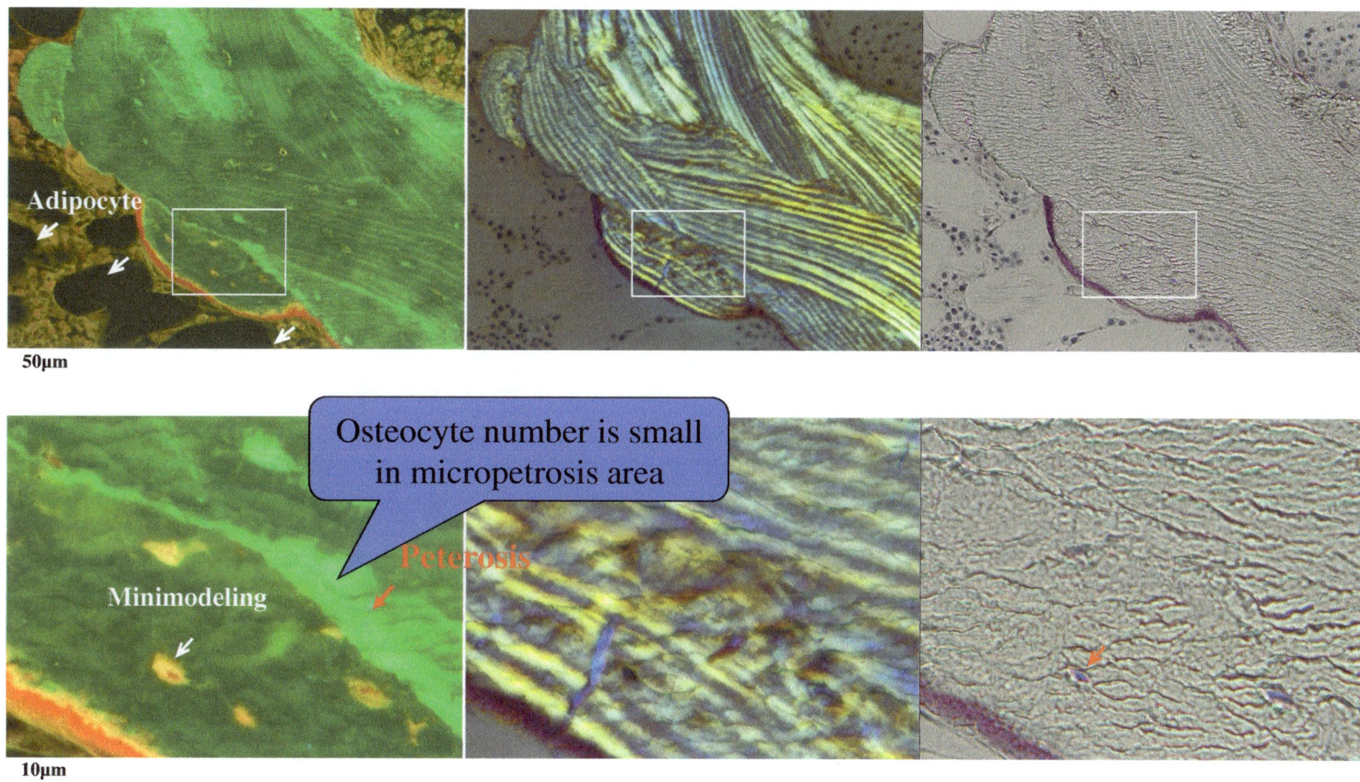

Fig. 19 Micropetrosis and minimodeling were demonstrated in hemodialysis patients with hypoparathyroidism.

mineralization by osteocytes induce the hypomineralized matrix, which is followed by mineralization by osteocytes (Yajima et al., 2018). In addition, PTHrP increases osteocytic osteolysis to release calcium and phosphorus into the circulation in people. Then, it is likely that hypomineralized matrix is observed as a result of erosion of mineralized matrix and/or defective mineralization of the matrix, which means the past demineralization by osteocytes in subjects with and without CKD (Lam et al., 2014; Nango et al., 2016; Yajima et al., 2010, 2018, 2019). Increased PTH and PTHrP of the serum increase osteocytic osteolysis in these subjects. And vitamin D receptor and PTH receptor were detected on the osteocyte (Lam et al., 2014). But whether mineralization by osteocytes is activated if vitamin D sterols are administered in spite of vitamin D expression on the osteocyte has not been demonstrated until our papers have been published (Lam et al., 2014; Nango et al., 2016; Yajima et al., 2010, 2018). In addition, RANKL is expressed by osteocyte to control bone turnover (Nakashima et al., 2011; Wilson, 2011). Sclerostin is also expressed on the osteocyte to suppresses excessive bone formation by osteoblasts (Collette et al., 2012; Ito et al., 2021; Kogawa et al., 2013; Stegen et al., 2018; Wijenayaka et al., 2011). and can also directly stimulate FGF 23 synthesis by osteocytes (Ito et al., 2021). Moreover, sclerostin stimulates osteocyte support of osteoclast activity through a RANKL-dependent pathway (Ito et al., 2021; Wijenayaka et al., 2011).

However, osteocyte maturation is impaired in CKD subjects (Pereira et al., 2018), which may affect these pathways. Recently, we found that mineralization by osteocytes is activated if vitamin D is administered to the patients after or during the treatment of renal hyperparathyroidism. Thus, special attention should be paid in evaluating osteocytic perilacunar/canalicular turnover in these patients. In the research published by *Bone*, we took bone samples from non-CKD patients whose bone remodeling was coupled and observed osteocyte lacunar walls by normal light microscopy and backscattered electron microscopy. The number of both eroded-surface predominant osteocyte lacunae and osteoid-surface predominant osteocyte lacunae was 10× higher in high-PTH group compared to that of these two osteocyte-lacunae in non-CKD group. As a result, it was revealed that osteocytic perilacunar/canalicular turnover in dialysis patients was coupled (Yajima et al., 2018).

We stated characteristics of hyperparathyroid bone disease caused by renal hyperparathyroidism. In addition, we stated pathology of cortical bone, low turnover bone disease and osteocytic perilacunar/canalicular turnover in renal bone disease because few papers regarding these three categories were published in the field of CKD. Overall, pathophysiology of renal hyperparathyroidism, treatment of renal hyperparathyroidism and renal bone disease were described in this review paper.

Acknowledgments

We would like to express our appreciation for the understanding of the Towa Hospital Director, Dr. Tsunamasa Inou and Professor Dr. Haruki Kume at Department of Urology, Graduate School of Medicine, The University of Tokyo, Tokyo, Japan. And we would like to express our appreciation for the contribution to this review paper, especially, Renal Bone Disease from Dr. David B. Burr at Indiana Center for Musculoskeletal Health and Department of Anatomy, Cell Biology and Physiology, Indiana University, School of Medicine, Indianapolis, IN, United States.

References

Akiyama, K. I., Miura, Y., Hayashi, H., Sakata, A., Matsumura, Y., Kojima, M., et al. (2020). Calciprotein particles regulate fibroblast growth factor-23 expression in osteoblasts. *Kidney International*, *97*(4), 702–712. https://doi.org/10.1016/j.kint.2019.10.019.

Barczynski, M., Konturek, A., Hubalewska-Dydejczyk, A., Cichon, S., & Nowak, W. (2009). Evaluation of Halle, Miami, Rome, and Vienna intraoperative iPTH assay criteria in guiding minimally invasive parathyroidectomy. *Langenbeck's Archives of Surgery*, *394*(5), 843–849. https://doi.org/10.1007/s00423-009-0510-z.

Behets, G. J., Spasovski, G., Sterling, L. R., Goodman, W. G., Spiegel, D. M., Broe, D., et al. (2015). Bone histomorphometry before and after long-term treatment with cinacalcet in dialysis patients with secondary hyperparathyroidism. *Kidney International*, *87*(7), 846–856. https://doi.org/10.1038/ki.2014.349.

Block, G. A., Spiegel, D. M., Ehrlich, J., Mehta, R., Lindbergh, J., Dreisbach, A., et al. (2005). Effects of sevelamer and calcium on coronary artery calcification in patients new to hemodialysis. *Kidney International*, *68*(4), 1815–1824. https://doi.org/10.1111/j.1523-1755.2005.00600.x.

Bonewald, L. F. (2010). The amazing osteocyte. *Journal of Bone and Mineral Research*, *26*(2), 229–238. https://doi.org/10.1002/jbmr.320.

Bover, J., Bailone, L., López-Báez, V., Benito, S., Ciceri, P., Galassi, A., et al. (2017). Osteoporosis, bone mineral density and CKD-MBD: Treatment considerations. *Journal of Nephrology*, *30*(5), 677–687, (Review). https://doi.org/10.1007/s40620-017-0404-z.

Bover, J., Urena, P., Brandenburg, V., Goldsmith, D., Ruiz, C., DaSilva, I., et al. (2014). Adynamic bone disease: From bone to vessels in chronic kidney disease. *Seminars in Nephrology, (Review)*, *34*(6), 626–640. https://doi.org/10.1016/j.semnephrol.2014.09.008.

Brossard, J. H., Cloutier, M., Roy, L., Lepage, R., Gascon-Barrĕ, M., & D'Amour, P. (1996). Accumulation of a non-(1-84) molecular form of parathyroid hormone (PTH) detected by intact PTH assay in renal failure: Importance in the interpretation of PTH values. *The Journal of Clinical Endocrinology and Metabolism*, *81*(11), 3923–3929. https://doi.org/10.1210/jcem.81.11.8923839.

Brown, E. M., Gamba, G., Riccardi, D., Lombardi, M., Butters, R., Kifor, O., et al. (1993). Cloning and characterization of an extracellular ca (2+)-sensing receptor from bovine parathyroid. *Nature*, *366*(6455), 575–580. https://doi.org/10.1038/366575a0.

Busse, B., Djonic, D., Milovanovic, P., Hahn, M., Püschel, K., Ritchie, R. O., et al. (2010). Decrease in the osteocyte lacunar density accompanied by hypermineralized lacunar

occlusion reveals failure and delay of remodeling in aged human bone. *Aging Cell, 9*(6), 1065–1075. https://doi.org/10.1111/j.1474-9726.2010.00633.x.
Carneiro, D. M., Solorzano, C. C., Nader, M. C., Ramirez, M., & Irvin, G. L., 3rd. (2003). Comparison of intraoperative iPTH assay (QPTH) criteria in guiding parathyroidectomy: Which criterion is the most accurate? *Surgery, 134*(6), 973–979. discussion 979 https://doi.org/10.1016/j.surg.2003.06.001.
Carpentier, V. T., Wong, J., Yeap, Y., Gan, C., Sutton-Smith, P., Badiel, A., et al. (2012). Increased proportion of hypermineralized osteocyte lacunae in osteoporotic and osteoarthritic human trabecular bone: Implications for bone remodeling. *Bone, 50*(3), 688–694. https://doi.org/10.1016/j.bone.2011.11.021.
Coco, M., & Rush, H. (2000). Increased incidence of hip fractures in dialysis patients with low serum parathyroid hormone. *American Journal of Kidney Diseases, 36*(6), 1115–1121. https://doi.org/10.1053/ajkd.2000.19812.
Collette, N. M., Genetos, D. C., Economides, A. N., Xie, L., Shahnazari, M., Yao, W., et al. (2012). Targeted deletion of Sost distal enhancer increases bone formation and bone mass. *Proceedings of the National Academy of Sciences of the United States of America, 109*(35), 14092–14097. https://doi.org/10.1073/pnas.1207188109.
Danese, M. D., Kim, J., Doan, Q. V., Dylan, M., Griffiths, R., & Chertow, G. M. (2006). PTH and the risks for hip, vertebral, and pelvic fractures among patients on dialysis. *American Journal of Kidney Disease, 47*(1), 149–156. https://doi.org/10.1053/j.ajkd.2005.09.024.
EVOLVE Trial Investigators, Chertow, G. M., Block, G. A., Correa-Rotter, R., Drüeke, T. B., Floege, J., et al. (2012). Effect of cinacalcet on cardiovascular disease in patients undergoing dialysis. *The New England Journal of Medicine, 367*(26), 2482–2494. https://doi.org/10.1056/NEJMoa1205624.
Frost, H. M. (1960). Micropetrosis. *Journal of Bone and Joint Surgery (American), 42-A*, 144–150.
Fukagawa, M., Yokoyama, K., Koiwa, F., Taniguchi, M., Shoji, T., Kazama, J. J., et al. (2013). CKD-MBD guideline working group: Japanese Society for Dialysis Therapy. Clinical practice guideline for the management of chronic kidney disease-mineral and bone disorder. *Therapeutic Apheresis and Dialysis, 17*(3), 247–288. https://doi.org/10.1111/1744-9987.12058.
Fukuda, N., Tanaka, H., Tominaga, Y., Fukagawa, M., Kurokawa, K., & Seino, Y. (1993). Decreased 1,25-dihydroxyvitamin D3 receptor density is associated with a more severe form of parathyroid hyperplasia in chronic uremic patients. *Journal of Clinical Investigation, 92*(3), 1436–1443. https://doi.org/10.1172/JCI116720.
Galitzer, H., Ben-Dov, I. Z., Silvet, J., & Naveh-Many, T. (2010). Parathyroid cell resistance to fibroblast growth factor 23 in secondary hyperparathyroidism of chronic kidney disease. *Kidney International*, 77(3), 211–218. http://dx.doi:https://doi.org/10.1038/ki.2009.464.
Geng, Y., Mosyak, L., Kurinov, I., Zuo, H., Sturchler, E., Cheng, T. C., et al. (2016). Structural mechanism of ligand activation in human calcium-sensing receptor. *eLife, 19*(5), e13662. https://doi.org/10.7554/eLife.13662.
Goodman, W. G. (2004). Calcimimetic agents for the treatment of secondary hyperparathyroidism. *Seminars in Nephrology, 24*(5), 460–463. https://doi.org/10.1016/j.semnephrol.2004.06.006.
Goodman, W. G., Ramirez, J. A., Belin, T. R., Chon, Y., Gales, B., Segre, G. V., et al. (1994). Development of adynamic bone in patients with secondary hyperparathyroidism after intermittent calcitriol therapy. *Kidney International, 46*(4), 1160–1166. https://doi.org/10.1038/ki.1994.380.
Guitérrez, O. M. (2010). Fibroblast growth factor 23 and disordered vitamin D metabolism in chronic kidney disease: Updating the "trade-off" hypothesis. *Clinical Journal of the American Society of Nephrology, 5*(9), 1710–1716. https://doi.org/10.1093/ndt/gfv229.

Hibi, Y., Tominaga, Y., Sato, T., Katayama, A., Haba, T., Uchida, K., et al. (2002). Reoperation for renal hyperparathyroidism. *World Journal of Surgery, 26*(10), 1301–1307. https://doi.org/10.1007/s00268-002-6731-8.

Hiramitsu, T., Tominaga, Y., Okada, M., Yamamoto, T., & Kobayashi, T. (2015). A retrospective study of the impact of intraoperative intact parathyroid hormone monitoring during total parathyroidectomy for secondary hyperparathyroidism: *STARD study. Medicine (Baltimore), 94*(29), e1213. https://doi.org/10.1097.MD.0000000000001213.

Hodsman, A. B., & Steer, B. M. (1992). Serum aluminum levels as a reflection of renal osteodystrophy status and bone surface aluminum staining. *Journal of the American Society of Nephrology, 2*(8), 1318–1327. https://doi.org/10.1681/ASN.V281318.

Hundahl, S. A., Fleming, I. D., Fremgen, A. M., & Menck, H. R. (1999). Two hundred eighty six cases of parathyroid carcinoma treated in the U.S. between 1985-1995. *Cancer, 86*(3), 538–544. https://doi.org/10.1002/(sici)1097-0142(19990801)86:3<538::aid-cncr25->3.0.co;2-k.

Ikeda, Y., Kurihara, H., Morita, N., Miyabe, R., & Takami, H. (2007). The role of quick bio-intact PTH(1-84) assay during parathyroidectomy for secondary hyperparathyroidism. *Journal of Surgical Research, 141*(2), 306–310. https://doi.org/10.1016/j.jss.2006.11.025.

Imanishi, Y., Hosokawa, Y., Yoshimoto, K., Schipani, E., Mallya, S., Papanikolaou, A., et al. (2001). Primary hyperparathyroidism caused by parathyroid-targeted overexpression of cyclin D1 in transgenic mice. *Journal of Clinical Investigation, 107*(9), 1093–1102. https://doi.org/10.1172/JCI10523.

Irvin, G. L., 3rd, Dembrow, V. D., & Prudhomme, D. L. (1993). Clinical usefulness of an intraoperative "quick parathyroid hormone" assay. *Surgery, 114*(6), 1019–1022 (discussion 1022-1023).

Isakova, T., Wahl, P., Vargas, G. S., Orlando, M., Gutiērrez, O. M., Scialla, J., et al. (2011). Fibroblast growth factor 23 is elevated before parathyroid hormone and phosphate in chronic kidney disease. *Kidney International, 79*(12), 1370–1378. https://doi.org/10.1038/ki.2011.47.

Ito, N., Prideaux, M., Wijenayaka, A. R., Yang, D., Ormsby, R. T., Bonewald, L. F., et al. (2021). Sclerostin directly stimulates osteocyte synthesis of fibroblast growth factor-23. *Calcified Tissue International, 109*(1), 66–76. https://doi.org/10.1007/s00223-021-00823-6.

Jadoul, M., Albert, J. M., Akiba, T., Akizawa, T., Arab, L., Bragg-Gresham, J. L., et al. (2006). Incidence and risk factors for hip or other bone fractures among hemodialysis patients in the dialysis outcomes and practice patterns study. *Kidney International, 70*(7), 1358–1366. https://doi.org/10.1038/sj.ki.5001754.

Jamal, S. A., Gilbert, J., Gordon, C., & Bauer, D. C. (2006). Cortical pQCT measures are associated with fractures in dialysis patients. *Journal of Bone and Mineral Research, 21*(4), 543–548. https://doi.org/10.1359/jbmr.060105.

Jamal, S. A., Vandermeer, B., Raggi, P., Mendelssohn, D. C., Chatterley, T., Dorgan, M., et al. (2013). Effect of calcium-based versus non-calcium-based phosphate binders on mortality in patients with chronic kidney disease: An updated systematic review and meta-analysis. *Lancet, 382*(9900), 1268–1277, (Review). https://doi.org/10.1016/S0140-6736(13)60897-1.

Jorgetti, V., dos Reis, L. M., & Ott, S. M. (2014). Ethnic differences in bone and mineral metabolism in healthy people and patients with CKD. *Kidney International, 85*(6), 1283–1289. https://doi.org/10.1038/ki.2013.443.

Kalantar-Zadeh, K., Gutekunst, L., Mehrotra, R., Mehrotra, R., Kovesdy, C. P., Bross, R., et al. (2010). Understanding sources of dietary phosphorus in the treatment of patients with chronic kidney disease. *Clinical Journal of American Society of Nephrology*, 5(3), 519–530, (Review). https://doi.org/10.2215/CJN.06080809.

Kawata, T., Tokunaga, S., Murai, M., Masuda, N., Haruyama, W., Shoukei, Y., et al. (2018). A novel calcimimetic agent, evocalcet (MT-4580/KHK7580), suppresses the

parathyroid cell function with little effect on the gastrointestinal tract or CYP isozymes in vivo and in vitro. *PLoS One*, *13*(4), e0195316. https://doi.org/10.1371/journal.pone.0195316.
Kazama, J. J., Omori, K., Higuchi, N., Takahashi, N., Ito, Y., Maruyama, H., et al. (2004). Intact PTH assay overestimates true 1-84 PTH levels after maxacalcitol therapy in dialysis patients with secondary hyperparathyroidism. *Nephrology, Dialysis, Transplantation*, *19*(4), 892–897. https://doi.org/10.1093/ndt/ghf038.
Keutmann, H. T., Niall, H. D., O'Riordan, J. L., & Potts, J. T., Jr. (1975). A reinvestigation of the amino-terminal sequence of human parathyroid hormone. *Biochemistry*, *14*(9), 1842–1947. https://doi.org/10.1021/bi00680a006.
Keutman, H. T., Sauer, M. M., Hendy, G. N., O'Riodan, L. H., & Potts, J. T., Jr. (1978). Complete amino acid sequence of human parathyroid hormone. *Biochemistry*, *17*(26), 5723–5729. https://doi.org/10.1021/bi00619a019.
Kifor, O., Moore, F. D., Jr., Wang, P., Goldstein, M., Vassilev, V., Kifor, I., et al. (1996). Reduced immunostaining for the extracellular Ca2+sensing receptor in primary and uremic secondary hyperparathyroidism. *Journal of Clinical Endocrinology and Metabolism*, *81*(4), 1598–1606. https://doi.org/10.1210/jcem.81.4.8636374.
Koea, J. B., & Shaw, J. H. (1999). Parathyroid cancer: Biology and management. *Surgical Oncology*, *8*(3), 155–165. https://doi.org/10.1016/s0960-7404(00)00037-7.
Kogawa, M., Wijenayaka, A. R., Ormsby, R. T., Thomas, G. P., Anderson, P. H., Bonewald, L. F., et al. (2013). Sclerostin regulates release of bone mineral by osteocytes by induction of carbonic anhydrase 2. *Journal of Bone and Mineral Research*, *28*(12), 2436–2448. https://doi.org/10.1002/jbmr.2003.
Komaba, H., Goto, S., Fujii, H., Hamada, Y., Kobayashi, A., Shibuya, K., et al. (2010). Depressed expression of klotho and FGFR1 in hyperplastic parathyroid glands from uremic patients. *Kidney Inemational.*, 77(3), 232–238. https://doi.org/10.1038/ki.2009.414.
Kuro-o, M. (2019). The klotho proteins in health and disease. *Nature Reviews Nephrology*, *15*(1), 27–44. https://doi.org/10.1038/s41581-018-0078-3.
Kuro-o, M., Matsumura, Y., Aizawa, H., Kawaguchi, H., Suga, T., Utsugi, T., et al. (1997). Mutation of the mouse klotho gene leads to a syndrome resembling ageing. *Nature*, *390*(6655), 45–51. http//dx.doi.org/10.1038/36285.
LaClaire, R. E., Hellman, R. N., Karp, S. L., Kraus, M., Ofner, S., Li, Q., et al. (2005). Prevalence of calcidiol deficiency in CKD: A cross-sectional study across latitudes in the United States. *American Journal of Kidney Diseases*, *45*(6), 1026–1033. https://doi.org/10.1053/j.ajkd.2005.02.029.
Lam, N. N., Triliana, R., Sawyer, R. K., Atkins, G. J., Morris, H. A., O'Loughlin, P. D., et al. (2014). Vitamin D receptor overexpression in osteoblasts and osteocytes prevents bone loss during vitamin D-deficiency. *Journal of Steroid Biochemistry and Molecular Biology*, *144*(Pt A), 128–131. https://doi.org/10.1016/j.jsbmb.2014.01.002.
Lokey, J., Pattou, F., Mondragon-Sanchez, A., Minuto, M., Mullineris, B., Wambergue, F., et al. (2000). Intraoperative decay profile of intact (1-84) parathyroid hormone in surgery for renal hyperparathyroidism—A consecutive series of 80 patients. *Surgery*, *128*(6), 1029–1034. https://doi.org/10.1067/msy.2000.110431.
Machado, N. N., & Wilhelm, S. M. (2019). Parathyroid cancer: A review. *Cancers (Basel)*, *11*(11), 1676. https://doi.org/10.3390/cancers11111676.
Malluche, H. H., & Faugere, M. C. (1990). Renal bone disease 1990: An unmet challenge for the nephrologist. *Kidney International*, *38*, 193–211 (Review).
Man, N. K., & Jungers, P. (1998). Dialysis therapy. *The New England Journal of Medicine*, *339*(14), 1004 (author reply 1005).
Martin, L. N., Kayath, M. J., Vieira, J. G., & Nosē-Alberti, V. (1998). Parathyroid glands in uraemic patients with refractory hyperparathyroidism: Histopathology and p53 protein expression analysis. *Histopathology*, *33*(1), 46–51. https://doi.org/10.1046/j.1365-2559.1998.00459x.

Marie, P. J., & Glorieux, F. H. (1983). Relation between hypomineralized periosteocytic lesions and bone mineralization in vitamin D-resistant rickets. *Calcified Tissue International*, *35*(4–5), 443–448. https://doi.org/10.1007/BF02405074.

Martin, K. J., Pickthorn, K., Huang, S., Block, G. A., Vick, A., Mount, P. F., et al. (2014). AMG 416 (velcalcetide) is novel peptide for the treatment of secondary hyperparathyroidism in a single-dose study in hemodialysis patients. *Kidney International*, *85*(1), 191–197. https://doi.org/10.1038/ki.2013.289.

Matsuoka, S., Tominaga, Y., Sato, T., Uno, N., Goto, N., Katayama, A., et al. (2007). QuiCk-intra operative bio-intact PTH assay at parathyroidectomy for secondary hyperparathyroidism. *World Journal of Surgery*, *31*(4), 824–831. https://doi.org/10.1007/s00268-006-0601-8.

Metzger, M., Houillier, P., Gauci, C., Haymann, J. P., Flamant, M., Thervet, E., et al. (2013). Relation between circulating levels of 25(OH) vitamin D and parathyroid hormone in chronic kidney disease: Quest for a threshold. *The Journal of Clinical Endocrinology and Metabolism*, *98*(7), 2922–2928. http:/dx.doi.org/10.1210/jc.2013-1294.

Miki, H., Sumitomo, M., Inoue, H., Kita, S., & Monden, Y. (1996). Parathyroid carcinoma in patients with chronic renal failure on maintenance hemodialysis. *Surgery*, *120*(5), 897–901. https://doi.org/10.1016/s0039-6060(96)80101-2.

Nakamura, M., Aoyama, N., Yamaguchi, S., & Sasano, Y. (2021). Expression of tartrate-resistant acid phosphatase and cathepsin K during osteoclast differentiation in developing mouse mandibles. *Biomedical Research*, *42*(1), 13–21. https://doi.org/10.2220/biomedres.42.13.

Nakashima, T., Hayashi, M., Fukunaga, T., Kurata, K., Oh-Hora, M., Feng, J. Q., et al. (2011). Evidence for osteocyte regulation of bone homeostasis through RANKL expression. *Nature Medicine*, *17*(10), 1231–1234. https://doi.org/10.1038/nm.2452.

Nango, N., Kubota, S., Hasegawa, T., Yashiro, W., Momose, A., & Matsuo, K. (2016). Osteocyte-directed bone demineralization along canaliculi. *Bone*, *84*, 279–288. https://doi.org/10.1016/j.bone.2015.12.006.

Navey-Many, T., & Volovelsky, O. (2020). Parathyroid cell proliferation in secondary hyperparathyroidism of chronic kidney disease. *International of Journal of molecular. Science*, *21*(12), 4332, (Review). https://doi.org/10.3390/ijms21124332.

Noori, N., Kalantar-Zadeh, K., Kovesdy, C. P., Bross, R., Benner, D., & Kopple, J. D. (2010). Association of dietary phosphorus intake and phosphorus to protein ratio with mortality in hemodialysis patients. *Clinical Journal of the American Society of Nephrology*, *5*(4), 683–692. https://doi.org/10.2215/CJN.08601209.

Onishi, K., Miyake, M., Hori, S., Onishi, S., Iida, K., Morizawa, Y., et al. (2020). γ-Klotho is correlated with resistance to docetaxel in castration-resistant prostate cancer. *Oncology Letters*, *19*(3), 2306–2316. https://doi.org/10.3892/ol.2020.11308.

Pako, J., Bikov, A., Barta, I., Matsueda, H., Puskas, R., Galffy, G., et al. (2020). Assessment of the circulating klotho protein in lung cancer patients. *Pathology and Oncology Research*, *26*(1), 233–238. https://doi.org/10.1007/s12253-018-0441-5.

Parfitt, A. M. (1998). Perspective: A structural approach to renal bone disease. *Journal of Bone and Mineral Research*, *13*(8), 1213–1220. https://doi.org/10.1359/jbmr.1998.13.8.1213.

Pearce, S. H., Bai, M., Quinn, S. J., Kifor, O., Brown, E. M., & Thakker, R. V. (1996). Functional characterization of calcium-sensing receptor mutations expressed in human embryonic kidney cells. *Journal of Clinical Investigation*, *15: 98(8)*, 1860–1866. https://doi.org/10.1172/JCI118987.

Pereira, R. C., Salusky, I. B., Roschger, P., Klaushofer, K., Yadin, O., Freymiller, E. G., et al. (2018). Impaired osteocyte maturation in the pathogenesis of renal osteodystrophy. *Kidney International*, *94*(5), 1002–1012. https://doi.org/10.1016/j.kint.2018.08.011.

Petit, T. L. (1985). Aluminum in human dementia. *American Journal of Kidney Diseases*, *6*(5), 313–316. https://doi.org/10.1016/s0272-6386(85)80086-x.

Power, J., Loveridge, N., Lyon, A., Rushton, N., Parker, M., & Reeve, J. (2003). Bone remodeling at the endocortical surface of the human femoral neck: A mechanism for regional cortical thinning in cases of hip fracture. *Journal of Bone and Mineral Research*, *18*(10), 1775–1780. https://doi.org/10.1359/jbmr.2003.18.10.1775.

Quarles, L. D. (2012). Role of FGF23 in vitamin D and phosphate metabolism: Implications in chronic kidney disease. *Nature reviews. Endocrinology*, *17:8(5)*, 276–286. https://doi.org/10.1038/nrendo.2011.218.

Quarles, L. D., Lobaugh, B., & Murphy, G. (1992). Intact parathyroid hormone overestimates the presence and severity of parathyroid-mediated osseous abnormalities in uremia. *The Journal of Clinical Endocrinology and Metabolism*, *75*(1), 145–150. https://doi.org/10.1210/jcem.75.1.1619003.

Raggi, P., Chertow, G. M., Torres, P. U., Csiky, B., Naso, A., Nossuli, K., et al. (2011). The ADVANCE study: A randomized study to evaluate the effects of cinacalcet plus low-dose vitamin D on vascular calcification in patients on hemodialysis. *Nephrology, Dialysis, Transplantation*, *26*(4), 1327–1339. https://doi.org/10.1093/ndt/gfq725.

Ricci, G., Assenza, M., Barreca, M., Liotta, G., Paganelli, L., Serao, A., et al. (2012). Parathyroid carcinoma: The importance of high clinical suspicion for a correct management. *International Journal of Surgical Oncology*, *2012*, 639148. https://doi.org/10.1155/2012/649148.

Rubinstein, T. A., Shahmoon, S., Zigmond, E., Etan, T., Merenbakh-Lamin, K., Pasmanik-Chor, M., et al. (2019). Klotho suppresses colorectal cancer through modulation of the unfolded protein response. *Oncogene*, *38*(6), 794–807. https://doi.org/10.1038/s41388-018-0489-4.

Sabatino, A., Cuppari, L., Stenvinkel, P., Lindholm, B., & Avesani, C. M. (2021). Sarcopenia in chronic kidney disease: What have we learned so far? *Journal of Nephrology*, *34*(4), 1347–1372, (Review). https://doi.org/10.1007/s40620-020-00840-y.

Schantz, A., & Castleman, B. (1973). Parathyroid carcinoma. A study of 70 cases. *Cancer*, *31*(3), 600–605.

Schmidt, F. N., Delsmann, M. M., Mletzko, K., Yorgan, T. A., Hahn, M., Siebert, U., et al. (2018). Ultra-high matrix mineralization of sperm whale auditory ossicles facilitates high sound pressure and high-frequency underwater hearing. *Proceedings of the Royal Society B: Biological Sciences*, *285*(1893). 20181820. https://doi.org/10.1098/rspb.2018.1820.

Schober, H.-C., Han, Z.-H., Foldes, A. J., Shih, M. S., Rao, D. S., Balena, R., et al. (1998). Mineralized bone loss at different sites in dialysis patients: Implications for prevention. *Journal of the American Society of Nephrology*, *9*(7), 1225–1233. https://doi.org/10.1681/ASN.V971225.

Segre, G. V., D'Amour, P. D., Hultman, A., & Potts, J. T., Jr. (1981). Effects of hepatectomy, nephrectomy, and nephrectomy/uremia on the metabolism of parathyroid hormone in the rat. *Journal of Clinical Investigation*, *67*(2), 439–448. (https://doi.org/10.1172.JCI110052).

Shilo, V., Levi, I. M., Abel, R., Mihailovic, A., Wasserman, G., Naveh-Many, T., et al. (2017). *Let-7* and *MicroRNA-148* regulate parathyroid hormone levels in secondary hyperparathyroidism. *Journal of the American Society of Nephrology*, *28*(8), 2353–2363. https://doi.org/10.1681/ASN.2016050585.

Shmulevich, R., Nissim, T. B., Wolf, I., Merenbakh-Lamin, K., Fishman, D., Sekler, I., et al. (2020). Klotho rewires cellular metabolism of breast cancer cells through alteration of calcium shuttling and mitochondrial activity. *Oncogene*, *39*(24), 4636–4649. https://doi.org/10.1038/s41388-020-1313-5.

Silver, T., & Naveh-Manny, T. (2011). FGF23 and the parathyroid glands. *Pediatric Nephrology*, *25*(11), 2241–2245. https://doi.org/10.1007/s00467-010-1565-3.

Slatopolsky, E., Finch, J., Clay, P., Sicard, G., Singer, G., Gao, P., et al. (2000). A novel mechanism for skeletal resistance in uremia. *Kidney International*, *58*(2), 753–761. https://doi.org/10.1046/j.1523-1755.2000.00222.x.

Slatopolsky, E., Finch, J., Denda, M., Ritter, C., Zhong, M., Dusso, A., et al. (1996). Phosphorus restriction prevents parathyroid gland growth. High phosphorus directly stimulates PTH secretion in vitro. *Journal of Clinical Investigation, 97*(11), 2534–2540. https://doi.org/10.1172/JCI118701.

Stegen, S., Stockmans, I., Moermans, K., Thienpont, B., Maxwell, P. H., & Carmeliet, P. (2018). Osteocytic oxygen sensing controls bone mass through epigenetic regulation of sclerostin. *Nature Communications, 9*(1), 2557. https://doi.org/10.s41467-018-04679-7.

Tanno, Y., Yokoyama, K., Nakayama, M., Katoh, A., Yamamoto, H., Iwasaki, K., et al. (2003). IRMA (whole PTH) is a more useful assay for the effect of PTH on bone than the allegro intact PTH assay in CAPD patients with low bone turnover marker. *Nephrology, Dialysis, Transplantation, 18*(Suppl. 3), iii97-98. https://doi.org/10.1093/hdt/gfg1025.

Tassone, F., Gianotti, L., Emmolo, I., Ghio, M., & Borreta, G. (2009). Glomerular filtration rate and parathyroid hormone secretion in primary hyperparathyroidism. *Journal of Clinical Endocrinology and Metabolism, 94*(11), 4458–4461. http://dx.doi.10.1210/jc.2009-0587.

Tominaga, Y., Kakuta, T., Yasunaga, C., Nakamura, M., Kadokura, Y., & Tahara, H. (2016). Evaluation of parathyroidectomy for secondary and tertiary hyperparathyroidism by the parathyroid Surgeons' Society of Japan. *Therapeutic Apheresis and Dialysis, 20*(1), 6–11. https://doi.org/10.1111/1744-9987.12352.

Tominaga, Y., Kohara, S., Namii, Y., Nagasaka, T., Haba, T., Uchida, K., et al. (1996). Clonal analysis of nodular parathyroid hyperplasia in renal hyperparathyroidism. *World Journal of Surgery, 20*(7), 744–750, discussion 750-752. https://doi.org/10.1007/s002689900113.

Tominaga, Y., & Rundolph, G. W. (Eds.). (2021). *Surgical treatment of secondary and tertiary hyperparathyroidism. Surgery of thyroid and parathyroid glands* (3rd ed., pp. 564–575).

Tominaga, Y., Sato, K., Tanaka, Y., Numano, M., Uchida, K., & Takagi, H. (1995). Histopathology and pathophysiology of secondary hyperparathyroidism due to chronic renal failure. *Clinical Nephrology, 44*, S42–S47.

Tominaga, Y., Tanaka, Y., Sato, K., Numano, M., Uchida, K., Falkmer, U., et al. (1992). Recurrent renal hyperparathyroidism and DNA analysis of autografted parathyroid tissue. *World Journal of Surgery, 16*(4), 595–602, discussion 602-603. https://doi.org/10.1007/BF02067331.

Tominaga, Y., Tsuzuki, T., Uchida, K., Haba, T., Otsuka, S., Ichimori, T., et al. (1999). Expression of PRAD1/cyclinD1, retinoblastoma gene products, and Ki67 in parathyroid hyperplasia caused by chronic renal failure versus primary adenoma. *Kidney International, 55*(4), 1375–1383. https://doi.org/10.1046/j.1523-1755.1999.00396.x.

Torres, P. U., & Broe, M. D. (2012). Calcium-sensing receptor, calcimimetics, and cardiovascular calcifications in chronic kidney disease. *Kidney International*, 82(1), 19–25. (Review). https://doi.org/10.1038/ki.2012.69.

Torres, P. U., Prié, D., Beck, L., De Brauwere, D., Leroy, C., & Friedlander, G. (2009). Klotho gene, phosphocalcic metabolism, and survival in dialysis. *Journal of Renal Nutrition, 19*(1), 50–56. https://doi.org/10.1053/j.jrn.2008.10.018.

Wang, X., Matthews, B. G., Yu, J., Novak, S., Grcevic, D., Sanjay, A., et al. (2019). PDGF modulates BMP2-induced osteogenesis in periosteal progenitor cells. *JBMR Plus, 15*(3), e10127. https://doi.org/10.1002/jbm4.10127.

Wheeler, D. C., London, G. M., Parfrey, P. S., Block, G. A., Correa-Rotter, R., Dehmel, B., et al. (2014). Effects of cinacalcet on atherosclerotic and nonatherosclerotic cardiovascular events in patients receiving hemodialysis: The evaluation of cinacalcet HCl therapy to lower cardio vascular events (EVOLVE) trial. *Journal of the American Heart Association, 3*(6), e001363. https://doi.org/10.1161/JAHA.114.001363.

Wijenayaka, A. R., Kogawa, M., Lim, H. P., Bonewald, L. F., Findlay, D. M., & Atkins, G. J. (2011). Sclerostin stimulates osteocyte support of osteoclast activity by a RANKL-dependent pathway. *PLoS One, 6*(10), e25900. https://doi.org/10.1371/journal.pone.0025900.

Wilson, C. (2011). Osteocytes, RANKL and bone loss. *Nature Reviews. Endocrinology*, 7(12), 693, (Comment). https://doi.org/10.1038/nrendo.2011.176.

Xu, F., Huang, Y., Zeng, M., Zhang, L., Ren, W., Qian, H., et al. (2021). Diagnostic values of intraoperative (1-84) parathyroid hormone levels are superior to intact parathyroid hormone for successful parathyroidectomy in patients with chronic kidney disease. *Endocrine Practice*, *27*(11), 1065–1071. 2100163-4: S1530-891X. https://doi.org/10.1016/j.eprac.2021.04.006.

Yajima, A., Inaba, M., Ogawa, Y., Tominaga, Y., Tanizawa, T., Inou, T., et al. (2007). Significance of time-course changes of serum bone markers after parathyroidectomy in patients with uraemic hyperparathyroidism. *Nephrology, Dialysis, Transplantation*, *22*(6), 1645–1657. https://doi.org/10.1093/ndt/gfl837.

Yajima, A., Inaba, M., Tominaga, Y., & Ito, A. (2007). Minimodeling reduces the rate of cortical bone loss in patients with secondary hyperparathyroidism. *American Journal of Kidney Diseases*, *49*(3), 440–451. https://doi.org/10.1053/j.ajkd.2006.11.045.

Yajima, A., Inaba, M., Tominaga, Y., & Ito, A. (2008). Bone formation by minimodeling is more active than remodeling after parathyroidectomy. *Kidney International*, *74*(6), 774–781. https://doi.org/10.1038/ki.2008.242.

Yajima, A., Inaba, M., Tominaga, Y., Nishizawa, Y., Ikeda, K., & Ito, A. (2010). Increased osteocyte death and mineralization inside bone after parathyroidectomy in patients with secondary hyperparathyroidism. *Journal of Bone and Mineral Research*, *25*(11), 2374–2381. https://doi.org/10.1002/jbmr.126.

Yajima, A., Inaba, M., Tominaga, Y., Tanaka, M., Otsubo, S., Nitta, K., et al. (2013). Impact of lanthanum carbonate on cortical bone in dialysis patients with adynamic bone disease. *Therapeutic Apheresis and Dialysis*, *17*, 41–48. https://doi.org/10.1111/1744-9987.12038.

Yajima, A., Ogawa, Y., Takahashi, H. E., Tominaga, Y., Inou, T., & Otsubo, O. (2003). Changes of bone remodeling immediately after parathyroidectomy for secondary hyperparathyroidism. *American Journal of Kidney Diseases*, *42*(4), 729–738. https://doi.org/10.1016/s0272-6386(03)00909-0.

Yajima, A., Tanaka, K., Tominaga, Y., Ogawa, Y., Tanizawa, T., Inou, T., et al. (2001). Early changes of bone histology and circulating markers of bone turnover after parathyroidectomy in hemodialysis patients with severe hyperparathyroidism. *Clinical Nephrology*, *56*(1), 27–34.

Yajima, A., Tsuchiya, K., Bonewald, L. F., Inaba, M., Tominaga, Y., Tanizawa, T., et al. (2017). Case report: Electron microscopic evaluation of bone from a patient treated with cinacalcet hydrochloride, maxacalcitol, and alfacalcidol for hyperparathyroid bone disease with secondary hyperparathyroidism. *Osteoporosis International*, *29*(5), 1203–1209. https://doi.org/10.1007/s00198-018-4402-3.

Yajima, A., Tsuchiya, K., Burr, D. B., Minner, D. E., Condon, K. W., Miller, C. A., et al. (2018). Osteocytic perilacunar/canalicular turnover in hemodialysis patients with high and low serum PTH levels. *Bone*, *113*, 68–76. https://doi.org/10.1016/j.bone.2018.05.002.

Yajima, A., Tsuchiya, K., Burr, D. B., Murata, T., Nakamura, M., Inaba, M., et al. (2021). Micropetrosis in hemodialysis patients. *Bone Reports*, *15*, 101150. https://doi.org/10.1016/j.bonr.2021.101150.

Yajima, A., Tsuchiya, K., Burr, D. B., Wallace, J. M., Damrath, J. D., Inaba, M., et al. (2019). The importance of biologically active vitamin D for mineralization by osteocytes after parathyroidectomy for renal hyperparathyroidism. *JBMR Plus*, *3*(11), e10234. https://doi.org/10.1002/jbm4.10234.

Zhang, L., Xing, C., Shen, C., Zeng, M., Yang, G., Mao, H., et al. (2016). Diagnostic accuracy study of intraoperative and perioperative serum intact PTH level for successful parathyroidectomy in 501 secondary hyperparathyroidism patients. *Scientific Reports*, *27*(6), 26841. https://doi.org/10.1038/srep27918.

Zitt, E., Fouque, D., Jacobson, S. H., Malberti, F., Ryba, M., Ureña, P., et al. (2013). Serum phosphorus reduction in dialysis patients treated with cinacalcet for secondary hyperparathyroidism results mainly from parathyroid hormone reduction. *Clinical Kidney Journal, 6*(3), 287–294. https://doi.org/10.1093/ckj/sft026.

Further reading

Bonewald, L. F. (2007). (Review) Osteocytes as dynamic multifunctional cells. *Annals of the New York Academy of Sciences, 1116*, 281–290. https://doi.org/10.1196/annals.1402.018.

Taniguchi, M., Tanaka, M., Hamano, T., Nakanishi, S., Fujii, H., Kato, H., et al. (2011). Comparison between whole and intact parathyroid hormone assays. *Therapeutic Apheresis and Dialysis, 15*(Supplement 1), 42–49. https://doi.org/10.1111/j.1744-9987.2011.00926.x.

CHAPTER THIRTEEN

Primary cilia and PTH1R interplay in the regulation of osteogenic actions

Juan A. Ardura*, Eduardo Martín-Guerrero, Sara Heredero-Jiménez, and Arancha R. Gortazar
Bone Physiopathology Laboratory, Department of Basic Medical Sciences, CEU San Pablo University, CEU Universities, Madrid, Spain
*Corresponding author: e-mail address: juanantonio.ardurarodriguez@ceu.es

Contents

Abstract

Primary cilia are subcellular structures specialized in sensing different stimuli in a diversity of cell types. In bone, the primary cilium is involved in mechanical sensing and transduction of signals that regulate the behavior of mesenchymal osteoprogenitors, osteoblasts and osteocytes. To perform its functions, the primary cilium modulates a plethora of molecules including those stimulated by the parathyroid hormone (PTH) receptor type I (PTH1R), a master regulator of osteogenesis. Binding of the agonists PTH or PTH-related protein (PTHrP) to the PTH1R or direct agonist-independent stimulation of the receptor activate PTH1R signaling pathways. In turn, activation of PTH1R

Vitamins and Hormones, Volume 120
ISSN 0083-6729
https://doi.org/10.1016/bs.vh.2022.04.001

leads to regulation of bone formation and remodeling. Herein, we describe the structure, function and molecular partners of primary cilia in the context of bone, playing special attention to those signaling pathways that are mediated directly or indirectly by PTH1R in association with primary cilia during the process of osteogenesis.

1. Introduction to primary cilia structure, biogenesis, intra-ciliary transport and molecular mechanotransducers in bone

The primary cilium is a nonmotile organelle based on microtubules that extends as a single unit from the cell surface of most mammalian cell types (Satir & Christensen, 2007). Cells that present primary cilia include lung, kidney and heart cells (Li, Jin, & Jin, 2019), neurons (Ma, Peterson, & Turnbull, 2011), chondrocytes (Yuan, Serra, & Yang, 2015), mononuclear cells isolated from peripheral blood and bone marrow (Singh, Chaudhry, & Merchant, 2016) and bone cells (Martín-Guerrero et al., 2020). Primary cilia are assembled and disassembled in each division cycle of the cell upon regulation by different stimuli and signaling pathways (Anvarian, Mykytyn, Mukhopadhyay, Pedersen, & Christensen, 2019).

Primary cilia structure is composed of a core of primary cilium microtubules surrounded by a ciliary membrane that is continuous with the plasma membrane and presents a unique lipid and receptor composition, enabling the cilium to detect signals from the extracellular environment. The information received by molecular detectors at the ciliary membrane is transmitted to intracellular molecules in order to regulate diverse developmental and physiological cell processes such as proliferation or differentiation (Wheway, Nazlamova, & Hancock, 2018).

The biogenesis of the primary cilium starts in the G0-G1 phase of the cell cycle, when cells enter quiescence. In the case of bone cells, primary cilia biogenesis is probably induced by bone morphogenetic proteins (BMPs) signaling through activation of the BMP type 1 receptor, but it could also occur in response to environmental stimuli such as mechanical stress (Komatsu, Kaartinen, & Mishina, 2011; Nguyen & Jacobs, 2013). Once the cell has entered quiescence, it must go through a series of processes to develop a fully functional cilium (Uzbekov et al., 2012). These processes begin with the mother centriole. First, membrane vesicles are recruited to encapsulate the distal end of the mother centriole, and subsequent interactions between the centriole and vesicles are stabilized by the formation of a structure named ciliary necklace. The resulting structure is composed of a basal body that consists in a mature centriole, accessorized by sub-distal appendages, connected to an immature centriole that is surrounded by a dense protein-rich pericentriolar matrix (Basten & Giles, 2013). A primary vesicle is formed and it expands its membrane by fusion of secondary vesicles and simultaneously is deformatted into an invaginated sac, resulting in a double membrane sheath around the apical end that interacts with the transitional fibers of the centriole and provides the matrix for the ciliary pocket. This ciliary pocket along with the ciliary necklace and transitional fibers are known as the transition zone: a structure that acts as a barrier to prevent the free exchange of proteins between the plasma and ciliary membranes (Basten & Giles, 2013). The assembly of the axial filament or axoneme begins with the supply of lipids and proteins into the ciliary membrane to form a structure that contains a ring of nine microtubule doublets (a 9 + 0 axoneme). Finally, the distal tip of the enveloped cilium fuses with the apical plasma membrane, leading to the emergence of the cilium into the extracellular space (Fliegauf, Benzing, & Omran, 2007; Hoey, Downs, & Jacobs, 2012; Rohatgi & Snell, 2010).

The primary cilium is organized as a semi-closed system and does not have any protein synthesis ability. Therefore, every ciliary protein requires to be imported, transported along the cilia and retained (Garcia-Gonzalo & Reiter, 2012). Given these spatial limitations, the transport system necessary to regulate the presence of proteins in the primary cilium is particularly relevant for the appropriate function of the cilium and the activation of several signaling pathways (Martin et al., 2018). There are 22 identified intraflagellar transport (IFT) proteins that modulate the presence of proteins in specific locations along the primary cilium (Martin et al., 2018; Zaghloul & Brugmann, 2011). IFT proteins form two complexes: IFT-A contains six proteins which preferentially bind with dynein-2 to transport proteins from the cilium tip to the cell body; a process known as retrograde transport. In contrast, IFT-B, that contains 16 proteins (subdivided into core and peripheral sub-complexes containing 10 and 6 different IFT particle proteins, respectively), which bind with kinesin-II, transports proteins from the basal body to the cilium tip (anterograde transport) (Martin et al., 2018; Yuan & Yang, 2016). Despite this classification, some IFT-A and IFT-B complexes may participate in ciliary transport in other directions (Eguether et al., 2014). In any case, it can be assumed that signaling pathways of the primary cilium rely on the trafficking of proteins into or out of the cilium mediated by the aforementioned IFT and kinesin/dynein motor proteins (Martin et al., 2018). Among other IFT proteins, Kif3a, an essential subunit of the kinesin II IFT motor protein system, exerts a significant role on both primary cilia formation in bone cells and in bone formation (Qiu et al., 2012; Temiyasathit et al., 2012). On this subject, specific deletion of Kif3a in osteoblasts in vivo causes a reduction in the number and length of primary cilia in these cells. Increased cell proliferation, impaired osteoblastic differentiation and enhanced adipogenesis, as well as reduced intracellular calcium responses to fluid flow shear stress, decreased mineralization and low bone mass were associated with the deletion of Kif3a in osteoblasts (Qiu et al., 2012). In another study, mice with an osteoblast- and osteocyte-specific knockout of Kif3a, exhibited decreased formation of new bone in response to mechanical bone loading compared to control mice (Temiyasathit et al., 2012). These results suggest that IFT protein function may be critical for the regulation of bone physiology due to mechanical loading.

The extracellular location of the cilium makes it susceptible to receive external stimuli, such as morphogenic, olfactory, chemical and mechanical stimuli (Basten & Giles, 2013). Sensitivity to mechanical stimulation is a fundamental property in a tissue subjected to continuous loading such as bone (Middleton, Ma, & You, 2017). There is a wide range of mechanical forces that are part of daily bone loading, including bone deformation, fluid flow shear stress, vibrations and strain (Jacobs, Temiyasathit, & Castillo, 2010). In this regard, the primary cilium has revealed to be an essential mechanosensor that triggers osteogenic responses induced by fluid flow shear stress (Moore, Chen, & Jacobs, 2019). The structure of primary cilia, its association with the extracellular matrix and the localization of mechanotransduction molecules along the membrane of the cilium are crucial elements involved in the mechanical sensing abilities of this organelle (Fig. 1), as we will discuss below.

I. The microarchitecture of the primary cilium seems to be a key feature to translate external stimuli such as fluid flow into cilia deformation and subsequent cellular responses, allowing this organelle to act as a mechanical-sensing structure (Hoey et al., 2012) (Fig. 1A). The axonemes of primary cilia present certain resistance to bending by shear stress thanks to the spatial distribution of the nine microtubule doublets at their axis. However, primary cilia display less resistance to bending than motile cilia due to the lack of radial spokes and nexin links that reinforce the axoneme of motile cilia (Hoey et al., 2012). It has been hypothesized that bending of the cilium contributes to the activation of signaling molecules at the surface of primary cilia membranes, including the opening of ion channels (Hoey et al., 2012).

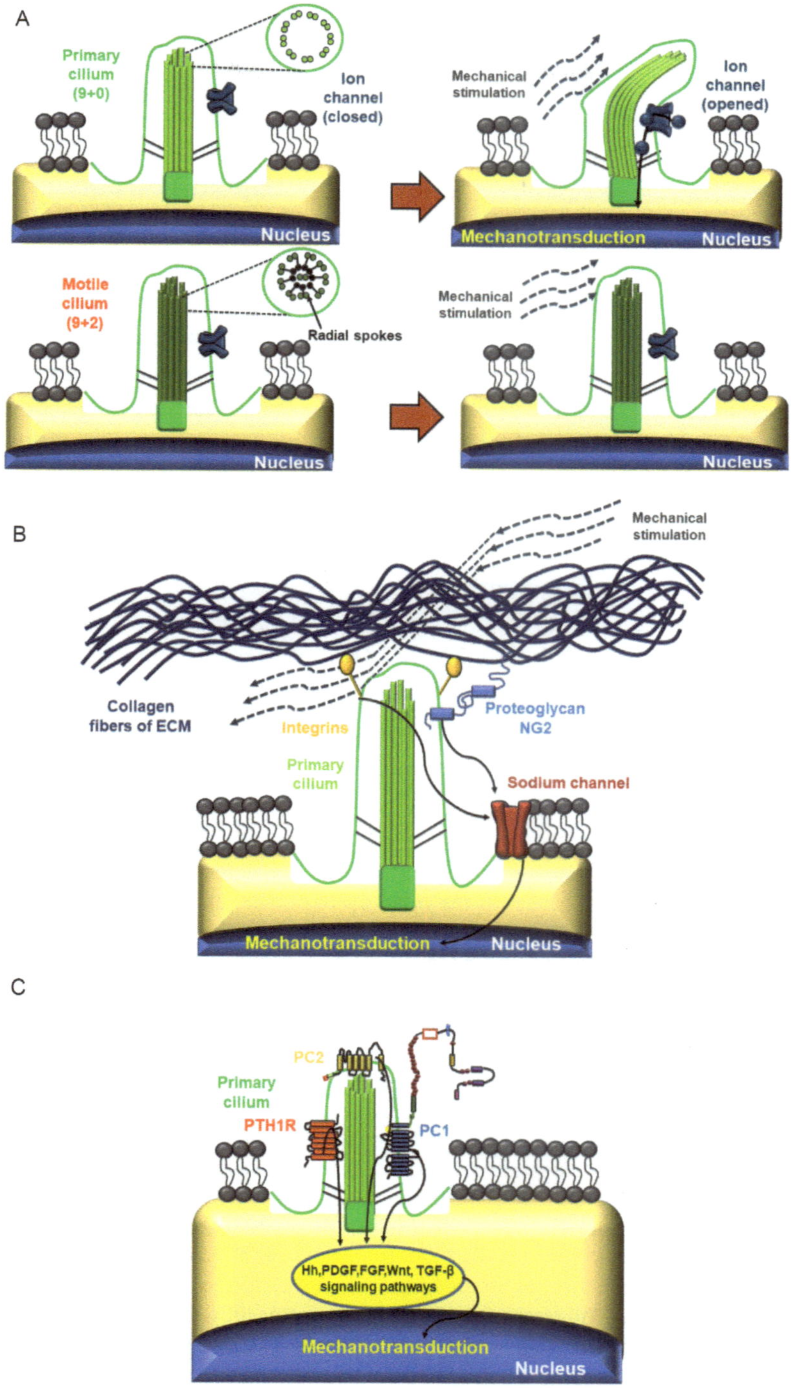

Fig. 1 See figure legend on next page.

If that were the case, a stiff cilium would experience lower membrane strains than a more flexible cilium when exposed to the same bending force. Therefore, a more flexible cilium such as the primary cilium, would be expected to be a greater sensitive mechanotransducer (Hoey et al., 2012).

II. The mechanosensory function of primary cilia also seems to rely on its interactions with the extracellular matrix (ECM) (Fig. 1B). Primary cilia can express at their membrane different type of integrins and NG2 proteoglycan that act as receptors for ECM molecules, allowing the direct binding of primary cilia to oriented collagen fibers of the matrix (McGlashan, Jensen, & Poole, 2006). These observations suggest that ciliary function is mediated, at least in part, by adhesion proteins. Nevertheless, the expression of NEDD9, a protein that localizes to focal adhesions and mediates integrin signaling and cell attachment to matrix, induces primary cilia resorption (Pugacheva, Jablonski, Hartman, Henske, & Golemis, 2007; Seeger-Nukpezah & Golemis, 2012). Moreover, a potential relationship between the cilium and the direction of secretion of ECM components and/or the maintenance of cell shape via cytoskeletal microtubules has also been proposed (Farnum & Wilsman, 2011). In this way, the primary cilium would act transferring mechanical stimuli from particularly oriented ECM to the cell cytoskeleton, subsequently regulating the activity of sodium channels mec-4 and mec-10 (EnaC superfamily) in order to convert mechanical

Fig. 1 Proposed mechanisms for mechanical sensing abilities of primary cilia. (A) The lack of nexins and radial spokes at the distal tip of primary cilia (upper A) make them less resistant to bending than motile cilia (lower A); bending of the cilium has been hypothesized as an activator of several signaling molecules at the surface of ciliary membrane, including ion channels. Microarchitecture of the primary cilium constitutes a key feature to translate external stimuli into subsequent cellular responses. (B) Interactions of ciliary integrins and NG2 proteoglycan with several ECM molecules allow direct binding of primary cilia to oriented collagen fibers of the matrix. Primary cilium would act transferring mechanical stimuli from particularly oriented ECM to the cell cytoskeleton, subsequently regulating the activity of several sodium channels in order to convert mechanical stimuli into biochemical responses. (C) Several molecules localize at the ciliary membrane, including G-coupled protein receptors (GPCRs) (i.e. PTH1R), stretch-activated ion channels (polycystin 2: PC2) and transcription factors (i.e. polycystin1: PC1), can translate mechanical stimuli into intracellular signals. Actions of these ciliary mechanotransducers are modulated by diverse signaling pathways like hedgehog (Hh), platelet-derived growth factor (PDGF), Wnt/β-catenin, transforming growth factor-β (TGF-β), BMP and fibroblast growth factor (FGF) signaling pathways, among others.

stimuli into an electrical response (Hoey et al., 2012). Therefore, the interplay among the ECM and primary cilia with the microtubule cytoskeleton within the cell involves a molecular mechanism that translates extracellular mechanical loading into intracellular responses.

III. In addition, the ciliary membrane contains molecules that can directly translate mechanical external stimuli into intracellular signals, including receptors for G-coupled protein receptors (GPCRs) (e.g. parathyroid hormone (PTH) receptor type 1 (PTH1R) (Martín-Guerrero et al., 2020; Zheng et al., 2018)), stretch-activated ion channels (e.g. polycystin 2 (Nauli et al., 2003)) and molecules that upon mechanical unloading suffer proteolysis and become transcription factors (e.g. polycystin 1 (Low et al., 2006)) (Fig. 1C). Several factors and signaling pathways mediate the actions of the aforementioned molecular mechanotransducers of the primary cilium, including hedgehog (Hh), platelet-derived growth factor (PDGF) (Anderson et al., 2008), Wnt/β-catenin, transforming growth factor-β (TGF-β), BMP and fibroblast growth factor (FGF) signaling pathways, among others.

2. Role of primary cilia in bone

Bone is a dynamic tissue that is continuously being remodeled by the coordinated actions of different bone cell populations (Florencio-Silva, Sasso, Sasso-Cerri, Simões, & Cerri, 2015). Osteocytes are the most abundant population of bone cells (Bonewald, 2007) and possess a network of interconnected dendritic-like processes that allow fast transduction of signals among other osteocytes and with osteoblasts and bone forming-cells derived from mesenchymal progenitors (Hemmatian, Bakker, Klein-Nulend, & van Lenthe, 2017). After differentiation from mesenchymal cells, active osteoblasts induce bone formation by controlling both synthesis and mineralization of bone extracellular matrix. In this regard, the primary cilium can directly regulate the activity of mesenchymal osteoprogenitors, osteoblasts and osteocytes upon mechanical stimulation. Mesenchymal progenitors of osteoblasts without functional primary cilia due to targeted genetic deletion are unable to proliferate in response to a mechanical stimulus, fail to deposit and orient collagen fibers to the strain field and do not differentiate into bone-forming osteoblasts (Leucht et al., 2013). Furthermore, both osteoblasts and osteocytes dramatically decrease their responsiveness to mechanical stimulation in the absence of primary cilia, hence reducing load-dependent bone formation (Qiu et al., 2012; Temiyasathit et al., 2012). Specific deletion of the primary cilium in osteoblasts also impairs their function, as reflected by reduction in mineral apposition rate and decreased expression of bone differentiation and formation markers Runx2, osterix, osteocalcin and Dmp1 in osteoblasts with non-functional primary cilia (Qiu et al., 2012). Interestingly, some cells such as chondrocytes present primary cilia preferentially colocalizing with the Golgi apparatus to potentially optimize secretion of extracellular matrix (Poole et al., 1997). These observations could imply that primary cilia detection of loading might be directly involved in the secretion of extracellular matrix also by osteoblasts. On the other hand, primary cilia are implicated in mechanical loading-induced bone formation by coordinating the proliferation and differentiation of bone forming osteoblasts, the continuous replenishment of osteoblasts from mesenchymal stem cells (Temiyasathit et al., 2012) and the activity of embedded fully mature osteocytes (Hoey, Kelly, & Jacobs, 2011; Moore et al., 2016). The coordination processes required for bone formation seem to be regulated in response to paracrine factors secreted by mechanosensitive osteocytes (Hoey et al., 2011; Leucht et al., 2013). Integrin attachments along the canalicular walls of the osteocyte network are highly stimulated by different mechanical forces during loading (Litzenberger, Kim, Tummala, & Jacobs, 2010; Marie, Haÿ, & Saidak, 2014). It has been proposed that primary cilia that attach to the lacunar wall can be activated, due to hydrodynamic pressure in the lacunocanalicular system and, thus, participate in bone mechanotransduction (Vaughan, Mullen, Verbruggen, & McNamara, 2015). Unlike other cell types, the primary cilium in bone translates mechanical stimulation into cellular responses independently of calcium flux and stretch-activated ion channels (Malone et al., 2007). The secretome obtained from conditioned media of mechanically stimulated osteocytes causes a significant upregulation of osteogenic genes, such as osteopontin (OPN) and cyclooxygenase-2 (COX-2) in mesenchymal stem cells (MSCs) (Hoey et al., 2011). This paracrine signaling mechanism between osteocytes and MSCs is primary cilia-dependent and potentially enhances mesenchymal cell differentiation towards an osteoblastic phenotype (Hoey et al., 2011). Thus, the primary cilium is also involved in controlling load-dependent paracrine signaling in bone by regulating the secretion of soluble factors by osteocytes in response to mechanical stimulation.

3. Primary cilia-depending signaling pathways associated with activation of PTH1R in osteogenesis: Hedgehog and TGF-β/BMPs pathways

3.1 PTH1R signaling pathway

PTH1R is a seven-transmembrane domain GPCR expressed in several tissues and organs including bone. This receptor can be activated by two naturally occurring ligands, PTH and PTHrP or by ligand-independent mechanical stimulation (Ardura, Portal-Núñez, Alonso, Bravo, & Gortazar, 2019; Maycas et al., 2015; Zhang, Frangos, & Chachisvilis, 2009). Activation of PTH1R by these stimuli leads to intracellular accumulation of second messengers, such as calcium, and triggers different signaling pathways, including those depending on protein kinase A, protein kinase C and Mitogen-Activated Protein Kinases ERK 1/2 (Ardura et al., 2019). In turn, stimulation of these pathways in mesenchymal osteoprogenitors, osteoblasts and osteocytes regulate bone cell proliferation and differentiation and bone formation (Ardura et al., 2019). The diverse signaling pathways promoted by PTH1R activation are also responsible for the secretion of factors that regulate bone remodeling, such as Receptor Activator for Nuclear Factor κ B Ligand (RANKL) (Ardura et al., 2019). These multiple effects on bone indicate that PTH1R is a key regulator of osteogenesis.

PTH1R actions have been related to primary cilia in bone mainly through the activation of hedgehog and TFG-β/BMP signaling pathways, described in the following paragraphs. Specific interactions of PTH1R and primary cilia will be detailed in Sections 4.1 and 4.2.

3.2 Hedgehog signaling

Hh signaling pathway is crucial during skeletal development. Mammals contain three Hh genes: Sonic Hedgehog (Shh)—a regulator of limb patterning during development, Indian Hedgehog (Ihh)—a regulator of endochondral bone formation, and Desert Hedgehog (Dhh)—involved in morphogenesis. Glioma-associated oncogene family members Gli1, Gli2 and Gli3 are three transcription factors that act as the principal mediators of Hh signaling (Chen & Jiang, 2013). Hh canonical signaling pathway is primarily activated in the primary cilium by a mechanism described as follows (Wheway et al., 2018). The Hh receptor Patched1 (PTC1), a 12-transmembrane protein, is localized at the primary cilium membrane in a basal off state. There, PTC1 receptor inhibits Hh signal transduction in the absence of Hh protein binding by repressing the seven-transmembrane receptor Smoothened (SMO) and keeping it out of the cilium (Yang, Andre, Ye, & Yang, 2015). Inactive SMO promotes proteolytic processing of Gli into a truncated form, GliR, that acts as a transcriptional repressor to block the expression of Hh target genes (Chen & Jiang, 2013). At the same time, Gli transcription factors are sequestered and suppressed by Suppressor of Fused (SuFu) at the primary cilium or targeted for degradation in this inactivated state (Wheway et al., 2018). In contrast, Hh ligands switch on this signaling pathway by binding to PTC1 and leading to the exit of PTC1 from the primary cilia structure. Subsequently, repression of SMO ceases and can translocate to the primary cilium where it is activated by endogenous SMO ligands—probably cholesterol-related molecules. At the ciliary membrane activated SMO can inhibit GliR production and block the SUFU–Gli association. Therefore, Hh signaling inhibits the production of GliR and allows the Gli active forms to be produced and translocate to the nucleus to stimulate Hh target gene expression (Chen & Jiang, 2013).

In addition to canonical pathways, Hh signaling can occur through non-canonical pathways, which involve either Gli-independent mechanisms or SMO-independent regulation of Gli activity (Anvarian et al., 2019). One of the non-canonical Hh signaling pathways activates small GTPases RhoA and Rac1 via Smo-coupled Gαi proteins (Polizio et al., 2011). Regarding this pathway, it has been described that promotion of osteogenesis is performed by balancing canonical Hh-Gli and noncanonical Hh-Gαi-RhoA pathways (Moran et al., 2018; Wei, Hu, Mishina, & Liu, 2016; Yuan & Yang, 2016).

Other non-canonical Hh signaling pathways involves activation of EVC ciliary proteins EVC1 and 2 located at the basal body and the base of the axoneme in the primary cilium. EVC1 and 2 proteins in chondrocytes of the cartilage can promote bone growth modulating Ihh signaling pathway activation (Ruiz-Perez et al., 2007; Zaghloul & Brugmann, 2011). The formation of EVC1/2 complexes has been reported to induce ECV1/2 interactions with SMO with the subsequent activation of this receptor, leading to SUFU/Gli dissociation and triggering translocation of active Gli factors to the nucleus (Humke, Dorn, Milenkovic, Scott, & Rohatgi, 2010). EVC1 also takes part in both parathyroid hormone related peptide (PTHrP)-dependent and -independent Ihh signaling pathways (Pacheco et al., 2012; Shao, Wang, Welter, & Ballock, 2012; Yuan et al., 2015). PTHrP, a peptide that binds PTH1R, has been related with Hh signaling during skeletal development. Ihh increases PTHrP expression during endochondral ossification (Karp et al., 2000; Whitfield, 2008; Xiang et al., 2014), a process that requires degradation of the cartilage matrix and substitution by osteoblasts to create bone matrix (see below at Section 4.1) (Gouttenoire et al., 2007).

Activation of Hh signaling through canonical or non-canonical pathways leads to bone formation and remodeling in response to mechanical stimulation, thanks to the abilities of primary cilia of bone progenitor cells to sense mechanical forces and subsequently induce osteogenesis (Barba et al., 2018; Shao et al., 2012). For example, Hh signaling is activated by primary cilium in response to low intensity pulsed ultrasounds to stimulate osteogenic differentiation and the maturation of osteoblasts to promote fracture healing (Matsumoto et al., 2018). Moreover, the Hh family members Ptc1, Gli1 and Gli3 associate with Zic1, a neural developmental transcription factor, to modulate shear flow mechanotransduction in osteocytes (Kalogeropoulos et al., 2010).

3.3 TGF-β/BMP signaling

Transforming Growth Factor Beta (TGF-β) signaling participates in bone development and maintenance through the control of cellular proliferation, differentiation, matrix deposition and migration. TGF-β ligands are a group of ligands that bind and activate hetero-tetrameric receptor complexes known as TGF-β receptors (TGFβR) (Wu, Chen, & Li, 2016). These complexes can be located and concentrated within the primary cilium (Labour, Riffault, Christensen, & Hoey, 2016). TGFβR1 and 2 are serine/threonine and tyrosine kinases that activate canonical or non-canonical signaling pathways (Vander Ark, Cao, & Li, 2018), being canonical signaling association with the primary cilium better studied (Aspera-Werz et al., 2019; Mönnich et al., 2018). In canonical signaling, TGF-β through the activation of TGFβR receptors, phosphorylate transcription factors of the R-SMAD family, predominantly SMAD 2/3. Phosphorylated R-SMADS form a complex with SMAD4 and translocate to the nucleus to regulate the expression of specific genes (Heldin & Moustakas, 2016). BMP signaling share common effectors with TGF-β. BMP ligand binding triggers the formation of TGFβR1 and 2 subsequently activating SMAD 1,4,5 and 8 (Anvarian et al., 2019; Takebayashi-Suzuki & Suzuki, 2020). In addition, BMP signaling has been described to cooperate with primary cilium in endothelial cells. In these cells, primary cilia enhance BMP9–SMAD 1,5 and 8 signaling, decreasing their migration speed in response to laminar shear stress (Vion et al., 2018). These observations suggest that a similar primary cilia-BMP interacting mechanism may occur in bone cells.

TGF-β1 is stored in the bone extracellular matrix as an inactive form, being activated during osteoclast-mediated resorption (Labour et al., 2016). Interactions between primary cilia and TGF-β signaling have been reported. For example, cigarette smoke perturbs the function of TGFβR1, disrupting primary cilia formation in mesenchymal stem cells causing inhibition of TGF-β signaling (Aspera-Werz et al., 2019). As a result, MSC migration to the fracture callus and the posterior differentiation to osteochondral cell progenitors is decreased (Aspera-Werz et al., 2019). MSCs recruitment and migration has been associated with primary cilia-dependent SMAD3 and TGFβRI activation in the ciliary base, with subsequent nuclear translocation of phosphorylated-SMAD3. This process requires the complete formation of primary cilia and functional IFT transport (Labour et al., 2016). Similarly, primary cilia association with BMP signaling has been observed. BMP signaling induce osteoblast differentiation and mineralization through SMAD 1, 5, and 8 activation in response to mechanical stimulation by using pulsed electromagnetic fields (Xie et al., 2016). In contrast, TGF-β1 can also regulate histone deacetylase activity (HDAC), blocking BMP-dependent maturation of osteoblasts. Thus, TGF-β1 can potentially impair osteoblast maturation and mechanosensation partially via HDAC6 regulation and reduction of the number of primary cilia (Ehnert et al., 2017; Xiang et al., 2017).

These findings suggest a wide spectrum of mechanisms by which TGF-β/BMP signaling modulate primary cilia function in the bone cell lineage. In addition, interactions of TGF-β/BMP signaling with ciliary PTH1R have also been shown (Zheng et al., 2018) and will be described in Section 4.2.

4. Primary cilia and PTH1R interactions in osteogenesis

4.1 Osteogenic actions during development: Endochondral ossification and postnatal growth plate

Signaling pathways and factors induced by primary cilia and PTH1R have been associated with regulation of osteogenesis during the development of the skeleton. In particular Ihh, PTHrP and BMP/TGF-β signaling pathways participate in the modulation of endochondral osteogenesis or ossification (Gao et al., 2020). This process that forms most of the human skeleton involves the formation of a hyaline cartilage template from condensed mesenchymal cells that differentiate into chondrocytes (Kronenberg, 2003). The cartilage templates, that resembles the future long bones, are surrounded by cartilage osteochondro-progenitor cells of mesenchymal origin (Mizuhashi, Nagata, Matsushita, Ono, & Ono, 2019), constituting the perichondrium (Hallett, Ono, & Ono, 2019).

Endochondral ossification arises at two distinct locations in the long bones. First, it starts at the bone diaphysis in primary ossification sites and continues at the bone epiphysis in secondary sites of ossification (Ortega, Behonick, & Werb, 2004). Chondrocytes at the center of the cartilage go through a series of differentiation phases becoming hypertrophic (Long & Ornitz, 2013). These hypertrophic chondrocytes promote the formation of the periosteal bone collar by inducing the differentiation of adjacent perichondrial cells into matrix-mineralizing osteoblasts via activation of Ihh and BMP signaling pathways (Long et al., 2004). During endochondral ossification at primary sites, the process of hypertrophy of chondrocytes is characterized by cell cycle exit, volume increase and secretion of an extracellular matrix based on type X collagen fibers (Haimov et al., 2020). Next, hypertrophic chondrocytes are substituted by either one of the following processes. One, type X collagen-based matrix mineralizes by calcification, preventing the nutrients from reaching the chondrocytes and causing them to undergo apoptotic cell death. Alternatively, hypertrophic chondrocytes could transdifferentiate into osteoblasts to form a bony matrix (Hallett et al., 2019; Wei et al., 2016; Yang, Tsang, Tang, Chan, & Cheah, 2014). In any case, the space once occupied by cartilage cells is subsequently invaded by blood vessels and replaced by a bone matrix, known as primary spongiosa, that will constitute the trabeculae and bone marrow of the primary ossification center (Hallett et al., 2019).

In addition, secondary ossification centers are formed at each end of the developing cartilage. The structure between the primary and secondary ossification centers is known as growth plate, which is responsible for longitudinal bone growth until the end of adolescence (Kronenberg, 2003). This region is mainly formed by chondrocytes arranged in three layers: irregularly scattered chondrocytes that form the resting zone, proliferating chondrocytes that form the column-shaped proliferative zone and chondrocytes with an increased cell volume that form the hypertrophic zone (Kronenberg, 2003). The columns formed by chondrocytes are oriented in such way that direct the lengthening of the bone along a distinct axis (Wei et al., 2016). The matrix that surrounds chondrocytes in this area contains several factors that mediate chondrocyte cell migration, proliferation and differentiation (Hallett et al., 2019). Chondrocytes derived from the hypertrophic zone undergo apoptosis or transdifferentiate into osteoblasts that produce extracellular matrix of the future bone tissue (Hallett et al., 2019; Jing, Wang, Li, Ma, & Feng, 2020). Chondrocyte to osteoblast transdifferentiation in the secondary ossification center may occur directly without an intermediate step of pluripotency or indirectly involving intermediate steps of dedifferentiation and redifferentiation (Wolff & Hartmann, 2019).

PTHrP, an agonist of PTH1R, is secreted from cells in the resting zone of the developing growth plate and maintain chondrocytes proliferating inhibiting their differentiation towards an hypertrophic phenotype (Kronenberg, 2003) (Fig. 2). Ihh -a molecule prioritarily activated by primary cilia- is mainly produced by prehypertrophic chondrocytes, in which promotes their proliferation, and also creates a negative feedback loop with

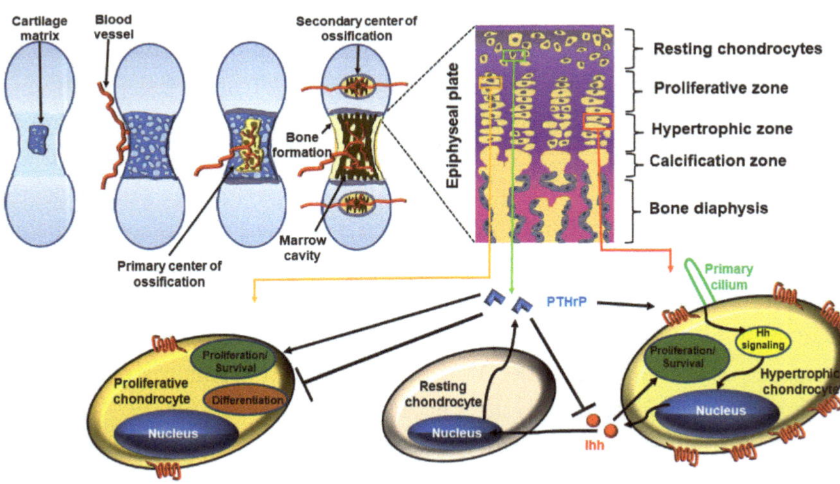

Fig. 2 Role of PTH1R and Hedgehog signaling in endochondral ossification. Schematic representation of endochondral ossification process, showing different states of chondrocyte differentiation at the epiphyseal plate. The initial step involves formation of a hyaline cartilage template from condensed mesenchymal cells that differentiate into chondrocytes that form a resting zone. Subsequently, chondrocytes become hypertrophic. Finally, hypertrophic chondrocytes die by apoptosis or transdifferentiate into osteoblasts to form bone matrix. Ihh/PTHrP negative feedback loop is essential in endochondral ossification: Ihh induces PTHrP expression of chondrocytes in the periarticular resting zone region, and Ihh-induced PTHrP maintains proliferation, induces survival and inhibits differentiation of chondrocytes in the proliferative zone, whereas represses proliferation and expression of Ihh in prehypertrophic chondrocytes.

PTHrP to maintain the length of the growth plate (Gao et al., 2020; Ohba, 2016). Ihh induces PTHrP expression of chondrocytes in the periarticular resting zone region and at the perichondrium. Ihh-induced PTHrP maintains the proliferation of chondrocytes at the proliferation/hypertrophic interface zone, promotes the survival of the proliferating chondrocytes by inducing the expression of the anti-apoptotic protein Bcl-2 and inhibits their differentiation towards hypertrophic chondrocytes (Amling et al., 1997; Kronenberg, 2003; Nishimori et al., 2019). In addition, PTHrP represses the expression of Ihh in prehypertrophic chondrocytes in which PTH1R expression is predominant—proliferating chondrocytes express low levels of PTH1R—causing a negative feedback loop (Schipani & Provot, 2003). Therefore, the suppression of hypertrophy keeps chondrocytes proliferating, and the distance between PTHrP-producing and Ihh-producing cells remains constant thanks to the Ihh-PTHrP negative feedback loop (Ohba, 2016). Thus, PTHrP primarily regulates hypertrophy and Ihh regulates proliferation of chondrocytes (van Donkelaar & Huiskes, 2007). Ultimately, chondrocytes stop proliferating driven by regulation of molecules such as TGFα/EFGR (epidermal growth factor receptor) or FGF (fibroblast growth factor, that represses Ihh expression (Schipani & Provot, 2003; Zhang et al., 2011). Once the proliferation program stops, chondrocytes activate an apoptotic process and die or transdifferentiate into osteoblasts (Jing et al., 2020; Kronenberg, 2003; Wolff & Hartmann, 2019). Ihh produced by pre- and hypertrophic chondrocytes has also been involved in osteoblastogenesis at the perichondrium and the primary spongiosa during endochondral ossification. In this regard hedgehog signaling is necessary for the differentiation of progenitors into Runx2-positive osteoblast precursors (Rodda & McMahon, 2006).

Primary cilia of chondrocytes at the growth plate and perichondrial progenitors have a crucial role during endochondral bone formation and skeletal development (Haycraft & Serra, 2008; Haycraft et al., 2007). A particular orientation of primary cilia in chondrocytes of articular cartilage subregions has been related to proper bone development. Primary cilia in chondrocytes of articular cartilage are oriented away from the articular surface to facilitate the appositional growth and maturation of cartilage (Farnum & Wilsman, 2011). Growth plate chondrocyte cilia orientation has been associated with decreased chondrocyte proliferation and increased hypertrophy (Waters & Beales, 2011). Also, proteins associated with primary cilia formation such as IFT88 and IFT4 are important for chondrocyte proliferation, existence of the growth plate and the expression of genes related to skeletogenesis

(Chang & Serra, 2013; Gouttenoire et al., 2007). Furthermore, chondrocytes are frequently exposed to compressive forces at the articular cartilage and the primary cilium has revealed to be essential for load-induced signaling responses (Shao et al., 2012). In this regard, Ihh signaling is mediated in growth plate chondrocytes by the primary cilium (Han et al., 2018; Shao et al., 2012), and given that Ihh controls PTHrP expression at periarticular chondrocytes it has been hypothesized that primary cilia indirectly modulate PTH1R actions on chondrocytes (Moore & Jacobs, 2018).

TGFβ/BMP signaling is another signaling pathway modulated by primary cilia (Labour et al., 2016). Receptors and signaling molecules activated by TGFβ localize to the cilium and regulate the migration of mesenchymal stem cells in bone (Labour et al., 2016). Moreover, TGFβ2 is activated by hedgehog signaling and, in turn, activates PTHrP signaling, thus regulating hypertrophic differentiation in the growth plate (Alvarez et al., 2002). On the other hand, primary cilia also concentrate BMP receptors during development (Lindbæk et al., 2015). BMP signaling is associated with promotion of proliferation in chondrocytes of the resting zone (Yoon et al., 2006), induction of Ihh/PTHrP expression and delaying the process of hypertrophic differentiation (Minina et al., 2001).

Other signaling pathways that impact endochondral ossification such as those triggered by Wnt and FGF factors could potentially interact with the primary cilium and PTH1R but this relationship is still not well defined.

4.2 Osteogenic actions in adult bone cells

As previously mentioned, several GPCRs are selectively targeted to primary cilia on many mammalian cell types (Mykytyn & Askwith, 2017). Among different GPCRs, the PTH receptor type I (PTH1R) has been described to co-localize at the primary cilium structure in cells of the nucleus pulposus of the intervertebral disc and in the osteoblast/osteocyte lineage (Martín-Guerrero et al., 2020; Zheng et al., 2018). It has been observed that specific transport of PTH1R to the primary cilium of nucleus pulposus cells is promoted by mechanical stimulation (Zheng et al., 2018). PTH1R relocalization to the primary cilium enhanced parathyroid hormone (PTH) signaling in nucleus pulposus cells by increasing the transcription of integrin αvβ6 and the activation of transforming growth factor (TGF)-β-connective tissue growth factor (CCN2)-matrix proteins signaling cascade (Zheng et al., 2018). This study also showed that PTH could attenuate intervertebral disc degeneration due to aging by a mechanism dependent on TGF-β activity, and CCN2 and aggrecan signaling. Interestingly, either PTH1R or primary cilia knockout in nucleus pulposus cells caused significant disc degeneration and inhibited PTH-dependent anti-aging actions on intervertebral discs (Zheng et al., 2018). Therefore, these results demonstrate that mechanical stimulus-dependent translocation of PTH1R to the primary cilium is a necessary process to further increase PTH-induced PTH1R signaling and anti-aging effects. In this regard, it has been previously shown that mechanical forces can directly stimulate changes in the structural conformation of PTH1R and trigger its signaling by an agonist-independent mechanism (Maycas et al., 2015; Zhang et al., 2009). Thus, the ability of PTH1R to be activated without stimulation by PTH/PTHrP suggest that the synergic response of PTH1R by mechanical stimulation and PTH could be caused by different possible mechanisms (Fig. 3). First, cells may have different pools of available PTH receptor. One pool that could be activated by PTH/PTHrP peptides and other pool that could be activated only by mechanical stimulation when translocated to the ciliary membrane. In this model, cells stimulated by both PTH/PTHrP and mechanical stimuli would have activation of both pools of PTH1R compared to cells stimulated just with the peptides. Activation of both pools of PTH1R may theoretically enhance the response of the cells by increasing the concentration of second messenger molecules or prolonging the activation of the receptors compared to an agonist-only stimulation (Fig. 3A). Other alternative mechanism could involve the induction of a structural conformation of the receptor by mechanical stimulation that might increase PTH/PTHrP peptide binding or its signaling. In this way, mechanical stimuli and PTH/PTHrP peptide could concomitantly activate an overresponsive receptor conformation that could not be activated by mechanical stimuli or PTH/PTHrP peptides independently (Fig. 3B). Other alternative theories, such as a better exposure of PTH1R to PTH/PTHrP peptide stimulation and subsequent enhanced receptor signaling when PTH1R is exposed at the primary ciliary cannot be excluded (Fig. 3C).

Co-localization of PTH1R with primary cilia has also been observed in the osteoblast/osteocyte lineage (Martín-Guerrero et al., 2020). In this study, activation of PTH1R by PTHrP increased cell survival, osteoblast gene expression (osteocalcin, runt-related transcription factor 2, and bone alkaline phosphatase) and the expression of the hedgehog transcription factor Gli-1 in osteocytes and osteoblasts. These effects were inhibited by primary cilia abrogation using specific siRNAs or inhibitors. In addition, a Gli-1

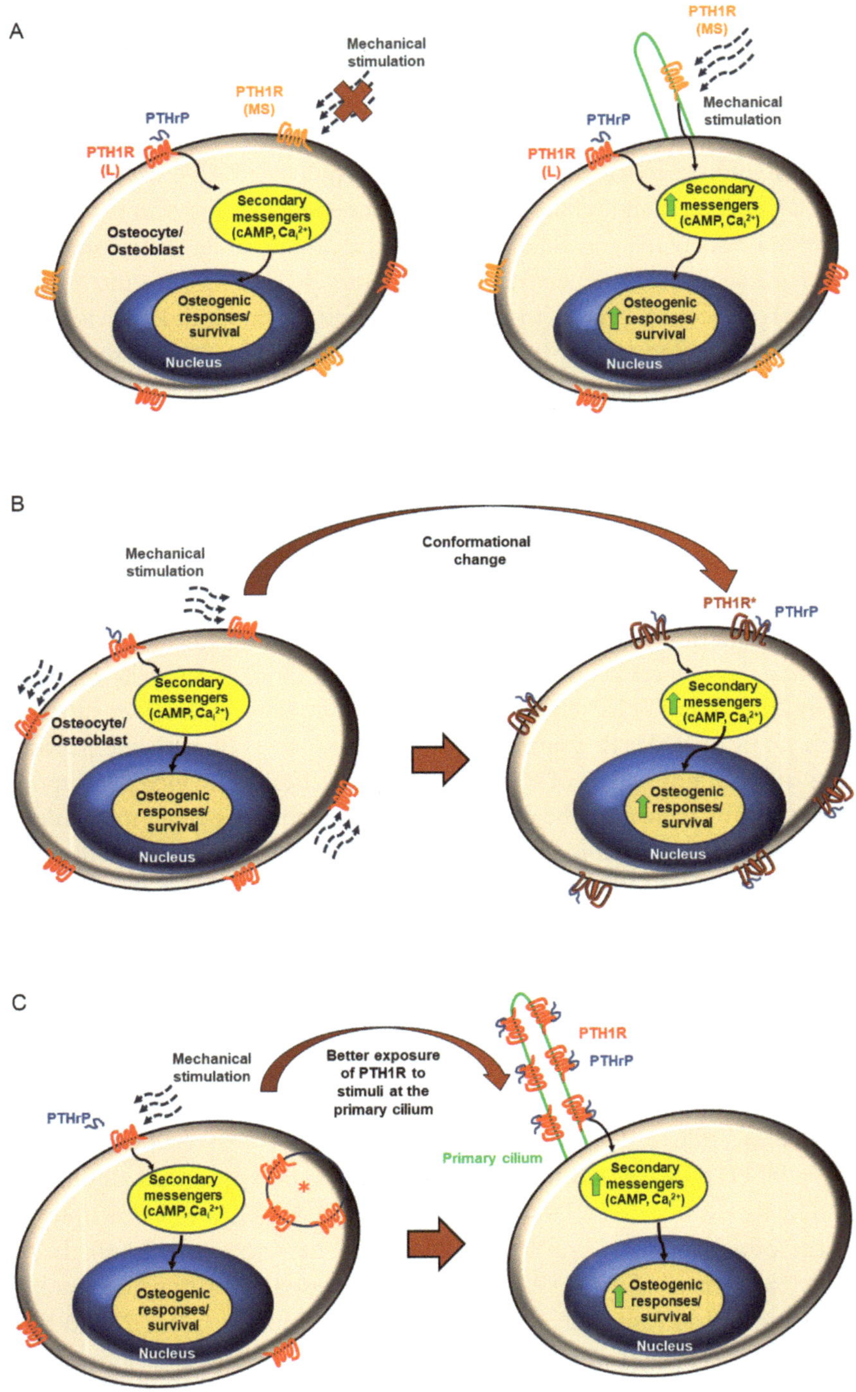

Fig. 3 Proposed mechanisms for PTH1R synergic activation. (A) Different pools of PTH receptor may exist in the cell, one activated by PTH/PTHrP ligands and other activated by mechanical stimulation only when translocated to the ciliary membrane. (B) Mechanical stimulation induces conformational reorganization of PTH receptor, which allows for increased ligand binding and/or signaling. (C) PTH1R location at the primary ciliary may increase its exposure to ligand peptide stimulation increasing its response.

antagonist showed to inhibit PTHrP prosurvival actions without affecting PTHrP-induced overexpression of osteogenic genes (Martín-Guerrero et al., 2020). These observations demonstrate that PTH1R activation induces prosurvival actions via primary cilia- and Gli-1-dependent mechanisms and modulates osteogenic responses via a primary cilia-dependent and Gli-1-independent pathway in osteocytes and osteoblasts (Martín-Guerrero et al., 2020). Even though these results were obtained in cells that were stimulated by an agonist but not by mechanical forces, an increase of the number of primary cilia-presenting cells was observed after mechanical stimulation of osteocytes and osteoblasts (Martín-Guerrero et al., 2020). Therefore, it may be possible that PTH/PTHrP peptides and mechanical stimulation may concomitantly activate the PTH1R in osteoblasts and osteocytes as observed in nucleus pulposus cells, by the mechanisms previously mentioned.

5. Conclusions

Primary cilia are subcellular structures that potentially activate several signaling pathways to regulate endochondral ossification during development and in the postnatal growth plate and during osteogenesis in adult bone cells. One molecular partner of primary cilia in the modulation of these processes of bone formation is the PTH1R. This receptor is a GPCR that can be activated by PTH/PTHrP ligands or by mechanical stimulation to regulate the differentiation of osteoprogenitors, chondrocytes and osteoblasts. Some studies suggest that PTH1R and primary cilium could associate in bone cells through direct physical interaction or by activation of common molecular transducers. Therefore, PTH1R might translocate and be activated at the primary cilium and also, primary cilia could activate signaling pathways upstream or downstream of PTH1R to regulate bone osteogenesis.

Acknowledgments

The author acknowledges the support of the Ministerio de Ciencia e Innovación (PID2019-109659RB-I00) from the Spanish Government, "Fundación Española de Investigación Ósea y del Metabolismo Mineral" (FEIOMM) and "Sociedad Española de Investigación Ósea y Metabolismo Mineral" (SEIOMM).

References

Alvarez, J., Sohn, P., Zeng, X., Doetschman, T., Robbins, D. J., & Serra, R. (2002). TGFbeta2 mediates the effects of hedgehog on hypertrophic differentiation and PTHrP expression. *Development (Cambridge, England)*, *129*(8), 1913–1924.

Amling, M., Neff, L., Tanaka, S., Inoue, D., Kuida, K., Weir, E., et al. (1997). Bcl-2 lies downstream of parathyroid hormone-related peptide in a signaling pathway that regulates chondrocyte maturation during skeletal development. *Journal of Cell Biology*, *136*(1), 205–213. https://doi.org/10.1083/jcb.136.1.205.

Anderson, C. T., Castillo, A. B., Brugmann, S. A., Helms, J. A., Jacobs, C. R., & Stearns, T. (2008). Primary cilia: Cellular sensors for the skeleton. *Anatomical Record*, *291*(9), 1074–1078. https://doi.org/10.1002/ar.20754.

Anvarian, Z., Mykytyn, K., Mukhopadhyay, S., Pedersen, L. B., & Christensen, S. T. (2019). Cellular signaling by primary cilia in development, organ function and disease. *Nature Reviews Nephrology*, *15*(4), 199–219. https://doi.org/10.1038/s41581-019-0116-9.

Ardura, J. A., Portal-Núñez, S., Alonso, V., Bravo, B., & Gortazar, A. R. (2019). Handling parathormone receptor type 1 in skeletal diseases: Realities and expectations of abaloparatide. *Trends in Endocrinology and Metabolism*, *30*(10), 756–766. https://doi.org/10.1016/j.tem.2019.07.014.

Aspera-Werz, R. H., Chen, T., Ehnert, S., Zhu, S., Fröhlich, T., & Nussler, A. K. (2019). Cigarette smoke induces the risk of metabolic bone diseases: Transforming growth factor beta signaling impairment via dysfunctional primary cilia affects migration, proliferation, and differentiation of human mesenchymal stem cells. *International Journal of Molecular Sciences*, *20*(12). https://doi.org/10.3390/ijms20122915.

Barba, M., Di Pietro, L., Massimi, L., Geloso, M. C., Frassanito, P., Caldarelli, M., et al. (2018). BBS9 gene in nonsyndromic craniosynostosis: Role of the primary cilium in the aberrant ossification of the suture osteogenic niche. *Bone*, *112*, 58–70. https://doi.org/10.1016/j.bone.2018.04.013.

Basten, S. G., & Giles, R. H. (2013). Functional aspects of primary cilia in signaling, cell cycle and tumorigenesis. *Cilia*, *2*(1). https://doi.org/10.1186/2046-2530-2-6.

Bonewald, L. F. (2007). Osteocytes as dynamic multifunctional cells. *Annals of the New York Academy of Sciences*, *1116*, 281–290. https://doi.org/10.1196/annals.1402.018.

Chang, C.-F., & Serra, R. (2013). Ift88 regulates hedgehog signaling, *Sfrp5* expression, and β-catenin activity in post-natal growth plate. *Journal of Orthopedic Research*, *31*(3), 350–356. https://doi.org/10.1002/jor.22237.

Chen, Y., & Jiang, J. (2013). Decoding the phosphorylation code in hedgehog signal transduction. *Cell Research*, *23*(2), 186–200. https://doi.org/10.1038/cr.2013.10.

Eguether, T., SanAgustin, J. T., Keady, B. T., Jonassen, J. A., Liang, Y., Francis, R., et al. (2014). IFT27 links the bbsome to ift for maintenance of the ciliary signaling compartment. *Developmental Cell*, *31*(3), 279–290. https://doi.org/10.1016/j.devcel.2014.09.011.

Ehnert, S., Sreekumar, V., Aspera-Werz, R. H., Sajadian, S. O., Wintermeyer, E., Sandmann, G. H., et al. (2017). TGF-β1 impairs mechanosensation of human osteoblasts via HDAC6-mediated shortening and distortion of primary cilia. *Journal of Molecular Medicine*, *95*(6), 653–663. https://doi.org/10.1007/s00109-017-1526-4.

Farnum, C. E., & Wilsman, N. J. (2011). Orientation of primary cilia of articular chondrocytes in three-dimensional space. *The Anatomical Record: Advances in Integrative Anatomy and Evolutionary Biology*, *294*(3), 533–549. https://doi.org/10.1002/ar.21330.

Fliegauf, M., Benzing, T., & Omran, H. (2007). When cilia go bad: Cilia defects and ciliopathies. *Nature Reviews Molecular Cell Biology*, *8*(11), 880–893. https://doi.org/10.1038/nrm2278.

Florencio-Silva, R., Sasso, G. R. S., Sasso-Cerri, E., Simões, M. J., & Cerri, P. S. (2015). Biology of bone tissue: Structure, function, and factors that influence bone cells. *BioMed Research International*, *2015*, 1–17. https://doi.org/10.1155/2015/421746.

Gao, Y., Gui, F., Li, D., Zhang, R., Sun, Q., & Guo, X. (2020). Fluoride regulates the expression of extracellular matrix HSPG and related signaling pathways FGFR3 and Ihh/PTHrP feedback loop during endochondral ossification. *Environmental Toxicology and Pharmacology*, *73*(103), 275. https://doi.org/10.1016/j.etap.2019.103275.

Garcia-Gonzalo, F. R., & Reiter, J. F. (2012). Scoring a backstage pass: Mechanisms of ciliogenesis and ciliary access. *Journal of Cell Biology*, *197*(6), 697–709. https://doi.org/10.1083/jcb.201111146.

Gouttenoire, J., Valcourt, U., Bougault, C., Aubert-Foucher, E., Arnaud, E., Giraud, L., et al. (2007). Knockdown of the intraflagellar transport protein IFT46 stimulates selective gene expression in mouse chondrocytes and affects early development in zebrafish. *Journal of Biological Chemistry*, *282*(42), 30960–30973. https://doi.org/10.1074/jbc.M705730200.

Haimov, H., Shimoni, E., Brumfeld, V., Shemesh, M., Varsano, N., Addadi, L., et al. (2020). Mineralization pathways in the active murine epiphyseal growth plate. *Bone*, *130*. https://doi.org/10.1016/j.bone.2019.115086.

Hallett, S. A., Ono, W., & Ono, N. (2019). Growth plate chondrocytes: Skeletal development, growth and beyond. *International Journal of Molecular Sciences*, *20*(23), 6009. https://doi.org/10.3390/ijms20236009.

Han, S., Park, H.-R., Lee, E.-J., Jang, J.-A., Han, M.-S., Kim, G.-W., et al. (2018). Dicam promotes proliferation and maturation of chondrocyte through Indian hedgehog signaling in primary cilia. *Osteoarthritis and Cartilage*, *26*(7), 945–953. https://doi.org/10.1016/j.joca.2018.04.008.

Haycraft, C. J., & Serra, R. (2008). *Chapter 11 Cilia involvement in patterning and maintenance of the skeleton* (pp. 303–332). https://doi.org/10.1016/S0070-2153(08)00811-9.

Haycraft, C. J., Zhang, Q., Song, B., Jackson, W. S., Detloff, P. J., Serra, R., et al. (2007). Intraflagellar transport is essential for endochondral bone formation. *Development*, *134*(2), 307–316. https://doi.org/10.1242/dev.02732.

Heldin, C. H., & Moustakas, A. (2016). Signaling receptors for TGF-β family members. *Cold Spring Harbor Perspectives in Biology*, *8*(8). https://doi.org/10.1101/cshperspect.a022053.

Hemmatian, H., Bakker, A. D., Klein-Nulend, J., & van Lenthe, G. H. (2017). Aging, osteocytes, and mechanotransduction. *Current Osteoporosis Reports*, *15*(5), 401–411. https://doi.org/10.1007/s11914-017-0402-z.

Hoey, D. A., Downs, M. E., & Jacobs, C. R. (2012). The mechanics of the primary cilium: An intricate structure with complex function. *Journal of Biomechanics*, *45*(1). https://doi.org/10.1016/j.jbiomech.2011.08.008.

Hoey, D. A., Kelly, D. J., & Jacobs, C. R. (2011). A role for the primary cilium in paracrine signaling between mechanically stimulated osteocytes and mesenchymal stem cells. *Biochemical and Biophysical Research Communications*, *412*(1), 182–187. https://doi.org/10.1016/j.bbrc.2011.07.072.

Humke, E. W., Dorn, K. V., Milenkovic, L., Scott, M. P., & Rohatgi, R. (2010). The output of hedgehog signaling is controlled by the dynamic association between suppressor of fused and the gli proteins. *Genes and Development*, *24*(7), 670–682. https://doi.org/10.1101/gad.1902910.

Jacobs, C. R., Temiyasathit, S., & Castillo, A. B. (2010). Osteocyte mechanobiology and pericellular mechanics. *Annual Review of Biomedical Engineering*, *12*(1), 369–400. https://doi.org/10.1146/annurev-bioeng-070909-105302.

Jing, Y., Wang, Z., Li, H., Ma, C., & Feng, J. (2020). Chondrogenesis defines future skeletal patterns via cell transdifferentiation from chondrocytes to bone cells. *Current Osteoporosis Reports*, *18*(3), 199–209. https://doi.org/10.1007/s11914-020-00586-3.

Kalogeropoulos, M., Varanasi, S. S., Olstad, O. K., Sanderson, P., Gautvik, V. T., Reppe, S., et al. (2010). Zic1 transcription factor in bone: Neural developmental protein regulates mechanotransduction in osteocytes. *The FASEB Journal*, *24*(8), 2893–2903. https://doi.org/10.1096/fj.09-148908.

Karp, S. J., Schipani, E., St-Jacques, B., Hunzelman, J., Kronenberg, H., & McMahon, A. P. (2000). Indian hedgehog coordinates endochondral bone growth and morphogenesis via parathyroid hormone related-protein-dependent and -independent pathways. *Development (Cambridge, England)*, *127*(3), 543–548.

Komatsu, Y., Kaartinen, V., & Mishina, Y. (2011). Cell cycle arrest in node cells governs ciliogenesis at the node to break left-right symmetry. *Development*, *138*(18), 3915–3920. https://doi.org/10.1242/dev.068833.

Kronenberg, H. M. (2003). Developmental regulation of the growth plate. *Nature*, *423*(6937), 332–336. https://doi.org/10.1038/nature01657.

Labour, M. N., Riffault, M., Christensen, S. T., & Hoey, D. A. (2016). TGFβ1-induced recruitment of human bone mesenchymal stem cells is mediated by the primary cilium in a SMAD3-dependent manner. *Scientific Reports*, *6*. https://doi.org/10.1038/srep35542.

Leucht, P., Monica, S. D., Temiyasathit, S., Lenton, K., Manu, A., Longaker, M. T., et al. (2013). Primary cilia act as mechanosensors during bone healing around an implant. *Medical Engineering & Physics*, *35*(3). https://doi.org/10.1016/j.medengphy.2012.06.005.

Li, S., Jin, S., & Jin, C. (2019). The correlative hypotheses between pitchfork and Kif3a in palate development. *Medical Hypotheses*, *126*, 23–25. https://doi.org/10.1016/j.mehy.2019.03.005.

Lindbæk, L., Warzecha, C., Koefoed, K., Mogensen, J., Schmid, F., Pedersen, L., et al. (2015). Coordination of TGFβ/BMP signaling is associated with the primary cilium. *Cilia*, *4*(S1), P17. https://doi.org/10.1186/2046-2530-4-S1-P17.

Litzenberger, J. B., Kim, J.-B., Tummala, P., & Jacobs, C. R. (2010). β1 integrins mediate mechanosensitive signaling pathways in osteocytes. *Calcified Tissue International*, *86*(4). https://doi.org/10.1007/s00223-010-9343-6.

Long, F., Chung, U., Ohba, S., McMahon, J., Kronenberg, H. M., & McMahon, A. P. (2004). Ihh signaling is directly required for the osteoblast lineage in the endochondral skeleton. *Development*, *131*(6), 1309–1318. https://doi.org/10.1242/dev.01006.

Long, F., & Ornitz, D. M. (2013). Development of the endochondral skeleton. *Cold Spring Harbor Perspectives in Biology*, *5*(1). https://doi.org/10.1101/cshperspect.a008334.

Low, S. H., Vasanth, S., Larson, C. H., Mukherjee, S., Sharma, N., Kinter, M. T., et al. (2006). Polycystin-1, STAT6, and P100 function in a pathway that transduces ciliary mechanosensation and is activated in polycystic kidney disease. *Developmental Cell*, *10*(1). https://doi.org/10.1016/j.devcel.2005.12.005.

Ma, X., Peterson, R., & Turnbull, J. (2011). Adenylyl cyclase type 3, a marker of primary cilia, is reduced in primary cell culture and in lumbar spinal cord in situ in G93A SOD1 mice. *BMC Neuroscience*, *12*. https://doi.org/10.1186/1471-2202-12-71.

Malone, A. M. D., Anderson, C. T., Tummala, P., Kwon, R. Y., Johnston, T. R., Stearns, T., et al. (2007). Primary cilia mediate mechanosensing in bone cells by a calcium-independent mechanism. *Proceedings of the National Academy of Sciences of the United States of America*, *104*(33), 13325–13330. https://doi.org/10.1073/pnas.0700636104.

Marie, P. J., Haÿ, E., & Saidak, Z. (2014). Integrin and cadherin signaling in bone: Role and potential therapeutic targets. *Trends in Endocrinology & Metabolism*, *25*(11). https://doi.org/10.1016/j.tem.2014.06.009.

Martin, L., Kaci, N., Estibals, V., Goudin, N., Garfa-Traore, M., Benoist-Lasselin, C., et al. (2018). Constitutively-active FGFR3 disrupts primary cilium length and IFT20 trafficking in various chondrocyte models of achondroplasia. *Human Molecular Genetics*, *27*(1), 1–13. https://doi.org/10.1093/hmg/ddx374.

Martín-Guerrero, E., Tirado-Cabrera, I., Buendía, I., Alonso, V., Gortázar, A. R., & Ardura, J. A. (2020). Primary cilia mediate parathyroid hormone receptor type 1 osteogenic actions in osteocytes and osteoblasts via Gli activation. *Journal of Cellular Physiology*, *235*(10), 7356–7369. https://doi.org/10.1002/jcp.29636.

Matsumoto, K., Shimo, T., Kurio, N., Okui, T., Ibaragi, S., Kunisada, Y., et al. (2018). Low-intensity pulsed ultrasound stimulation promotes osteoblast differentiation through hedgehog signaling. *Journal of Cellular Biochemistry, 119*(6), 4352–4360. https://doi.org/10.1002/jcb.26418.

Maycas, M., Ardura, J. A., de Castro, L. F., Bravo, B., Gortázar, A. R., & Esbrit, P. (2015). Role of the parathyroid hormone type 1 receptor (PTH1R) as a mechanosensor in osteocyte survival. *Journal of Bone and Mineral Research, 30*(7). https://doi.org/10.1002/jbmr.2439.

McGlashan, S. R., Jensen, C. G., & Poole, C. A. (2006). Localization of extracellular matrix receptors on the chondrocyte primary cilium. *The Journal of Histochemistry and Cytochemistry: Official Journal of the Histochemistry Society, 54*(9). https://doi.org/10.1369/jhc.5A6866.2006.

Middleton, K. A. J., Ma, Y.-H. V., & You, L. (2017). Measuring bone cell response to fluid shear stress and hydrostatic/dynamic pressure. *Experimental Methods in Orthopaedic Biomechanics*. https://doi.org/10.1016/B978-0-12-803802-4.00014-7.

Minina, E., Wenzel, H. M., Kreschel, C., Karp, S., Gaffield, W., McMahon, A. P., et al. (2001). BMP and Ihh/PTHrP signaling interact to coordinate chondrocyte proliferation and differentiation. *Development (Cambridge, England), 128*(22), 4523–4534.

Mizuhashi, K., Nagata, M., Matsushita, Y., Ono, W., & Ono, N. (2019). Growth plate borderline chondrocytes behave as transient mesenchymal precursor cells. *Journal of Bone and Mineral Research, 34*(8), 1387–1392. https://doi.org/10.1002/jbmr.3719.

Mönnich, M., Borgeskov, L., Breslin, L., Jakobsen, L., Rogowski, M., Doganli, C., et al. (2018). CEP128 localizes to the subdistal appendages of the mother centriole and regulates TGF-β/BMP signaling at the primary cilium. *Cell Reports, 22*(10), 2584–2592. https://doi.org/10.1016/j.celrep.2018.02.043.

Moore, E. R., Chen, J. C., & Jacobs, C. R. (2019). Prx1-expressing progenitor primary cilia mediate bone formation in response to mechanical loading in mice. *Stem Cells International, 2019*. https://doi.org/10.1155/2019/3094154.

Moore, E. R., & Jacobs, C. R. (2018). The primary cilium as a signaling nexus for growth plate function and subsequent skeletal development. *Journal of Orthopaedic Research: Official Publication of the Orthopedic Research Society, 36*(2). https://doi.org/10.1002/jor.23732.

Moore, B. S., Stepanchick, A. N., Tewson, P. H., Hartle, C. M., Zhang, J., Quinn, A. M., et al. (2016). Cilia have high cAMP levels that are inhibited by sonic hedgehog-regulated calcium dynamics. *Proceedings of the National Academy of Sciences of the United States of America, 113*(46), 13069–13074. https://doi.org/10.1073/pnas.1602393113.

Moran, J., Sanderson, K. G., Maynes, J., Vig, A., Batmanabane, V., Kannu, P., et al. (2018). IFT80 mutations cause a novel complex ciliopathy phenotype with retinal degeneration. *Clinical Genetics, 94*(3–4), 368–372. https://doi.org/10.1111/cge.13408.

Mykytyn, K., & Askwith, C. (2017). G-protein-coupled receptor signaling in cilia. *Cold Spring Harbor Perspectives in Biology, 9*(9). https://doi.org/10.1101/cshperspect.a028183.

Nauli, S. M., Alenghat, F. J., Luo, Y., Williams, E., Vassilev, P., Li, X., et al. (2003). Polycystins 1 and 2 mediate mechanosensation in the primary cilium of kidney cells. *Nature Genetics, 33*(2). https://doi.org/10.1038/ng1076.

Nguyen, A. M., & Jacobs, C. R. (2013). Emerging role of primary cilia as mechanosensors in osteocytes. *Bone, 54*(2), 196–204. https://doi.org/10.1016/j.bone.2012.11.016.

Nishimori, S., Lai, F., Shiraishi, M., Kobayashi, T., Kozhemyakina, E., Yao, T.-P., et al. (2019). PTHrP targets HDAC4 and HDAC5 to repress chondrocyte hypertrophy. *JCI Insight, 4*(5). https://doi.org/10.1172/jci.insight.97903.

Ohba, S. (2016). Hedgehog signaling in endochondral ossification. *Journal of Developmental Biology, 4*(2), 20. https://doi.org/10.3390/jdb4020020.

Ortega, N., Behonick, D. J., & Werb, Z. (2004). Matrix remodeling during endochondral ossification. *Trends in Cell Biology*, *14*(2), 86–93. https://doi.org/10.1016/j.tcb.2003.12.003.

Pacheco, M., Valencia, M., Caparrós-Martín, J. A., Mulero, F., Goodship, J. A., & Ruiz-Perez, V. L. (2012). Evc works in chondrocytes and osteoblasts to regulate multiple aspects of growth plate development in the appendicular skeleton and cranial base. *Bone*, *50*(1), 28–41. https://doi.org/10.1016/j.bone.2011.08.025.

Polizio, A. H., Chinchilla, P., Chen, X., Kim, S., Manning, D. R., & Riobo, N. A. (2011). Heterotrimeric Gi proteins link hedgehog signaling to activation of Rho small GTPases to promote fibroblast migration. *Journal of Biological Chemistry*, *286*(22), 19589–19596. https://doi.org/10.1074/jbc.M110.197111.

Poole, C. A., Jensen, C. G., Snyder, J. A., Gray, C. G., Hermanutz, V. L., & Wheatley, D. N. (1997). Confocal analysis of primary cilia structure and colocalization with the Golgi apparatus in chondrocytes and aortic smooth muscle cells. *Cell Biology International*, *21*(8), 483–494. https://doi.org/10.1006/cbir.1997.0177.

Pugacheva, E. N., Jablonski, S. A., Hartman, T. R., Henske, E. P., & Golemis, E. A. (2007). HEF1-dependent aurora a activation induces disassembly of the primary cilium. *Cell*, *129*(7). https://doi.org/10.1016/j.cell.2007.04.035.

Qiu, N., Xiao, Z., Cao, L., Buechel, M. M., David, V., Roan, E., et al. (2012). Disruption of Kif3a in osteoblasts results in defective bone formation and osteopenia. *Journal of Cell Science*. https://doi.org/10.1242/jcs.095893.

Rodda, S. J., & McMahon, A. P. (2006). Distinct roles for hedgehog and canonical Wnt signaling in specification, differentiation and maintenance of osteoblast progenitors. *Development*, *133*(16), 3231–3244. https://doi.org/10.1242/dev.02480.

Rohatgi, R., & Snell, W. J. (2010). The ciliary membrane. *Current Opinion in Cell Biology*, *22*(4), 541–546. https://doi.org/10.1016/j.ceb.2010.03.010.

Ruiz-Perez, V. L., Blair, H. J., Rodriguez-Andres, M. E., Blanco, M. J., Wilson, A., Liu, Y.-N., et al. (2007). Evc is a positive mediator of Ihh-regulated bone growth that localizes at the base of chondrocyte cilia. *Development*, *134*(16), 2903–2912. https://doi.org/10.1242/dev.007542.

Satir, P., & Christensen, S. T. (2007). Overview of structure and function of mammalian cilia. *Annual Review of Physiology*, *69*(1), 377–400. https://doi.org/10.1146/annurev.physiol.69.040705.141236.

Schipani, E., & Provot, S. (2003). PTHrP, PTH, and the PTH/PTHrP receptor in endochondral bone development. *Birth Defects Research Part C: Embryo Today: Reviews*, *69*(4), 352–362. https://doi.org/10.1002/bdrc.10028.

Seeger-Nukpezah, T., & Golemis, E. A. (2012). The extracellular matrix and ciliary signaling. *Current Opinion in Cell Biology*, *24*(5), 652–661. https://doi.org/10.1016/j.ceb.2012.06.002.

Shao, Y. Y., Wang, L., Welter, J. F., & Ballock, R. T. (2012). Primary cilia modulate Ihh signal transduction in response to hydrostatic loading of growth plate chondrocytes. *Bone*, *50*(1), 79–84. https://doi.org/10.1016/j.bone.2011.08.033.

Singh, M., Chaudhry, P., & Merchant, A. A. (2016). Primary cilia are present on human blood and bone marrow cells and mediate hedgehog signaling. *Experimental Hematology*, *44*(12), 1181–1187.e2. https://doi.org/10.1016/j.exphem.2016.08.009.

Takebayashi-Suzuki, K., & Suzuki, A. (2020). Intracellular communication among morphogen signaling pathways during vertebrate body. *Genes*, *11*, 341. https://doi.org/10.3390/genes11030341.

Temiyasathit, S., Tang, W. J., Leucht, P., Anderson, C. T., Monica, S. D., Castillo, A. B., et al. (2012). Mechanosensing by the primary cilium: Deletion of Kif3A reduces bone formation due to loading. *PLoS One*, 7(3). https://doi.org/10.1371/journal.pone.0033368.

Uzbekov, R. E., Maurel, D. B., Aveline, P. C., Pallu, S., Benhamou, C. L., & Rochefort, G. Y. (2012). Centrosome fine ultrastructure of the osteocyte mechanosensitive primary cilium. *Microscopy and Microanalysis*, *18*(6), 1430–1441. https://doi.org/10.1017/S1431927612013281.

van Donkelaar, C. C., & Huiskes, R. (2007). The PTHrP–Ihh feedback loop in the embryonic growth plate allows PTHrP to control hypertrophy and Ihh to regulate proliferation. *Biomechanics and Modeling in Mechanobiology*, *6*(1–2), 55–62. https://doi.org/10.1007/s10237-006-0035-0.

Vander Ark, A., Cao, J., & Li, X. (2018). TGF-β receptors: In and beyond TGF-β signaling. *Cellular Signaling*, *52*, 112–120. https://doi.org/10.1016/j.cellsig.2018.09.002.

Vaughan, T. J., Mullen, C. A., Verbruggen, S. W., & McNamara, L. M. (2015). Bone cell mechanosensation of fluid flow stimulation: A fluid–structure interaction model characterizing the role integrin attachments and primary cilia. *Biomechanics and Modeling in Mechanobiology*, *14*(4), 703–718. https://doi.org/10.1007/s10237-014-0631-3.

Vion, A.-C., Alt, S., Klaus-Bergmann, A., Szymborska, A., Zheng, T., Perovic, T., et al. (2018). Primary cilia sensitize endothelial cells to BMP and prevent excessive vascular regression. *Journal of Cell Biology*, *217*(5), 1651–1665. https://doi.org/10.1083/jcb.201706151.

Waters, A. M., & Beales, P. L. (2011). Ciliopathies: An expanding disease spectrum. *Pediatric Nephrology*, *26*(7), 1039–1056. https://doi.org/10.1007/s00467-010-1731-7.

Wei, X., Hu, M., Mishina, Y., & Liu, F. (2016). Developmental regulation of the growth plate and cranial synchondrosis. *Journal of Dental Research*, *95*(11), 1221–1229. https://doi.org/10.1177/0022034516651823.

Wheway, G., Nazlamova, L., & Hancock, J. T. (2018). Signaling through the primary cilium. *Frontiers in Cell and Developmental Biology*, *6*. https://doi.org/10.3389/fcell.2018.00008.

Whitfield, J. F. (2008). The solitary (primary) cilium—A mechanosensory toggle switch in bone and cartilage cells. *Cellular Signaling*, *20*(6), 1019–1024. https://doi.org/10.1016/j.cellsig.2007.12.001.

Wolff, L. I., & Hartmann, C. (2019). A second career for chondrocytes—Transformation into osteoblasts. *Current Osteoporosis Reports*, *17*(3), 129–137. https://doi.org/10.1007/s11914-019-00511-3.

Wu, M., Chen, G., & Li, Y. P. (2016). TGF-β and BMP signaling in osteoblast, skeletal development, and bone formation, homeostasis and disease. *Bone Research*, *4*. https://doi.org/10.1038/boneres.2016.9.

Xiang, W., Guo, F., Cheng, W., Zhang, J., Huang, J., Wang, R., et al. (2017). HDAC6 inhibition suppresses chondrosarcoma by restoring the expression of primary cilia. *Oncology Reports*, *38*(1), 229–236. https://doi.org/10.3892/or.2017.5694.

Xiang, W., Jiang, T., Guo, F., Xu, T., Gong, C., Cheng, P., et al. (2014). Evaluating the role of PTH in promotion of chondrosarcoma cell proliferation and invasion by inhibiting primary cilia expression. *International Journal of Molecular Sciences*, *15*(11), 19816–19831. https://doi.org/10.3390/ijms151119816.

Xie, Y. F., Shi, W. G., Zhou, J., Gao, Y. H., Li, S. F., Fang, Q. Q., et al. (2016). Pulsed electromagnetic fields stimulate osteogenic differentiation and maturation of osteoblasts by upregulating the expression of BMPRII localized at the base of primary cilium. *Bone*, *93*, 22–32. https://doi.org/10.1016/j.bone.2016.09.008.

Yang, J., Andre, P., Ye, L., & Yang, Y. Z. (2015). The hedgehog signaling pathway in bone formation. *International Journal of Oral Science*, 7(2), 73–79. https://doi.org/10.1038/ijos.2015.14.

Yang, L., Tsang, K. Y., Tang, H. C., Chan, D., & Cheah, K. S. E. (2014). Hypertrophic chondrocytes can become osteoblasts and osteocytes in endochondral bone formation. *Proceedings of the National Academy of Sciences of the United States of America*, *111*(33), 12097–12102. https://doi.org/10.1073/pnas.1302703111.

Yoon, B. S., Pogue, R., Ovchinnikov, D. A., Yoshii, I., Mishina, Y., Behringer, R. R., et al. (2006). BMPs regulate multiple aspects of growth-plate chondrogenesis through opposing actions on FGF pathways. *Development*, *133*(23), 4667–4678. https://doi.org/10.1242/dev.02680.

Yuan, X., Serra, R. A., & Yang, S. (2015). Function and regulation of primary cilia and intraflagellar transport proteins in the skeleton. *Annals of the New York Academy of Sciences*, *1335*(1), 78–99. https://doi.org/10.1111/nyas.12463.

Yuan, X., & Yang, S. (2016). Primary cilia and intraflagellar transport proteins in bone and cartilage. *Journal of Dental Research*, *95*(12), 1341–1349. https://doi.org/10.1177/0022034516652383.

Zaghloul, N. A., & Brugmann, S. A. (2011). The emerging face of primary cilia. *Genesis*, *49*(4), 231–246. https://doi.org/10.1002/dvg.20728.

Zhang, Y. L., Frangos, J. A., & Chachisvilis, M. (2009). Mechanical stimulus alters conformation of type 1 parathyroid hormone receptor in bone cells. *American Journal of Physiology. Cell Physiology*, *296*(6), C1391–C1399.

Zhang, X., Siclari, V. A., Lan, S., Zhu, J., Koyama, E., Dupuis, H. L., et al. (2011). The critical role of the epidermal growth factor receptor in endochondral ossification. *Journal of Bone and Mineral Research*, *26*(11), 2622–2633. https://doi.org/10.1002/jbmr.502.

Zheng, L., Cao, Y., Ni, S., Qi, H., Ling, Z., Xu, X., et al. (2018). Ciliary parathyroid hormone signaling activates transforming growth factor-β to maintain intervertebral disc homeostasis during aging. *Bone Research*, *6*(1), 21. https://doi.org/10.1038/s41413-018-0022-y.

Kolibri 360 GmbH
Ferdinand-Jühlke-Straße 7
99095 Erfurt, Deutschland
produktsicherheit@kolibri360.de